REAL ANALYSIS

Aimed at advanced undergraduates and beginning graduate students, *Real Analysis* offers a rigorous yet accessible course in the subject. Carothers, presupposing only a modest background in real analysis or advanced calculus, writes with an informal style and incorporates historical commentary as well as notes and references.

The book looks at metric and linear spaces, offering an introduction to general topology while emphasizing normed linear spaces. It addresses function spaces and provides familiar applications, such as the Weierstrass and Stone–Weierstrass approximation theorems, functions of bounded variation, Riemann–Stieltjes integration, and a brief introduction to Fourier analysis. Finally, it examines Lebesgue measure and integration on the line. Illustrations and abundant exercises round out the text.

Real Analysis will appeal to students in pure and applied mathematics as well as researchers in statistics, education, engineering, and economics.

N. L. Carothers is Professor of Mathematics at Bowling Green State University in Bowling Green, Ohio.

Real Analysis

N. L. CAROTHERS

Bowling Green State University

CAMBRIDGE UNIVERSITY PRESS
Cambridge, New York, Melbourne, Madrid, Cape Town, Singapore, São Paulo

Cambridge University Press
The Edinburgh Building, Cambridge CB2 2RU, UK

Published in the United States of America by Cambridge University Press, New York

www.cambridge.org
Information on this title: www.cambridge.org/9780521497497

First published 2000

A catalogue record for this publication is available from the British Library

Library of Congress Cataloguing in Publication data

Carothers, N. L., 1952–

Real analysis / N. L. Carothers.

p. cm.

Includes bibliographical references and index.

ISBN 0-521-49749-3. – ISBN 0-521-49756-6 (pbk.)

1. Mathematical analysis. I. Title.
QA300.C32 1999

515 – dc21 98-31982
CIP

ISBN-13 978-0-521-49749-7 hardback
ISBN-10 0-521-49749-3 hardback

ISBN-13 978-0-521-49756-5 paperback
ISBN-10 0-521-49756-6 paperback

Transferred to digital printing 2006

"Details are all that matters: God dwells there, and you never get to see Him if you don't struggle to get them right."

– Stephen J. Gould

". . . lots of things worth saying can only be said loosely."

– William Cooper

Contents

Preface

This book is based on a course in real analysis offered to advanced undergraduates and first-year graduate students at Bowling Green State University. In many respects it is a perfectly ordinary first course in analysis, but there are some important differences. For one, the typical audience for the class includes many nonspecialists, students of statistics, economics, and education, as well as students of pure and applied mathematics at the undergraduate and graduate levels. What's more, the students come from a wide variety of backgrounds. This makes the course something of a challenge to teach. The material must be presented efficiently, but without sacrificing the less well-prepared student. The course must be essentially self-contained, but not so pedestrian that the more experienced student is bored. And the course should offer something of value to both the specialist and the nonspecialist. The following pages contain my personal answer to this challenge.

To begin, I make a few compromises: Extra details are given on metric and normed linear spaces in place of general topology, and a thorough attack on Riemann–Stieltjes and Lebesgue integration on the line in place of abstract measure and integration. On the other hand, I avoid euphemisms and specialized notation and, instead, attempt to remain faithful to the terminology and notation used in more advanced settings. Next, to make the course more meaningful to the nonspecialist (and more fun for me), I toss in a few historical tidbits along the way.

By way of prerequisites, I assume that the reader has had at least one semester of advanced calculus or real analysis at the undergraduate level. For example, I assume that the reader has been exposed to (and is moderately comfortable with) an "ε-δ" presentation of convergence, completeness, and continuity on the real line; a few "name" theorems (Bolzano–Weierstrass, for one); and a rigorous definition of the Riemann integral, but I do not presuppose any real depth or breadth of understanding of these topics beyond their basics.

The writing style throughout is deliberately conversational. While I have tried to be as precise as possible, the odd detail here and there is sometimes left to the reader, which is reflected by the use of a parenthetical (Why?) or (How?). The decision to omit these few details is motivated by the hope that the student who can successfully navigate through this "guided tour" of analysis, who is willing to get involved with the mathematics at hand, will come away with something valuable in the process.

You will notice, too, that I don't try to keep secrets. Important ideas are often broached long before they are needed in the formal presentation. A particular theme may be repeated in several different forms before it is made flesh. This repetition is necessary if new definitions and new ideas are to seem natural and appropriate. Once such an idea is finally made formal, there is usually a real savings in the "definition–theorem–proof" cycle. The student who has held on to the thread can usually see the connections without difficulty or fanfare.

The book is divided, rather naturally, into three parts. The first part concerns general metric and normed spaces. This serves as a beginner's guide to general topology. The second part serves as a transition from the discussion of abstract spaces to concrete spaces of functions. The emphasis here is on the space of continuous real-valued functions and a few of its relatives. A discussion of Riemann–Stieltjes integration is included to set the stage for the later transition to Lebesgue measure and integration in the third and last part. A more detailed description of the contents is given below.

Where to start is always problematic; a certain amount of review is arguably necessary. Chapters One, Two, and Ten, along with their references, provide a source for such review (albeit incomplete at times). These chapters serve as a rather long introduction to Parts One and Two, primarily spelling out notation and recalling facts from advanced calculus, but also making the course somewhat self-contained.

The "real" course begins in Chapter Three, with metric and normed spaces, with frequent emphasis on normed spaces. From there we collect "C" words: convergence, continuity, connectedness, completeness, compactness, and category.

Part Two concerns spaces of functions. The reader will find a particularly heavy emphasis on the interplay between algebra, topology, and analysis here, which serves as a transition from the "sterile" abstraction of metric spaces to the "practical" abstraction of such results as the Weierstrass theorem and the Riesz representation theorem.

Part Three concerns Lebesgue measure and integration on the real line, culminating in Lebesgue's differentiation theorem. While I have opted for a "hands-on" approach to Lebesgue measure on the line, I have not been shy about using the machinery developed in the first two parts of the book. In other words, rather than presenting measure theory from an abstract point of view, with Lebesgue measure as a special case, I have chosen to concentrate solely on Lebesgue measure on the line, but from as lofty a viewpoint as I can muster. This approach is intended to keep the discussion down to earth while still easing the transition to abstract measure theory and functional analysis in subsequent courses.

This is an ambitious list of topics for two semesters. In actual practice, several topics can safely be left for the interested and ambitious reader to discover independently. For example, the sections on completions, equivalent metrics, infinitely differentiable functions, equicontinuity, continuity and category, and the Riesz representation theorem (among others) could be omitted.

A few words are in order about the exercises. I included as many as I could manage without undermining the text. They come in all shapes and sizes. And, like the text itself, there is a fair amount of built-in repetition. But the exercises are intended to be part of the presentation, not just a few stray thoughts appended to the end of a chapter. For this reason, the exercises are peppered throughout the text; each is placed near what I consider to be its natural position in the flow of ideas.

The beginner is encouraged to at least read through the exercises – those that look too difficult at first may seem easier on their third or fourth appearance. And the key ideas come up at least that often. A word of warning to the instructor in this regard: Some restraint is needed in assigning certain problems too early. There are occasional "sleepers" (deceptively difficult problems), intended to serve more as brainwashing than as homework. A veteran will have little trouble spotting them. And a word of warning to the student, too: Since the exercises are part of the text, a few important notions make their first appearance in an exercise. Be on the lookout for **bold** type; it's used to highlight key words and will help you spot these important exercises.

You will notice that certain of the exercises are marked with a small triangle (▷) in the margin. For a variety of reasons, I have deemed these exercises important for a full understanding of the material. Many are straightforward "computations," some are simple detail checking, and at least a few unveil the germs of ideas essential for later developments. Again, a veteran will find it easy to distinguish one from the other. In my own experience, the marked exercises provide a reasonable source for assignments as well as topics for in-class discussion.

To encourage independent study (and because I enjoyed doing it), I have included a short section of "Notes and Remarks" at the end of each chapter. Here I discuss additional or peripheral topics of interest, alternate presentations, and historical commentary. The references cited here include not only primary sources, both technical and historical, but also various secondary sources, such as survey or expository articles.

A word or two about organization: Exercises are numbered consecutively within a given chapter. However, when referring to a given exercise from outside its home chapter, a chapter number is also included. Thus, Exercise 14 refers to the fourteenth exercise in the current chapter, while Exercise 3.26 refers to the twenty-sixth exercise in Chapter Three. The various lemmas, theorems, corollaries, and examples are likewise numbered consecutively within a chapter, without regard to label, and always carry the number of the chapter where they reside. This means that the lemma immediately following Proposition 10.5 is labeled Lemma 10.6, even if it is the first lemma to appear in the chapter, and Lemma 10.6 may well be followed by Theorem 10.7, the second theorem in the chapter. In any case, all three items appear in Chapter Ten.

Many people endured this project with me, and quite a few helped along the way. I would not have survived the process had it not been for the constant encouragement and expert guidance offered by my friends Patrick Flinn and Stephen Dilworth. Equally important were my colleagues Steven Seubert and Kit Chan, who graciously agreed to field-test the notes, and who patiently entertained endless discussions of minutiae. Of course, a large debt of gratitude is also owed to the many students who suffered through early versions of these notes. You have them to thank for each passage that "works" (and only me to blame for those that don't). Finally, copious thanks to my wife Cheryl, who, with good humor and affection, indulged my musings and maintained my sanity.

– N. C.

PART ONE

METRIC SPACES

CHAPTER ONE

Calculus Review

Our goal in this chapter is to provide a quick review of a handful of important ideas from advanced calculus (and to encourage a bit of practice on these fundamentals). We will make no attempt to be thorough. Our purpose is to set the stage for later generalizations and to collect together in one place some of the notation that should already be more or less familiar. There are sure to be missing details, unexplained terminology, and incomplete proofs. On the other hand, since much of this material will reappear in later chapters in a more general setting, you will get to see some of the details more than once. In fact, you may find it entertaining to refer to this chapter each time an old name is spoken in a new voice. If nothing else, there are plenty of keywords here to assist you in looking up any facts that you have forgotten.

The Real Numbers

First, let's agree to use a standard notation for the various familiar sets of numbers. $\mathbb{R}$ denotes the set of all real numbers; $\mathbb{C}$ denotes the set of all complex numbers (although our major concern here is $\mathbb{R}$, we will use complex numbers from time to time); $\mathbb{Z}$ stands for the integers (negative, zero, and positive); $\mathbb{N}$ is the set of natural numbers (positive integers); and $\mathbb{Q}$ is the set of rational numbers. We won't give the set of irrational numbers its own symbol; rather we'll settle for writing $\mathbb{R}\setminus\mathbb{Q}$ (the set-theoretic difference of $\mathbb{R}$ and $\mathbb{Q}$).

We will assume most of the basic algebraic and order properties of these sets, but we will review a few important ideas. Of greatest importance to us is that the set $\mathbb{R}$ of real numbers is **complete** – in more than one sense! First, recall that a subset A of $\mathbb{R}$ is said to be **bounded above** if there is some $x \in \mathbb{R}$ such that $a \leq x$ for all $a \in A$. Any such number x is called an **upper bound** for A. The real numbers are constructed so that any nonempty set with an upper bound has, in fact, a least upper bound (l.u.b.). We won't give the details of this construction; instead we'll take this property as an axiom:

The Least Upper Bound Axiom (sometimes called **the completeness axiom).** *Any nonempty set of real numbers with an upper bound has a least upper bound.*

That is, if $A \subset \mathbb{R}$ is nonempty and bounded above, then there is a number $s \in \mathbb{R}$ satisfying: (i) s is an upper bound for A; and (ii) if x is any upper bound for A, then

$s \le x$. In other words, if $y < s$, then we must have $y < a \le s$ for some $a \in A$. (Why?) We even have a notation for this: In this case we write $s = \text{l.u.b.}\, A = \sup A$ (for **supremum**). If A fails to be bounded above, we set $\sup A = +\infty$, and if $A = \emptyset$, we put $\sup A = -\infty$ since, after all, every real number is an upper bound for A.

Example 1.1

$\sup(-\infty, 1) = 1$ and $\sup\{2 - (1/n) : n = 1, 2, \dots\} = 2$. Notice, please, that $\sup A$ is *not* necessarily an element of A.

An immediate consequence of the least upper bound axiom is that we also have **greatest lower bounds** (g.l.b.), just by turning things around. The details are left as Exercise 1.

EXERCISE

▷ **1.** If A is a nonempty subset of $\mathbb{R}$ that is bounded below, show that A has a greatest lower bound. That is, show that there is a number $m \in \mathbb{R}$ satisfying: (i) m is a lower bound for A; and (ii) if x is a lower bound for A, then $x \le m$. [Hint: Consider the set $-A = \{-a : a \in A\}$ and show that $m = -\sup(-A)$ works.]

We have a notation for greatest lower bounds, too, of course: We write $m = \text{g.l.b.}\, A = \inf A$ (for **infimum**). It follows from Exercise 1 that $\inf A = -\sup(-A)$. Thus, $\inf A = -\infty$ if A isn't bounded below, and $\inf \emptyset = +\infty$. In case a set A is both bounded above and bounded below, we simply say that A is **bounded**.

EXERCISES

2. Let A be a bounded subset of $\mathbb{R}$ containing at least two points. Prove:

(a) $-\infty < \inf A < \sup A < +\infty$.

(b) If B is a nonempty subset of A, then $\inf A \le \inf B \le \sup B \le \sup A$.

(c) If B is the set of all upper bounds for A, then B is nonempty, bounded below, and $\inf B = \sup A$.

▷ **3.** Establish the following apparently different (but "fancier") characterization of the supremum. Let A be a nonempty subset of $\mathbb{R}$ that is bounded above. Prove that $s = \sup A$ if and only if (i) s is an upper bound for A, and (ii) for every $\varepsilon > 0$, there is an $a \in A$ such that $a > s - \varepsilon$. State and prove the corresponding result for the infimum of a nonempty subset of $\mathbb{R}$ that is bounded below.

Recall that a sequence (x_n) of real numbers is said to **converge** to $x \in \mathbb{R}$ if, for every $\varepsilon > 0$, there is a positive integer N such that $|x_n - x| < \varepsilon$ whenever $n \ge N$. In this case, we call x the **limit** of the sequence (x_n) and write $x = \lim_{n\to\infty} x_n$.

▷ **4.** Let A be a nonempty subset of $\mathbb{R}$ that is bounded above. Show that there is a sequence (x_n) of elements of A that converges to $\sup A$.

5. Suppose that $a_n \le b$, for all n, and that $a = \lim_{n\to\infty} a_n$ exists. Show that $a \le b$. Conclude that $a \le \sup_n a_n = \sup\{a_n : n \in \mathbb{N}\}$.

▷ **6.** Prove that every convergent sequence of real numbers is bounded. Moreover, if (a_n) is convergent, show that $\inf_n a_n \le \lim_{n\to\infty} a_n \le \sup_n a_n$.

As an application of the least upper bound axiom, we next establish *the Archimedean property in* $\mathbb{R}$.

Lemma 1.2. *If x and y are positive real numbers, then there is some positive integer n such that $nx > y$.*

PROOF. Suppose that no such n existed; that is, suppose that $nx \le y$ for all $n \in \mathbb{N}$. Then $A = \{nx : n \in \mathbb{N}\}$ is bounded above by y, and so $s = \sup A$ is finite. Now, since $s - x < s$, we must have some element of A in between, that is, $s - x < nx \le s$ for some $n \in \mathbb{N}$. But then $s < (n+1)x$. And what's wrong? Well, since $(n+1)x \in A$, we should instead have $(n+1)x \le s$. This contradiction tells us that it is unacceptable to have $nx \le y$ for all n, and so we must have $nx > y$ for some n. □

This simple observation does a lot of good:

Theorem 1.3. *If a and b are real numbers with $a < b$, then there is a rational $r \in \mathbb{Q}$ with $a < r < b$.*

PROOF. Since $b - a > 0$, we may apply Lemma 1.2 to get a positive integer q such that $q(b - a) > 1$. But if qa and qb differ by more than 1, there must be some integer in between. That is, there is some $p \in \mathbb{Z}$ with $qa < p < qb$. Thus $a < p/q < b$. □

EXERCISES

7. If $a < b$, then there is also an irrational $x \in \mathbb{R} \setminus \mathbb{Q}$ with $a < x < b$. [Hint: Find an irrational of the form $p\sqrt{2}/q$.]

8. Given $a < b$, show that there are, in fact, infinitely many distinct rationals between a and b. The same goes for irrationals, too.

9. Show that the least upper bound axiom also holds in $\mathbb{Z}$ (i.e., each nonempty subset of $\mathbb{Z}$ with an upper bound in $\mathbb{Z}$ has a least upper bound in $\mathbb{Z}$), but that it *fails* to hold in $\mathbb{Q}$.

It follows from Theorem 1.3 that every real number is the limit of a monotone (i.e., increasing or decreasing) sequence of rationals (or irrationals). We'll want to take full advantage of this fact, and we'll see at least one more reason why it's true. First, though, let's give a formal statement of the property behind it.

Theorem 1.4. *A monotone, bounded sequence of real numbers converges.*

PROOF. Let $(x_n) \subset \mathbb{R}$ be monotone and bounded. We first suppose that (x_n) is increasing. Now, since (x_n) is bounded, we may set $x = \sup_n x_n$ (a real number). We will show that $x = \lim_{n\to\infty} x_n$.

Let $\varepsilon > 0$. Since $x - \varepsilon < x = \sup_n x_n$, we must have $x_N > x - \varepsilon$ for some N. But then, for any $n \ge N$, we have $x - \varepsilon < x_N \le x_n \le x$. (Why?) That is, $|x - x_n| < \varepsilon$ for all $n \ge N$. Consequently, (x_n) converges and $x = \sup_n x_n = \lim_{n\to\infty} x_n$.

Finally, if (x_n) is decreasing, consider the increasing sequence $(-x_n)$. From the first part of the proof, $(-x_n)$ converges to $\sup_n(-x_n) = -\inf_n x_n$. It then follows that (x_n) converges to $\inf_n x_n$. □

In subsequent chapters we will consider certain properties of the real line that may be defined either in terms of sequences or in terms of subsets of $\mathbb{R}$. To better appreciate the connection between sequences and sets, we will show how Theorem 1.4 gives a quick proof of the nested interval theorem. Later in this chapter we will use the nested interval theorem to define a strange and beautiful subset of $\mathbb{R}$ called the Cantor set.

The Nested Interval Theorem 1.5. *If (I_n) is a sequence of closed, bounded, nonempty intervals in $\mathbb{R}$ with $I_1 \supset I_2 \supset I_3 \supset \cdots$, then $\bigcap_{n=1}^{\infty} I_n \ne \emptyset$. If, in addition,* length $(I_n) \to 0$, *then $\bigcap_{n=1}^{\infty} I_n$ contains precisely one point.*

PROOF. Write $I_n = [\,a_n, b_n\,]$. Then $I_n \supset I_{n+1}$ means that $a_n \le a_{n+1} \le b_{n+1} \le b_n$ for all n. Thus, $a = \lim_{n\to\infty} a_n = \sup_n a_n$ and $b = \lim_{n\to\infty} b_n = \inf_n b_n$ both exist (as finite real numbers) and satisfy $a \le b$. (Why?) Thus we must have $\bigcap_{n=1}^{\infty} I_n = [\,a, b\,]$. Indeed, if $x \in I_n$ for all n, then $a_n \le x \le b_n$ for all n, and hence $a \le x \le b$. Conversely, if $a \le x \le b$, then $a_n \le x \le b_n$ for all n. That is, $x \in I_n$ for all n. Finally, if $b_n - a_n = \text{length}\,(I_n) \to 0$, then $a = b$ and so $\bigcap_{n=1}^{\infty} I_n = \{a\}$. □

Examples 1.6

(a) Please note that it is essential that the intervals used in the nested interval theorem be both closed and bounded. Indeed, $\bigcap_{n=1}^{\infty}[n, \infty) = \emptyset$ and $\bigcap_{n=1}^{\infty}(0, 1/n] = \emptyset$.

(b) Suppose that (I_n) is a sequence of closed intervals with $I_n \supset I_{n+1}$, for all n and with length $(I_n) \to 0$ as $n \to \infty$. If $\bigcap_{n=1}^{\infty} I_n = \{x\}$, then any sequence of points (x_n), with $x_n \in I_n$ for all n, must converge to x. (Why?)

A sequence of sets (I_n) with $I_n \supset I_{n+1}$ for all n is often said to be a *decreasing* sequence of sets. Thus, the nested interval theorem might be paraphrased by saying that a decreasing sequence of closed, bounded, nonempty intervals "converges" to a nonempty set. In this language, the nested interval theorem is at least reminiscent of the fact that a monotone bounded sequence of real numbers is convergent. And with good reason: The fact that monotone bounded sequences converge is actually equivalent to the least upper bound axiom, as is the nested interval theorem. That is, we might just as well have assumed the conclusion of either Theorem 1.4 or Theorem 1.5 as an axiom for $\mathbb{R}$ and deduced the existence of least upper bounds as a corollary. As evidence, here is a proof that the nested interval theorem implies the existence of least upper bounds (this is similar in spirit to Bolzano's original proof):

Let A be a nonempty subset of $\mathbb{R}$ that is bounded above. Specifically, let $a_1 \in A$ and let b_1 be an upper bound for A. For later reference, set $I_1 = [\, a_1, b_1 \,]$. Now consider the point $x_1 = (a_1 + b_1)/2$, halfway between a_1 and b_1. If x_1 is an upper bound for A, we set $I_2 = [\, a_1, x_1 \,]$; otherwise, there is an element $a_2 \in A$ with $a_2 > x_1$. In this case, set $I_2 = [\, a_2, b_1 \,]$. In either event, I_2 is a closed subinterval of I_1 of the form $[\, a_2, b_2 \,]$, where $a_2 \in A$ and b_2 is an upper bound for A. Moreover, length$(I_2) \le$ length$(I_1)/2$. We now start the process all over again, using I_2 in place of I_1, and obtain a closed subinterval $I_3 = [\, a_3, b_3 \,] \subset I_2$, where $a_3 \in A$ and b_3 is an upper bound for A, with length$(I_3) \le$ length$(I_2)/2 \le$ length$(I_1)/4$. By induction, we get a sequence of nested closed intervals $I_n = [\, a_n, b_n \,]$, where $a_n \in A$ and b_n is an upper bound for A, with length$(I_n) \le$ length$(I_1)/2^{n-1} \to 0$ as $n \to \infty$. The single point $b \in \bigcap_{n=1}^{\infty} I_n$ is the least upper bound for A. (Why?)

EXERCISES

10. Let $a_1 = \sqrt{2}$ and let $a_{n+1} = \sqrt{2a_n}$ for $n \ge 1$. Show that (a_n) converges and find its limit. [Hint: Show that (a_n) is increasing and bounded.]

11. Fix $a > 0$ and let $x_1 > \sqrt{a}$. For $n \ge 1$, define

$$x_{n+1} = \frac{1}{2}\left(x_n + \frac{a}{x_n}\right).$$

Show that (x_n) converges and that $\lim_{n\to\infty} x_n = \sqrt{a}$.

12. Suppose that $s_1 > s_2 > 0$ and let $s_{n+1} = \frac{1}{2}(s_n + s_{n-1})$ for $n \ge 2$. Show that (s_n) converges. [Hint: Show that (s_{2n-1}) decreases and (s_{2n}) increases.]

▷ **13.** Let $a_n \ge 0$ for all n, and let $s_n = \sum_{i=1}^{n} a_i$. Show that (s_n) converges if and only if (s_n) is bounded.

Recall that a sequence of real numbers (x_n) is said to be **Cauchy** if, for every $\varepsilon > 0$, there is an integer $N \ge 1$ such that $|x_n - x_m| < \varepsilon$ whenever $n, m \ge N$.

▷ **14.** Prove that a convergent sequence is Cauchy, and that any Cauchy sequence is bounded.

▷ **15.** Show that a Cauchy sequence with a convergent subsequence actually converges.

16.

(a) Why is $0.4999\ldots = 0.5$? (Try to give more than one reason.)

(b) Write $0.234234234\ldots$ as a fraction.

(c) Precisely which real numbers between 0 and 1 have more than one decimal representation? Explain.

Our second approach to describing the elements of $\mathbb{R}$ as limits of sequences of rational numbers is to consider **decimals**. We might as well do this in some generality.

Proposition 1.7. *Fix an integer $p \geq 2$, and let (a_n) be any sequence of integers satisfying $0 \leq a_n \leq p-1$ for all n. Then, $\sum_{n=1}^{\infty} a_n/p^n$ converges to a number in $[0, 1]$.*

PROOF. Since $a_n \geq 0$, the partial sums $\sum_{n=1}^{N} a_n/p^n$ are nonnegative and increase with N. Thus, to show that the series converges to some number in $[0, 1]$, we just need to show that 1 is an upper bound for the sequence of partial sums. But this is easy:

$$\sum_{n=1}^{N} \frac{a_n}{p^n} \leq \sum_{n=1}^{N} \frac{p-1}{p^n} \leq (p-1)\sum_{n=1}^{\infty} \frac{1}{p^n} = 1.$$

(Why? What does this say when $p = 10$?) □

Conversely, each x in $[0, 1]$ can be so represented:

Proposition 1.8. *Let p be an integer, $p \geq 2$, and let $0 \leq x \leq 1$. Then there is a sequence of integers (a_n) with $0 \leq a_n \leq p-1$ for all n such that $x = \sum_{n=1}^{\infty} a_n/p^n$.*

PROOF. Certainly the case $x = 0$ causes no real strain, so let us suppose that $0 < x \leq 1$. We will construct (a_n) by induction.

Choose a_1 to be the largest integer satisfying $a_1/p < x$. (How?) Since $x > 0$, it follows that $a_1 \geq 0$; and since $x \leq 1$, we have $a_1 < p$. Because a_1 is an integer, this means that $a_1 \leq p-1$. Also, since a_1 is largest, we must have $a_1/p < x \leq (a_1+1)/p$.

Next, choose a_2 to be the largest integer satisfying $a_1/p + a_2/p^2 < x$. Check that $0 \leq a_2 \leq p-1$ and that $a_1/p + a_2/p^2 < x \leq a_1/p + (a_2+1)/p^2$.

By induction we get a sequence of integers (a_n) with $0 \leq a_n \leq p-1$ such that

$$\frac{a_1}{p} + \cdots + \frac{a_n}{p^n} < x \leq \frac{a_1}{p} + \cdots + \frac{a_n+1}{p^n}.$$

Obviously, $x = \sum_{n=1}^{\infty} a_n/p^n$. (Why?) □

The series $\sum_{n=1}^{\infty} a_n/p^n$ is called a base p (or p-adic) decimal expansion for x. It is sometimes written in the shorter form $x = 0.a_1a_2a_3\cdots$ (base p). It does not have to be unique (even for ordinary base 10 decimals: $0.5 = 0.4999\cdots$). One problem is that our construction is designed to produce nonterminating decimal expansions. In the particular case where $x = a_1/p + \cdots + (a_n+1)/p^n = q/p^n$, for some integer $0 < q \leq p^n$, the construction will give us a repeating string of $p-1$'s in the decimal expansion for x since $1/p^n = \sum_{k=n+1}^{\infty}(p-1)/p^k$. That is, any such x has two distinct base p decimal expansions:

$$x = \frac{a_1}{p} + \cdots + \frac{a_n+1}{p^n} = \frac{a_1}{p} + \cdots + \frac{a_n}{p^n} + \sum_{k=n+1}^{\infty} \frac{p-1}{p^k}.$$

We now have several methods for finding a sequence of rationals that increase or decrease to a given real number. An application of this fact can be used to define expressions such as a^x for real exponents x. For example, if $a > 1$, and if x is any real number, then we set $a^x = \sup\{a^r : r \in \mathbb{Q},\ r < x\}$. We get away with this because a^r is well defined and increasing for $r \in \mathbb{Q}$.

You may have been tempted to use logarithms or exponentials to define a^x, but we would need a similar line of reasoning to define, say, e^x (or even e itself!), and we would need quite a bit more machinery to define $\log x$. As long as we've already digressed from decimals, let's construct e. For this we'll use a simple (but extremely useful) inequality.

Bernoulli's Inequality 1.9. *If $a > -1$, $a \neq 0$, then $(1+a)^n > 1 + na$ for any integer $n > 1$.*

The proof of Bernoulli's inequality is left as an exercise. We'll apply it to prove:

Proposition 1.10.

(i) $\left(1+\frac{1}{n}\right)^n$ *is strictly increasing.*

(ii) $\left(1+\frac{1}{n}\right)^{n+1}$ *is strictly decreasing.*

(iii) $2 \le \left(1+\frac{1}{n}\right)^n < \left(1+\frac{1}{n}\right)^{n+1} \le 4.$

(iv) *Both sequences converge to the same limit $e = \lim_{n\to\infty}(1+(1/n))^n$, where $2 < e < 4$.*

PROOF. (i) We need to show that $(1+1/(n+1))^{n+1}/(1+(1/n))^n > 1$. For this we rewrite and apply Bernoulli's inequality:

$$\begin{aligned}
\frac{(1+\frac{1}{n+1})^{n+1}}{(1+\frac{1}{n})^n} &= \left(1+\frac{1}{n}\right)\cdot\left(\frac{1+\frac{1}{n+1}}{1+\frac{1}{n}}\right)^{n+1} \\
&= \left(1+\frac{1}{n}\right)\cdot\left(\frac{n^2+2n}{(n+1)^2}\right)^{n+1} \\
&= \left(1+\frac{1}{n}\right)\cdot\left(1-\frac{1}{(n+1)^2}\right)^{n+1} \\
&> \left(1+\frac{1}{n}\right)\cdot\left(1-\frac{1}{n+1}\right) = 1 \qquad \text{(by Bernoulli).}
\end{aligned}$$

(ii) This case is very similar to (i).

$$\begin{aligned}
\frac{(1+\frac{1}{n})^{n+1}}{(1+\frac{1}{n+1})^{n+2}} &= \frac{1}{(1+\frac{1}{n})}\cdot\left(\frac{1+\frac{1}{n}}{1+\frac{1}{n+1}}\right)^{n+2} \\
&= \left(\frac{n}{n+1}\right)\cdot\left(\frac{(n+1)^2}{n^2+2n}\right)^{n+2} \\
&= \left(\frac{n}{n+1}\right)\cdot\left(1+\frac{1}{n(n+2)}\right)^{n+2} \\
&> \left(\frac{n}{n+1}\right)\cdot\left(1+\frac{1}{n}\right) = 1 \qquad \text{(by Bernoulli).}
\end{aligned}$$

(iii) Since $1+(1/n) > 1$, we have $(1+(1/n))^n < (1+(1/n))^{n+1}$. Since $(1+(1/n))^n$ increases, the left-hand side is at least 2 (the first term); and since $(1+(1/n))^{n+1}$ decreases, the right-hand side is at most 4 (the first term).

(iv) Finally, we define $e = \lim_{n\to\infty}(1+(1/n))^n$, and conclude that

$$\lim_{n\to\infty}\left(1+\frac{1}{n}\right)^{n+1} = \lim_{n\to\infty}\left(1+\frac{1}{n}\right)\lim_{n\to\infty}\left(1+\frac{1}{n}\right)^n = e. \quad \square$$

The same proof applies to the sequence $(1+(x/n))^n$ for any $x \in \mathbb{R}$, and we may define $e^x = \lim_{n\to\infty}(1+(x/n))^n$. The full details of this last conclusion are best left for another day. See Exercise 18(b).

EXERCISES

▷ **17.** Given real numbers a and b, establish the following formulas: $|a+b| \le |a|+|b|$, $|\,|a|-|b|\,| \le |a-b|$, $\max\{a,b\} = \frac{1}{2}(a+b+|a-b|)$, and $\min\{a,b\} = \frac{1}{2}(a+b-|a-b|)$.

18.

(a) Given $a > -1$, $a \ne 0$, use induction to show that $(1+a)^n > 1+na$ for any integer $n > 1$.

(b) Use (a) to show that, for any $x > 0$, the sequence $(1+(x/n))^n$ increases.

(c) If $a > 0$, show that $(1+a)^r > 1+ra$ holds for any *rational* exponent $r > 1$. [Hint: If $r = p/q$, then apply (a) with $n = q$ and (b) with $x = ap$.]

(d) Finally, show that (c) holds for any *real* exponent $r > 1$.

19. If $0 < c < 1$, show that $c^n \to 0$; and if $c > 0$, show that $c^{1/n} \to 1$. [Hint: Use Bernoulli's inequality for each, once with $c = 1/(1+x)$, $x > 0$ and once with $c^{1/n} = 1+x_n$, where $x_n > 0$.]

20. Given $a, b > 0$, show that $\sqrt{ab} \le \frac{1}{2}(a+b)$ (this is the arithmetic-geometric mean inequality). Generalize this to $(a_1 \cdot a_2 \cdots a_n)^{1/n} \le (1/n)(a_1+a_2+\cdots+a_n)$. [Hint: Induction and Bernoulli's inequality.]

▷ **21.** Let $p \ge 2$ be a fixed integer, and let $0 < x < 1$. If x has a finite-length base p decimal expansion, that is, if $x = a_1/p + \cdots + a_n/p^n$ with $a_n \ne 0$, prove that x has precisely *two* base p decimal expansions. Otherwise, show that the base p decimal expansion for x is unique. Characterize the numbers $0 < x < 1$ that have *repeating* base p decimal expansions. How about *eventually repeating*?

As long as we are on the subject of sequences, this is a good time to outline part of the master plan! Virtually everything that we need to know about the real line $\mathbb{R}$ and about functions $f : \mathbb{R} \to \mathbb{R}$ can be described in terms of convergent sequences. Indeed, a continuous function $f : \mathbb{R} \to \mathbb{R}$ could be defined as a function that "preserves" convergent sequences: $f(\lim_{n\to\infty} x_n) = \lim_{n\to\infty} f(x_n)$. If we hope to understand continuous functions (and we do!), then it is of great importance to us to know precisely which real sequences converge. So far we know that monotone, bounded sequences converge, and that any convergent sequence is necessarily bounded. (Why?) These two facts together raise the question: Does every bounded sequence converge? Of course not. But just how "far" from convergent is a typical bounded sequence? To answer this, we will want to broaden our definition of limit. First a few easy observations.

Let (a_n) be a bounded sequence of real numbers, and consider the sequences:

$$t_n = \inf\{a_n, a_{n+1}, a_{n+2}, \dots\} \quad \text{and} \quad T_n = \sup\{a_n, a_{n+1}, a_{n+2}, \dots\}.$$

Then (t_n) *increases*, (T_n) *decreases*, and $\inf_k a_k \le t_n \le T_n \le \sup_k a_k$ for all n. (Why?) Thus we may speak of $\lim_{n\to\infty} t_n$ as the "lower limit" and $\lim_{n\to\infty} T_n$ as the "upper limit" of our original sequence (a_n). And that is exactly what we will do.

Now these same considerations are meaningful even if we start with an *unbounded* sequence (a_n), although in that case we will have to allow the values $\pm\infty$ for at least some of the t_n's or T_n's (possibly both). That is, if we permit comparisons to $\pm\infty$, then the t_n's still increase and the T_n's still decrease. Of course we will want to use $\sup_n t_n$ and $\inf_n T_n$ in place of $\lim_{n\to\infty} t_n$ and $\lim_{n\to\infty} T_n$, since "sup" and "inf" have more or less obvious extensions to subsets of the extended real number system $[-\infty, +\infty]$ whereas "lim" does not. Even so, we are sure to get caught saying something like "(t_n) *converges* to $+\infty$." But we will pay a stiff penalty for too much rigor here; even a simple fact could have a tediously long description. For the remainder of this section you are encouraged to interpret words such as "limit" and "converges" in this looser sense.

Given any sequence of real numbers (a_n), we define

$$\liminf_{n\to\infty} a_n = \varliminf_{n\to\infty} a_n = \sup_{n\ge 1}\left(\inf\{a_n, a_{n+1}, a_{n+2}, \dots\}\right)$$

and

$$\limsup_{n\to\infty} a_n = \varlimsup_{n\to\infty} a_n = \inf_{n\ge 1}\big(\sup\{a_n, a_{n+1}, a_{n+2}, \dots\}\big).$$

That is, $\liminf_{n\to\infty} a_n = \sup_n t_n$ ($=\lim_{n\to\infty} t_n$ if (a_n) is bounded from below) and $\limsup_{n\to\infty} a_n = \inf_n T_n$ ($=\lim_{n\to\infty} T_n$ if (a_n) is bounded from above). The name "lim inf" is short for "limit inferior," while "lim sup" is short for "limit superior."

EXERCISES

22. Show that $\inf_n a_n \le \liminf_{n\to\infty} a_n \le \limsup_{n\to\infty} a_n \le \sup_n a_n$.

23. If (a_n) is convergent, show that $\liminf_{n\to\infty} a_n = \limsup_{n\to\infty} a_n = \lim_{n\to\infty} a_n$.

▷ **24.** Show that $\limsup_{n\to\infty}(-a_n) = -\liminf_{n\to\infty} a_n$.

▷ **25.** If $\limsup_{n\to\infty} a_n = -\infty$, show that (a_n) diverges to $-\infty$. If $\limsup_{n\to\infty} a_n = +\infty$, show that (a_n) has a *subsequence* that diverges to $+\infty$. What happens if $\liminf_{n\to\infty} a_n = \pm\infty$?

If we start with a *bounded* sequence (a_n), then

$$M = \limsup_{n\to\infty} a_n = \lim_{n\to\infty}\left(\sup\{a_k : k \ge n\}\right) \ne \pm\infty,$$

and hence:

$$\begin{cases} \text{for every } \varepsilon > 0, \text{ there is an integer } N \geq 1 \text{ such that} \\ M - \varepsilon < \sup\{a_k : k \geq n\} < M + \varepsilon \text{ for all } n \geq N. \end{cases}$$

Thus, the number $M = \limsup_{n\to\infty} a_n$ is *characterized* by the following:

$$(*) \begin{cases} \text{for every } \varepsilon > 0, \text{ we have } a_n < M + \varepsilon \text{ for all but finitely} \\ \text{many } n, \text{ and } M - \varepsilon < a_n \text{ for infinitely many } n. \end{cases}$$

EXERCISES

▷ **26.** Prove the characterization of lim sup given above. That is, given a bounded sequence (a_n), show that the number $M = \limsup_{n\to\infty} a_n$ satisfies $(*)$ and, conversely, that any number M satisfying $(*)$ must equal $\limsup_{n\to\infty} a_n$. State and prove the corresponding result for $m = \liminf_{n\to\infty} a_n$.

▷ **27.** Prove that every sequence of real numbers (a_n) has a subsequence (a_{n_k}) that converges to $\limsup_{n\to\infty} a_n$. [Hint: If $M = \limsup_{n\to\infty} a_n = \pm\infty$, we must interpret the conclusion loosely; this case is handled in Exercise 25. If $M \neq \pm\infty$, use $(*)$ to choose (a_{n_k}) satisfying $|a_{n_k} - M| < 1/k$, for example.] There is necessarily also a subsequence that converges to $\liminf_{n\to\infty} a_n$. Why?

28. By modifying the argument in the previous exercise, show that every sequence of real numbers has a *monotone* subsequence.

29. If (a_{n_k}) is a convergent subsequence of (a_n), show that $\liminf_{n\to\infty} a_n \leq \lim_{k\to\infty} a_{n_k} \leq \limsup_{n\to\infty} a_n$.

30. If $a_n \leq b_n$ for all n, and if (a_n) converges, show that $\lim_{n\to\infty} a_n \leq \liminf_{n\to\infty} b_n$.

31. If (a_n) is convergent and (b_n) is bounded, show that $\limsup_{n\to\infty} (a_n + b_n) \leq \lim_{n\to\infty} a_n + \limsup_{n\to\infty} b_n$.

32. Given a sequence (a_n) of real numbers, let $\mathcal{S}$ be the set of all limits of convergent subsequences of (a_n) (including, possibly, $\pm\infty$). For example, it follows from Exercise 27 that $\limsup_{n\to\infty} a_n$ and $\liminf_{n\to\infty} a_n$ are both elements of $\mathcal{S}$. Show that, in fact, $\limsup_{n\to\infty} a_n = \sup \mathcal{S}$ and $\liminf_{n\to\infty} a_n = \inf \mathcal{S}$.

The ability to find a convergent subsequence of an arbitrary sequence, as in Exercise 27, leads to a whole slew of corollaries. See if you can supply proofs for the following:

The Bolzano–Weierstrass Theorem 1.11. *Every bounded sequence of real numbers has a convergent subsequence.*

Corollary 1.12. *If (a_n) is a convergent sequence, then* $\liminf_{n\to\infty} a_n = \limsup_{n\to\infty} a_n = \lim_{n\to\infty} a_n$.

Corollary 1.13. *Every Cauchy sequence of real numbers converges.*

Corollary 1.14. *Every bounded sequence of real numbers has a Cauchy subsequence.*

[Hint: See Exercises 14 and 15 for more on Cauchy sequences.]

Finally, we come full circle:

Proposition 1.15. *If* (a_n) *is bounded, and if* $\liminf_{n\to\infty} a_n = \limsup_{n\to\infty} a_n$, *then* (a_n) *converges and* $\lim_{n\to\infty} a_n = \limsup_{n\to\infty} a_n$.

PROOF. Let $a = \liminf_{n\to\infty} a_n = \limsup_{n\to\infty} a_n$, and let $\varepsilon > 0$. From our characterizations of lim inf and lim sup, there is an $N_1 \geq 1$ such that $a - \varepsilon < a_n$ for all $n \geq N_1$ (since $a = \liminf_{n\to\infty} a_n$), and there is an $N_2 \geq 1$ such that $a_n < a + \varepsilon$ for all $n \geq N_2$ (since $a = \limsup_{n\to\infty} a_n$). Thus, for $n \geq \max\{N_1, N_2\}$ we have $|a - a_n| < \varepsilon$. □

You may recall that a sequence of real numbers converges if and only if it is Cauchy. Although one approach to this fact has already been suggested in the exercises, it is such an important property of the real numbers that it is well worth the effort to give a second proof!

First recall that if a sequence converges, then it is Cauchy; and if a sequence is Cauchy, then it is also bounded. (See the exercises for more details.) We want to reverse the first implication, and so we may assume that we have a bounded sequence to start with. This helps, since for a bounded sequence (a_n) both $\limsup_{n\to\infty} a_n$ and $\liminf_{n\to\infty} a_n$ are (finite) real numbers. Given a Cauchy sequence, then, we only need to check that these two numbers are equal, which is easier than it might sound.

Theorem 1.16. *A sequence of real numbers converges if (and only if) it is Cauchy.*

PROOF. Let (a_n) be Cauchy, and let $\varepsilon > 0$. Choose $N \geq 1$ such that $|a_n - a_m| < \varepsilon$ for all $m, n \geq N$. Then, in particular, we have $a_N - \varepsilon < a_n < a_N + \varepsilon$ for all $n \geq N$; thus, (a_n) is bounded. But $a_N - \varepsilon < a_n$ for $n \geq N$ implies that $a_N - \varepsilon \leq \liminf a_n$, while $a_n < a_N + \varepsilon$ for $n \geq N$ implies that $\limsup_{n\to\infty} a_n \leq a_N + \varepsilon$. (Why?) Since $-\infty < \liminf a_n \leq \limsup_{n\to\infty} a_n < \infty$, we may subtract these results and conclude that $\limsup_{n\to\infty} a_n - \liminf_{n\to\infty} a_n < 2\varepsilon$. Since $\varepsilon > 0$ is arbitrary, we get that $\limsup_{n\to\infty} a_n = \liminf_{n\to\infty} a_n$. □

EXERCISES

▷ **33.** Show that (x_n) converges to $x \in \mathbb{R}$ if and only if every subsequence (x_{n_k}) of (x_n) has a *further* subsequence $(x_{n_{k_l}})$ that converges to x.

34. Suppose that $a_n \geq 0$ and that $\sum_{n=1}^{\infty} a_n < \infty$.

(i) Show that $\liminf_{n\to\infty} n a_n = 0$.

(ii) Give an example showing that $\limsup_{n\to\infty} n a_n > 0$ is possible.

35. (The ratio test): Let $a_n \geq 0$.

(i) If $\limsup_{n\to\infty} a_{n+1}/a_n < 1$, show that $\sum_{n=1}^{\infty} a_n < \infty$.

(ii) If $\liminf_{n\to\infty} a_{n+1}/a_n > 1$, show that $\sum_{n=1}^{\infty} a_n$ diverges.

(iii) Find examples of both a convergent and a divergent series having $\lim_{n\to\infty} a_{n+1}/a_n = 1$.

36. (The root test): Let $a_n \geq 0$.

(i) If $\limsup_{n\to\infty} \sqrt[n]{a_n} < 1$, show that $\sum_{n=1}^{\infty} a_n < \infty$.

(ii) If $\liminf_{n\to\infty} \sqrt[n]{a_n} > 1$, show that $\sum_{n=1}^{\infty} a_n$ diverges.

(iii) Find examples of both a convergent and a divergent series having $\lim_{n\to\infty} \sqrt[n]{a_n} = 1$.

▷ **37.** If (E_n) is a sequence of subsets of a fixed set S, we define

$$\limsup_{n\to\infty} E_n = \bigcap_{n=1}^{\infty}\left(\bigcup_{k=n}^{\infty} E_k\right) \quad \text{and} \quad \liminf_{n\to\infty} E_n = \bigcup_{n=1}^{\infty}\left(\bigcap_{k=n}^{\infty} E_k\right).$$

Show that

$$\liminf_{n\to\infty} E_n \subset \limsup_{n\to\infty} E_n \quad \text{and that} \quad \liminf_{n\to\infty}\left(E_n^c\right) = \left(\limsup_{n\to\infty} E_n\right)^c.$$

38. Show that

$$\limsup_{n\to\infty} E_n = \{x \in S : x \in E_n \text{ for infinitely many } n\}$$

and that

$$\liminf_{n\to\infty} E_n = \{x \in S : x \in E_n \text{ for all but finitely many } n\}.$$

39. How would you define the limit (if it exists) of a sequence of sets? What should the limit be if $E_1 \supset E_2 \supset \cdots$? If $E_1 \subset E_2 \subset \cdots$? Compute $\liminf_{n\to\infty} E_n$ and $\limsup_{n\to\infty} E_n$ in both cases and test your conjecture.

Limits and Continuity

In this section we present a brief refresher course on limits and continuity for real-valued functions. With any luck, much of what we have to say will be very familiar. To begin, let f be a real-valued function defined (at least) for all points in some open interval containing the point $a \in \mathbb{R}$ except, possibly, at a itself. We will refer to such a set as a **punctured neighborhood** of a. Given a number $L \in \mathbb{R}$, we write $\lim_{x\to a} f(x) = L$ to mean:

$$\left\{\begin{array}{l}\text{for every } \varepsilon > 0, \text{ there is some } \delta > 0 \text{ such that } |f(x) - L| < \varepsilon \\ \text{whenever } x \text{ satisfies } 0 < |x - a| < \delta.\end{array}\right.$$

We say that $\lim_{x\to a} f(x)$ *exists* if there is some number $L \in \mathbb{R}$ that satisfies the requirements spelled out above. The proof of our first result is left as an exercise.

Theorem 1.17. *Let f be a real-valued function defined in some punctured neighborhood of $a \in \mathbb{R}$. Then, the following are equivalent:*

(i) *There exists a number L such that $\lim_{x \to a} f(x) = L$ (by the ε-δ definition).*

(ii) *There exists a number L such that $f(x_n) \to L$ whenever $x_n \to a$, where $x_n \neq a$ for all n.*

(iii) *$(f(x_n))$ converges (to* something) *whenever $x_n \to a$, where $x_n \neq a$ for all n.*

The point to item (iii) is that if $\lim_{n \to \infty} f(x_n)$ always exists, then it must actually be independent of the choice of (x_n). This is not as mystical as it might sound; indeed, if $x_n \to a$ and $y_n \to a$, then the sequence $x_1, y_1, x_2, y_2, \ldots$ also converges to a. (How does this help?) This particular phrasing is interesting because it does not refer to L. That is, we can test for the *existence* of a limit without knowing its value.

Now suppose that f is defined in a neighborhood of a, this time including the point a itself. We say that f **is continuous at** a if $\lim_{x \to a} f(x) = f(a)$. That is, if:

$$\begin{cases} \text{for every } \varepsilon > 0, \text{ there is a } \delta > 0 \text{ (that depends on } f, a, \text{ and } \varepsilon\text{)} \\ \text{such that } |f(x) - f(a)| < \varepsilon \text{ whenever } x \text{ satisfies } |x - a| < \delta. \end{cases}$$

Notice that we replaced L by $f(a)$ and we dropped the requirement that $x \neq a$. Theorem 1.17 has an obvious extension to this case (and its proof is also left as an exercise).

Theorem 1.18. *Let f be a real-valued function defined in some neighborhood of $a \in \mathbb{R}$. Then, the following are equivalent:*

(i) *f is continuous at a (by the ε-δ definition);*

(ii) *$f(x_n) \to f(a)$ whenever $x_n \to a$;*

(iii) *$(f(x_n))$ converges (to* something) *whenever $x_n \to a$.*

Notice that we dropped the requirement that $x_n \neq a$. Thus, if $\lim_{n \to \infty} f(x_n)$ always exists, then it must equal $f(a)$. (Why?)

You might also recall that we have a notation for left- and right-hand limits and left and right continuity. For example, if we define

$$f(a-) = \lim_{x \to a^-} f(x) \qquad \text{and} \qquad f(a+) = \lim_{x \to a^+} f(x)$$

(provided that these limits exist, of course), then we could add another equivalence to Theorem 1.18:

1.18. (iv) *$f(a-)$ and $f(a+)$ both exist, and both are equal to $f(a)$.*

One-sided limits are peculiar to functions defined on $\mathbb{R}$, and they do not generalize very well (because they are tied to the order in $\mathbb{R}$). But they are very good at what they do: They permit the cataloguing of very refined types of discontinuities. For example, we say that f is **right-continuous** at a if $f(a+)$ exists and equals $f(a)$, and we say that f has a **jump discontinuity** at a if $f(a-)$ and $f(a+)$ both exist but at least one is different from $f(a)$. A function having only jump discontinuities is not so very bad. In particular, monotone functions are rather well behaved:

Proposition 1.19. *Let $f : (a, b) \to \mathbb{R}$ be monotone and let $a < c < b$. Then, $f(c-)$ and $f(c+)$ both exist. Thus, f can have only jump discontinuities.*

PROOF. We might as well suppose that f is increasing (otherwise, consider $-f$). In that case, $f(c)$ is an upper bound for $\{f(t) : a < t < c\}$ and a lower bound for $\{f(t) : c < t < b\}$. All that remains is to check that $\sup\{f(t) : a < t < c\} = \lim_{x\to c^-} f(x)$ and $\inf\{f(t) : c < t < b\} = \lim_{x\to c^+} f(x)$. We will sketch the proof of the first of these.

Given $\varepsilon > 0$, there is some x_0 with $a < x_0 < c$ such that $\sup_{t<c} f(t) - \varepsilon < f(x_0) \le \sup_{t<c} f(t)$. Now let $\delta = c - x_0 > 0$. Then, if $c - \delta < x < c$, we get $x_0 < x < c$, and so $f(x_0) \le f(x) \le \sup_{t<c} f(t)$. Thus, $|f(x) - \sup_{t<c} f(t)| < \varepsilon$. □

EXERCISES

40. Prove Theorem 1.17.

41. Prove Theorem 1.18, including 1.18 (iv) as one of the equivalent conditions.

42. Given $f : (a, b) \to \mathbb{R}$ and $x \in (a, b)$, consider the statements: (i) $\lim_{h\to 0} |f(x + h) - f(x)| = 0$ and (ii) $\lim_{h\to 0} |f(x + h) - f(x - h)| = 0$. Show that (i) always implies (ii). Give an example where (ii) holds but not (i).

43. Modify Theorem 1.17 to characterize the statement $\lim_{x\to a^+} f(x) = L$, and check your new version by providing a proof!

44. If $f : \mathbb{R} \to \mathbb{R}$ is increasing and bounded, show that $\lim_{x\to\infty} f(x)$ and $\lim_{x\to-\infty} f(x)$ both exist.

▷ **45.** Let $f : [a, b] \to \mathbb{R}$ be continuous and suppose that $f(x) = 0$ whenever x is rational. Show that $f(x) = 0$ for every x in $[a, b]$.

▷ **46.** Let $f : \mathbb{R} \to \mathbb{R}$ be continuous.

(a) If $f(0) > 0$, show that $f(x) > 0$ for all x in some open interval $(-a, a)$.

(b) If $f(x) \ge 0$ for every rational x, show that $f(x) \ge 0$ for all real x. Will this result hold with "≥ 0" replaced by "> 0"? Explain.

47. Let f, g, h, and k be defined on $[0, 1]$ as follows:

$$f(x) = \begin{cases} 0 & \text{if } x \notin \mathbb{Q} \\ 1 & \text{if } x \in \mathbb{Q} \end{cases} \qquad h(x) = \begin{cases} 1 - x & \text{if } x \notin \mathbb{Q} \\ x & \text{if } x \in \mathbb{Q} \end{cases}$$

$$g(x) = \begin{cases} 0 & \text{if } x \notin \mathbb{Q} \\ x & \text{if } x \in \mathbb{Q} \end{cases} \qquad k(x) = \begin{cases} 0 & \text{if } x \notin \mathbb{Q} \\ 1/n & \text{if } x = m/n \in \mathbb{Q} \\ & \text{(in lowest terms).} \end{cases}$$

Prove that f is not continuous at any point in $[0, 1]$, that g is continuous only at $x = 0$, that h is continuous only at $x = 1/2$, and that k is continuous only at the irrational points in $[0, 1]$.

48. Give an example of a one-to-one, onto function $f : [0, 1] \to [0, 1]$ that is not monotone. Can you find a monotone, one-to-one function that is not onto? Or a monotone, onto function that is not one-to-one?

49. Let $f : (a, b) \to \mathbb{R}$ be monotone and let $a < x < b$. Show that f is continuous at x if and only if $f(x-) = f(x+)$.

50. Let D denote the set of rationals in $[0, 1]$ and suppose that $f : D \to \mathbb{R}$ is increasing. Show that there is an increasing function $g : [0, 1] \to \mathbb{R}$ such that $g(x) = f(x)$ whenever x is rational. [Hint: For $x \in [0, 1]$, define $g(x) = \sup\{f(t) : 0 \le t \le 1,\ t \in \mathbb{Q}\}$.]

51. Let $f : [a, b] \to \mathbb{R}$ be increasing and define $g : [a, b] \to \mathbb{R}$ by $g(x) = f(x+)$ for $a \le x < b$ and $g(b) = f(b)$. Prove that g is increasing and right-continuous.

Notes and Remarks

Although we cannot claim to have reviewed every last detail that you might need for an untroubled reading of these pages, we have managed to at least recall several important issues. Bartle [1964] and Fulks [1969] are good sources for a review of advanced calculus; Apostol [1975] and Stromberg [1981] are good sources for further details on the topics discussed in this chapter.

Full details of the construction of the real numbers "from scratch" can be found in Birkhoff and MacLane [1965], Goffman [1953a], Hewitt and Stromberg [1965], and Sprecher [1970]. For more on the various equivalent notions of completeness for the real numbers, see the aptly titled article "Completeness of the real numbers" in Goffman [1974]. For more on the history of rigorous analysis, see Boyer [1968], Edwards [1979], Grabiner [1983], Grattan-Guinness [1970], Kitcher [1983], Kleiner [1989], and Kline [1972]. As an interesting tidbit in this vein, Dudley [1989] points out that no proof of the so-called Bolzano–Weierstrass theorem (Corollary 1.11) has ever been found among Bolzano's writings. For a curious observation about real numbers with "ambiguous" decimal representations, see Petkovšek [1990].

Exercise 42 is taken from Apostol [1975].

CHAPTER TWO

Countable and Uncountable Sets

Equivalence and Cardinality

We have seen that the rational numbers are densely distributed on the real line in the sense that there is always a rational between any two distinct real numbers. But even more is true. In fact, it follows that there must be infinitely many rational numbers between any two distinct reals. (Why?) In sharp contrast to this picture of the rationals as a "dense" set, we will show in this section that the rational numbers are actually rather sparsely represented among the real numbers. We will do so by "counting" the rationals!

We say that two sets A and B are **equivalent** if there is a one-to-one correspondence between them. That is, A and B are equivalent if there exists some function $f : A \to B$ that is both one-to-one and onto. As a quick example, you might recall from calculus that the map $x \mapsto \arctan x$ is a strictly increasing (hence one-to-one) function from $\mathbb{R}$ onto the open interval $(-\pi/2, \pi/2)$. Thus, $\mathbb{R}$ is equivalent to $(-\pi/2, \pi/2)$. For convenience we may occasionally write $A \sim B$ in place of the phrase "A is equivalent to B." Please note that the relation "is equivalent to" is an equivalence relation.

The notion of equivalence is supposed to lead us to a notion of the relative sizes of sets. Equivalent sets should, by rights, have the same "number" of elements. For this reason we sometimes say that equivalent sets have the same *cardinality*. (A *cardinal number* is a number that indicates size without regard to order; we will have more to say about cardinal numbers later.) We put this to immediate use: A set A is called **finite** if $A = \emptyset$ or if A is equivalent to the set $\{1, 2, \ldots, n\}$ for some $n \in \mathbb{N}$; otherwise, we say that A is **infinite**. It follows that an infinite set must contain finite subsets of all orders. (Why?)

An infinite set A is said to be **countable** (or **countably infinite**) if A is equivalent to $\mathbb{N}$. That is, the elements of a countable set A can be **enumerated**, or counted, according to their correspondence with the natural numbers: $A = \{x_1, x_2, x_3, \ldots\}$, where the x_i are distinct. Note that this is not quite the same as a sequence. Here A is the range of a *one-to-one* function $f : \mathbb{N} \to A$ and we are simply displaying the elements of A in the order inherited from $\mathbb{N}$; that is, $A = \{f(1), f(2), \ldots\}$. Let us look at a few specific examples.

Examples 2.1

(a) $\mathbb{Z} \sim \mathbb{N}$. To see this, define $f : \mathbb{Z} \to \mathbb{N}$ by $f(n) = 2n$ if $n \geq 1$ and $f(n) = -2n+1$ if $n \leq 0$. The positive integers in $\mathbb{Z}$ are mapped to the even numbers in $\mathbb{N}$, while 0 and the negative integers in $\mathbb{Z}$ are mapped to the odd numbers in $\mathbb{N}$. That f is both one-to-one and onto is easy to check. Notice, please, that $\mathbb{Z}$ is equivalent to a *proper* subset of itself! This is typical of infinite sets.

(b) $\mathbb{N} \times \mathbb{N} \sim \mathbb{N}$. A quick proof is supplied by the fundamental theorem of arithmetic: Each positive integer $k \in \mathbb{N}$ can be uniquely written as $k = 2^{m-1}(2n - 1)$ for some $m, n \in \mathbb{N}$. (Factor out the largest power of 2 from k and what remains is necessarily an odd number.) Here is our map: Define $f : \mathbb{N} \times \mathbb{N} \to \mathbb{N}$ by $f(m, n) = 2^{m-1}(2n - 1)$. That f is both one-to-one and onto is obvious. We will give a second proof shortly.

In actual practice it makes life easier if we simply lump finite and countably infinite sets together under the heading of countable sets or, to be precise, *at-most-countable* sets. After all, the elements of a finite set can surely be counted. The easiest way to perform this consolidation is by modifying our definition of a countable set. *Henceforth, we will say that a countable set is one that is equivalent to some* subset *of* $\mathbb{N}$. This obviously now includes finite sets, but does it include any new, inappropriate sets? To see that this gives us just what we wanted, we prove:

Lemma 2.2. *An infinite subset of* $\mathbb{N}$ *is countable; that is, if* $A \subset \mathbb{N}$ *and if* A *is infinite, then* A *is equivalent to* $\mathbb{N}$.

PROOF. Recall that $\mathbb{N}$ is well ordered. That is, each nonempty subset of $\mathbb{N}$ has a smallest element. Thus, since $A \neq \emptyset$, there is a smallest element $x_1 \in A$. Then $A \setminus \{x_1\} \neq \emptyset$, and there must be a smallest $x_2 \in A \setminus \{x_1\}$. But now $A \setminus \{x_1, x_2\} \neq \emptyset$, and so we continue, setting $x_3 = \min(A \setminus \{x_1, x_2\})$. By induction we can find $x_1, x_2, x_3, \ldots, x_n, \ldots \in A$, where $x_n = \min(A \setminus \{x_1, \ldots, x_{n-1}\})$.

How do we know that this process exhausts A? Well, suppose that $x \in A \setminus \{x_1, x_2, \ldots\} \neq \emptyset$. Then the set $\{k : x_k > x\}$ must be nonempty (otherwise we would have $x \in A$ and $x < x_1 = \min A$), and hence it has a least element. That is, there is some n with $x_1 < \cdots < x_{n-1} < x < x_n$. But this contradicts the choice of x_n as the first element in $A \setminus \{x_1, \ldots, x_{n-1}\}$. Consequently, A is countable. □

It follows from Lemma 2.2 that a subset of $\mathbb{N}$ is either finite or is infinite and equivalent to $\mathbb{N}$. Please be forewarned: Not all authors agree with the convention that we have adopted. We have chosen to group finite and countably infinite sets together under the heading of countable sets to avoid the nuisance of providing two separate statements for each of our results.

The proof of Lemma 2.2 shows that an infinite subset S of $\mathbb{N}$ can be written as a strictly increasing subsequence of $\mathbb{N}$; that is, $S = \{n_1 < n_2 < n_3 < \cdots\}$. This, together with the order properties of the real line $\mathbb{R}$, make short work of finding monotone subsequences.

Theorem 2.3. *Every sequence of real numbers has a monotone subsequence.*

PROOF. Given a sequence (a_n), let $S = \{n : a_m > a_n \text{ for all } m > n\}$. If S is infinite, with elements $n_1 < n_2 < n_3 < \cdots$, then $a_{n_1} < a_{n_2} < a_{n_3} < \cdots$ is a (strictly) increasing subsequence.

If, on the other hand, S is finite, then $\mathbb{N} \setminus S$ is a nonempty subset of $\mathbb{N}$. Thus, there is a least element $n_1 \in \mathbb{N} \setminus S$ such that $n \notin S$ for all $n \geq n_1$. Since $n_1 \notin S$, there is some $n_2 > n_1$ such that $a_{n_2} \leq a_{n_1}$. But $n_2 \notin S$, and so there is some $n_3 > n_2$ such that $a_{n_3} \leq a_{n_2}$. And so on. Thus, $a_{n_1} \geq a_{n_2} \geq a_{n_3} \geq \cdots$ is a decreasing subsequence. □

We cannot pass up a chance to drop a few names:

Corollary 2.4. (The Bolzano–Weierstrass Theorem) *Every bounded sequence of real numbers has a convergent subsequence.*

Corollary 2.5. *Every Cauchy sequence of real numbers converges.*

EXERCISES

1. Check that the relation "is equivalent to" defines an equivalence relation. That is, show that (i) $A \sim A$, (ii) $A \sim B$ if and only if $B \sim A$, and (iii) if $A \sim B$ and $B \sim C$, then $A \sim C$.

2. If A is an infinite set, prove that A contains a subset of size n for any $n \geq 1$.

3. Given finitely many countable sets $A_1, \ldots, A_n$, show that $A_1 \cup \cdots \cup A_n$ and $A_1 \times \cdots \times A_n$ are countable sets.

▷ **4.** Show that any infinite set has a countably infinite subset.

5. Prove that a set is infinite if and only if it is equivalent to a proper subset of itself. [Hint: If A is infinite and $x \in A$, show that A is equivalent to $A \setminus \{x\}$.]

▷ **6.** If A is infinite and B is countable, show that A and $A \cup B$ are equivalent. [Hint: No containment relation between A and B is assumed here.]

7. Let A be countable. If $f : A \to B$ is onto, show that B is countable; if $g : C \to A$ is one-to-one, show that C is countable. [Hint: Be careful!]

8. Show that $(0, 1)$ is equivalent to $[\,0, 1\,]$ and to $\mathbb{R}$.

9. Show that $(0, 1)$ is equivalent to the unit square $(0, 1) \times (0, 1)$. [Hint: "Interlace" decimals – but carefully!]

10. Prove that $(0, 1)$ can be put into one-to-one correspondence with the set of all *functions* $f : \mathbb{N} \to \{0, 1\}$.

To motivate our next several results, we present a second proof that $\mathbb{N} \times \mathbb{N}$ is equivalent to $\mathbb{N}$. We begin by arranging the elements of $\mathbb{N} \times \mathbb{N}$ in a matrix (see Figure 2.1).

The arrows have been added to show how we are going to enumerate $\mathbb{N} \times \mathbb{N}$. We will count the pairs in the order indicated by the arrows: $(1, 1)$, $(2, 1)$, $(1, 2)$, $(3, 1)$, $(2, 2)$, and so on, accounting for each upward slanting diagonal in succession.

Notice that all of the pairs along a given diagonal have the same sum. The entries of $(1, 1)$ add to 2, the entries of both $(2, 1)$ and $(1, 2)$ add to 3, each pair of entries on the

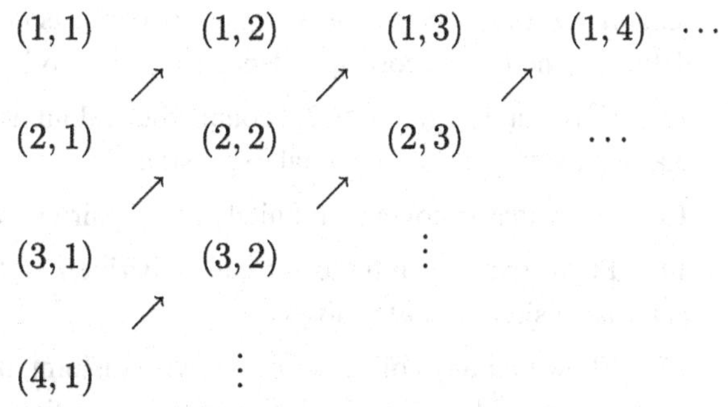

$$\begin{array}{ccccccc} (1,1) & & (1,2) & & (1,3) & & (1,4) & \cdots \\ & \nearrow & & \nearrow & & \nearrow & \\ (2,1) & & (2,2) & & (2,3) & & \cdots \\ & \nearrow & & \nearrow & & & \\ (3,1) & & (3,2) & & \vdots & & \\ & \nearrow & & & & & \\ (4,1) & & \vdots & & & & \end{array}$$

Figure 2.1

next diagonal add to 4, and so on. Moreover, for any given n, there are exactly n pairs whose entries sum to $n + 1$. Said in other words, there are exactly n pairs on the nth diagonal. Based on these observations, it is possible to give an explicit formula for this correspondence between $\mathbb{N}$ and $\mathbb{N} \times \mathbb{N}$. We leave the details as Exercise 11.

Now the fact that $\mathbb{N} \times \mathbb{N} \sim \mathbb{N}$ actually gives us a ton of new information. For example:

Theorem 2.6. *The countable union of countable sets is countable; that is, if A_i is countable for $i = 1, 2, 3, \ldots$, then $\bigcup_{i=1}^{\infty} A_i$ is countable.*

PROOF. Since each A_i is countable, we can arrange their elements collectively in a matrix:

$$\begin{array}{llll} A_1 : & a_{1,1} & a_{1,2} & a_{1,3} & \cdots \\ A_2 : & a_{2,1} & a_{2,2} & a_{2,3} & \cdots \\ A_3 : & a_{3,1} & a_{3,2} & a_{3,3} & \cdots, \end{array}$$

and so $\bigcup_{i=1}^{\infty} A_i$ is the range of a map on $\mathbb{N} \times \mathbb{N}$. (How?) That is, $\bigcup_{i=1}^{\infty} A_i$ is equivalent to a subset of $\mathbb{N} \times \mathbb{N}$ and hence to a subset of $\mathbb{N}$. □

Corollary 2.7. *$\mathbb{Q}$ is countable.* (Why?)

Example 2.8

While we are at it, let us make an observation about decimals. Given an integer $p \geq 2$, recall that the real numbers having a nonunique base p decimal expansion are of the form a/p^n, where $a \in \mathbb{Z}$ and $n = 0, 1, 2, \ldots$. Thus, only countably many reals have a nonunique base p decimal expansion. (Why?) In fact, because there are only countably many bases p to consider, the set of real numbers having a nonunique decimal expansion relative to *some* base is still a countable set.

EXERCISES

11. Here is an explicit correspondence between $\mathbb{N} \times \mathbb{N}$ and $\mathbb{N}$ (based on the "diagonal" argument preceding Corollary 2.6). Let $a_1 = 0$, and for $n = 2, 3, \ldots$, let $a_n = \sum_{i=1}^{n-1} i = n(n-1)/2$. Show that the correspondence $(m, n) \mapsto a_{m+n-1} + n$, from $\mathbb{N} \times \mathbb{N}$ to $\mathbb{N}$, is both one-to-one and onto. Said in another way, show that the map $m \mapsto (a_n - m + 1, m - a_{n-1})$, where n is chosen so that $a_{n-1} < m \leq a_n$, defines a one-to-one correspondence from $\mathbb{N}$ onto $\mathbb{N} \times \mathbb{N}$.

12. Given an integer $p \geq 2$, "count" the real numbers in (0, 1) that have an eventually repeating base p decimal expansion.

▷ **13.** Show that $\mathbb{N}$ contains infinitely many pairwise disjoint infinite subsets.

14. Prove that any infinite set can be written as the countably infinite union of pairwise disjoint infinite subsets.

▷ **15.** Show that any collection of pairwise disjoint, nonempty open intervals in $\mathbb{R}$ is at most countable. [Hint: Each one contains a rational!]

16. The *algebraic numbers* are those real or complex numbers that are the roots of polynomials having *integer* coefficients. Prove that the set of algebraic numbers is countable. [Hint: First show that the set of polynomials having integer coefficients is countable.]

Any infinite set that is not countable is called **uncountable**, for obvious reasons. Countably infinite sets are considered "small" infinite sets, while uncountable sets are "big" infinite sets (see the exercises for more on this). From this point of view, $\mathbb{Q}$ is "small" relative to $\mathbb{R}$:

Theorem 2.9. *$\mathbb{R}$ is uncountable.*

PROOF. To begin, first note that it is enough to show that $\mathbb{R}$ has an uncountable subset. (Why?) Thus, it is enough to show that (0, 1) is uncountable. To accomplish this we will show that any countable subset of (0, 1) is proper.

Given any sequence (a_n) in (0, 1), we construct an element x in (0, 1) with $x \neq a_n$ for any n. We begin by listing the decimal expansions of the a_n; for example:

$$\begin{aligned}
a_1 &= 0\,.\,\boxed{3}\;1\;5\;7\;2\ldots\\
a_2 &= 0\,.\,0\;\boxed{4}\;2\;6\;8\ldots\\
a_3 &= 0\,.\,9\;1\;\boxed{5}\;3\;6\ldots\\
a_4 &= 0\,.\,7\;5\;9\;\boxed{9}\;9\ldots
\end{aligned}$$

(If any a_n has two representations, just include both – the resulting list is still countable.)

Now let $x = 0.533353\ldots$, where the nth digit in the expansion for x is taken to be 3, unless a_n happens to have 3 as its nth digit, in which case we take 5. (This is why we highlighted the nth digit in the expansion of a_n. The choices 3 and 5 are more or less arbitrary here – we just want to avoid the troublesome digits 0 and 9.) Then, the decimal representation of x is unique because it does not end in all 0s or all 9s, and $x \neq a_n$ for any n because the decimal expansions for x and a_n differ in the nth place. Thus we have shown that (a_n) is a proper subset of (0, 1) and hence that (0, 1) is uncountable. □

Corollary 2.10. *$\mathbb{R} \setminus \mathbb{Q}$, the set of irrational numbers, is uncountable.* (Why?)

Examples 2.11

(a) Returning to an earlier observation, recall that the set of real numbers having a nonunique decimal expansion relative to *some* base is a countable set. Thus, "most" real numbers have a unique decimal expansion relative to *every* base!

(b) A real number that is not algebraic is called *transcendental.* It follows from Exercise 16 that "most" real numbers are transcendental, although it is not at all clear how we would *find* even one such number! This example demonstrates the curious power of cardinality in existential arguments. Other notions of "big" versus "small" sets will lend themselves equally well to similar sorts of existence proofs. We will repeat this theme several times before we are finished.

EXERCISES

17. If A is uncountable and B is countable, show that A and $A \setminus B$ are equivalent. In particular, conclude that $A \setminus B$ is uncountable.

18. Show that the set of all real numbers in the interval $(0, 1)$ whose base 10 decimal expansion contains no 3s or 7s is uncountable.

19. Show that the set of all functions $f : A \to \{0, 1\}$ is equivalent to $\mathcal{P}(A)$, the power set of A (i.e., the set of all subsets of A).

20. Prove that $\mathbb{N}$ contains uncountably many infinite subsets $(N_\alpha)_{\alpha\in\mathbb{R}}$ such that $N_\alpha \cap N_\beta$ is *finite* if $\alpha \neq \beta$. (This one's hard!)

Here is what we have so far: A countably infinite set is small in the sense that every subset is either finite or else the same "size" as the whole set. An uncountable set, on the other hand, is certainly bigger than any countable set because a countable subset of an uncountable set is necessarily proper. From this point of view, countably infinite sets are the "smallest" infinite sets; a "smaller" subset of a countably infinite set must be finite. But while there is a "smallest infinity," there is no largest – we can always build bigger and bigger sets.

Given a set A, we write $\mathcal{P}(A)$ for the **power set** of A – the set of all subsets of A. Now A is clearly equivalent to a subset of $\mathcal{P}(A)$ (namely, the collection of all singletons $\{a\}$, where $a \in A$) but, as it happens, $\mathcal{P}(A)$ is always "bigger" than A:

Cantor's Theorem 2.12. *No map $F : A \to \mathcal{P}(A)$ can be onto.*

PROOF. Given $F : A \to \mathcal{P}(A)$, consider $B = \{x \in A : x \notin F(x)\} \in \mathcal{P}(A)$. We claim that $B \neq F(y)$ for any $y \in A$. Indeed, if $B = F(y)$, then we are faced with the following alternatives:

$$y \in F(y) = B \quad\Longrightarrow\quad y \notin F(y) \qquad \text{or} \qquad y \notin F(y) = B \quad\Longrightarrow\quad y \in F(y),$$

and both lead to contradictions! □

While we won't take the time to fully justify the notation, each set has a **cardinal number** assigned to it, written card(A) and read "the cardinality of A," that uniquely specifies the number of elements of A. For finite sets the cardinality is literally the number of elements, as in card$\{1, \ldots, n\} = n$. For countably infinite sets we use the cardinal $\aleph_0$ (read "aleph-nought"), as in card$(\mathbb{N}) = \aleph_0$. And for $\mathbb{R}$ we write card$(\mathbb{R}) = \mathfrak{c}$ (for "continuum").

We will not pursue this notation much further, but it does provide a convenient shorthand and can actually clarify certain arguments. For example, we might write card$(A) =$ card(B) to mean that the sets A and B are equivalent. And we might use the formula card$(A) \leq$ card(B) to mean that there is a one-to-one map $f : A \to B$ from A into B. (Why is this a good choice?) But this raises the question of whether the order that we have imposed on cardinal numbers is reasonable. In other words, if card$(A) \leq$ card(B) and card$(B) \leq$ card(A) both hold, is it the case that card$(A) =$ card(B)? The answer is "yes" and is given in the following celebrated theorem.

F. Bernstein's Theorem 2.13. *Let A and B be nonempty sets. If there exist a one-to-one map $f : A \to B$, from A into B, and a one-to-one map $g : B \to A$, from B into A, then there is a map $h : A \to B$ that is both one-to-one and onto.*

PROOF. First, consider Figure 2.2. We would like to find a subset S of A so that

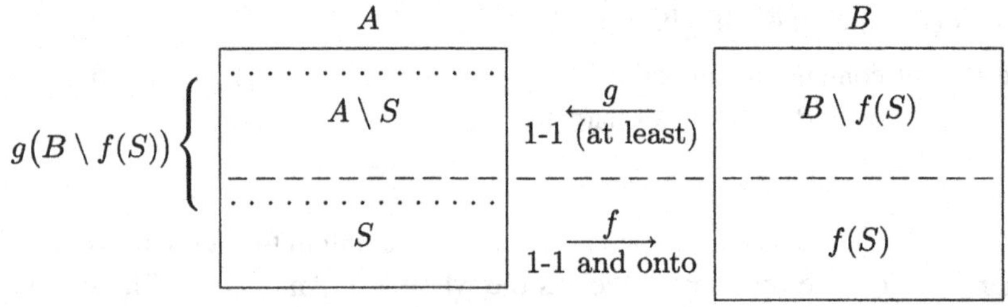

A B
$g(B \setminus f(S))$ $A \setminus S$ $\xleftarrow{g}$ 1-1 (at least) $B \setminus f(S)$
S $\xrightarrow{f}$ 1-1 and onto $f(S)$

Figure 2.2

we may define h to be f on S and g^{-1} on $A \setminus S$. As the figure suggests, for this to work we will need a subset S satisfying $g(B \setminus f(S)) = A \setminus S$. To this end, define a map $H : \mathcal{P}(A) \to \mathcal{P}(A)$ by

$$H(S) = A \setminus g(B \setminus f(S)).$$

In this notation, the problem is to find a "fixed point" for H, that is, a set S such that $H(S) = S$.

Claim. H is "increasing"; that is, $S \subset T \Longrightarrow H(S) \subset H(T)$. (Just check.)

Now to see that H must fix some set, let $\mathcal{C} = \{S \subset A : S \subset H(S)\}$, and let $\bar{S} = \bigcup \mathcal{C}$. ($\bar{S}$ is the least upper bound of the sets with $S \subset H(S)$. We do not exclude the possibility that $\mathcal{C} = \emptyset$ here; in that case we take $\bar{S} = \emptyset$.) We will show that $H(\bar{S}) = \bar{S}$.

First, $\bar{S} \subset H(\bar{S})$. Indeed, because $S \subset \bar{S}$ for all $S \in \mathcal{C}$, we have $S \subset H(S) \subset H(\bar{S})$ for all $S \in \mathcal{C}$ and hence $\bar{S} \subset H(\bar{S})$.

It now follows that $H(\bar{S}) \subset H(H(\bar{S}))$. That is, $H(\bar{S}) \in \mathcal{C}$ and hence $H(\bar{S}) \subset \bar{S}$. Consequently, $H(\bar{S}) = \bar{S}$. $\square$

What we have actually been doing in this section is developing an "arithmetic" for cardinal numbers. For example, it turns out that $\text{card}(A \times B) = \text{card}(A) \cdot \text{card}(B)$, which works just as you would suspect for *finite* sets. For infinite sets A and B, we instead use the equation to *define* the product of cardinal numbers. For instance, Example 2.1 (b) tells us that $\aleph_0 \cdot \aleph_0 = \aleph_0$. How might you justify the formula: $\mathfrak{c} \cdot \aleph_0 = \mathfrak{c}$?

A few more examples will help to explain this "arithmetic" with cardinal numbers.

Examples 2.14

(a) The collection of all sequences of 0s and 1s is uncountable. How so? Well, if (a_n) is a sequence of 0s and 1s, then $\sum_{n=1}^{\infty} a_n/2^n$ represents an element of $[0, 1]$ and, conversely, each element of $[0, 1]$ can be so represented. That is, the map $(a_n) \mapsto 0.a_1a_2a_3\cdots$ (base 2) is onto. Hence the set of all 0–1 sequences, written $\{0, 1\}^{\mathbb{N}}$, has cardinality at least that of $[0, 1]$. But, in fact, the two sets are equivalent. (Why?)

(b) We next note that the set of all 0–1 sequences is equivalent to $\mathcal{P}(\mathbb{N})$. This is easy: If $A \subset \mathbb{N}$, we define a sequence (a_n) by $a_n = 1$ if $n \in A$ and $a_n = 0$ if $n \notin A$. The correspondence $A \mapsto (a_n)$ is clearly both one-to-one and onto.

With the help of these two examples, we can make a rather fanciful calculation:

$$\mathfrak{c} = \text{card}([0, 1]) = \text{card}(\mathcal{P}(\mathbb{N})) = \text{card}\left(\{0, 1\}^{\mathbb{N}}\right) = 2^{\text{card}\,\mathbb{N}} = 2^{\aleph_0}.$$

Here we used a variation on the formula $\text{card}(A \times B) = \text{card}(A) \cdot \text{card}(B)$, namely, $\text{card}(A^B) = \text{card}(A)^{\text{card}(B)}$.

Occasionally it is convenient to use a shorthand for certain sets that mirrors their cardinality. For example, if we use "2" as a shorthand for the two-point set $\{0, 1\}$, then we might write $2^{\mathbb{N}}$ in place of $\mathcal{P}(\mathbb{N})$, or, more generally, 2^A in place of $\mathcal{P}(A)$. Along similar lines, we can prove that $\mathbb{R}^\infty$, the collection of all real sequences, has the same cardinality as $\mathbb{R}$. Of course, $\mathbb{R}^\infty$ is the same as $\mathbb{R}^{\mathbb{N}}$, the product of countably many copies of $\mathbb{R}$, and so

$$\text{card}\left(\mathbb{R}^{\mathbb{N}}\right) = \mathfrak{c}^{\aleph_0} = \left(2^{\aleph_0}\right)^{\aleph_0} = 2^{\aleph_0 \cdot \aleph_0} = 2^{\aleph_0} = \mathfrak{c}.$$

The Cantor Set

We next examine an intriguing and unusual subset of $\mathbb{R}$ called the Cantor set (or, sometimes, Cantor's *ternary* set). Our investigations here should provide us with a natural lead-in to several of the topics that are ahead of us. We will construct an uncountable (hence "large") subset of $[0, 1]$ that is somehow also "meager." We begin by applying the nested interval theorem to a particular batch of intervals.

Consider the process of successively removing "middle thirds" from the interval $[0, 1]$ (Figure 2.3).

We continue this process inductively. At the nth stage we construct I_n from I_{n-1} by removing 2^{n-1} disjoint, open, "middle thirds" intervals from I_{n-1}, each of length 3^{-n}; we will call this discarded set J_n. Thus, I_n is the union of 2^n closed subintervals of I_{n-1},

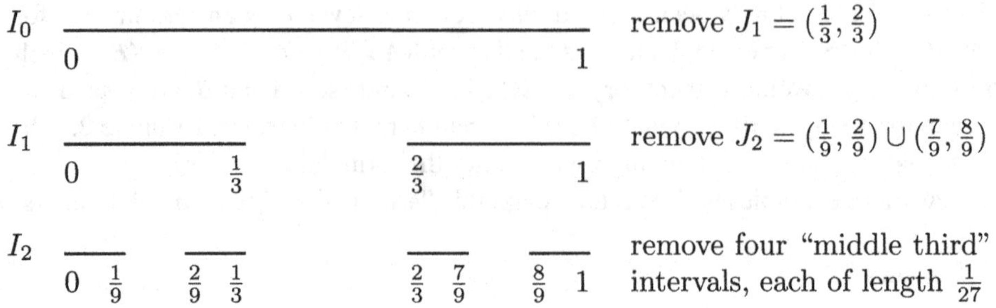

I_0 — 0 to 1 — remove $J_1 = (\frac{1}{3}, \frac{2}{3})$

I_1 — 0, $\frac{1}{3}$; $\frac{2}{3}$, 1 — remove $J_2 = (\frac{1}{9}, \frac{2}{9}) \cup (\frac{7}{9}, \frac{8}{9})$

I_2 — 0, $\frac{1}{9}$; $\frac{2}{9}$, $\frac{1}{3}$; $\frac{2}{3}$, $\frac{7}{9}$; $\frac{8}{9}$, 1 — remove four "middle third" intervals, each of length $\frac{1}{27}$

Figure 2.3

and the complement of I_n in $[0, 1]$ is $J_1 \cup \cdots \cup J_n$. The Cantor set Δ is defined as the set of points that still remain at the end of this process, in other words, the "limit" of the sets I_n. More precisely, $\Delta = \bigcap_{n=1}^{\infty} I_n$. It follows from the nested interval theorem that $\Delta \neq \emptyset$, but notice that Δ *is at least countably infinite*. The endpoints of each I_n are in Δ:

$$0,\ 1,\ 1/3,\ 2/3,\ 1/9,\ 2/9,\ \ldots \in \Delta.$$

We will refer to these points as **the endpoints of** Δ, that is, all of the points in Δ of the form $a/3^n$ for some integers a and n.

As we shall see presently, Δ is actually uncountable! This is more than a little surprising. Just try to imagine how terribly sparse the next few levels of the "middle thirds" diagram would look on the page. Adding even a few more levels defies the limits of typesetting! For good measure we will give two proofs that Δ is uncountable, the first being somewhat combinatorial.

Notice that each subinterval of I_{n-1} results in two subintervals of I_n (after discarding a middle third). We label these two new intervals L and R (for left and right) as in Figure 2.4.

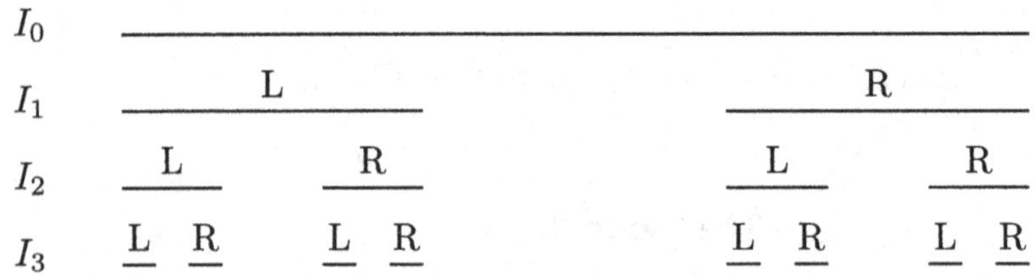

I_0

I_1 — L, R

I_2 — L, R, L, R

I_3 — L R, L R, L R, L R

Figure 2.4

As we progress down through the levels of the diagram toward the Cantor set (somewhere far below), imagine that we "step down" from one level to the next by repeatedly choosing either a step to the left (landing on an L interval in the next level below) or a step to the right (landing on an R interval). At each stage we are only allowed to step down to a subinterval of the interval we are presently on – jumping across "gaps" is not allowed! Thus, each string of choices, LRLRRLLRLLLR..., describes a unique "path" from the top level I_0 down to the bottom level Δ. The Cantor set, then, is quite literally the "dust" at the end of the trail. Said another way, each such "path" determines a unique sequence of nested subintervals, one from each level, whose intersection is a single point of Δ.

Conversely, each point $x \in \Delta$ lies at the end of exactly one such path, because at any given level there is only one possible subinterval of I_n on our diagram, call it $\tilde{I}_n$, that contains x. The resulting sequence of intervals $(\tilde{I}_n)$ is clearly nested. (Why?) Thus, the Cantor set Δ is in one-to-one correspondence with the set of all paths, that is, the set of all sequences of Ls and Rs. Of course, any two choices would have done just as well, so we might also say that Δ is equivalent to the set of all sequences of 0s and 1s – a set we already know to be uncountable. Here is what this means:

$$\operatorname{card}(\Delta) = \operatorname{card}\left(2^{\mathbb{N}}\right) = \operatorname{card}([0,1]).$$

Absolutely amazing! The Cantor set is just as "big" as $[0,1]$ and yet it strains the imagination to picture such a sparse set of points.

Before we give our second proof that Δ is uncountable, let's see why Δ is "small" (in at least one sense). We will show that Δ has "measure zero"; that is, the "measure" or "total length" of all of the intervals in its complement $[0,1] \setminus \Delta$ is 1. Here's why: By induction, the total length of the 2^{n-1} disjoint intervals comprising J_n (the set we discard at the nth stage) is $2^{n-1}/3^n$, and so the total length of $[0,1] \setminus \Delta$ must be

$$\sum_{n=1}^{\infty} \frac{2^{n-1}}{3^n} = \frac{1}{3} \sum_{n=1}^{\infty} \left(\frac{2}{3}\right)^{n-1} = \frac{1}{3} \cdot \frac{1}{1 - \frac{2}{3}} = 1.$$

We have discarded everything!? And left uncountably many points behind!? How bizarre! This simultaneous "bigness" and "smallness" is precisely what makes the Cantor set so intriguing. The exercises will supply even more ways to say that Δ is both "big" and "small."

Our second proof that Δ is uncountable is based on an equivalent characterization of Δ in terms of *ternary* (base 3) decimals. Recall that each x in $[0,1]$ can be written, in possibly more than one way, as: $x = 0.a_1a_2a_3\cdots$ (base 3), where each $a_n = 0$, 1, or 2. This three-way choice for decimal digits (base 3) corresponds to the three-way splitting of intervals that we saw earlier. To see this, let us consider a few specific examples. For instance, the three cases $a_1 = 0$, 1, or 2 correspond to the three intervals $[0, 1/3]$, $(1/3, 2/3)$, and $[2/3, 1]$, as in Figure 2.5.

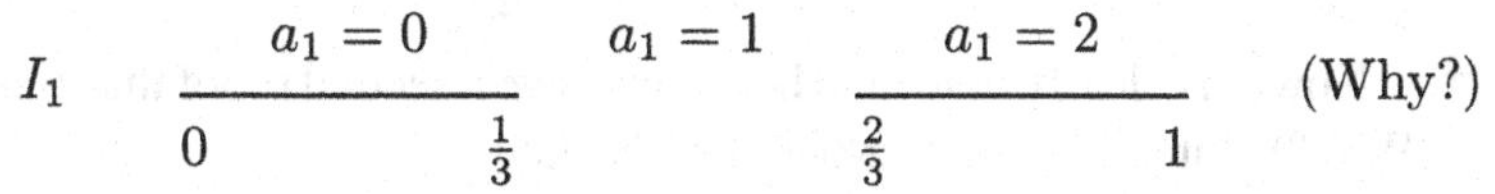

$$I_1 \quad \underset{0 \qquad\qquad \frac{1}{3}}{\overline{\quad a_1 = 0 \quad}} \qquad a_1 = 1 \qquad \underset{\frac{2}{3} \qquad\qquad 1}{\overline{\quad a_1 = 2 \quad}} \qquad \text{(Why?)}$$

Figure 2.5

There is some ambiguity at the endpoints:

$$\begin{aligned} 1/3 &= 0.1 \text{ (base 3)} = 0.0222\ldots \text{ (base 3)}, \\ 2/3 &= 0.2 \text{ (base 3)} = 0.1222\ldots \text{ (base 3)}, \\ 1 &= 1.0 \text{ (base 3)} = 0.2222\ldots \text{ (base 3)}, \end{aligned}$$

but each of these ambiguous cases has at least one representation with a_1 in the proper range. Next, Figure 2.6 shows the situation for I_2 (but this time ignoring the discarded

	$a_1 = 0$ and		$a_1 = 2$ and		
	$a_2 = 0$	$a_2 = 2$	$a_2 = 0$	$a_2 = 2$	
I_2	0 —— $\frac{1}{9}$	$\frac{2}{9}$ —— $\frac{1}{3}$	$\frac{2}{3}$ —— $\frac{7}{9}$	$\frac{8}{9}$ —— 1	(Why?)

Figure 2.6

intervals). Again, some confusion is possible at the endpoints:

$$1/9 = 0.01 \text{ (base 3)} = 0.00222\ldots \text{ (base 3)},$$
$$8/9 = 0.22 \text{ (base 3)} = 0.21222\ldots \text{ (base 3)}.$$

We will take these few examples as proof of the following

Theorem 2.15. *$x \in \Delta$ if and only if x can be written as $\sum_{n=1}^{\infty} a_n/3^n$, where each a_n is either 0 or 2.*

Thus the Cantor set consists of those points in $[0, 1]$ having *some* base 3 decimal representation that excludes the digit 1. Knowing this we can list all sorts of elements of Δ. For example, $1/4 \in \Delta$ because $1/4 = 0.020202\ldots$ (base 3). Theorem 2.15 also leads to another proof that Δ is uncountable; or, rather, it gives us a new way of writing the old proof. The first proof used sequences of 0s and 1s, and now we find ourselves with sequences of 0s and 2s; the connection isn't hard to guess.

Corollary 2.16. *Δ is uncountable; in fact, Δ is equivalent to $[0, 1]$.*

PROOF. By altering our notation we can easily display a correspondence between Δ and $[0, 1]$. Each $x \in \Delta$ may be written $x = \sum_{n=1}^{\infty} 2b_n/3^n$, where $b_n = 0$ or 1, and now we define **the Cantor function** $f : \Delta \to [0, 1]$ by

$$f\left(\sum_{n=1}^{\infty} \frac{2b_n}{3^n}\right) = \sum_{n=1}^{\infty} \frac{b_n}{2^n} \qquad (b_n = 0,\ 1).$$

That is,

$$f\big(0.a_1a_2a_3\cdots \text{ (base 3)}\big) = 0.\frac{a_1}{2}\frac{a_2}{2}\frac{a_3}{2}\cdots \text{ (base 2)} \qquad (a_n = 0,\ 2).$$

Now f is clearly onto, and hence we have a second proof that Δ is uncountable. (Why?) But f isn't one-to-one; here's why:

$$\begin{aligned} f(1/3) &= f\big(0.0222\ldots \text{ (base 3)}\big) = 0.0111\ldots \text{ (base 2)} \\ &= 0.1 \text{ (base 2)} = f\big(0.2 \text{ (base 3)}\big) = f(2/3). \end{aligned}$$

The same phenomenon occurs at each pair of endpoints of any discarded "middle third" interval (i.e., a subinterval of J_n):

$$\begin{aligned} f(1/9) &= f\big(0.00222\ldots \text{ (base 3)}\big) = 0.00111\ldots \text{ (base 2)} \\ &= 0.01 \text{ (base 2)} = f\big(0.02 \text{ (base 3)}\big) = f(2/9). \end{aligned}$$

It is easy to see that f is increasing; that is, if $x, y \in \Delta$ with $x < y$, then $f(x) \leq f(y)$. We leave it as an exercise to check that $f(x) = f(y)$ if and only if x and y are endpoints of a discarded "middle third" interval (see Exercise 26). Thus, f is one-to-one except at the endpoints of Δ (a countable set), where it's two-to-one. It follows that Δ is equivalent to $[0, 1]$. (How?) □

EXERCISES

▷ **21.** Show that any ternary decimal of the form $0.a_1a_2 \cdots a_n11$ (base 3), i.e., any finite-length decimal ending in two (or more) 1s, is *not* an element of Δ.

▷ **22.** Show that Δ contains no (nonempty) open intervals. In particular, show that if $x, y \in \Delta$ with $x < y$, then there is some $z \in [0, 1] \setminus \Delta$ with $x < z < y$. (It follows from this that Δ is *nowhere dense*, which is another way of saying that Δ is "small.")

▷ **23.** The endpoints of Δ are those points in Δ having a finite-length base 3 decimal expansion (not necessarily in the proper form), that is, all of the points in Δ of the form $a/3^n$ for some integers n and $0 \leq a \leq 3^n$. Show that the endpoints of Δ other than 0 and 1 can be written as $0.a_1a_2 \cdots a_{n+1}$ (base 3), where each a_k is 0 or 2, except a_{n+1}, which is either 1 or 2. That is, the discarded "middle third" intervals are of the form $(0.a_1a_2 \cdots a_n1,\ 0.a_1a_2 \cdots a_n2)$, where both entries are points of Δ written in base 3.

24. Show that Δ is *perfect*; that is, every point in Δ is the limit of a sequence of distinct points from Δ. In fact, show that every point in Δ is the limit of a sequence of distinct endpoints.

25. Define $g : \mathbb{R} \to \mathbb{R}$ by $g(x) = 1$ if $x \in \Delta$, and $g(x) = 0$ otherwise. At which points of $\mathbb{R}$ is g continuous?

▷ **26.** Let $f : \Delta \to [0, 1]$ be the Cantor function (defined above) and let $x, y \in \Delta$ with $x < y$. Show that $f(x) \leq f(y)$. If $f(x) = f(y)$, show that x has two distinct binary decimal expansions. Finally, show that $f(x) = f(y)$ if and only if x and y are "consecutive" endpoints of the form $x = 0.a_1a_2 \cdots a_n1$ and $y = 0.a_1a_2 \cdots a_n2$ (base 3).

27. Fix $n \geq 1$, and let $I_{n,k}$, $k = 1, \ldots, 2^{n-1}$ be the component subintervals of the nth level Cantor set I_n. If $x, y \in \Delta$ with $|x - y| < 3^{-n}$, show that x and y are in the same component $I_{n,k}$. For this same pair of points show that $|f(x) - f(y)| \leq 2^{-n}$.

The observation made in Exercise 26 enables us to extend the definition of the Cantor function f to all of $[0, 1]$ in an obvious way: We take f to be an appropriate constant on each of the open intervals that make up $[0, 1] \setminus \Delta$. For example, we would set $f(x) = f(1/3) = 1/2$ for each x in the interval $(1/3, 2/3)$ and $f(x) = f(1/9) = 1/4$ for each x in $(1/9, 2/9)$. See Figure 2.7.

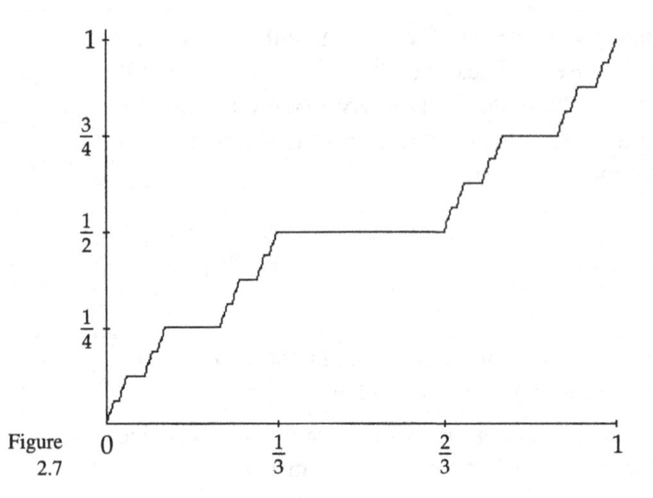

Figure 2.7

Formally, we define $f(x) = \sup\{f(y) : y \in \Delta,\ y \leq x\}$ for $x \in [0, 1] \setminus \Delta$. The new function $f : [0, 1] \to [0, 1]$ is still increasing (why?) and is actually continuous! (We will prove this in the next section.) Some authors refer to this extension as *the Cantor–Lebesgue function* or *Lebesgue's singular function*. We will simply call it the Cantor function. It is called a singular function because $f' = 0$ at almost every point in $[0, 1]$. That is, $f' = 0$ on $[0, 1] \setminus \Delta$, a set of measure 1. But we are getting ahead of ourselves.

EXERCISES

28. Let $f : \Delta \to [0, 1]$ be the Cantor function (as originally defined). Check that $f(x) = \sup\{f(y) : y \in \Delta,\ y \leq x\}$ for any $x \in \Delta$.

▷ **29.** Prove that the extended Cantor function $f : [0, 1] \to [0, 1]$ (as defined above) is increasing. [Hint: Consider cases.]

The construction of the Cantor set admits all sorts of generalizations. For example, suppose that we fix α with $0 < \alpha < 1$ and we repeat our "middle thirds" construction except that at the nth stage each of the open intervals we remove is now taken to have length $\alpha\, 3^{-n}$. (And we still want these to be in the "middle" of an interval from the current level – it is important that the remaining closed intervals turn out to be nested.) Figure 2.8 shows the first few levels of this generalized construction in the case $\alpha = 3/5$.

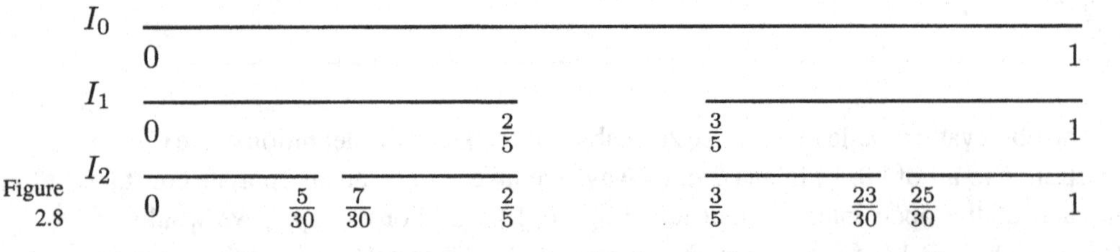

Figure 2.8

The limit of this process, called a **generalized Cantor set**, is very much like the ordinary Cantor set. It is uncountable, perfect, nowhere dense, and so on, but this one now has nonzero measure. We leave it as an exercise to check that the generalized Cantor set with parameter α has measure $\beta = 1-\alpha$. We label these sets according to their measure; that is, we write Δ_β to mean the generalized Cantor set with measure β.

EXERCISES

30. Check that the construction of the generalized Cantor set with parameter α, as described above, leads to a set of measure $1 - \alpha$; that is, check that the discarded intervals now have total length α.

31. Now that we know the description of Δ in terms of ternary decimals, it might be interesting to consider a similar construction using another base. For example, fix an integer $p > 3$ (to use as the base) and an integer $0 < d < p$ (as the omitted digit). Describe the set of all points in $[0, 1]$ that have some base p decimal expansion that excludes the digit d. Is it uncountable? Does it have measure zero?

The Cantor set satisfies another rather curious property: The set of all possible differences of pairs of elements of Δ fills up the interval $[-1, 1]$; in symbols, $\Delta - \Delta = \{y - x : x,\ y \in \Delta\} = [-1, 1]$. The original proof, due to Steinhaus, is based on a clever geometric observation. The claim is that the equation $y - x = b$ has a solution $x, y \in \Delta$ for any $-1 \le b \le 1$. That is, for any $-1 \le b \le 1$, the line $y = x + b$ must pass through the set $\Delta \times \Delta$.

Now the set $\Delta \times \Delta$ can be constructed inside the square $[0, 1] \times [0, 1]$ in much the same way that Δ is constructed inside $[0, 1]$. We begin with the full square A_0, remove "middle thirds" both horizontally and vertically, and arrive at the set of four subsquares $A_1 = ([0, 1/3] \cup [2/3, 1]) \times ([0, 1/3] \cup [2/3, 1])$. Next "cross out" the middle thirds, both horizontally and vertically, from these four squares to arrive at 16 smaller subsquares, a set that we will call A_2. And continue. The "limit" of this process is the set $\Delta \times \Delta = \bigcap_{n=1}^{\infty} A_n$.

To see that a line of the form $y = x + b$, where $-1 \le b \le 1$, must pass through $\Delta \times \Delta$, it is enough to show that $y = x + b$ always hits any A_n, for then we could apply a version of the nested interval theorem in $\mathbb{R}^2$ to finish the proof. (We will see just such a theorem in Chapter Seven.) For now we will settle for the following "visual" proof. Convince yourself that any line of slope 1 that passes through the square $[0, 1] \times [0, 1]$ must also pass through each A_n by considering the following pictures (showing A_1 on the left, A_2 on the right, and a "worst-case" line drawn through each square). Note that by "scaling" it is enough to understand just the first case; see Figure 2.9.

Monotone Functions

As we saw in the first chapter, monotone functions are reasonably well behaved. In particular, a monotone function has (at worst) only jump discontinuities. It follows that a monotone function must have lots of points of continuity. Here's why:

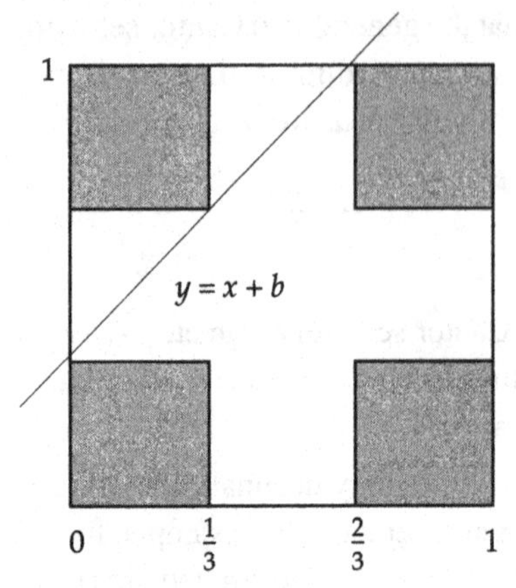

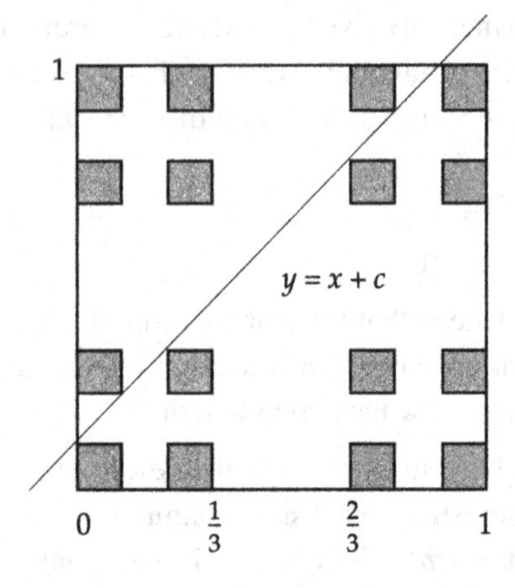

Figure 2.9

Theorem 2.17. *If $f : (a, b) \to \mathbb{R}$ is monotone, then f has at most countably many points of discontinuity in (a, b), all of which are jump discontinuities.*

PROOF. That f has only jump discontinuities follows from Proposition 1.19. Now we just need to count the points of discontinuity.

Let's reflect on the situation. If $f : (a, b) \to \mathbb{R}$ is, say, increasing, and if $c \in (a, b)$, then the left- and right-hand limits of f at c satisfy $f(c-) \le f(c) \le f(c+)$. (Why?) In particular, f is discontinuous at c if and only if $f(c-) < f(c+)$. Consequently, if c and d are two different points of discontinuity for f, then the intervals $(f(c-), f(c+))$ and $(f(d-), f(d+))$ are nonempty and *disjoint*. (Why?) Thus,

$$\{(f(c-), f(c+)) : c \text{ is a point of discontinuity for } f\}$$

is a collection of nonempty, disjoint open intervals in $\mathbb{R}$, and any such collection must be *countable* (see Exercise 15). □

Theorem 2.17 allows us to clean up a few details from the last section:

Corollary 2.18. *If $f : [a, b] \to [c, d]$ is both monotone and onto, then f is continuous.*

Corollary 2.19. *The Cantor function $f : [0, 1] \to [0, 1]$ is continuous!*

Theorem 2.17 has a *converse* (see Exercise 34 for a detailed statement). Given any countable set D in $\mathbb{R}$, we can construct an increasing function $f : \mathbb{R} \to \mathbb{R}$ that is discontinuous precisely at the points of D. Here is a brief sketch:

Let $D = \{x_1, x_2, \dots\}$, and let (ε_n) be a sequence of positive numbers with $\sum_{n=1}^{\infty} \varepsilon_n < \infty$. We define $f(x) = \sum_{x_n \le x} \varepsilon_n$, where the sum is over the set $\{n : x_n \le x\}$ and where $f(x) = 0$ if this set is empty. Notice that $0 \le f(x) \le \sum_{n=1}^{\infty} \varepsilon_n < \infty$ in any case.

Now, if $x < y$, then

$$f(y) = \sum_{x_n \le y} \varepsilon_n = \sum_{x_n \le x} \varepsilon_n + \sum_{x < x_n \le y} \varepsilon_n = f(x) + \sum_{x < x_n \le y} \varepsilon_n \ge f(x).$$

Thus, f is increasing. Next we consider this formula in each of the cases $x = x_k$ and $y = x_k$. First,

$$x = x_k < y \quad \Longrightarrow \quad f(y) = f(x_k) + \sum_{x_k < x_n \le y} \varepsilon_n.$$

Claim. $f(x_k+) = f(x_k)$; i.e.,

$$\lim_{y \to x_k^+} \sum_{x_k < x_n \le y} \varepsilon_n = 0 \qquad \text{because} \qquad \sum_{n=N}^{\infty} \varepsilon_n \to 0 \quad \text{as} \quad N \to \infty.$$

And, in the second case,

$$x < x_k = y \quad \Longrightarrow \quad f(x_k) = f(x) + \sum_{x < x_n \le x_k} \varepsilon_n \ge f(x) + \varepsilon_k.$$

Claim. $f(x_k-) = f(x_k) - \varepsilon_k$; i.e.,

$$\lim_{x \to x_k^-} \sum_{x < x_n \le x_k} \varepsilon_n = \varepsilon_k.$$

Thus, $f(x_k-) + \varepsilon_k = f(x_k) = f(x_k+)$ and $f(x_k+) - f(x_k-) = \varepsilon_k$.

The proof that f is continuous at each $x \in \mathbb{R} \setminus D$ is similar.

EXERCISES

32. Deduce from Theorem 2.17 that a monotone function $f : \mathbb{R} \to \mathbb{R}$ has points of continuity in every open interval.

33. Let $f : [\,a, b\,] \to \mathbb{R}$ be monotone. Given n distinct points $a < x_1 < x_2 < \cdots < x_n < b$, show that $\sum_{i=1}^{n} |f(x_i+) - f(x_i-)| \le |f(b) - f(a)|$. Use this to give another proof that f has at most countably many (jump) discontinuities.

34. Let $D = \{x_1, x_2, \ldots\}$, and let $\varepsilon_n > 0$ with $\sum_{n=1}^{\infty} \varepsilon_n < \infty$. Define $f(x) = \sum_{x_n \le x} \varepsilon_n$ (as above). Check the following: (i) f is discontinuous at the points of D; (ii) f is right-continuous everywhere; and (iii) f is continuous at each point $x \in \mathbb{R} \setminus D$. How might this construction be modified so as to yield a *strictly* increasing function with these same properties?

35. Let $f : [\,a, b\,] \to \mathbb{R}$ be increasing, and let (x_n) be an enumeration of the discontinuities of f. For each n, let $a_n = f(x_n) - f(x_n-)$ and $b_n = f(x_n+) - f(x_n)$ be the left and right "jumps" in the graph of f, where $a_n = 0$ if $x_n = a$ and $b_n = 0$ if $x_n = b$. Show that $\sum_{n=1}^{\infty} a_n \le f(b) - f(a)$ and $\sum_{n=1}^{\infty} b_n \le f(b) - f(a)$.

36. In the notation of Exercise 35, define $h(x) = \sum_{x_n \le x} a_n + \sum_{x_n < x} b_n$. Show that h is increasing and that $g = f - h$ is both continuous and increasing. Thus, each increasing function f can be written as the sum of a continuous increasing function g and a "pure jump" function h.

Notes and Remarks

For an infinitely enjoyable discussion of the infinite, see the article "Infinity" by Hans Hahn [1956a]. The clever proof of Theorem 2.3, and more, can be found in Newman and Parsons [1988]. For an alternate proof of Corollary 2.7, see Campbell [1986].

Countable (and uncountable) sets were introduced by Cantor. Indeed, most of the results in this chapter are due to Cantor himself. In particular, Corollary 2.7, Theorem 2.9, and Theorem 2.12 are due to Cantor; see Dunham [1990]. The statement of Theorem 2.13 originated as an open question in one of Cantor's seminars. You will often see it referred to variously as the Cantor–Bernstein theorem or as the Schröder–Bernstein theorem. According to Dudley [1989], full credit should go to Felix Bernstein, who was a 19-year-old student at the time! At any rate, Hausdorff [1937] refers to it as Bernstein's theorem. The proof given here is taken from an exercise in Willard [1970] but is probably much older.

The proof of Theorem 2.12 may remind you of Russell's paradox. Briefly, Russell's paradox demonstrates that there are limitations on what may be regarded as a *set*. As Russell would ask, is the collection $\mathcal{U}$ of all sets again a set? If so, then we might consider the set $\mathcal{B} = \{A \in \mathcal{U} : A \notin A\}$. Now if we accept $\mathcal{U}$ as a set, then the "rules" of set operations say that we are stuck with accepting $\mathcal{B}$ as a set, too. With that decision made, the "rules" likewise permit us to ask the question, is $\mathcal{B} \in \mathcal{B}$? A moment's reflection on what this means will have your head spinning! Evidently not everyone gets to be a set. We have taken the easy way out and left the concept of "set" as a primitive, undefined notion. Not to worry, though; we are on solid ground. Trust me!

Although Example 2.11 (b) might suggest that it is impossible to construct a single transcendental number, that is not entirely true. Since the algebraic numbers are countable, the "diagonalization" technique used in the proof of Theorem 2.9, if carefully applied, would yield a specific transcendental number. Better still, it is actually possible to display uncountably many transcendental numbers: In 1844, Liouville first proved that transcendental numbers exist by showing that any number of the form $\sum_{n=1}^{\infty} a_n/10^{n!}$, where the a_n are integers with $1 \le a_n \le 9$, is transcendental. However, not all transcendental numbers are of this form. Following this discovery, Hermite showed in 1873 that e is transcendental, and Lindemann showed in 1882 that π is transcendental. For more details, see Oxtoby [1971], Stromberg [1981], and Kline [1972]. For more on what mathematicians mean by the word "impossible," see Davis [1986].

In addition to the books by Dudley [1989] and Hausdorff [1937], you can find more abstract set theory in the books by Boas [1960], Folland [1984], Hewitt and Stromberg [1965], Halmos [1960], Kaplansky [1977], Kolmogorov and Fomin [1970], and Torchinsky [1988]. Several of the references include a bit of history, too. The books by Willard and Dudley, for example, have copious notes and references to original works. Kline [1972] is a mammoth source of information about mathematics in general. Hawkins [1970] has a detailed exposition of the events leading to the great "revolution" in analysis following the work of Riemann, Weierstrass, and Cantor (roughly speaking, the years 1875–1925). Manheim [1964] traces the early development of abstract set theory and point set topology.

Cantor first mentioned "the" Cantor set in connection with the concept of "perfect" sets in Cantor [1883], but the set itself was not discovered by Cantor. Examples of this type, including "the" Cantor set, had already been introduced by H. J. S. Smith in connection with two constructions for nowhere dense sets in H. J. S. Smith [1875]. According to Hawkins [1970], Smith's results did not become well known until the 1880s. The title of Smith's paper, "On the integration of discontinuous functions," highlights the connection between abstract set theory and integration. Interest in infinite sets and "pathological" sets was born out of the study of Riemann's integrability condition and its relation to Fourier's work on trigonometric series. (See Hawkins [1970], Manheim [1964], Rogosinski [1950], and the series of articles by Dauben [1971, 1974, 1983].) We will have more to say about this in Part Three; in any event, the Cantor set will remain an important example throughout this course.

The "visual" proof that $\Delta - \Delta = [-1, 1]$ is originally due to Steinhaus [1917]. For a proof based on the ternary decimal representation of Δ, see Randolph [1940]. For more on the Cantor set and generalized Cantor sets, see Chae [1980], Randolph [1968], Coppel [1983], and Majumder [1965]. For more on the Cantor function, see Chalice [1991] and Hille and Tamarkin [1929].

The construction used in the converse to Theorem 2.17 is based on the presentation in Rudin [1953]. Our results about monotone functions will turn out to be very useful later in the course when we discuss "the problem of moments." This famous problem has its roots in mathematical physics, but it is of consequence to probability and statistics as well. We will postpone further discussion of the problem; for more details and a few clues about what is ahead, see the short note "Stieltjes on the Stieltjes integral" in Birkhoff [1973].

CHAPTER THREE

Metrics and Norms

In the beginning there were operations – hundreds of them – limits, derivatives, integrals, sums; all of the many operations on functions, sequences, sets, vectors, matrices, and whatever else you might have encountered in calculus. The hallmark of twentieth-century mathematics is that we now view these operations as functions defined on entire collections of "abstract" objects rather than as specific actions taken on individual objects, one at a time. Maurice Fréchet, in a short expository article from 1950, had this to say (the italics are his own):

> In modern times it has been recognized that it is possible to elaborate full mathematical theories dealing with elements of which the nature is not specified, that is, with abstract elements. A collection of these abstract elements will be called an *abstract set*. If to this set there is added some rule of association of these elements, or some relation between them, the set will be called an *abstract space*. A natural generalization of function consists in associating with any element x of an abstract set E a number $f(x)$. Functional analysis is the study of such "functionals" $f(x)$. More generally, *general analysis* is the theory of the transformations $y = F[x]$ of an element x of an abstract set E into an element y of another (or the same) abstract set F. It is obvious that the study of general analysis should be preceded by a discussion of abstract spaces.
>
> It is necessary to keep in mind that these notions are *not of a metaphysical nature*; that when we speak of an abstract element we mean that the nature of this element is indifferent, but *we do not mean at all that this element is unreal*. Our theory will apply to all elements; in particular, applications of it may be made to the natural sciences. Of course, due attention must be paid to any properties which depend essentially on the nature of any special category of elements under investigation.

Early examples of this type of abstraction appeared in 1906 in Fréchet's thesis, "Sur quelques points du calcul functionnel," in which he introduced a notion of distance defined on abstract sets of points. In particular, Fréchet considered the collection $C[0, 1]$, consisting of all continuous real-valued functions defined on the closed interval $[0, 1]$, where we measure the distance between two functions by taking the maximum vertical distance between their graphs; that is, $\text{dist}(f, g) = \max_{0 \le t \le 1} |f(t) - g(t)|$. (This distance function was actually well known in 1906, but Fréchet was the first to view it as a small part of a much bigger picture.) Given a notion of distance between elements of $C[0, 1]$, it makes sense to ask questions like: Is integration continuous? That is, are the numbers $\int_0^1 f(t)\,dt$ and $\int_0^1 g(t)\,dt$ "close" whenever f and g are "close"?

This new point of view proved to have immediate applications; in that same year Friedrich Riesz used Fréchet's ideas to give a new proof of a result of Erhardt Schmidt, stating that any orthonormal system in $C[0,1]$ must be countable. In fact, Riesz extended this result to another collection of functions and in so doing introduced the L_p spaces. Riesz's techniques revolutionized the study of trigonometric series. To say that Fréchet's ideas caught on would be an understatement; the study of modern analysis would be lost without them. By 1928, Fréchet had compiled a monograph on his research on abstract spaces entitled *Les Espaces Abstraits*. (The word "space" has come to connote an abstract set of points that carries with it some additional structure.) Much of the terminology we will use, and certainly most of our examples of abstract spaces, can be found in Fréchet's monograph. By mathematical standards, 1928 is not so very long ago.

Metric Spaces

Given a set M, how might we define a distance function on M? What would we want a "reasonable" distance to do? Certainly we would want our distance to be (defined and) nonnegative for any pair of points in M. Let's start there: Let $d : M \times M \to [0, \infty)$ be a nonnegative, real-valued function defined on all pairs of elements from M. We would probably expect to have $d(x, x) = 0$ for any $x \in M$. And $d(x, y) = 0$ should mean that $x = y$. We would most likely want our distance to also satisfy $d(x, y) = d(y, x)$ for all pairs of points $x, y \in M$. Anything else? Well, in the hope of preserving at least a bit of the geometry granted by the familiar distances in $\mathbb{R}$ and $\mathbb{R}^n$, we might also require one last property. The distance function should satisfy **the triangle inequality**: For each triple of points x, y, z in M, we ask that $d(x, y) \le d(x, z) + d(z, y)$. The triangle inequality is the embodiment of that old saw, "The shortest distance between two points is a straight line." This timid little inequality will turn out to be immensely valuable.

A function d on $M \times M$ satisfying the following properties is called a **metric** on M.

(i) $0 \le d(x, y) < \infty$ for all pairs $x, y \in M$.
(ii) $d(x, y) = 0$ if and only if $x = y$.
(iii) $d(x, y) = d(y, x)$ for all pairs $x, y \in M$.
(iv) $d(x, y) \le d(x, z) + d(z, y)$ for all $x, y, z \in M$.

A function d on $M \times M$ that satisfies all of the above save item (ii) is sometimes called a *pseudometric*. Thus, a pseudometric will permit distinct points to be 0 distance apart.

The couple (M, d), consisting of a set M together with a metric d defined on M, is called a **metric space**. If a particular metric on M is understood, or if the argument at hand works equally well for any metric, we may forego this formality and simply refer to the set M as a metric space, with the tacit understanding that a metric d is available on demand.

Examples 3.1

(a) Every set M admits at least one metric. For example, check that the function defined by $d(x, y) = 1$ for any $x \neq y$ in M, and $d(x, x) = 0$ for all x in M, is a metric. This mundane, but always available, metric is called the **discrete metric** on M. It will prove to be much more interesting than first appearances suggest. A set supplied with its discrete metric will be called a **discrete space**.

(b) An important example for our purposes is the real line $\mathbb{R}$ together with its usual metric $d(a,b) = |a-b|$. Any time we refer to $\mathbb{R}$ without explicitly naming a metric, the absolute value metric is always understood to be the one that we have in mind.

(c) Any subset of a metric space is again a metric space in a very natural way. If d is a metric on M, and if A is a subset of M, then $d(x, y)$ is defined for any pair of points $x, y \in A$. Moreover, the restriction of d to $A \times A$ obviously still satisfies properties (i)–(iv). That is, the metric that is defined on M automatically defines a metric on A by restriction. We will even use the same letter d and simply refer to the metric space (A, d). Of particular interest in this regard is that $\mathbb{N}$, $\mathbb{Z}$, $\mathbb{Q}$, and $\mathbb{R} \setminus \mathbb{Q}$ each come already supplied with a natural metric, namely, the restriction of the usual metric on $\mathbb{R}$. In each case, we will refer to this restriction as the usual metric.

EXERCISES

1. Show that

$$d(x, y) = \left| \frac{1}{x} - \frac{1}{y} \right|$$

defines a metric on $(0, \infty)$.

▷ **2.** If d is a metric on M, show that $|d(x,z) - d(y,z)| \le d(x,y)$ for any $x, y, z \in M$.

3. As it happens, some of our requirements for a metric are redundant. To see why this is so, let M be a set and suppose that $d : M \times M \to \mathbb{R}$ satisfies $d(x,y) = 0$ if and only if $x = y$, and $d(x,y) \le d(x,z) + d(y,z)$ for all $x, y, z \in M$. Prove that d is a metric; that is, show that $d(x,y) \ge 0$ and $d(x,y) = d(y,x)$ hold for all x, y.

4. Let M be a set and suppose that $d : M \times M \to [0, \infty)$ satisfies properties (i), (ii), and (iii) for a metric on M and the triangle inequality *reversed*: $d(x,y) \ge d(x,z) + d(z,y)$. Prove that M has at most one point.

▷ **5.** There are other, albeit less natural, choices for a metric on $\mathbb{R}$. For instance, check that $\rho(a,b) = \sqrt{|a-b|}$, $\sigma(a,b) = |a-b|/(1+|a-b|)$, and $\tau(a,b) = \min\{|a-b|, 1\}$ each define metrics on $\mathbb{R}$. [Hint: To show that σ is a metric, you might first show that the function $F(t) = t/(1+t)$ is increasing and satisfies $F(s+t) \le F(s) + F(t)$ for $s, t \ge 0$. A similar approach will also work for ρ and τ.]

▷ **6.** If d is any metric on M, show that $\rho(x,y) = \sqrt{d(x,y)}$, $\sigma(x,y) = d(x,y)/(1 + d(x,y))$, and $\tau(x,y) = \min\{d(x,y), 1\}$ are also metrics on M. [Hint: $\sigma(x,y) = F(d(x,y))$, where F is as in Exercise 5.]

7. Here is a generalization of Exercises 5 and 6. Let $f : [0,\infty) \to [0,\infty)$ be increasing and satisfy $f(0) = 0$, and $f(x) > 0$ for all $x > 0$. If f also satisfies

$f(x+y) \le f(x) + f(y)$ for all $x, y \ge 0$, then $f \circ d$ is a metric whenever d is a metric. Show that each of the following conditions is sufficient to ensure that $f(x+y) \le f(x) + f(y)$ for all $x, y \ge 0$:
(a) f has a second derivative satisfying $f'' \le 0$;
(b) f has a decreasing first derivative;
(c) $f(x)/x$ is decreasing for $x > 0$.
[Hint: First show that (a) $\Longrightarrow$ (b) $\Longrightarrow$ (c).]

8. If d_1 and d_2 are both metrics on the same set M, which of the following yield metrics on M: $d_1 + d_2$? $\max\{d_1, d_2\}$? $\min\{d_1, d_2\}$? If d is a metric, is d^2 a metric?

9. Recall that $2^{\mathbb{N}}$ denotes the set of all sequences (or "strings") of 0s and 1s. Show that $d(a, b) = \sum_{n=1}^{\infty} 2^{-n}|a_n - b_n|$, where $a = (a_n)$ and $b = (b_n)$ are sequences of 0s and 1s, defines a metric on $2^{\mathbb{N}}$.

10. The *Hilbert cube* H^∞ is the collection of all real sequences $x = (x_n)$ with $|x_n| \le 1$ for $n = 1, 2, \ldots$.
(i) Show that $d(x, y) = \sum_{n=1}^{\infty} 2^{-n}|x_n - y_n|$ defines a metric on H^∞.
(ii) Given $x, y \in H^\infty$ and $k \in \mathbb{N}$, let $M_k = \max\{|x_1 - y_1|, \ldots, |x_k - y_k|\}$. Show that $2^{-k}M_k \le d(x, y) \le M_k + 2^{-k}$.

11. Let $\mathbb{R}^\infty$ denote the collection of all real sequences $x = (x_n)$. Show that the expression

$$d(x, y) = \sum_{n=1}^{\infty} \frac{1}{n!} \frac{|x_n - y_n|}{1 + |x_n - y_n|}$$

defines a metric on $\mathbb{R}^\infty$.

12. Check that $d(f, g) = \max_{a \le t \le b} |f(t) - g(t)|$ defines a metric on $C[a, b]$, the collection of all continuous, real-valued functions defined on the closed interval $[a, b]$.

13. Fréchet's metric on $C[0, 1]$ is by no means the only choice (although we will see later that it is a good one). For example, show that $\rho(f, g) = \int_0^1 |f(t) - g(t)|\,dt$ and $\sigma(f, g) = \int_0^1 \min\{|f(t) - g(t)|, 1\}\,dt$ also define metrics on $C[0, 1]$.

▷ **14.** We say that a subset A of a metric space M is **bounded** if there is some $x_0 \in M$ and some constant $C < \infty$ such that $d(a, x_0) \le C$ for all $a \in A$. Show that a finite union of bounded sets is again bounded.

▷ **15.** We define the **diameter** of a nonempty subset A of M by $\operatorname{diam}(A) = \sup\{d(a, b) : a, b \in A\}$. Show that A is bounded if and only if $\operatorname{diam}(A)$ is finite.

Normed Vector Spaces

A large and important class of metric spaces are also vector spaces (over $\mathbb{R}$ or $\mathbb{C}$). Notice, for example, that $C[0, 1]$ is a vector space (and even a ring). An easy way to build a metric on a vector space is by way of a length function or norm. A **norm** on a vector space V is a function $\|\cdot\| : V \to [0, \infty)$ satisfying:

(i) $0 \le \|x\| < \infty$ for all $x \in V$;
(ii) $\|x\| = 0$ if and only if $x = 0$ (the zero vector in V);
(iii) $\|\alpha x\| = |\alpha|\,\|x\|$ for any scalar α and any $x \in V$; and
(iv) **the triangle inequality**: $\|x + y\| \le \|x\| + \|y\|$ for all $x, y \in V$.

A function $\|\cdot\| : V \to [0, \infty)$ satisfying all of the above properties except (ii) is called a *pseudonorm* on V; that is, a pseudonorm permits nonzero vectors to have 0 length.

The pair $(V, \|\cdot\|)$, consisting of a vector space V together with a norm on V, is called a **normed vector space** (or normed *linear* space). Just as with metric spaces, we may be a bit lax with this formality. Phrases such as "let V be a normed vector space" carry the tacit understanding that a norm is lurking about in the background.

It is easy to see that any norm induces a metric on V by setting $d(x, y) = \|x - y\|$. We will refer to this particular metric as the **usual** metric on $(V, \|\cdot\|)$. We may even be so bold as to refer to $(V, \|\cdot\|)$ as a metric space with the clear understanding that the usual metric induced by the norm is the one that we have in mind. Not all metrics on a vector space come from norms, however, so we cannot afford to be totally negligent (see Exercise 16).

Examples 3.2

(a) The absolute value function $|\cdot|$ clearly defines a norm on $\mathbb{R}$.

(b) Each of the following defines a norm on $\mathbb{R}^n$:

$$\|x\|_1 = \sum_{i=1}^{n} |x_i|, \qquad \|x\|_2 = \left(\sum_{i=1}^{n} |x_i|^2\right)^{1/2},$$

and $\|x\|_\infty = \max_{1 \le i \le n} |x_i|$, where $x = (x_1, \ldots, x_n) \in \mathbb{R}^n$. The first and last expressions are very easy to check while the second takes a bit more work. (Although this is probably familiar from calculus, we will supply a proof shortly.) The function $\|\cdot\|_2$ is often called *the Euclidean norm* and is generally accepted as the norm of choice on $\mathbb{R}^n$. As it happens, for any $1 \le p < \infty$, the expression $\|x\|_p = \left(\sum |x_i|^p\right)^{1/p}$ defines a norm on $\mathbb{R}^n$; see Theorem 3.8.

(c) Each of the following defines a norm on $C[a, b]$:

$$\|f\|_1 = \int_a^b |f(t)|\,dt, \qquad \|f\|_2 = \left(\int_a^b |f(t)|^2\,dt\right)^{1/2},$$

$$\text{and} \qquad \|f\|_\infty = \max_{a \le t \le b} |f(t)|.$$

Again, the second expression is hardest to check (and we will do so later; for now, see Exercise 25). The last expression is generally taken as "the" norm on $C[a, b]$.

(d) If $(V, \|\cdot\|)$ is a normed vector space, and if W is a *linear subspace* of V, then W is also normed by $\|\cdot\|$. That is, the restriction of $\|\cdot\|$ to W defines a norm on W.

(e) We might also consider the sequence space analogues of the "scale" of norms on $\mathbb{R}^n$ given in (b). For $1 \le p < \infty$, we define ℓ_p to be the collection of all real sequences $x = (x_n)$ for which $\sum_{n=1}^{\infty} |x_n|^p < \infty$, and we define ℓ_∞ to be the collection of all bounded real sequences. Each ℓ_p is a vector space under "coordinatewise" addition and scalar multiplication. Moreover, the expression $\|x\|_p = \left(\sum |x_n|^p\right)^{1/p}$ if $1 \le p < \infty$ or $\|x\|_\infty = \sup_n |x_n|$ if $p = \infty$ defines a norm on ℓ_p. The cases $p = 1$ and $p = \infty$ are easy to check (see Exercise 21), the case $p = 2$ is given as Theorem 3.4, while the case $1 < p < \infty$ is given as Theorem 3.8.

We can complete the details of several of our examples if we prove that ℓ_2 is a vector space and that $\|\cdot\|_2$ is a norm on ℓ_2. Now it is easy to see that if $\|x\|_2 = 0$, then $x_n = 0$ for all n and hence that $x = 0$ (the zero vector in ℓ_2). Also, given $x \in \ell_2$ and $\alpha \in \mathbb{R}$, it is easy to see that $\alpha x \in \ell_2$, where $\alpha x = (\alpha x_n)$, and that $\|\alpha x\|_2 = |\alpha|\|x\|_2$. What is not so clear is whether $x + y = (x_n + y_n)$ is in ℓ_2 whenever x and y are in ℓ_2. In other words, if x and y are square-summable, does it follow that $x + y$ is square-summable? A moment's reflection will convince you that to answer this question we will need to know something about the "dot product" $\sum x_n y_n$. This extra bit of information is supplied by the following lemma.

Lemma 3.3. (The Cauchy–Schwarz Inequality) $\sum_{i=1}^{\infty} |x_i y_i| \le \|x\|_2 \|y\|_2$ *for any* $x, y \in \ell_2$.

PROOF. To simplify our notation a bit, let's agree to write $\langle x, y\rangle = \sum x_i y_i$. We first consider the case where $x, y \in \mathbb{R}^n$ (that is, $x_i = 0 = y_i$ for all $i > n$). In this case, $\langle x, y\rangle$ is the usual "dot product" in $\mathbb{R}^n$. Also notice that we may suppose that $x, y \ne 0$. (There is nothing to show if either is 0.)

Now let $t \in \mathbb{R}$ and consider

$$0 \le \|x + ty\|_2^2 = \langle x + ty, x + ty\rangle = \|x\|_2^2 + 2t\langle x, y\rangle + t^2\|y\|_2^2.$$

Since this (nontrivial) quadratic in t is always nonnegative, it must have a nonpositive discriminant. (Why?) Thus, $\left(2\langle x, y\rangle\right)^2 - 4\|x\|_2^2\,\|y\|_2^2 \le 0$ or, after simplifying, $|\langle x, y\rangle| \le \|x\|_2\,\|y\|_2$. That is, $\left|\sum_{i=1}^{n} x_i y_i\right| \le \|x\|_2\,\|y\|_2$.

Now this isn't quite what we wanted, but it actually implies the stronger inequality in the statement of the lemma. Why? Because the inequality that we have shown must also hold for the vectors $(\,|x_i|\,)$ and $(\,|y_i|\,)$. That is,

$$\sum_{i=1}^{n} |x_i|\,|y_i| \;\le\; \|(\,|x_i|\,)\|_2\,\|(\,|y_i|\,)\|_2 \;=\; \|x\|_2\,\|y\|_2.$$

Finally, let $x, y \in \ell_2$. Then for each n we have

$$\sum_{i=1}^{n} |x_i y_i| \le \left(\sum_{i=1}^{n} |x_i|^2\right)^{1/2} \left(\sum_{i=1}^{n} |y_i|^2\right)^{1/2} \le \|x\|_2\,\|y\|_2.$$

Thus, $\sum_{i=1}^{\infty} x_i y_i$ must be absolutely convergent and satisfy $\sum_{i=1}^{\infty} |x_i y_i| \le \|x\|_2\,\|y\|_2$. $\square$

Now we are ready to prove the triangle inequality for the ℓ_2-norm.

Theorem 3.4. (Minkowski's Inequality) *If $x, y \in \ell_2$, then $x + y \in \ell_2$. Moreover,* $\|x + y\|_2 \le \|x\|_2 + \|y\|_2$.

PROOF. It follows from the Cauchy–Schwarz inequality that for each n we have

$$\sum_{i=1}^{n} |x_i + y_i|^2 = \sum_{i=1}^{n} |x_i|^2 + 2\sum_{i=1}^{n} x_i y_i + \sum_{i=1}^{n} |y_i|^2$$
$$\le \|x\|_2^2 + 2\|x\|_2 \|y\|_2 + \|y\|_2^2 = (\|x\|_2 + \|y\|_2)^2 .$$

Thus, since n is arbitrary, we have $x + y \in \ell_2$ and $\|x + y\|_2 \le \|x\|_2 + \|y\|_2$. □

We have now shown that ℓ_2 is a vector space and that $\|\cdot\|_2$ is a norm on ℓ_2. As you have no doubt already surmised, the proof is essentially identical to the one used to show that $\|\cdot\|_2$ is a norm on $\mathbb{R}^n$. In the next section a variation on this theme will be used to prove that ℓ_p is a vector space and that $\|\cdot\|_p$ is a norm.

EXERCISES

16. Let V be a vector space, and let d be a metric on V satisfying $d(x, y) = d(x - y, 0)$ and $d(\alpha x, \alpha y) = |\alpha|\, d(x, y)$ for every $x, y \in V$ and every scalar α. Show that $\|x\| = d(x, 0)$ defines a norm on V (that has d as its "usual" metric). Give an example of a metric on the vector space $\mathbb{R}$ that fails to be associated with a norm in this way.

17. Recall that for $x \in \mathbb{R}^n$ we have defined $\|x\|_1 = \sum_{i=1}^{n} |x_i|$ and $\|x\|_\infty = \max_{1 \le i \le n} |x_i|$. Check that each of these is indeed a norm on $\mathbb{R}^n$.

▷ **18.** Show that $\|x\|_\infty \le \|x\|_2 \le \|x\|_1$ for any $x \in \mathbb{R}^n$. Also check that $\|x\|_1 \le n\|x\|_\infty$ and $\|x\|_1 \le \sqrt{n}\, \|x\|_2$.

19. Show that we have $\sum_{i=1}^{n} x_i y_i = \|x\|_2 \|y\|_2$ (equality in the Cauchy–Schwarz inequality) if and only if x and y are *proportional*, that is, if and only if either $x = \alpha y$ or $y = \alpha x$ for some $\alpha \ge 0$.

20. Show that $\|A\| = \max_{1 \le i \le n} \left(\sum_{j=1}^{m} |a_{i,j}|^2 \right)^{1/2}$ is a norm on the vector space $\mathbb{R}^{n \times m}$ of all $n \times m$ real matrices $A = [a_{i,j}]$.

21. Recall that we defined ℓ_1 to be the collection of all absolutely summable sequences under the norm $\|x\|_1 = \sum_{n=1}^{\infty} |x_n|$, and we defined ℓ_∞ to be the collection of all bounded sequences under the norm $\|x\|_\infty = \sup_{n \ge 1} |x_n|$. Fill in the details showing that each of these spaces is in fact a normed vector space.

22. Show that $\|x\|_\infty \le \|x\|_2$ for any $x \in \ell_2$, and that $\|x\|_2 \le \|x\|_1$ for any $x \in \ell_1$.

23. The subset of ℓ_∞ consisting of all sequences that converge to 0 is denoted by c_0. (Note that c_0 is actually a linear subspace of ℓ_∞; thus c_0 is also a normed vector space under $\|\cdot\|_\infty$.) Show that we have the following *proper* set inclusions: $\ell_1 \subset \ell_2 \subset c_0 \subset \ell_\infty$.

More Inequalities

We next supply the promised extension of Theorem 3.4 to the spaces ℓ_p, $1 < p < \infty$. Just as in the case of ℓ_2, notice that several facts are easy to check. For example, it is clear that $\|x\|_p = 0$ implies that $x = 0$, and it is easy to see that $\|\alpha x\|_p = |\alpha|\,\|x\|_p$ for any scalar α. Thus we lack only the triangle inequality. We begin with a few classical inequalities that are of interest in their own right. The first shows that ℓ_p is at least a vector space:

Lemma 3.5. *Let $1 < p < \infty$ and let $a, b \geq 0$. Then, $(a+b)^p \leq 2^p(a^p + b^p)$. Consequently, $x + y \in \ell_p$ whenever $x, y \in \ell_p$.*

PROOF. $(a+b)^p \leq (2\max\{a,b\})^p = 2^p \max\{a^p, b^p\} \leq 2^p(a^p + b^p)$. Thus, if $x, y \in \ell_p$, then $\sum_{n=1}^{\infty} |x_n + y_n|^p \leq 2^p \sum_{n=1}^{\infty} |x_n|^p + 2^p \sum_{n=1}^{\infty} |y_n|^p < \infty$. □

Lemma 3.6. (Young's Inequality) *Let $1 < p < \infty$ and let q be defined by $1/p + 1/q = 1$. Then, for any $a, b \geq 0$, we have $ab \leq a^p/p + b^q/q$, with equality occurring if and only if $a^{p-1} = b$.*

PROOF. Since the inequality trivially holds if either a or b is 0, we may certainly suppose that $a, b > 0$. Next notice that $q = p/(p-1)$ also satisfies $1 < q < \infty$ and $p - 1 = p/q = 1/(q-1)$. Thus, the functions $f(t) = t^{p-1}$ and $g(t) = t^{q-1}$ are *inverses* for $t \geq 0$.

The proof of the inequality follows from a comparison of areas (see Figure 3.1). The area of the rectangle with sides of lengths a and b is at most the sum of the areas under the graphs of the functions $y = x^{p-1}$ for $0 \leq x \leq a$ and $x = y^{q-1}$ for

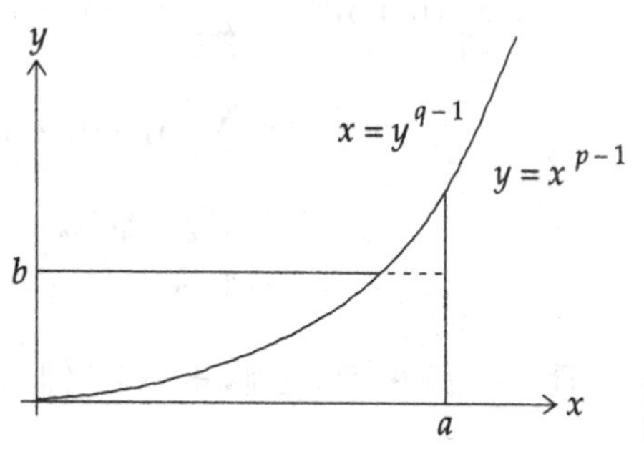

Figure 3.1

$0 \leq y \leq b$. That is,

$$ab \leq \int_0^a x^{p-1}\,dx + \int_0^b y^{q-1}\,dy = \frac{a^p}{p} + \frac{b^q}{q}.$$

Clearly, equality can occur only if $a^{p-1} = b$. □

When $p = q = 2$, Young's inequality reduces to the *arithmetic–geometric mean inequality* (although it is usually stated in the form $\sqrt{ab} \leq (a+b)/2$). Young's inequality will supply the extension of the Cauchy–Schwarz inequality that we need.

Lemma 3.7. (Hölder's Inequality) *Let* $1 < p < \infty$ *and let* q *be defined by* $1/p + 1/q = 1$. *Given* $x \in \ell_p$ *and* $y \in \ell_q$, *we have* $\sum_{i=1}^{\infty} |x_i y_i| \le \|x\|_p \, \|y\|_q$.

PROOF. We may suppose that $\|x\|_p > 0$ and $\|y\|_q > 0$ (since, otherwise, there is nothing to show). Now, for $n \ge 1$ we use Young's inequality to estimate:

$$\sum_{i=1}^{n} \left| \frac{x_i y_i}{\|x\|_p \, \|y\|_q} \right| \le \frac{1}{p} \sum_{i=1}^{n} \left| \frac{x_i}{\|x\|_p} \right|^p + \frac{1}{q} \sum_{i=1}^{n} \left| \frac{y_i}{\|y\|_q} \right|^q \le \frac{1}{p} + \frac{1}{q} = 1.$$

Thus, $\sum_{i=1}^{n} |x_i y_i| \le \|x\|_p \, \|y\|_q$ for any $n \ge 1$, and the result follows. □

Our proof of the triangle inequality will be made easier if we first isolate one of the key calculations. Notice that if $x \in \ell_p$, then the sequence $(\,|x_n|^{p-1})_{n=1}^{\infty} \in \ell_q$, because $(p-1)q = p$. Moreover,

$$\|(\,|x_n|^{p-1})\|_q = \left(\sum_{i=1}^{\infty} |x_i|^p \right)^{1/q} = \|x\|_p^{p-1}.$$

Theorem 3.8. (Minkowski's Inequality) *Let* $1 < p < \infty$. *If* $x, y \in \ell_p$, *then* $x + y \in \ell_p$ *and* $\|x + y\|_p \le \|x\|_p + \|y\|_p$.

PROOF. We have already shown that $x + y \in \ell_p$ (Lemma 3.5). To prove the triangle inequality, we once again let q be defined by $1/p + 1/q = 1$, and we now use Hölder's inequality to estimate:

$$\begin{aligned}
\sum_{i=1}^{\infty} |x_i + y_i|^p &= \sum_{i=1}^{\infty} |x_i + y_i| \cdot |x_i + y_i|^{p-1} \\
&\le \sum_{i=1}^{\infty} |x_i| \cdot |x_i + y_i|^{p-1} + \sum_{i=1}^{\infty} |y_i| \cdot |x_i + y_i|^{p-1} \\
&\le \|x\|_p \cdot \|(\,|x_n + y_n|^{p-1})\|_q + \|y\|_p \cdot \|(\,|x_n + y_n|^{p-1})\|_q \\
&= \|x + y\|_p^{p-1} \left(\|x\|_p + \|y\|_p \right).
\end{aligned}$$

That is, $\|x + y\|_p^p \le \|x + y\|_p^{p-1} \left(\|x\|_p + \|y\|_p \right)$, and the triangle inequality follows. □

EXERCISES

24. The conclusion of Lemma 3.7 also holds in the case $p = 1$ and $q = \infty$. Why?

25. The same techniques can be used to show that $\|f\|_p = \left(\int_0^1 |f(t)|^p \, dt \right)^{1/p}$ defines a norm on $C[0, 1]$ for any $1 < p < \infty$. State and prove the analogues of Lemma 3.7 and Theorem 3.8 in this case. (Does Lemma 3.7 still hold in this setting for $p = 1$ and $q = \infty$?)

26. Given $a, b > 0$, show that $\lim_{p\to\infty}(a^p + b^p)^{1/p} = \max\{a, b\}$. [Hint: If $a < b$ and $r = a/b$, show that $(1/p)\log(1 + r^p) \to 0$ as $p \to \infty$.] What happens as $p \to 0$? as $p \to -1$? as $p \to -\infty$?

Limits in Metric Spaces

Now that we have generalized the notion of distance, we can easily define the notions of convergence and continuity in metric spaces. It will help a bit, though, if we first generate some notation for "small" sets. Throughout this section, unless otherwise specified, we will assume that we are always dealing with a generic metric space (M, d).

Given $x \in M$ and $r > 0$, the set $B_r(x) = \{y \in M : d(x, y) < r\}$ is called the **open ball** about x of radius r. If we also need to refer to the metric d, then we write $B_r^d(x)$. We may occasionally refer to the set $\{y \in M : d(x, y) \le r\}$ as the *closed* ball about x of radius r, but we will not bother with any special notation for closed balls.

Examples 3.9

(a) In $\mathbb{R}$ we have $B_r(x) = (x - r, x + r)$, the open interval of radius r about x, while in $\mathbb{R}^2$ the set $B_r(x)$ is the open disk of radius r centered at x.

(b) In a *discrete* space $B_1(x) = \{x\}$ and $B_2(x) = M$.

(c) In a normed vector space $(V, \|\cdot\|)$ the balls centered at 0 play a special role (see Exercise 32); in this setting $B_r(0) = \{x : \|x\| < r\}$.

A subset A of M is said to be **bounded** if it is contained in some ball, that is, if $A \subset B_r(x)$ for some $x \in M$ and some $r > 0$. But exactly which x and r does not much matter. In fact, A is bounded if and only if for *any* $x \in M$ we have $\sup_{a\in A} d(x, a) < \infty$. (Why?) Related to this is the **diameter** of A, defined by $\text{diam}(A) = \sup\{d(a, b) : a, b \in A\}$. The diameter of A is a convenient measure of size because it does not refer to points outside of A.

EXERCISES

Each of the following exercises is set in a generic metric space (M, d).

27. Show that $\text{diam}(B_r(x)) \le 2r$, and give an example where strict inequality occurs.

28. If $\text{diam}(A) < r$, show that $A \subset B_r(a)$ for some $a \in A$.

▷ **29.** Prove that A is bounded if and only if $\text{diam}(A) < \infty$.

▷ **30.** If $A \subset B$, show that $\text{diam}(A) \le \text{diam}(B)$.

31. Give an example where $\text{diam}(A \cup B) > \text{diam}(A) + \text{diam}(B)$. If $A \cap B \ne \emptyset$, show that $\text{diam}(A \cup B) \le \text{diam}(A) + \text{diam}(B)$.

▷ **32.** In a normed vector space $(V, \|\cdot\|)$ show that $B_r(x) = x + B_r(0) = \{x + y : \|y\| < r\}$ and that $B_r(0) = rB_1(0) = \{rx : \|x\| < 1\}$.

A **neighborhood** of x is any set containing an open ball about x. You should think of a neighborhood of x as a "thick" set of points near x. We will use this new terminology to streamline our definition of convergence.

We say that a sequence of points (x_n) in M **converges** to a point $x \in M$ if $d(x_n, x) \to 0$. Now, since this definition is stated in terms of the sequence of *real numbers* $\big(d(x_n, x)\big)_{n=1}^{\infty}$, we can easily derive the following equivalent reformulations:

$$\begin{cases} (x_n) \text{ converges to } x \text{ if and only if, given any } \varepsilon > 0, \text{ there is} \\ \text{an integer } N \geq 1 \text{ such that } d(x_n, x) < \varepsilon \text{ whenever } n \geq N, \end{cases}$$

or

$$\begin{cases} (x_n) \text{ converges to } x \text{ if and only if, given any } \varepsilon > 0, \text{ there is} \\ \text{an integer } N \geq 1 \text{ such that } \{x_n : n \geq N\} \subset B_\varepsilon(x). \end{cases}$$

If it should happen that $\{x_n : n \geq N\} \subset A$ for some N, we say that the sequence (x_n) is **eventually** in A. Thus, our last formulation can be written

$$\begin{cases} (x_n) \text{ converges to } x \text{ if and only if, given any } \varepsilon > 0, \\ \text{the sequence } (x_n) \text{ is eventually in } B_\varepsilon(x) \end{cases}$$

or, in yet another incarnation,

$$\begin{cases} (x_n) \text{ converges to } x \text{ if and only if the sequence} \\ (x_n) \text{ is eventually in every neighborhood of } x. \end{cases}$$

This final version is blessed by a total lack of Ns and εs! In any event, just as with real sequences, we usually settle for the shorthand $x_n \to x$ in place of the phrase (x_n) converges to x. On occasion we will want to display the set M, or d, or both, and so we may also write $x_n \xrightarrow{d} x$ or $x_n \to x$ in (M, d). We also define Cauchy (or d-Cauchy, if we need to specify d) in the obvious way: A sequence (x_n) is **Cauchy** in (M, d) if, given any $\varepsilon > 0$, there is an integer $N \geq 1$ such that $d(x_m, x_n) < \varepsilon$ whenever $m, n \geq N$. We can reword this just a bit to read: (x_n) is Cauchy if and only if, given $\varepsilon > 0$, there is an integer $N \geq 1$ such that $\operatorname{diam}(\{x_n : n \geq N\}) \leq \varepsilon$. (How?)

Much of what we already know about sequences of real numbers will carry over to this new setting – but not everything! The reader is strongly encouraged to test the limits of this transition by supplying proofs for the following easy results.

EXERCISES

Each of the following exercises is set in a metric space M with metric d.

33. Limits are unique. [Hint: $d(x, y) \leq d(x, x_n) + d(x_n, y)$.]

▷ **34.** If $x_n \to x$ in (M, d), show that $d(x_n, y) \to d(x, y)$ for any $y \in M$. More generally, if $x_n \to x$ and $y_n \to y$, show that $d(x_n, y_n) \to d(x, y)$.

35. If $x_n \to x$, then $x_{n_k} \to x$ for any subsequence (x_{n_k}) of (x_n).

▷ **36.** A convergent sequence is Cauchy, and a Cauchy sequence is bounded (that is, the set $\{x_n : n \geq 1\}$ is bounded).

▷ **37.** A Cauchy sequence with a convergent subsequence converges.

38. A sequence (x_n) has a Cauchy subsequence if and only if it has a subsequence (x_{n_k}) for which $d\left(x_{n_k}, x_{n_{k+1}}\right) < 2^{-k}$ for all k.

▷ **39.** If every subsequence of (x_n) has a *further* subsequence that converges to x, then (x_n) converges to x.

Now, while several familiar results about sequences in $\mathbb{R}$ have carried over successfully to the "abstract" setting of metric spaces, at least a few will not survive the journey. Two especially fragile cases are: Cauchy sequences *need not* converge and bounded sequences *need not* have convergent subsequences. A few specific examples might help your appreciation of their delicacy.

Examples 3.10

(a) Consider the sequence $(1/n)_{n=1}^{\infty}$ living in the space $M = (0, 1]$ under its usual metric. Then, $(1/n)$ is Cauchy but, annoyingly, does not converge to any point in M. (Why?) Notice too that $(1/n)$ is a bounded sequence with no convergent subsequence.

(b) Consider $M = \mathbb{R}$ supplied with the discrete metric. Then, $(n)_{n=1}^{\infty}$ is a bounded sequence with no Cauchy subsequence!

(c) At least one good thing happens in any discrete space: Cauchy sequences always converge. But for a simple reason. In a discrete space, a sequence (x_n) is Cauchy if and only if it is *eventually constant*; that is, if and only if $x_n = x$ for some (fixed) x and all n sufficiently large. (Why?)

Let's take a closer look at $\mathbb{R}^n$ (with its usual metric). Since $d(x, y) = \|x - y\|_2 = \left(\sum_{i=1}^{n} |x_i - y_i|^2\right)^{1/2} \geq |x_j - y_j|$ for any $j = 1, \ldots, n$, it follows that a sequence of vectors $x^{(k)} = (x_1^k, \ldots, x_n^k)$ in $\mathbb{R}^n$ converges (is Cauchy) if and only if each of the coordinate sequences $(x_j^k)_{k=1}^{\infty}$ converges (is Cauchy) in $\mathbb{R}$. (Why?) Thus, nearly every fact about convergent sequences in $\mathbb{R}$ "lifts" successfully to $\mathbb{R}^n$. For example, any Cauchy sequence in $\mathbb{R}^n$ converges in $\mathbb{R}^n$, and any bounded sequence in $\mathbb{R}^n$ has a convergent subsequence.

How much of this has to do with the particular metric that we chose for $\mathbb{R}^n$? And will this same result "lift" to the spaces ℓ_1, ℓ_2, or ℓ_∞, for example? We cannot hope for much, but each of these spaces shares at least one thing in common with $\mathbb{R}^n$. Since all three of the norms $\|\cdot\|_1$, $\|\cdot\|_2$, and $\|\cdot\|_\infty$ satisfy $\|x\| \geq |x_j|$ for any j, it follows that convergence in ℓ_1, ℓ_2, or ℓ_∞ will imply "coordinatewise" convergence. That is, if $x^{(k)} = (x_n^k)_{n=1}^{\infty}$, $k = 1, 2, \ldots$, is a sequence (of sequences!) in, say, ℓ_1, and if $x^{(k)} \to x$ in ℓ_1, then we must have $x_n^k \to x_n$ (as $k \to \infty$) for each $n = 1, 2, \ldots$. A simple example will convince you that the converse does not hold, in general, in this new setting. The sequence $e^{(k)} = (0, \ldots, 0, 1, 0, \ldots)$, where the kth entry is 1 and the rest are 0s, converges "coordinatewise" to $0 = (0, 0, \ldots)$, but $(e^{(k)})$ does *not* converge to 0 in any of the metric spaces ℓ_1, ℓ_2, or ℓ_∞. Why? Because in each of the three spaces we have $d(e^{(k)}, 0) = \|e^{(k)}\| = 1$. In fact, $(e^{(k)})$ is not even Cauchy because in each case we also have $\|e^{(k)} - e^{(m)}\| \geq 1$ for any $k \neq m$.

EXERCISES

40. Here is a positive result about ℓ_1 that may restore your faith in intuition. Given any (fixed) element $x \in \ell_1$, show that the sequence $x^{(k)} = (x_1, \ldots, x_k, 0, \ldots) \in \ell_1$ (i.e., the first k terms of x followed by all 0s) converges to x in ℓ_1-norm. Show that the same holds true in ℓ_2, but give an example showing that it fails (in general) in ℓ_∞.

41. Given $x, y \in \ell_2$, recall that $\langle x, y \rangle = \sum_{i=1}^\infty x_i y_i$. Show that if $x^{(k)} \to x$ and $y^{(k)} \to y$ in ℓ_2, then $\langle x^{(k)}, y^{(k)} \rangle \to \langle x, y \rangle$.

▷ **42.** Two metrics d and ρ on a set M are said to be **equivalent** if they generate the same convergent sequences; that is, $d(x_n, x) \to 0$ if and only if $\rho(x_n, x) \to 0$. If d is any metric on M, show that the metrics ρ, σ, and τ, defined in Exercise 6, are all equivalent to d.

▷ **43.** Show that the usual metric on $\mathbb{N}$ is equivalent to the discrete metric. Show that any metric on a *finite* set is equivalent to the discrete metric.

▷ **44.** Show that the metrics induced by $\|\cdot\|_1$, $\|\cdot\|_2$, and $\|\cdot\|_\infty$ on $\mathbb{R}^n$ are all equivalent. [Hint: See Exercise 18.]

45. We say that two norms on the same vector space X are equivalent if the metrics they induce are equivalent. Show that $\|\cdot\|$ and $|||\cdot|||$ are equivalent on X if and only if they generate the same sequences tending to 0; that is, $\|x_n\| \to 0$ if and only if $|||x_n||| \to 0$.

▷ **46.** Given two metric spaces (M, d) and (N, ρ), we can define a metric on the product $M \times N$ in a variety of ways. Our only requirement is that a sequence of pairs (a_n, x_n) in $M \times N$ should converge precisely when both coordinate sequences (a_n) and (x_n) converge (in (M, d) and (N, ρ), respectively). Show that each of the following define metrics on $M \times N$ that enjoy this property and that all three are equivalent:

$$d_1\big((a, x), (b, y)\big) = d(a, b) + \rho(x, y),$$
$$d_2\big((a, x), (b, y)\big) = \big(d(a, b)^2 + \rho(x, y)^2\big)^{1/2},$$
$$d_\infty\big((a, x), (b, y)\big) = \max\{d(a, b), \rho(x, y)\}.$$

Henceforth, any implicit reference to "the" metric on $M \times N$, sometimes called **the product metric**, will mean one of d_1, d_2, or d_∞. Any one of them will serve equally well; use whichever looks most convenient for the argument at hand.

While we are not yet ready for an all-out attack on continuity, it couldn't hurt to give a hint as to what is ahead. Given a function $f : (M, d) \to (N, \rho)$ between two metric spaces, and given a point $x \in M$, we have at least two plausible sounding definitions for the continuity of f at x. Each definition is derived from its obvious counterpart for real-valued functions by replacing absolute values with an appropriate metric.

For example, we might say that f is continuous at x if $\rho(f(x_n), f(x)) \to 0$ whenever $d(x_n, x) \to 0$. That is, f should send sequences converging to x into sequences converging to $f(x)$. This says that f "commutes" with limits: $f(\lim_{n\to\infty} x_n) = \lim_{n\to\infty} f(x_n)$. Sounds like a good choice.

Or we might try doctoring the familiar ε-δ definition from a first course in calculus. In this case we would say that f is continuous at x if, given any $\varepsilon > 0$, there always exists a $\delta > 0$ such that $\rho(f(x), f(y)) < \varepsilon$ whenever $d(x, y) < \delta$. Written in slightly different terms, this definition requires that $f\left(B_\delta^d(x)\right) \subset B_\varepsilon^\rho(f(x))$. That is, f maps a sufficiently small neighborhood of x into a given neighborhood of $f(x)$.

We will rewrite the definition once more, but this time we will use an inverse image. Recall that the *inverse image* of a set A, under a function $f : X \to Y$, is defined to be the *set* $\{x \in X : f(x) \in A\}$ and is usually written $f^{-1}(A)$. (The inverse image of any set under any function always makes sense. Although the notation is similar, inverse images have nothing whatever to do with inverse *functions*, which don't always make sense.) Stated in terms of an inverse image, our condition reads: $B_\delta^d(x) \subset f^{-1}\left(B_\varepsilon^\rho(f(x))\right)$. Look a bit imposing? Well, it actually tells us quite a bit. It says that the inverse image of a "thick" set containing $f(x)$ must still be "thick" near x. Curious. Figure 3.2 may help you with these new definitions. Better still, draw a few pictures of your own!

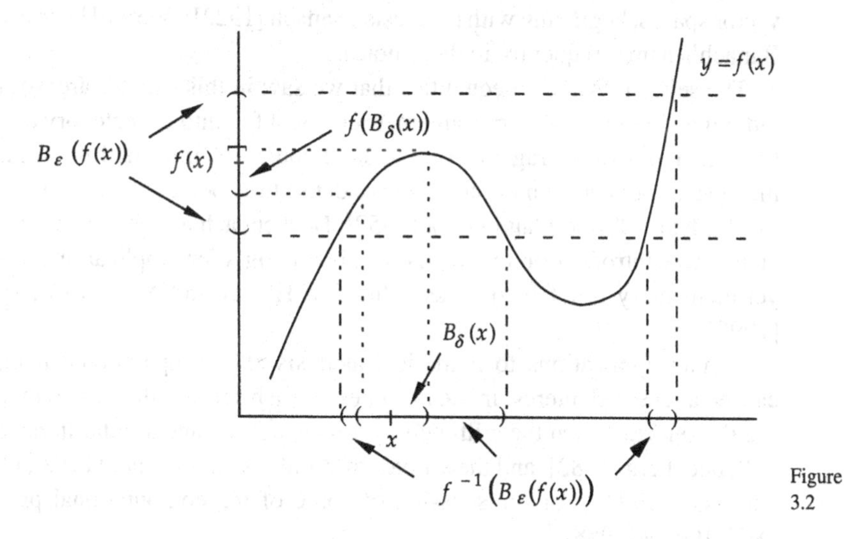

Figure 3.2

This sets the stage for what is ahead. Each of the two possible definitions for continuity seems perfectly reasonable. Certainly we would hope that the two turn out to be equivalent. But what do convergent sequences have to do with "thick" sets? And just what is a "thick" set anyway?

Notes and Remarks

The quotation at the start of this chapter is taken from Fréchet [1950]; his thesis appears in Fréchet [1906]. His book, Fréchet [1928], was published as one of the volumes in a series of monographs edited by Émile Borel. The authors in this series include every "name" French mathematician of that time: Baire, Borel, Lebesgue, Lévy, de La Vallée Poussin, and many others. The full title of Fréchet's book, including subtitle, is enlightening: *Les espaces abstraits et leur théorie considérée comme introduction a l'analyse générale* (Abstract spaces and their theory considered as an introduction to general analysis). The paper by Riesz mentioned in the introductory passage is Riesz [1906].

It was Hausdorff who gave us the name "metric space." Indeed, his classic work *Grundzüge der Mengenlehre*, Leipzig, 1914, is the source for much of our terminology regarding abstract sets and abstract spaces. An English translation of Hausdorff's book is available as *Set Theory* (Hausdorff [1937]). If we had left it up to Fréchet, we would be calling metric spaces "spaces of type (D)."

For more on metric spaces, normed spaces, and $\mathbb{R}^n$, see Copson [1968], Goffman and Pedrick [1965], Goldberg [1976], Hoffman [1975], Kaplansky [1977], Kasriel [1971], Kolmogorov and Fomin [1970], and Kuller [1969]. For a look at modern applications of metric space notions, see Barnsley [1988] and Edgar [1990].

Normed vector spaces were around for some time before anyone bothered to formalize their definition. Quite often you will see the great Polish mathematician Stefan Banach mentioned as the originator of normed vector spaces, but this is only partly true. In any case, it is fair to say that Banach gave the first *thorough* treatment of normed vector spaces, beginning with his thesis (Banach [1922]). We will have cause to mention Banach's name frequently in these notes.

The several "name" inequalities that we saw in this chapter are, for the most part, older than the study of norms and metrics. Most fall into the category of "mean values" (various types of averages). An excellent source of information on inequalities and mean values of every shape and size is a dense little book with the apt title *Inequalities*, by Hardy, Littlewood, and Pólya [1952]. Beckenbach and Bellman [1961] provide an elementary introduction to inequalities, including a few applications. For a very slick, yet elementary proof of the inequalities of Hölder and Minkowski, see Maligranda [1995].

Certain applications to numerical analysis and computational mathematics have caused a renewed interest in mean values. For a brief introduction to this exciting area, see the selection "On the arithmetic-geometric mean and similar iterative algorithms" in Schoenberg [1982], and the articles by Almkvist and Berndt [1988], Carlson [1971], and Miel [1983]. For a discussion of some of the computational practicalities, see D. H. Bailey [1988].

CHAPTER FOUR

Open Sets and Closed Sets

Open Sets

One of the themes of this (or any other) course in real analysis is the curious interplay between various notions of "big" sets and "small" sets. We have seen at least one such measure of size already: Uncountable sets are big, whereas countable sets are small. In this chapter we will make precise what was only hinted at in Chapter Three – the rather vague notion of a "thick" set in a metric space. For our purposes, a "thick" set will be one that contains an entire neighborhood of each of its points. But perhaps we can come up with a better name.... Throughout this chapter, unless otherwise specified, we live in a generic metric space (M, d).

A set U in a metric space (M, d) is called an **open set** if U contains a neighborhood of each of its points. In other words, U is an open set if, given $x \in U$, there is some $\varepsilon > 0$ such that $B_\varepsilon(x) \subset U$.

Examples 4.1

(a) In any metric space, the whole space M is an open set. The empty set $\emptyset$ is also open (by default).

(b) In $\mathbb{R}$, any open interval is an open set. Indeed, given $x \in (a, b)$, let $\varepsilon = \min\{x - a, b - x\}$. Then, $\varepsilon > 0$ and $(x - \varepsilon, x + \varepsilon) \subset (a, b)$. The cases (a, ∞) and $(-\infty, b)$ are similar. While we're at it, notice that the interval $[0, 1)$, for example, is *not* open in $\mathbb{R}$ because it does not contain an entire neighborhood of 0.

(c) In a discrete space, $B_1(x) = \{x\}$ is an open set for any x. (Why?) It follows that *every* subset of a discrete space is open.

Before we get too carried away, we should follow the lead suggested by our last two examples and check that every open ball is in fact an open set.

Proposition 4.2. *For any $x \in M$ and any $\varepsilon > 0$, the open ball $B_\varepsilon(x)$ is an open set.*

PROOF. Let $y \in B_\varepsilon(x)$. Then $d(x, y) < \varepsilon$ and hence $\delta = \varepsilon - d(x, y) > 0$. We will show that $B_\delta(y) \subset B_\varepsilon(x)$ (as in Figure 4.1). Indeed, if $d(y, z) < \delta$, then, by the triangle inequality, $d(x, z) \leq d(x, y) + d(y, z) < d(x, y) + \delta = d(x, y) + \varepsilon - d(x, y) = \varepsilon$. $\square$

Let's collect our thoughts. First, every open ball is open. Next, it follows from the definition of open sets that an open set must actually be a *union* of open balls. In fact,

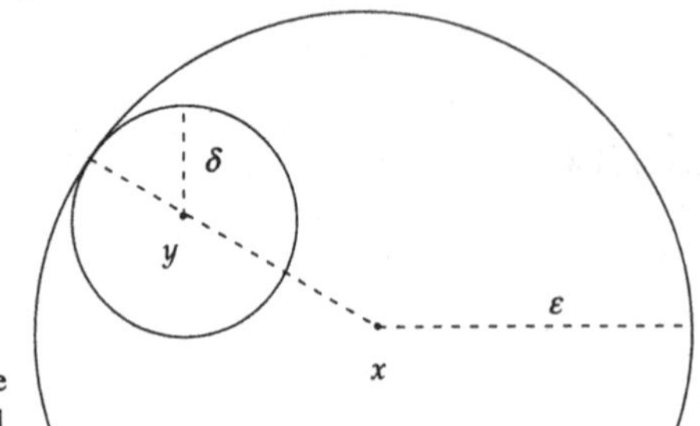

Figure 4.1

if U is open, then $U = \bigcup\{B_\varepsilon(x) : B_\varepsilon(x) \subset U\}$. Moreover, *any* arbitrary union of open balls is again an open set. (Why?) Here's what all of this means:

Theorem 4.3. *An* arbitrary union *of open sets is again open; that is, if* $(U_\alpha)_{\alpha \in A}$ *is any collection of open sets, then* $V = \bigcup_{\alpha \in A} U_\alpha$ *is open.*

PROOF. If $x \in V$, then $x \in U_\alpha$ for some $\alpha \in A$. But then, since U_α is open, $B_\varepsilon(x) \subset U_\alpha \subset V$ for some $\varepsilon > 0$. □

Intersections aren't nearly as generous:

Theorem 4.4. *A* finite intersection *of open sets is open; that is, if each of* $U_1, \ldots, U_n$ *is open, then so is* $V = U_1 \cap \cdots \cap U_n$.

PROOF. If $x \in V$, then $x \in U_i$ for all $i = 1, \ldots, n$. Thus, for each i there is an $\varepsilon_i > 0$ such that $B_{\varepsilon_i}(x) \subset U_i$. But then, setting $\varepsilon = \min\{\varepsilon_1, \ldots, \varepsilon_n\} > 0$, we have $B_\varepsilon(x) \subset \bigcap_{i=1}^n B_{\varepsilon_i}(x) \subset \bigcap_{i=1}^n U_i = V$. □

Example 4.5

The word "finite" is crucial in Theorem 4.4 because $\bigcap_{n=1}^\infty(-1/n, 1/n) = \{0\}$, and $\{0\}$ is not open in $\mathbb{R}$. (Why?)

Now, since the real line $\mathbb{R}$ is of special interest to us, let's characterize the open subsets of $\mathbb{R}$. This will come in handy later. But it should be stressed that while this characterization holds for $\mathbb{R}$, it does not have a satisfactory analogue even in $\mathbb{R}^2$. (As we will see in Chapter Six, not every open set in the plane can be written as a union of *disjoint* open disks.)

Theorem 4.6. *If U is an open subset of $\mathbb{R}$, then U may be written as a* **countable** *union of* **disjoint** *open intervals. That is,* $U = \bigcup_{n=1}^\infty I_n$, *where* $I_n = (a_n, b_n)$ *(these may be unbounded) and* $I_n \cap I_m = \emptyset$ *for* $n \neq m$.

PROOF. We know that U can be written as a union of open intervals (because each $x \in U$ is in some open interval I with $I \subset U$). What we need to show is that U is a union of *disjoint* open intervals – such a union, as we know, must be countable (see Exercise 2.15).

We first claim that each $x \in U$ is contained in a *maximal* open interval $I_x \subset U$ in the sense that if $x \in I \subset U$, where I is an open interval, then we must have

$I \subset I_x$. Indeed, given $x \in U$, let

$$a_x = \inf\{a : (a, x] \subset U\} \qquad \text{and} \qquad b_x = \sup\{b : [x, b) \subset U\}.$$

Then, $I_x = \big(a_x, b_x\big)$ satisfies $x \in I_x \subset U$, and I_x is clearly maximal. (Check this!)

Next, notice that for any $x, y \in U$ we have either $I_x \cap I_y = \emptyset$ or $I_x = I_y$. Why? Because if $I_x \cap I_y \neq \emptyset$, then $I_x \cup I_y$ is an open interval containing both I_x and I_y. By maximality we would then have $I_x = I_y$. It follows that U is the union of disjoint (maximal) intervals: $U = \bigcup_{x \in U} I_x$. □

Now any time we make up a new definition in a metric space setting, it is usually very helpful to find an equivalent version stated exclusively in terms of sequences. To motivate this in the particular case of open sets, let's recall:

$$x_n \to x \iff (x_n) \text{ is eventually in } B_\varepsilon(x), \text{ for any } \varepsilon > 0$$

and hence

$$x_n \to x \iff (x_n) \text{ is eventually in } U, \text{ for any open set } U \text{ containing } x.$$

(Why?) This last statement essentially characterizes open sets:

Theorem 4.7. *A set U in (M, d) is open if and only if, whenever a sequence (x_n) in M converges to a point $x \in U$, we have $x_n \in U$ for all but finitely many n.*

PROOF. The forward implication is clear from the remarks preceding the theorem. Let's see why the new condition implies that U is open:

If U is *not* open, then there is an $x \in U$ such that $B_\varepsilon(x) \cap U^c \neq \emptyset$ for all $\varepsilon > 0$. In particular, for each n there is some $x_n \in B_{1/n}(x) \cap U^c$. But then $(x_n) \subset U^c$ and $x_n \to x$. (Why?) Thus, the new condition also fails. □

In slightly different language, Theorem 4.7 is saying that the only way to reach a member of an open set is by traveling well inside the set; there are no inhabitants on the "frontier." In essence, you cannot visit a single resident without seeing a whole neighborhood!

Closed Sets

What good would "open" be without "closed"? A set F in a metric space (M, d) is said to be a **closed set** if its complement $F^c = M \setminus F$ is open.

We can draw several immediate (although not terribly enlightening) conclusions:

Examples 4.8

(a) $\emptyset$ and M are always closed. (And so it is possible for a set to be *both* open and closed!)

(b) An arbitrary intersection of closed sets is closed. A finite union of closed sets is closed.

(c) Any finite set is closed. Indeed, it is enough to show that $\{x\}$ is always closed. (Why?) Given any $y \in M \setminus \{x\}$ (that is, any $y \neq x$), note that $\varepsilon = d(x, y) > 0$, and hence $B_\varepsilon(y) \subset M \setminus \{x\}$.

(d) In $\mathbb{R}$, each of the intervals $[a, b]$, $[a, \infty)$, and $(-\infty, b]$ is closed. Also, $\mathbb{N}$ and Δ are closed sets. (Why?)

(e) In a discrete space, *every* subset is closed.

(f) *Sets are not "doors"!* $(0, 1]$ is neither open nor closed in $\mathbb{R}$!

As yet, our definition is not terribly useful. It would be nice if we had an intrinsic characterization of closed sets – something that did not depend on a knowledge of open sets – something in terms of sequences, for example. For this let's first make an observation: F is closed if and only if F^c is open, and so F is closed if and only if

$$x \in F^c \Longrightarrow B_\varepsilon(x) \subset F^c \qquad \text{for some } \varepsilon > 0.$$

But this is the same as saying: F is closed if and only if

$$B_\varepsilon(x) \cap F \neq \emptyset \qquad \text{for every } \varepsilon > 0 \Longrightarrow x \in F. \tag{4.1}$$

This is our first characterization of closed sets. (Compare this with the phrase "F is not open," as in the proof of Theorem 4.7. They are similar, but not the same!)

Notice, please, that if $x \in F$, then $B_\varepsilon(x) \cap F \neq \emptyset$ necessarily follows; we are interested in the reverse implication here. In general, a point x that satisfies $B_\varepsilon(x) \cap F \neq \emptyset$ for every $\varepsilon > 0$ is evidently "very close" to F in the sense that x cannot be separated from F by any positive distance. At worst, x might be on the "boundary" of F. Thus condition (4.1) is telling us that a set is closed if and only if it contains all such "boundary" points. Exercises 33, 40, and 41 make these notions more precise. For now, let's translate condition (4.1) into a sequential characterization of closed sets.

Theorem 4.9. *Given a set F in (M, d), the following are equivalent:*

(i) *F is closed; that is, $F^c = M \setminus F$ is open.*

(ii) *If $B_\varepsilon(x) \cap F \neq \emptyset$ for every $\varepsilon > 0$, then $x \in F$.*

(iii) *If a sequence $(x_n) \subset F$ converges to some point $x \in M$, then $x \in F$.*

PROOF. (i) $\Longleftrightarrow$ (ii): This is clear from our observations above and the definition of an open set.

(ii) $\Longrightarrow$ (iii): Suppose that $(x_n) \subset F$ and $x_n \xrightarrow{d} x \in M$. Then $B_\varepsilon(x)$ contains infinitely many x_n for any $\varepsilon > 0$, and hence $B_\varepsilon(x) \cap F \neq \emptyset$ for any $\varepsilon > 0$. Thus $x \in F$, by (ii).

(iii) $\Longrightarrow$ (ii): If $B_\varepsilon(x) \cap F \neq \emptyset$ for all $\varepsilon > 0$, then for each n there is an $x_n \in B_{1/n}(x) \cap F$. The sequence (x_n) satisfies $(x_n) \subset F$ and $x_n \to x$. Hence, by (iii), $x \in F$. $\square$

Condition (iii) of Theorem 4.9 is just a rewording of our sequential characterization of open sets (Theorem 4.7) applied to $U = F^c$. Most authors take (iii) as the definition of a closed set. In other words, condition (iii) says that a closed set must contain all of its *limit points*. That is, "closed" means closed under the operation of taking of limits. (Exercise 33 explores a slightly different, but more precise, notion of limit point.)

EXERCISES

1. Show that an "open rectangle" $(a, b) \times (c, d)$ is an open set in $\mathbb{R}^2$. More generally, if A and B are open in $\mathbb{R}$, show that $A \times B$ is open in $\mathbb{R}^2$. If A and B are closed in $\mathbb{R}$, show that $A \times B$ is closed in $\mathbb{R}^2$.

2. If F is a closed set and G is an open set in a metric space M, show that $F \setminus G$ is closed and that $G \setminus F$ is open.

▷ **3.** Some authors say that two metrics d and ρ on a set M are equivalent if they generate the same open sets. Prove this. (Recall that we have defined equivalence to mean that d and ρ generate the same convergent sequences. See Exercise 3.42.)

4. Prove that *every* subset of a metric space M can be written as the intersection of open sets.

▷ **5.** Let $f : \mathbb{R} \to \mathbb{R}$ be continuous. Show that $\{x : f(x) > 0\}$ is an open subset of $\mathbb{R}$ and that $\{x : f(x) = 0\}$ is a closed subset of $\mathbb{R}$.

6. Give an example of an infinite closed set in $\mathbb{R}$ containing only irrationals. Is there an open set consisting entirely of irrationals?

7. Show that every open set in $\mathbb{R}$ is the union of (countably many) open intervals with *rational* endpoints. Use this to show that the collection $\mathcal{U}$ of all open subsets of $\mathbb{R}$ has the same cardinality as $\mathbb{R}$ itself.

▷ **8.** Show that every open interval (and hence every open set) in $\mathbb{R}$ is a countable union of closed intervals and that every closed interval in $\mathbb{R}$ is a countable intersection of open intervals.

9. Let d be a metric on an infinite set M. Prove that there is an open set U in M such that both U and its complement are infinite. [Hint: Either (M, d) is discrete or it's not. . . .]

10. Given $y = (y_n) \in H^\infty$, $N \in \mathbb{N}$, and $\varepsilon > 0$, show that $\{x = (x_n) \in H^\infty : |x_k - y_k| < \varepsilon,\ k = 1, \ldots, N\}$ is open in H^∞ (see Exercise 3.10).

▷ **11.** Let $e^{(k)} = (0, \ldots, 0, 1, 0, \ldots)$, where the kth entry is 1 and the rest are 0s. Show that $\{e^{(k)} : k \geq 1\}$ is closed as a subset of ℓ_1.

12. Let F be the set of all $x \in \ell_\infty$ such that $x_n = 0$ for all but finitely many n. Is F closed? open? neither? Explain.

13. Show that c_0 is a closed subset of ℓ_∞. [Hint: If $(x^{(n)})$ is a sequence (of sequences!) in c_0 converging to $x \in \ell_\infty$, note that $|x_k| \leq \left|x_k - x_k^{(n)}\right| + \left|x_k^{(n)}\right|$ and now choose n so that $\left|x_k - x_k^{(n)}\right|$ is small *independent* of k.]

14. Show that the set $A = \{x \in \ell_2 : |x_n| \leq 1/n,\ n = 1, 2, \ldots\}$ is a closed set in ℓ_2 but that $B = \{x \in \ell_2 : |x_n| < 1/n,\ n = 1, 2, \ldots\}$ is *not* an open set. [Hint: Does $B \supset B_\varepsilon(0)$?]

Now, as we've seen, some sets are neither open nor closed. However, it is possible to describe the "open part" of a set and the "closure" of a set. Here's what we'll do:

Given a set A in (M, d), we define the **interior** of A, written $\text{int}(A)$ or A°, to be the largest open set contained in A. That is,

$$\begin{aligned}\text{int}(A) = A^\circ &= \bigcup\{U : U \text{ is open and } U \subset A\}\\ &= \bigcup\{B_\varepsilon(x) : B_\varepsilon(x) \subset A \text{ for some } x \in A,\ \varepsilon > 0\} \qquad \text{(Why?)}\\ &= \{x \in A : B_\varepsilon(x) \subset A \text{ for some } \varepsilon > 0\}.\end{aligned}$$

Note that A° is clearly an *open subset* of A.

We next define the **closure** of A, written $\text{cl}(A)$ or $\bar{A}$, to be the smallest closed set containing A. That is,

$$\text{cl}(A) = \bar{A} = \bigcap\{F : F \text{ is closed and } A \subset F\}.$$

Please take note of the "dual" nature of our two new definitions.

Now it is clear that $\bar{A}$ is a *closed* set containing A – and necessarily the smallest one. But it's not so clear which points are in $\bar{A}$ or, more precisely, which points are in $\bar{A} \setminus A$. We could use a description of $\bar{A}$ that is a little easier to "test" on a given set A. It follows from our last theorem that $x \in \bar{A}$ if and only if $B_\varepsilon(x) \cap \bar{A} \neq \emptyset$ for every $\varepsilon > 0$. The description that we are looking for simply removes this last reference to $\bar{A}$.

Proposition 4.10. *$x \in \bar{A}$ if and only if $B_\varepsilon(x) \cap A \neq \emptyset$ for every $\varepsilon > 0$.*

PROOF. One direction is easy: If $B_\varepsilon(x) \cap A \neq \emptyset$ for every $\varepsilon > 0$, then $B_\varepsilon(x) \cap \bar{A} \neq \emptyset$ for every $\varepsilon > 0$, and hence $x \in \bar{A}$ by Theorem 4.9.

Now, for the other direction, let $x \in \bar{A}$ and let $\varepsilon > 0$. If $B_\varepsilon(x) \cap A = \emptyset$, then A is a subset of $\left(B_\varepsilon(x)\right)^c$, a closed set. Thus, $\bar{A} \subset \left(B_\varepsilon(x)\right)^c$. (Why?) But this is a contradiction, because $x \in \bar{A}$ while $x \notin \left(B_\varepsilon(x)\right)^c$. □

Corollary 4.11. *$x \in \bar{A}$ if and only if there is a sequence $(x_n) \subset A$ with $x_n \to x$.*

That is, $\bar{A}$ is the set of all limits of convergent sequences in A (including limits of constant sequences).

Example 4.12

Here are a few easy examples in $\mathbb{R}$. (Check the details!)

(a) $\text{int}\big((0, 1]\big) = (0, 1)$ and $\text{cl}\big((0, 1\,]\big) = [\,0, 1\,]$,

(b) $\text{int}\left(\{(1/n) : n \geq 1\}\right) = \emptyset$ and $\text{cl}\left(\{(1/n) : n \geq 1\}\right) = \{(1/n) : n \geq 1\} \cup \{0\}$,

(c) $\text{int}(\mathbb{Q}) = \emptyset$ and $\text{cl}(\mathbb{Q}) = \mathbb{R}$,

(d) $\text{int}(\Delta) = \emptyset$ and $\text{cl}(\Delta) = \Delta$.

EXERCISES

Unless otherwise specified, each of the following exercises is set in a generic metric space (M, d).

15. The set $A = \{y \in M : d(x, y) \leq r\}$ is sometimes called the *closed ball* about x of radius r. Show that A is a closed set, but give an example showing that A need not equal the closure of the open ball $B_r(x)$.

16. If $(V, \|\cdot\|)$ is any normed space, prove that the closed ball $\{x \in V : \|x\| \leq 1\}$ is always the closure of the open ball $\{x \in V : \|x\| < 1\}$.

▷ **17.** Show that A is open if and only if $A^\circ = A$ and that A is closed if and only if $\bar{A} = A$.

▷ **18.** Given a nonempty bounded subset E of $\mathbb{R}$, show that $\sup E$ and $\inf E$ are elements of $\bar{E}$. Thus $\sup E$ and $\inf E$ are elements of E whenever E is *closed*.

▷ **19.** Show that $\operatorname{diam}(A) = \operatorname{diam}(\bar{A})$.

20. If $A \subset B$, show that $\bar{A} \subset \bar{B}$. Does $\bar{A} \subset \bar{B}$ imply $A \subset B$? Explain.

21. If A and B are any sets in M, show that $\overline{A \cup B} = \bar{A} \cup \bar{B}$ and $\overline{A \cap B} \subset \bar{A} \cap \bar{B}$. Give an example showing that this last inclusion can be proper.

22. True or false? $(A \cup B)^\circ = A^\circ \cup B^\circ$.

23. If $x \neq y$ in M, show that there are disjoint open sets U, V with $x \in U$ and $y \in V$. Moreover, show that U and V can be chosen so that even $\bar{U}$ and $\bar{V}$ are disjoint.

24. Show that $\bar{A} = (\operatorname{int}(A^c))^c$ and that $A^\circ = (\operatorname{cl}(A^c))^c$.

25. A set that is simultaneously open and closed is sometimes called a **clopen** set. Show that $\mathbb{R}$ has no nontrivial clopen sets. [Hint: If U is a nontrivial open subset of $\mathbb{R}$, show that $\bar{U}$ is strictly bigger than U.]

26. We define the distance from a point $x \in M$ to a nonempty set A in M by $d(x, A) = \inf\{d(x, a) : a \in A\}$. Prove that $d(x, A) = 0$ if and only if $x \in \bar{A}$.

27. Show that $|d(x, A) - d(y, A)| \leq d(x, y)$ and conclude that the map $x \mapsto d(x, A)$ is continuous.

28. Given a set A in M and $\varepsilon > 0$, show that $\{x \in M : d(x, A) < \varepsilon\}$ is an open set and that $\{x \in M : d(x, A) \leq \varepsilon\}$ is a closed set (and each contains A).

29. Show that every closed set in M is the intersection of countably many open sets and that every open set in M is the union of countably many closed sets. [Hint: What is $\bigcap_{n=1}^\infty \{x \in M : d(x, A) < (1/n)\}$?]

30.

(a) For each $n \in \mathbb{Z}$, let F_n be a closed subset of $(n, n+1)$. Show that $F = \bigcup_{n \in \mathbb{Z}} F_n$ is a closed set in $\mathbb{R}$. [Hint: For each fixed n, first show that there is a $\delta_n > 0$ so that $|x - y| \geq \delta_n$ whenever $x \in F_n$ and $y \in F_m$, $m \neq n$.]

(b) Find a sequence of disjoint closed sets in $\mathbb{R}$ whose union is *not* closed.

31. If $x \notin F$, where F is closed, show that there are disjoint open sets U, V with $x \in U$ and $F \subset V$. (This extends the first result in Exercise 23 since $\{y\}$ is closed.) Is it possible to find U and V so that $\bar{U}$ and $\bar{V}$ are disjoint? Is it possible to extend this result further to read: Any two disjoint closed sets are contained in disjoint open sets?

32. We define the distance between two (nonempty) subsets A and B of M by $d(A, B) = \inf\{d(a, b) : a \in A, b \in B\}$. Give an example of two disjoint closed sets A and B in $\mathbb{R}^2$ with $d(A, B) = 0$.

▷ **33.** Let A be a subset of M. A point $x \in M$ is called a **limit point** of A if every neighborhood of x contains a point of A that is different from x itself, that is, if $(B_\varepsilon(x) \setminus \{x\}) \cap A \neq \emptyset$ for every $\varepsilon > 0$. If x is a limit point of A, show that every neighborhood of x contains infinitely many points of A.

▷ **34.** Show that x is a limit point of A if and only if there is a sequence (x_n) in A such that $x_n \to x$ and $x_n \neq x$ for all n.

35. Let A' be the set of limit points of a set A. Show that A' is closed and that $\bar{A} = A' \cup A$. Show that $A' \subset A$ if and only if A is closed. (A' is called the *derived set* of A.)

36. Suppose that $x_n \xrightarrow{d} x \in M$, and let $A = \{x\} \cup \{x_n : n \geq 1\}$. Prove that A is closed.

37. Prove the Bolzano–Weierstrass theorem: Every bounded infinite subset of $\mathbb{R}$ has a limit point. [Hint: Use the nested interval theorem. If A is a bounded infinite subset of $\mathbb{R}$, then A is contained in some closed bounded interval I_1. At least one of the left or right halves of I_1 contains infinitely many points of A. Call this new closed interval I_2. Continue.]

38. A set P is called **perfect** if it is empty or if it is a closed set and every point of P is a limit point of P. Show that Δ is perfect. Show that $\mathbb{R}$ is perfect when considered as a subset of $\mathbb{R}^2$.

39. Show that a nonempty perfect subset P of $\mathbb{R}$ is uncountable. This gives yet another proof that the Cantor set is uncountable. [Hint: First convince yourself that P is infinite, and assume that P is countable, say $P = \{x_1, x_2, \ldots\}$. Construct a decreasing sequence of nested closed intervals $[\, a_n, b_n \,]$ such that $(a_n, b_n) \cap P \neq \emptyset$ but $x_n \notin [\, a_n, b_n \,]$. Use the nested interval theorem to get a contradiction.]

40. If $x \in A$ and x is *not* a limit point of A, then x is called an **isolated point** of A. Show that a point $x \in A$ is an isolated point of A if and only if $(B_\varepsilon(x) \setminus \{x\}) \cap A = \emptyset$ for some $\varepsilon > 0$. Prove that a subset of $\mathbb{R}$ can have at most countably many isolated points, thus showing that every uncountable subset of $\mathbb{R}$ has a limit point.

41. Related to the notion of limit points and isolated points are boundary points. A point $x \in M$ is said to be a **boundary point** of A if each neighborhood of x hits both A and A^c. In symbols, x is a boundary point of A if and only if $B_\varepsilon(x) \cap A \neq \emptyset$ and $B_\varepsilon(x) \cap A^c \neq \emptyset$ for every $\varepsilon > 0$. Verify each of the following formulas, where $\text{bdry}(A)$ denotes the set of boundary points of A:
(a) $\text{bdry}(A) = \text{bdry}(A^c)$,
(b) $\text{cl}(A) = \text{bdry}(A) \cup \text{int}(A)$,
(c) $M = \text{int}(A) \cup \text{bdry}(A) \cup \text{int}(A^c)$.
Notice that the first and last equations tell us that each set A partitions M into three regions: the points "well inside" A, the points "well outside" A, and the points on the common boundary of A and A^c.

42. If E is a nonempty bounded subset of $\mathbb{R}$, show that $\sup E$ and $\inf E$ are both boundary points of E. Hence, if E is also closed, then $\sup E$ and $\inf E$ are elements of E.

43. Show that bdry(A) is always a closed set; in fact, bdry$(A) = \bar{A} \setminus A^\circ$.

44. Show that A is closed if and only if bdry$(A) \subset A$.

45. Give examples showing that bdry$(A) = \emptyset$ and bdry$(A) = M$ are both possible.

▷ **46.** A set A is said to be **dense** in M (or, as some authors say, *everywhere dense*) if $\bar{A} = M$. For example, both $\mathbb{Q}$ and $\mathbb{R} \setminus \mathbb{Q}$ are dense in $\mathbb{R}$. Show that A is dense in M if and only if any of the following hold:
(a) Every point in M is the limit of a sequence from A.
(b) $B_\varepsilon(x) \cap A \neq \emptyset$ for every $x \in M$ and every $\varepsilon > 0$.
(c) $U \cap A \neq \emptyset$ for every nonempty open set U.
(d) A^c has empty interior.

47. Let G be open and let D be dense in M. Show that $\overline{G \cap D} = \bar{G}$. Give an example showing that this equality may fail if G is not open.

▷ **48.** A metric space is called **separable** if it contains a countable dense subset. Find examples of countable dense sets in $\mathbb{R}$, in $\mathbb{R}^2$, and in $\mathbb{R}^n$.

49. Prove that ℓ_2 and H^∞ are separable. [Hint: Consider finitely nonzero sequences of the form $(r_1, \ldots, r_n, 0, 0, \ldots)$, where each r_k is rational.]

50. Show that ℓ_∞ is *not* separable. [Hint: Consider the set $2^{\mathbb{N}}$, consisting of all sequences of 0s and 1s, as a subset of ℓ_∞. We know that $2^{\mathbb{N}}$ is uncountable. Now what?]

51. Show that a separable metric space has at most countably many isolated points.

52. If M is separable, show that any collection of disjoint open sets in M is at most countable.

53. Can you find a countable dense subset of $C[0, 1]$?

54. A set A is said to be **nowhere dense** in M if $\text{int}(\text{cl}(A)) = \emptyset$. Show that $\{x\}$ is nowhere dense if and only if x is *not* an isolated point of M.

55. Show that every finite subset of $\mathbb{R}$ is nowhere dense. Is every countable subset of $\mathbb{R}$ nowhere dense? Show that the Cantor set is nowhere dense in $\mathbb{R}$.

56. If A and B are nowhere dense in M, show that $A \cup B$ is nowhere dense. Give an example showing that an infinite union of nowhere dense sets need not be nowhere dense.

57. If A is closed, show that A is nowhere dense if and only if A^c is dense if and only if A has an empty interior.

58. Let (r_n) be an enumeration of $\mathbb{Q}$. For each n, let I_n be the open interval centered at r_n of radius 2^{-n}, and let $U = \bigcup_{n=1}^\infty I_n$. Prove that U is a proper, open, dense subset of $\mathbb{R}$ and that U^c is nowhere dense in $\mathbb{R}$.

59. If A is closed, show that bdry(A) is nowhere dense.

60. Show that each of the following is equivalent to the statement "A is nowhere dense":

(a) $\bar{A}$ contains no nonempty open set.

(b) Each nonempty open set in M contains a nonempty open subset that is disjoint from A.

(c) Each nonempty open set in M contains an open ball that is disjoint from A.

The Relative Metric

Although it is a digression at this point, we need to generate some terminology for later use. First, given a nontrivial subset A of a metric space (M, d), recall that A "inherits" the metric d by restriction. Thus, the metric space (A, d) has open sets, closed sets, convergent sequences, and so on, of its own. How are these related to the open sets, closed sets, convergent sequences, and so on, of (M, d)? The answer comes from examining the open balls in (A, d). Note that for $x \in A$ we have

$$B_\varepsilon^A(x) = \{a \in A : d(x, a) < \varepsilon\} = A \cap \{y \in M : d(x, y) < \varepsilon\} = A \cap B_\varepsilon^M(x),$$

where superscripts have been used to distinguish between a ball in A and a ball in M. Thus, a subset G of A is open in (A, d), or open *relative* to A, if, given $x \in G$, there is some $\varepsilon > 0$ such that

$$G \supset B_\varepsilon^A(x) = A \cap B_\varepsilon^M(x).$$

This observation leads us to the following:

Proposition 4.13. *Let $A \subset M$.*

(i) *A set $G \subset A$ is open in (A, d) if and only if $G = A \cap U$, where U is open in (M, d).*

(ii) *A set $F \subset A$ is closed in (A, d) if and only if $F = A \cap C$, where C is closed in (M, d).*

(iii) $\text{cl}_A(E) = A \cap \text{cl}_M(E)$ *for any subset E of A (where the subscripts distinguish between the closure of E in (A, d) and the closure of E in (M, d)).*

PROOF. We will prove (i) and leave the rest as exercises.

First suppose that $G = A \cap U$, where U is open in (M, d). If $x \in G \subset U$, then $x \in B_\varepsilon^M(x) \subset U$ for some $\varepsilon > 0$. But since $G \subset A$, we have $x \in A \cap B_\varepsilon^M(x) = B_\varepsilon^A(x) \subset A \cap U = G$. Thus, G is open in (A, d).

Next suppose that G is open in (A, d). Then, for each $x \in G$, there is some $\varepsilon_x > 0$ such that $x \in B_{\varepsilon_x}^A(x) = A \cap B_{\varepsilon_x}^M(x) \subset G$. But now it is clear that $U = \bigcup\{B_{\varepsilon_x}^M(x) : x \in G\}$ is an open set in (M, d) satisfying $G = A \cap U$. □

We paraphrase the statement "G is open in (A, d)" by saying that "G is open in A," or "G is open relative to A," or perhaps "G is relatively open in A." The same goes for closed sets. In the case of closures, the symbols $\mathrm{cl}_A(E)$ are read "the closure of E in A." Another notation for $\mathrm{cl}_A(E)$ is $\bar{E}^A$.

Examples 4.14

(a) Let $A = (0, 1] \cup \{2\}$, considered as a subset of $\mathbb{R}$. Then, $(0, 1]$ is open in A and $\{2\}$ is both open and closed in A. (Why?)

(b) We may consider $\mathbb{R}$ as a subset of $\mathbb{R}^2$ in an obvious way – all pairs of the form $(x, 0)$, $x \in \mathbb{R}$. The metric that $\mathbb{R}$ inherits from $\mathbb{R}^2$ in this way is nothing but the usual metric on $\mathbb{R}$. (Why?) Similarly, $\mathbb{R}^2$ may be considered as a natural subset of $\mathbb{R}^3$ (as the xy-plane, for instance). What happens in this case? Figure 4.2 might help.

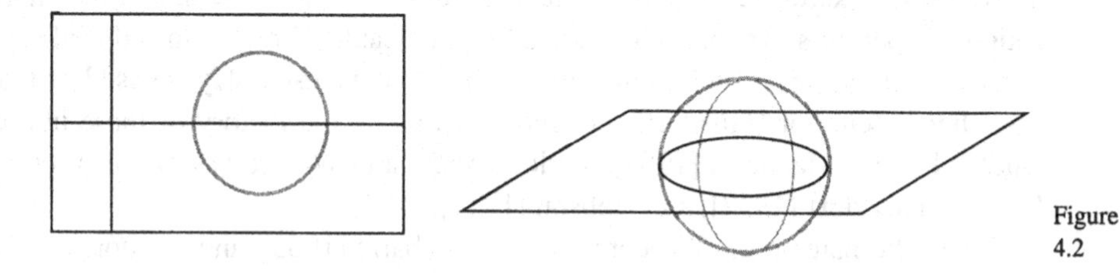

Figure 4.2

EXERCISES

Throughout, M denotes an arbitrary metric space with metric d.

▷ **61.** Complete the proof of Proposition 4.13.

▷ **62.** Suppose that A is open in (M, d) and that $G \subset A$. Show that G is open in A if and only if G is open in M. Is the result still true if "open" is replaced everywhere by "closed"? Explain.

63. Is there a nonempty subset of $\mathbb{R}$ that is open when considered as a subset of $\mathbb{R}^2$? closed?

64. Show that the analogue of part (iii) of Proposition 4.13 for relative interiors is *false*. Specifically, find sets $E \subset A \subset \mathbb{R}$ such that $\mathrm{int}_A(E) = A$ while $\mathrm{int}_{\mathbb{R}}(E) = \emptyset$.

65. Let A be a subset of M. If G and H are disjoint open sets in A, show that there are disjoint open sets U and V in M such that $G = U \cap A$ and $H = V \cap A$. [Hint: Let $U = \bigcup\{B^M_{\varepsilon/2}(x) : x \in G \text{ and } B^A_\varepsilon(x) \subset G\}$. Do the same for V and H.]

66. Let $A \subset B \subset M$. If A is dense in B (how would you define this?), and if B is dense in M, show that A is dense in M.

67. Let G be open and let D be dense in M. Show that $G \cap D$ is dense in G. Give an example showing that this may fail if G is not open.

68. If A is a separable subset of M (that is, if A has a countable dense subset of its own), show that $\bar{A}$ is also separable.

69. A collection (U_α) of open sets is called an *open base* for M if every open set in M can be written as a union of U_α. For example, the collection of all open intervals in $\mathbb{R}$ with *rational* endpoints is an open base for $\mathbb{R}$ (and this is even a countable collection). (Why?) Prove that M has a countable open base if and only if M is separable. [Hint: If $\{x_n\}$ is a countable dense set in M, consider the collection of open balls with rational radii centered at the x_n.]

◊

Notes and Remarks

For sets of real numbers, the concepts of neighborhoods, limit points (Exercise 33), derived sets (Exercise 35), perfect sets (Exercise 38), closed sets, and the characterization of open sets (Theorem 4.6) are all due to Cantor. Fréchet introduced separable spaces (Exercise 48). Much of the terminology that we use today is based on that used by either Fréchet or Hausdorff. For more details on the history of these notions see Dudley [1989], Manheim [1964], Taylor [1982], and Willard [1970]; also see Fréchet [1928], Haussdorf [1937], and Hobson [1927].

For an alternate proof of Theorem 4.6, see Labarre [1965], and for more on "Cantor-like" nowhere dense subsets of $\mathbb{R}$ (as in Exercise 58), see the short note in Wilansky [1953b].

CHAPTER FIVE

Continuity

Continuous Functions

Throughout this chapter, unless otherwise specified, (M, d) and (N, ρ) are arbitrary metric spaces and $f : M \to N$ is a function mapping M into N. We say that f is **continuous at a point** $x \in M$ if:

$$\begin{cases} \text{for every } \varepsilon > 0, \text{ there is a } \delta > 0 \text{ (which depends on } f, x, \text{ and } \varepsilon) \text{ such} \\ \text{that } \rho(f(x), f(y)) < \varepsilon \text{ whenever } y \in M \text{ satisfies } d(x, y) < \delta. \end{cases}$$

Recall from our earlier discussions that we may rephrase this definition (how?) to read:

$$\begin{cases} f \text{ is continuous at } x \text{ if, for any } \varepsilon > 0, \text{ there is a } \delta > 0 \text{ such that} \\ f\left(B_\delta^d(x)\right) \subset B_\varepsilon^\rho(f(x)) \text{ or, equivalently, } B_\delta^d(x) \subset f^{-1}\left(B_\varepsilon^\rho(f(x))\right). \end{cases}$$

If f is continuous at every point of M, we simply say that f is **continuous on** M, or often just that f is continuous.

By now it should be clear that any statement concerning arbitrary open balls will translate into a statement concerning arbitrary open sets. Thus, there is undoubtedly a characterization of continuity available that may be stated exclusively in terms of open sets. Of course, any statement concerning open sets probably has a counterpart using closed sets. And don't forget sequences! Open sets and closed sets can each be characterized in terms of convergent sequences, and so we would expect to find a characterization of continuity in terms of convergent sequences, too. At any rate, we've done enough hinting around about reformulations of the definition of continuity. It's time to put our cards on the table.

Theorem 5.1. *Given $f : (M, d) \to (N, \rho)$, the following are equivalent:*

(i) *f is continuous on M (by the ε-δ definition).*

(ii) *For any $x \in M$, if $x_n \to x$ in M, then $f(x_n) \to f(x)$ in N.*

(iii) *If E is closed in N, then $f^{-1}(E)$ is closed in M.*

(iv) *If V is open in N, then $f^{-1}(V)$ is open in M.*

PROOF. (i) $\Longrightarrow$ (ii): (Compare this with the case $f : \mathbb{R} \to \mathbb{R}$.) Suppose that $x_n \xrightarrow{d} x$. Given $\varepsilon > 0$, let $\delta > 0$ be such that $f\left(B_\delta^d(x)\right) \subset B_\varepsilon^\rho(f(x))$. Then, since $x_n \xrightarrow{d} x$, we have that (x_n) is eventually in $B_\delta^d(x)$. But this implies that $(f(x_n))$ is eventually in $B_\varepsilon^\rho(f(x))$. Since ε is arbitrary, this means that $f(x_n) \xrightarrow{\rho} f(x)$.

(ii) $\Longrightarrow$ (iii): Let E be closed in (N, ρ). Given $(x_n) \subset f^{-1}(E)$ such that $x_n \xrightarrow{d} x \in M$, we need to show that $x \in f^{-1}(E)$. But $(x_n) \subset f^{-1}(E)$ implies that $(f(x_n)) \subset E$, while $x_n \xrightarrow{d} x \in M$ tells us that $f(x_n) \xrightarrow{\rho} f(x)$ from (ii). Thus, since E is closed, we have that $f(x) \in E$ or $x \in f^{-1}(E)$.

(iii) $\Longleftrightarrow$ (iv) is obvious, since $f^{-1}(A^c) = \left(f^{-1}(A)\right)^c$. See Exercise 1.

(iv) $\Longrightarrow$ (i): Given $x \in M$ and $\varepsilon > 0$, the set $B_\varepsilon^\rho(f(x))$ is open in (N, ρ) and so, by (iv), the set $f^{-1}\left(B_\varepsilon^\rho(f(x))\right)$ is open in (M, d). But then $B_\delta^d(x) \subset f^{-1}\left(B_\varepsilon^\rho(f(x))\right)$, for some $\delta > 0$, because $x \in f^{-1}\left(B_\varepsilon^\rho(f(x))\right)$. □

Example 5.2

(a) Define $\chi_{\mathbb{Q}} : \mathbb{R} \to \mathbb{R}$ by $\chi_{\mathbb{Q}}(x) = 1$, if $x \in \mathbb{Q}$, and $\chi_{\mathbb{Q}}(x) = 0$, if $x \notin \mathbb{Q}$. Then, $\chi_{\mathbb{Q}}^{-1}(B_{1/3}(1)) = \mathbb{Q}$ and $\chi_{\mathbb{Q}}^{-1}(B_{1/3}(0)) = \mathbb{R} \setminus \mathbb{Q}$. Thus $\chi_{\mathbb{Q}}$ cannot be continuous at any point of $\mathbb{R}$ because neither $\mathbb{Q}$ nor $\mathbb{R} \setminus \mathbb{Q}$ contains an interval.

(b) A function $f : M \to N$ between metric spaces is called an **isometry** (into) if f preserves distances: $\rho(f(x), f(y)) = d(x, y)$ for all $x, y \in M$. Obviously, an isometry is continuous. The natural inclusions from $\mathbb{R}$ into $\mathbb{R}^2$ (i.e., $x \mapsto (x, 0)$) and from $\mathbb{R}^2$ into $\mathbb{R}^3$ (this time $(x, y) \mapsto (x, y, 0)$) are isometries. (Why?)

(c) Let $f : \mathbb{N} \to \mathbb{R}$ be *any* function. Then f is continuous! Why? Because $\{n\}$ is an open ball in $\mathbb{N}$. Specifically, $\{n\} = B_{1/2}(n) \subset f^{-1}(B_\varepsilon(f(n)))$ for any $\varepsilon > 0$.

(d) $f : \mathbb{R} \to \mathbb{N}$ is continuous if and only if f is *constant*! Why? [Hint: See Exercise 4.25.]

(e) Relative continuity can sometimes be counterintuitive. From (a) we know that $\chi_{\mathbb{Q}}$ has *no* points of continuity relative to $\mathbb{R}$, but the restriction of $\chi_{\mathbb{Q}}$ to $\mathbb{Q}$ is everywhere continuous *relative to* $\mathbb{Q}$! Why? (See Exercise 9 for more details.)

(f) If y is any fixed element of (M, d), then the real-valued function $f(x) = d(x, y)$ is continuous on M. As we will see, even more is true (see Exercises 20 and 34).

EXERCISES

Throughout, M denotes an arbitrary metric space with metric d.

▷ **1.** Given a function $f : S \to T$ and sets $A, B \subset S$ and $C, D \subset T$, establish the following:

(i) $A \subset f^{-1}(f(A))$, with equality for all A if and only if f is one-to-one.

(ii) $f\left(f^{-1}(C)\right) \subset C$, with equality for all C if and only if f is onto.

(iii) $f(A \cup B) = f(A) \cup f(B)$.

(iv) $f^{-1}(C \cup D) = f^{-1}(C) \cup f^{-1}(D)$.

(v) $f(A \cap B) \subset f(A) \cap f(B)$, with equality for all A and B if and only if f is one-to-one.

(vi) $f^{-1}(C \cap D) = f^{-1}(C) \cap f^{-1}(D)$.

(vii) $f(A) \setminus f(B) \subset f(A \setminus B)$.

(viii) $f^{-1}(C \setminus D) = f^{-1}(C) \setminus f^{-1}(D)$.

Generalize, wherever possible, to arbitrary unions and intersections.

▷ **2.** Given a subset A of some "universal" set S, we define $\chi_A : S \to \mathbb{R}$, the **characteristic function** of A, by $\chi_A(x) = 1$ if $x \in A$ and $\chi_A(x) = 0$ if $x \notin A$. Prove or disprove the following formulas: $\chi_{A \cup B} = \chi_A + \chi_B$, $\chi_{A \cap B} = \chi_A \cdot \chi_B$, $\chi_{A \setminus B} = \chi_A - \chi_B$. What corrections are necessary?

3. If $f : A \to B$ and $C \subset B$, what is $\chi_C \circ f$ (as a characteristic function)?

4. Show that $\chi_\Delta : \mathbb{R} \to \mathbb{R}$, the characteristic function of the Cantor set, is discontinuous at each point of Δ.

5. Is there a continuous characteristic function on $\mathbb{R}$? If $A \subset \mathbb{R}$, show that χ_A is continuous at each point of $\operatorname{int}(A)$. Are there any other points of continuity?

6. Let $f : \mathbb{R} \to \mathbb{R}$ be continuous. Show that $\{x : f(x) > 0\}$ is an open subset of $\mathbb{R}$ and that $\{x : f(x) = 0\}$ is a closed subset of $\mathbb{R}$. If $f(x) = 0$ whenever x is rational, show that $f(x) = 0$ for every real x.

7.

(a) If $f : M \to \mathbb{R}$ is continuous and $a \in \mathbb{R}$, show that the sets $\{x : f(x) > a\}$ and $\{x : f(x) < a\}$ are open subsets of M.

(b) Conversely, if the sets $\{x : f(x) > a\}$ and $\{x : f(x) < a\}$ are open for every $a \in \mathbb{R}$, show that f is continuous.

(c) Show that f is continuous even if we assume only that the sets $\{x : f(x) > a\}$ and $\{x : f(x) < a\}$ are open for every *rational* a.

▷ **8.** Let $f : \mathbb{R} \to \mathbb{R}$ be continuous.

(a) If $f(0) > 0$, show that $f(x) > 0$ for all x in some interval $(-a, a)$.

(b) If $f(x) \geq 0$ for every rational x, show that $f(x) \geq 0$ for all real x. Will this result hold with "≥ 0" replaced by ">0"? Explain.

▷ **9.** Let $A \subset M$. Show that $f : (A, d) \to (N, \rho)$ is continuous at $a \in A$ if and only if, given $\varepsilon > 0$, there is a $\delta > 0$ such that $\rho(f(x), f(a)) < \varepsilon$ whenever $d(x, a) < \delta$ *and* $x \in A$. We paraphrase this statement by saying that "f has a point of continuity relative to A."

10. Let $A = (0, 1\,] \cup \{2\}$, considered as a subset of $\mathbb{R}$. Show that every function $f : A \to \mathbb{R}$ is continuous, relative to A, at 2.

11. Let A and B be subsets of M, and let $f : M \to \mathbb{R}$. Prove or disprove the following statements:

(a) If f is continuous at each point of A and f is continuous at each point of B, then f is continuous at each point of $A \cup B$.

(b) If $f\,|_A$ is continuous, relative to A and $f\,|_B$ is continuous, relative to B, then $f\,|_{A \cup B}$ is continuous, relative to $A \cup B$.

If either statement is not true in general, what modifications are necessary to make it so?

12. Let $I = (\mathbb{R} \setminus \mathbb{Q}) \cap [\,0, 1\,]$ with its usual metric. Prove that there is a continuous function g mapping I *onto* $\mathbb{Q} \cap [\,0, 1\,]$.

13. Let (r_n) be an enumeration of the rationals in $[\,0, 1\,]$ and define f on $[\,0, 1\,]$ by $f(x) = \sum_{r_n < x} 2^{-n}$. Show that f is everywhere discontinuous on $[\,0, 1\,]$ but that f is everywhere continuous when considered as a function on only $[\,0, 1\,] \setminus \mathbb{Q}$.

14. A continuous function on $\mathbb{R}$ is completely determined by its values on $\mathbb{Q}$. Use this to "count" the continuous functions $f : \mathbb{R} \to \mathbb{R}$.

15. Suppose that $f : \mathbb{R} \to \mathbb{R}$ satisfies $f(x + y) = f(x) + f(y)$ for every x, $y \in \mathbb{R}$. If f is continuous at some point $x_0 \in \mathbb{R}$, prove that there is some constant $a \in \mathbb{R}$ such that $f(ax) = ax$ for all $x \in \mathbb{R}$. That is, an additive function that is continuous at even one point is linear – and hence continuous on all of $\mathbb{R}$.

16. Let $f : \mathbb{R} \to \mathbb{R}$, and define $G : \mathbb{R} \to \mathbb{R}^2$ by $G(x) = (x, f(x))$, so that the range of G is the *graph* of f. Show that f is continuous if and only if G is continuous if and only if both of the sets $A = \{(x, y) : y \leq f(x)\}$ and $B = \{(x, y) : y \geq f(x)\}$ are closed in $\mathbb{R}^2$. In particular, if f is continuous, then the graph of f is closed in $\mathbb{R}^2$.

▷ **17.** Let $f, g : (M, d) \to (N, \rho)$ be continuous, and let D be a dense subset of M. If $f(x) = g(x)$ for all $x \in D$, show that $f(x) = g(x)$ for all $x \in M$. If f is onto, show that $f(D)$ is dense in N.

18. Let $f : (M, d) \to (N, \rho)$ be continuous, and let A be a separable subset of M. Prove that $f(A)$ is separable.

▷ **19.** A function $f : \mathbb{R} \to \mathbb{R}$ is said to satisfy a **Lipschitz condition** if there is a constant $K < \infty$ such that $|f(x) - f(y)| \leq K|x - y|$ for all $x, y \in \mathbb{R}$. More economically, we may say that f is Lipschitz (or Lipschitz with constant K if a particular constant seems to matter). Show that $\sin x$ is Lipschitz with constant $K = 1$. Prove that a Lipschitz function is (uniformly) continuous.

▷ **20.** If d is a metric on M, show that $|d(x, z) - d(y, z)| \leq d(x, y)$ and conclude that the function $f(x) = d(x, z)$ is continuous on M for any fixed $z \in M$. This says that $d(x, y)$ is *separately continuous* – continuous in each variable separately.

21. If $x \neq y$ in M, show that there are disjoint open sets U, V with $x \in U$ and $y \in V$. Moreover, U and V can be chosen so that $\bar{U}$ and $\bar{V}$ are disjoint.

22. Define $E : \mathbb{N} \to \ell_1$ by $E(n) = (1, \ldots, 1, 0, \ldots)$, where the first n entries are 1 and the rest 0. Show that E is an isometry (into).

23. Define $S : c_0 \to c_0$ by $S(x_1, x_2, \ldots) = (0, x_1, x_2, \ldots)$. That is, S shifts the entries forward and puts 0 in the empty slot. Show that S is an isometry (into).

24. Let V be a normed vector space. If $y \in V$ is fixed, show that the maps $\alpha \mapsto \alpha\, y$, from $\mathbb{R}$ into V, and $x \mapsto x + y$, from V into V, are continuous.

▷ **25.** A function $f : (M, d) \to (N, \rho)$ is called **Lipschitz** if there is a constant $K < \infty$ such that $\rho(f(x), f(y)) \leq K d(x, y)$ for all x, $y \in M$. Prove that a Lipschitz mapping is continuous.

26. Provide the answer to a question raised in Chapter Three by showing that integration is continuous. Specifically, show that the map $L(f) = \int_a^b f(t)\,dt$ is Lipschitz with constant $K = b - a$ for $f \in C[a, b]$.

27. Fix $k \geq 1$ and define $f : \ell_\infty \to \mathbb{R}$ by $f(x) = x_k$. Is f continuous? [Hint: f is Lipschitz.]

28. Define $g : \ell_2 \to \mathbb{R}$ by $g(x) = \sum_{n=1}^{\infty} x_n/n$. Is g continuous?

29. Fix $y \in \ell_\infty$ and define $h : \ell_1 \to \ell_1$ by $h(x) = (x_n y_n)_{n=1}^{\infty}$. Show that h is continuous.

▷ **30.** Let $f : (M, d) \to (N, \rho)$. Prove that f is continuous if and only if $f\left(\bar{A}\right) \subset \overline{f(A)}$ for every $A \subset M$ if and only if $f^{-1}(B^\circ) \subset \left(f^{-1}(B)\right)^\circ$ for every $B \subset N$. Give an example of a continuous f such that $f\left(\bar{A}\right) \neq \overline{f(A)}$ for some $A \subset M$.

31. Let $f : (M, d) \to (N, \rho)$.

(a) If $M = \bigcup_{n=1}^{\infty} U_n$, where each U_n is an open set in M, and if f is continuous on each U_n, show that f is continuous on M.

(b) If $M = \bigcup_{n=1}^{N} E_n$, where each E_n is a closed set in M, and if f is continuous on each E_n, show that f is continuous on M.

(c) Give an example showing that f can fail to be continuous on all of M if, instead, we use a countably infinite union of closed sets $M = \bigcup_{n=1}^{\infty} E_n$ in (b).

32. A real-valued function f on a metric space M is called *lower semicontinuous* if, for each real α, the set $\{x \in M : f(x) \leq \alpha\}$ is closed in M. (For example, if $g : M \to \mathbb{R}$ is continuous and $x_0 \in M$, then the function f defined by $f(x) = g(x)$ for $x \neq x_0$, and $f(x_0) = g(x_0) - 1$ is lower semicontinuous.) Prove that f is lower semicontinuous if and only if $f(x) \leq \liminf_{n\to\infty} f(x_n)$ whenever $x_n \to x$ in M. [Hint: For the forward implication, suppose that $x_n \to x$ and $m = \liminf_{n\to\infty} f(x_n) < \infty$. Then, for every $\varepsilon > 0$, the set $\{t \in M : f(t) \leq m + \varepsilon\}$ is closed and contains infinitely many x_n.]

33. A function $f : M \to \mathbb{R}$ is called *upper semicontinuous* if $-f$ is lower semicontinuous. Formulate the analogue of Exercise 32 for upper semicontinuous functions.

Theorem 5.1 characterizes continuous functions in terms of open sets and closed sets. As it happens, we can use these characterizations "in reverse" to derive information about open and closed sets. In particular, we can characterize closures in terms of certain continuous functions.

Given a nonempty set A and a point $x \in M$, we define **the distance from** x **to** A by:

$$d(x, A) = \inf\{d(x, a) : a \in A\}.$$

Clearly, $0 \leq d(x, A) < \infty$ for any x and any A, but it is not necessarily true that $d(x, A) > 0$ when $x \notin A$. For example, $d(x, \mathbb{Q}) = 0$ for any $x \in \mathbb{R}$.

Proposition 5.3. *$d(x, A) = 0$ if and only if $x \in \bar{A}$.*

PROOF. $d(x, A) = 0$ if and only if there is a sequence of points (a_n) in A such that $d(x, a_n) \to 0$. But this means that $a_n \to x$ and, hence, $x \in \bar{A}$ by Corollary 4.10. □

Note that Proposition 5.3 has given us another connection between limits in M and limits in $\mathbb{R}$. Loosely speaking, Proposition 5.3 shows that 0 is a limit point of $\{d(x, a) : a \in A\}$ if and only if x is a limit point of A. We can get even more mileage out of this observation by checking that the map $x \mapsto d(x, A)$ is actually *continuous*. For this it suffices to establish the following inequality:

Proposition 5.4. $|d(x, A) - d(y, A)| \leq d(x, y)$.

PROOF. $d(x, a) \le d(x, y) + d(y, a)$ for any $a \in A$. But $d(x, A)$ is a lower bound for $d(x, a)$; hence $d(x, A) \le d(x, y) + d(y, a)$. Now, by taking the infimum over $a \in A$, we get $d(x, A) \le d(x, y) + d(y, A)$. Since the roles of x and y are interchangeable, we're done. □

To appreciate what this has done for us, let's make two simple observations. First, if $f : M \to \mathbb{R}$ is a continuous function, then the set $E = \{x \in M : f(x) = 0\}$ is closed. (Why?) Conversely, if E is a closed set in M, then E is the "zero set" of some continuous real-valued function on M; in particular, $E = \{x \in M : d(x, E) = 0\}$. Thus a set E is closed if and only if $E = f^{-1}(\{0\})$ for some continuous function $f : M \to \mathbb{R}$. Conclusion: If you know all of the closed (or open) sets in a metric space M, then you know all of the continuous real-valued functions on M (Theorem 5.1). Conversely, if you know all of the continuous real-valued functions on M, then you know all of the closed (or open) sets in M.

EXERCISES

Unless otherwise stated, each of the following exercises is set in a general metric space (M, d).

▷ **34.** Show that d is continuous on $M \times M$, where $M \times M$ is supplied with "the" product metric (see Exercise 3.46). This says that d is *jointly* continuous, that is, continuous as a function of two variables. [Hint: If $x_n \to x$ and $y_n \to y$, show that $d(x_n, y_n) \to d(x, y)$.]

35. Show that a set U is open in M if and only if $U = f^{-1}(V)$ for some continuous function $f : M \to \mathbb{R}$ and some open set V in $\mathbb{R}$.

▷ **36.** Suppose that we are given a point x and a sequence (x_n) in a metric space M, and suppose that $f(x_n) \to f(x)$ for every continuous, real-valued function f on M. Does it follow that $x_n \to x$ in M? Explain.

37. If F is closed and $x \notin F$, show that there are disjoint open sets U, V with $x \in U$ and $F \subset V$. Can U and V be chosen so that $\bar{U}$ and $\bar{V}$ are disjoint?

38. Given disjoint nonempty closed sets E, F, define $f : M \to \mathbb{R}$ by $f(x) = d(x, E)/[d(x, E) + d(x, F)]$. Show that f is a continuous function on M with $0 \le f \le 1$, $f^{-1}(\{0\}) = E$, and $f^{-1}(\{1\}) = F$. Use this to find disjoint open sets U and V with $E \subset U$ and $F \subset V$. Can U and V be chosen so that $\bar{U}$ and $\bar{V}$ are disjoint?

39. Show that every open set in M is the union of countably many closed sets, and that every closed set is the intersection of countably many open sets.

40. We define the distance between two nonempty subsets A and B of M by $d(A, B) = \inf\{d(a, b) : a \in A, b \in B\}$. Give an example of two disjoint closed sets A and B in $\mathbb{R}^2$ with $d(A, B) = 0$.

41. Let C be a closed set in $\mathbb{R}$ and let $f : C \to \mathbb{R}$ be continuous. Show that there is a continuous function $g : \mathbb{R} \to \mathbb{R}$ with $g(x) = f(x)$ for every $x \in C$. We say that g is a continuous *extension* of f to all of $\mathbb{R}$. In particular, every continuous function on the Cantor set Δ extends continuously to all of $\mathbb{R}$. [Hint: The complement of C is the countable union of disjoint open intervals. Define g by "connecting the dots" across each of these open intervals.]

42. Suppose that $f : \mathbb{Q} \to \mathbb{R}$ is Lipschitz. Show that f extends to a continuous function $h : \mathbb{R} \to \mathbb{R}$. Is h unique? Explain. [Hint: Given $x \in \mathbb{R}$, choose a sequence of rationals (r_n) converging to x and argue that $h(x) = \lim_{n\to\infty} f(r_n)$ exists and is actually independent of the sequence (r_n).]

Homeomorphisms

By now we have seen how the convergent sequences in a metric space determine all of its open (or closed) sets and all of its continuous functions. We have also seen how the open sets determine which sequences converge and which functions are continuous. And we have seen that the continuous functions, in turn, determine the open sets in a metric space and so too, indirectly, its convergent sequences.

Any one of these three – the convergent sequences, the open sets, or the continuous functions – forms the "soul" of a metric space, the essence that distinguishes one metric space from another in "spirit," if not in "body." As a concrete example of this "gestalt," consider $\mathbb{Z}$ and $\mathbb{N}$. The algebraic and order properties of $\mathbb{Z}$ and $\mathbb{N}$ are surely different, but as metric spaces $\mathbb{Z}$ and $\mathbb{N}$ are essentially the same: countably infinite discrete spaces. Every subset is open, every real-valued function is continuous, and only (eventually) constant sequences converge. From this point of view, $\mathbb{Z}$ and $\mathbb{N}$ are indistinguishable as metric spaces.

All of this suggests an idea: Two metric spaces might be considered "similar" if there is a "similarity" between their open sets, or their convergent sequences, or their continuous functions. Not necessarily "identical," mind you, just "similar." But how do we make this precise? The answer comes from examining our notion of equivalence for metrics.

Suppose that we are handed two metrics, d and ρ, on the same set M. How do we compare (M, d) and (M, ρ)? Well, consider the following list of observations (see Exercises 3.42 and 4.3):

(M, d) and (M, ρ) are "similar"
$\iff$ d and ρ are equivalent metrics on M
$\iff$ d and ρ generate the same convergent sequences
$\iff$ d and ρ generate the same open (closed) sets.

Now let's bring continuous functions into the picture:

d and ρ are equivalent metrics on M
$\iff$ { d and ρ generate the same continuous real-valued functions on M
$\iff$ { d and ρ generate the same continuous functions (with any range) on M.

And, finally, let's consolidate all of these observations into one:

$$d \text{ and } \rho \text{ are equivalent metrics on } M \iff \begin{cases} \text{The identity map } i : (M, d) \to (M, \rho) \text{ and its inverse } i^{-1} : \\ (M, \rho) \to (M, d) \text{ (also the identity) are both } \textit{continuous}. \text{ (Why?)} \end{cases}$$

Generalizing on this last observation, we say that two metric spaces (M, d) and (N, ρ) are **homeomorphic** ("similar-shape") if there is a one-to-one and onto map $f : M \to N$ such that *both* f and f^{-1} are continuous. Such a map f is called a **homeomorphism** from M onto N. Note that f is a homeomorphism if and only if f^{-1} is a homeomorphism (from N onto M). You should think of homeomorphic spaces as *essentially* identical. In particular, if d and ρ are equivalent metrics on M, then (M, d) and (M, ρ) are homeomorphic.

Theorem 5.5. *Let $f : (M, d) \to (N, \rho)$ be one-to-one and onto. Then the following are equivalent:*

(i) *f is a homeomorphism.*
(ii) $x_n \xrightarrow{d} x \iff f(x_n) \xrightarrow{\rho} f(x)$.
(iii) *G is open in $M \iff f(G)$ is open in N.*
(iv) *E is closed in $M \iff f(E)$ is closed in N.*
(v) *$\hat{d}(x, y) = \rho\big(f(x), f(y)\big)$ defines a metric on M equivalent to d.*

The proof of Theorem 5.5 is left as an exercise. The conclusion to be drawn from this rather long statement is that a homeomorphism provides a correspondence not just between the points of M and N, but also between the convergent sequences in M and N, as well as between the open and closed sets in M and N. There is also a correspondence between the continuous real-valued functions on M and N; see Exercise 54.

Let's look at a few specific examples.

Examples 5.6

(a) Note that the relation "is homeomorphic to" is an equivalence relation. In particular, every metric space is homeomorphic to itself (by way of the identity map). More generally, note that $f : M \to N$ is a homeomorphism if and only if $f^{-1} : N \to M$ is a homeomorphism.

(b) From our earlier discussion, we know that if d and ρ are equivalent metrics on M, then (M, d) and (M, ρ) are homeomorphic (under the identity map). However, if (M, d) and (M, ρ) are homeomorphic, it does *not* follow that d and ρ are equivalent; see Exercise 50.

(c) $(\mathbb{R}, \text{usual})$ is *not* homeomorphic to $(\mathbb{R}, \text{discrete})$. Why? (Try to think of more than one reason.) But $(\mathbb{N}, \text{usual})$ is homeomorphic to $(\mathbb{N}, \text{discrete})$. Check this!

(d) All three of the spaces $(\mathbb{R}^n, \|\cdot\|_1)$, $(\mathbb{R}^n, \|\cdot\|_2)$, and $(\mathbb{R}^n, \|\cdot\|_\infty)$ are homeomorphic. See Exercises 3.18 and 3.44.

(e) Suppose that $f : M \to N$ is an isometry from M onto N; that is, an onto map satisfying $\rho(f(x), f(y)) = d(x, y)$ for all $x, y \in M$. Now an isometry is evidently one-to-one; hence f has an inverse that satisfies $\rho(a, b) = d(f^{-1}(a), f^{-1}(b))$ for all $a, b \in N$. (Why?) That is, f^{-1} is also an isometry. Clearly, then, f is a homeomorphism. In this case, however, we would emphasize the fact that M and N are more than merely "alike" by saying that M and N are **isometric**. Isometric spaces are exact replicas of one another; they are identical in every feature save the "names" of their elements. For example, the interval $[0, 1]$ is isometric to the interval $[4, 5]$; indeed, it is isometric to any closed interval of length 1.

(f) In $\mathbb{R}$, any two intervals that look alike are homeomorphic. $[0, 1]$ and $[a, b]$ are homeomorphic, as are $(0, 1)$ and (a, b). The interval $(0, 1)$ is also homeomorphic to $\mathbb{R}$, and $(0, 1]$ is homeomorphic to $[a, b)$. Why? [Hint: The map $x \mapsto 2 - 3x$ is a homeomorphism from $(0, 1]$ onto $[-1, 2)$, while $x \mapsto \arctan x$ is a homeomorphism from $\mathbb{R}$ onto $(-\pi/2, \pi/2)$.]

(g) Any two intervals that look different *are* different. For example, $(0, 1]$ is *not* homeomorphic to (a, b). The argument may be a bit hard to follow, so hang on! Suppose that $(0, 1]$ is homeomorphic to (a, b) under some homeomorphism f. Then, by removing 1 from $(0, 1]$ and its image $c = f(1)$ from (a, b), we would have that $(0, 1)$ is homeomorphic to $(a, c) \cup (c, b)$. (Why should this work?) But $(0, 1)$ is homeomorphic to $\mathbb{R}$, and so $\mathbb{R}$ would be have to be homeomorphic to $(a, c) \cup (c, b)$, too. From this it follows that $\mathbb{R}$ could be written as the disjoint union of two nontrivial open sets, which is impossible (see Exercise 4.25). The arguments in the various other cases are similar in spirit.

(h) Although it will take some time before we can explain all of the details, you might find it comforting to know that $\mathbb{R}$ is not homeomorphic to $\mathbb{R}^2$ and that the unit interval $[0, 1]$ is not homeomorphic to the unit square $[0, 1] \times [0, 1]$. More generally, if $m \neq n$, then $\mathbb{R}^n$ and $\mathbb{R}^m$ are not homeomorphic. In other words, spaces with different "dimensions" are apparently different.

EXERCISES

43. If you are not already convinced, prove that two metrics d and ρ on a set M are equivalent if and only if the identity map on M is a homeomorphism from (M, d) to (M, ρ).

44. Check that the relation "is homeomorphic to" is an equivalence relation on pairs of metric spaces.

45. Prove that $\mathbb{N}$ (with its usual metric) is homeomorphic to $\{(1/n) : n \geq 1\}$ (with its usual metric).

▷ **46.** Show that every metric space is homeomorphic to one of finite diameter. [Hint: Every metric is equivalent to a bounded metric.]

47. Define $E : \mathbb{N} \to \ell_1$ by $E(n) = (1, \ldots, 1, 0, \ldots)$, where the first n entries are 1 and the rest are 0. Show that E is an isometry (into).

▷ **48.** Prove that $\mathbb{R}$ is homeomorphic to $(0, 1)$ and that $(0, 1)$ is homeomorphic to $(0, \infty)$. Is $\mathbb{R}$ *isometric* to $(0, 1)$? to $(0, \infty)$? Explain.

49. Let V be a normed vector space. Given a fixed vector $y \in V$, show that the map $f(x) = x + y$ (*translation* by y) is an isometry on V. Given a nonzero scalar $\alpha \in \mathbb{R}$, show that the map $g(x) = \alpha x$ (*dilation* by α) is a homeomorphism on V.

50. Let (M, d) denote the set $\{0\} \cup \{(1/n) : n \geq 1\}$ under its usual metric. Define a second metric ρ on M by setting $\rho(1/n, 1/m) = |1/n - 1/m|$ for $m, n \geq 2$, $\rho(1/n, 1) = 1/n$ for $n \geq 2$, $\rho(1/n, 0) = 1 - 1/n$ for $n \geq 2$, and $\rho(0, 1) = 1$. Show that (M, d) and (M, ρ) are homeomorphic but that the identity map from (M, d) to (M, ρ) is *not* continuous.

51. Let (M, ρ) be a separable metric space and assume that $\rho(x, y) \leq 1$ for every x, $y \in M$. Given a countable dense set $\{x_n : n \geq 1\}$ in M, define a map $f : M \to H^\infty$, from M into the Hilbert cube (Exercise 3.10), by $f(x) = \big(\rho(x, x_n)\big)_{n=1}^{\infty}$.

(i) Prove that f is one-to-one and continuous. In fact, f satisfies $d\big(f(x), f(y)\big) \leq \rho(x, y)$, where d is the metric on H^∞.

(ii) Fix $\varepsilon > 0$ and $x \in H^\infty$. Find $\delta > 0$ such that $\rho(x, y) < \varepsilon$ whenever $d\big(f(x), f(y)\big) < \delta$. Conclude that f is a homeomorphism into H^∞.

You may find the following simple lemma useful in working the subsequent batch of exercises.

Lemma 5.7. *Let $f : L \to M$ and $g : M \to N$, where L, M, and N are metric spaces. If f is continuous at $x \in L$, and if g is continuous at $f(x) \in M$, then $g \circ f : L \to N$ is continuous at $x \in L$.*

PROOF. $x_n \to x$ in $L \Longrightarrow f(x_n) \to f(x)$ in $M \Longrightarrow g(f(x_n)) \to g(f(x))$ in N. □

EXERCISES

Throughout, M denotes a generic metric space with metric d.

▷ **52.** Prove Theorem 5.5.

▷ **53.** Suppose that we are given a point x and a sequence (x_n) in a metric space M, and suppose that $f(x_n) \to f(x)$ for every continuous real-valued function f on M. Prove that $x_n \to x$ in M.

▷ **54.** Let $f : (M, d) \to (N, \rho)$ be one-to-one and onto. Prove that the following are equivalent: (i) f is a homeomorphism and (ii) $g : N \to \mathbb{R}$ is continuous if and only if $g \circ f : M \to \mathbb{R}$ is continuous. [Hint: Use the characterization given in Theorem 5.5 (ii).]

55. Let $f : (M, d) \to (N, \rho)$ be a homeomorphism. Prove that M is separable if and only if N is separable.

▷ **56.** Let $f : (M, d) \to (N, \rho)$.

(i) We say that f is an **open** map if $f(U)$ is open in N whenever U is open in M; that is, f maps open sets to open sets. Give examples of a continuous map that is not open and an open map that is not continuous. [Hint: Please note that the definition depends on the target space N.]

(ii) Similarly, f is called **closed** if it maps closed sets to closed sets. Give examples of a continuous map that is not closed and a closed map that is not continuous.

▷ **57.** Let $f : (M, d) \to (N, \rho)$ be one-to-one and onto. Show that the following are equivalent: (i) f is open; (ii) f is closed; and (iii) f^{-1} is continuous. Consequently, f is a homeomorphism if and only if both f and f^{-1} are open (closed).

58. Let $f : (M, d) \to (N, \rho)$ be one-to-one and onto. Prove that f is a homeomorphism if and only if $f(\bar{A}) = \overline{f(A)}$ for every subset A of M.

59.

(a) Show that an open, continuous map need not be closed, even if it is onto. [Hint: Consider the map $\pi(x, y) = x$ from $\mathbb{R}^2$ onto $\mathbb{R}$.]

(b) Show that a closed, continuous map need not be open, even if it is onto. [Hint: Consider the map $x \mapsto \cos x$ from $[0, 2\pi]$ onto $[-1, 1]$.]

60. Let (M, d) be a metric space, and let τ be the discrete metric on M. Then, (M, d) and (M, τ) are homeomorphic if and only if every subset of M is open in (M, d) if and only if every function $f : (M, d) \to \mathbb{R}$ is continuous.

61. Show that $\mathbb{N}$ is homeomorphic to the set $\{e^{(n)} : n \geq 1\}$ when considered as a subset of any one of the spaces c_0, ℓ_1, ℓ_2, or ℓ_∞. [Hint: The map $n \mapsto e^{(n)}$ is continuous and open. Why?] If we instead take the discrete metric on $\mathbb{N}$, show that the map $n \mapsto e^{(n)}$ is an isometry into c_0.

Perhaps you have heard the word *topology*? Well, now you know something about it! Topology is the study of continuous transformations or, what amounts to the same thing, the study of open sets. This rather loose description will have to do for now. In any case, a property that can be characterized solely in terms of open sets is usually referred to as a *topological property*. In other words, a topological property is one that is preserved by homeomorphisms. For example, separability (having a countable dense subset) is a topological property, while boundedness is not (see Exercises 55 and 46). And Example 5.6 (h) would seem to suggest that the "dimension" of a space is a topological property. The word topology is also used as the name for the collection of all open sets. For example, we might say that convergence and continuity in M depend on the topology of M. This description is more to the point than saying that either depends on the metric of M.

From this point on we will be very much concerned with whether or not a given property is preserved by homeomorphisms. Such properties are invariant under slight changes in the metric and so are typically more "forgiving" than those that depend intimately on a particular metric.

The Space of Continuous Functions

We write $C(M)$ for the collection of all continuous, real-valued functions on (M, d). As we have seen, the collection $C(M)$ contains a wealth of information about the metric space (M, d) itself. This being a course in analysis (or had you forgotten?), we want to know everything possible about continuous functions on metric spaces. Since we are allowed to focus our attention on real-valued functions, $C(M)$ is the space that we need to master. We will find that $C(M)$ comes equipped with an incredible amount of *algebraic* structure – all inherited from $\mathbb{R}$. We will show that $C(M)$ is a *vector space*, an *algebra*, and a *lattice*. One of our goals will be to find a metric (or norm) on $C(M)$ that is compatible with its algebraic structure. While this will take no small effort on our part, it is well worth it. The scenery alone more than justifies the trip; analysis, algebra, and topology all flourish in $C(M)$.

Given real-valued functions $f, g : M \to \mathbb{R}$, we define all of the usual algebraic operations on f and g "pointwise." That is, we define $c \cdot f$, $c \in \mathbb{R}$, $f + g$, and $f \cdot g$ by $(c \cdot f)(x) = cf(x)$, $(f + g)(x) = f(x) + g(x)$, and $(f \cdot g)(x) = f(x)g(x)$, for all $x \in M$. In this way, the ring structure of $\mathbb{R}$ "lifts" to the real-valued functions on M. The order structure of $\mathbb{R}$ also lifts: We define $f \le g$ to mean that $f(x) \le g(x)$ for all $x \in M$. From here we can make sense out of all sorts of expressions, for example, $|f|(x) = |f(x)|$, $\max\{f, g\}(x) = \max\{f(x), g(x)\}$, and $\min\{f, g\}(x) = \min\{f(x), g(x)\}$.

Now if M is a metric space, what we would like to know is whether the space $C(M)$ is "closed" under all of these various operations. You won't be surprised to learn that the answer is: Yes. For example, it follows from Lemma 5.7 that if $f : M \to \mathbb{R}$ is continuous, then so are cf, $|f|$, f^2, $\sin(f)$, and so on (How?) The other cases that we want to consider are slightly more elaborate compositions involving two functions at a time, such as $x \mapsto (f(x), g(x)) \mapsto f(x) + g(x)$. Another easy lemma will make short work of the details.

Lemma 5.8. *If $f, g : M \to \mathbb{R}$ are continuous, then so is the function $h : M \to \mathbb{R}^2$, defined by $h(x) = \big(f(x), g(x)\big)$ for $x \in M$.*

PROOF. $x_n \to x$ in $M \implies f(x_n) \to f(x)$ and $g(x_n) \to g(x)$ in $\mathbb{R} \implies h(x_n) \to h(x)$ in $\mathbb{R}^2$. (Why?) □

Here's the plan of attack: Each of the functions $f + g$, $f \cdot g$, $\max\{f, g\}$, and $\min\{f, g\}$ is the composition of two functions. First, $x \mapsto (f(x), g(x))$, and then the pair $(f(x), g(x))$ in $\mathbb{R}^2$ is mapped to $f(x) + g(x)$, or $f(x)g(x)$, or $\max\{f(x), g(x)\}$, or $\min\{f(x), g(x)\}$. If f and g are continuous, then the first map is always continuous by Lemma 5.8, and so we only need to know whether the second map is continuous from $\mathbb{R}^2$ into $\mathbb{R}$ in each of the four cases. Here are some of the details (you may want to supply a few more).

Examples 5.9

(a) The map $(x, y) \mapsto x + y$ is continuous: If $x_n \to x$ and $y_n \to y$ in $\mathbb{R}$, then $x_n + y_n \to x + y$ because $|(x_n + y_n) - (x + y)| \le |x_n - x| + |y_n - y|$. Alternatively, you might show that the set $\{(x, y) : |(x + y) - (a + b)| < \varepsilon\}$ is open in $\mathbb{R}^2$.

(b) The map $(x, y) \mapsto \max\{x, y\}$ is continuous; an easy way to see this is to write $\max\{x, y\} = \frac{1}{2}(x + y + |x - y|)$. (How does this help?) For $(x, y) \mapsto \min\{x, y\}$, use the fact that $\min\{x, y\} = \frac{1}{2}(x + y - |x - y|)$.

(c) The map $(x, y) \mapsto xy$ is continuous since $xy = \frac{1}{4}\left[(x + y)^2 - (x - y)^2\right]$. (How would a "direct" proof go?)

Combining these observations with Lemma 5.8 gives:

Theorem 5.10. *Let $f, g : M \to \mathbb{R}$ be continuous. Then, $f \pm g$, $f \cdot g$, $\max\{f, g\}$, and $\min\{f, g\}$ are all continuous.*

If we use the pointwise definitions for algebraic operations in $C(M)$, then $C(M)$ becomes a vector space (it is closed under addition and scalar multiplication), an algebra (or ring – it is also closed under products), and a lattice (each pair of functions has a max and a min back in $C(M)$). The most important observation for now is that $C(M)$ is a vector space; we will have much more to say about the lattice and ring structures later.

Our next task is to determine, if possible, a metric or a norm on $C(M)$ that will be compatible with these algebraic operations. We have been given a hint as to how to do this by Fréchet himself. The norm of choice on $C[\,a, b\,]$ is apparently $\|f\|_\infty = \max_{a \le t \le b} |f(t)|$. We have already checked that this is, in fact, a norm on $C[\,a, b\,]$ (that is, it "respects" the vector space operations in $C[\,a, b\,]$). That this norm does still more is outlined in the following exercise.

EXERCISE

62. If $f, g \in C[\,a, b\,]$, show that $\|fg\|_\infty \le \|f\|_\infty \|g\|_\infty$. Also show that $\|\max\{f, g\}\|_\infty \le \max\{\|f\|_\infty, \|g\|_\infty\}$, and that $\|f\|_\infty \le \|g\|_\infty$ whenever $|f| \le |g|$.

We know that homeomorphic spaces are supposed to have (essentially) the same collection of continuous functions. Let's make this even more precise in at least one special case.

EXERCISE

63. Let $[\,a, b\,]$ be any closed, bounded interval in $\mathbb{R}$, and let $\sigma : [\,0, 1\,] \to [\,a, b\,]$ be defined by $\sigma(t) = a + t(b - a)$. Prove that:

(i) σ is a homeomorphism.

(ii) $f \in C[\,a, b\,]$ if (and only if) $f \circ \sigma \in C[\,0, 1\,]$.

(iii) The map $f \mapsto f \circ \sigma$ is an isometry from $C[\,a, b\,]$ onto $C[\,0, 1\,]$. The map $T(f) = f \circ \sigma$ actually does much more; it is both an algebra and a lattice isomorphism. That is, it also preserves the algebraic and order structures. Specifically, given any $f, g \in C[\,a, b\,]$, check that:

(iv) $T(\alpha f + \beta g) = \alpha T(f) + \beta T(g)$ for all $\alpha, \beta \in \mathbb{R}$.

(v) $T(fg) = T(f)\,T(g)$.

(vi) $T(f) \le T(g)$ if and only if $f \le g$.

Thus, for all practical purposes, $C[\,a, b\,]$ and $C[\,0, 1\,]$ are identical.

But will the norm on $C[\,a, b\,]$ give any clues to a possible norm on $C(\mathbb{R})$? Since the elements of $C(\mathbb{R})$ need not be bounded (let alone actually attain a maximum value), we cannot expect to use $\sup_{t\in\mathbb{R}} |f(t)|$, for example. A norm may be too much to hope for, but it is easy enough to define a metric on $C(\mathbb{R})$. This, too, comes to us from Fréchet.

EXERCISE

64. Given $n \in \mathbb{N}$ and $f, g \in C(\mathbb{R})$, let $d_n(f, g) = \max_{|t|\le n} |f(t) - g(t)|$. Then d_n defines a pseudometric on $C(\mathbb{R})$. (Why?) Show that $d(f, g) = \sum_{n=1}^{\infty} 2^{-n} d_n(f, g)/(1 + d_n(f, g))$ defines a metric on $C(\mathbb{R})$.

It will take quite a bit more work before we can settle the issue of a reasonable metric on $C(M)$ – even in a few special cases. But at least one case is easy to describe. If M is a *finite* set, say $M = \{x_1, \ldots, x_n\}$ (under *any* metric), then we may identify $C(M)$ with $\mathbb{R}^n$ by identifying each $f \in C(M)$ with its range $(f(x_1), \ldots, f(x_n)) \in \mathbb{R}^n$. Why does this work? Recall that any metric on a finite set is necessarily equivalent to the discrete metric, and so every function $f : M \to \mathbb{R}$ is continuous. Thus, each $y \in \mathbb{R}^n$ defines an element $f \in C(M)$ by setting $f(x_k) = y_k$, for $k = 1, \ldots, n$.

If we use coordinatewise operations on $\mathbb{R}^n$, this correspondence even preserves the algebraic structure on $C(M)$. For example, check that if $f, g \in C(M)$, then $f + g$ corresponds to the vector $(f(x_1) + g(x_1), \ldots, f(x_n) + g(x_n))$. Similarly, $f \cdot g$ corresponds to $(f(x_1)g(x_1), \ldots, f(x_n)g(x_n))$ and $|f|$ corresponds to $(|f(x_1)|, \ldots, |f(x_n)|)$. Finally, we can induce a suitable norm on $C(M)$ by taking the "max" norm on $\mathbb{R}^n$. Specifically, check that $\|f\|_\infty = \max_{1\le i\le n} |f(x_i)|$, the norm induced on $C(M)$ by this correspondence, satisfies $\|fg\|_\infty \le \|f\|_\infty \|g\|_\infty$ and $\|f\|_\infty \le \|g\|_\infty$ whenever $|f| \le |g|$.

Our goal is now a little clearer: To define a norm on $C(M)$, we want M to be "like" a finite set. Whatever that might mean, we would certainly hope that $[a, b]$ turns out to be "like" a finite set (after all, that case works just fine already). We will put these issues aside for now, but they will resurface in Chapter Eight when we finally arrive at a plausible generalization of finite sets (which really will include $[a, b]$ as a special case).

Notes and Remarks

The so-called Lipschitz condition of Exercises 19 and 25 was introduced by Rudolph Lipschitz in 1876 (for more on this, see the discussion in Chapter Seven).

The definition of continuity in terms of open sets is due to Hausdorff. For various notions of "almost" continuous and "nearly" continuous functions, based on variations of Exercise 30, see Beslin [1992] and Tong [1992].

Semicontinuity (Exercises 32 and 33) was introduced by René Baire in his thesis, Baire [1899]. Also see Radó [1942]. Related to Exercises 7 and 32 is the intoxicating article by Foster Brooks [1971], where sets of the form $\{x : f(x) > a\}$ are called "cut sets."

For more on the notion of "dimension," which was referred to in passing in Example 5.6 (h), see Menger [1943].

The algebraic and lattice structures of the space $C(M)$ have been the topic of a great deal of research during the last 50 years. For much more on this, see Birkhoff [1948], Gillman and Jerison [1960], Goffman and Pedrick [1965], Jameson [1974], Kuller [1969], Schaefer [1980], and Simmons [1963]. The short note by Aron and Fricke [1986] provides an elementary proof of the fact that a linear, multiplicative map $\varphi : C(\mathbb{R}) \to \mathbb{R}$ (i.e., an algebra homomorphism) is given by point evaluation. Compare this with Exercise 63.

CHAPTER SIX

Connectedness

Connected Sets

We have a few details to clean up before we move on to other things; these concern the special role of intervals in $\mathbb{R}$ and their use in characterizing the open sets in $\mathbb{R}$ given in Chapter Four (see Theorem 4.6 and Exercise 4.25). As we'll see in this section, a better understanding of the special nature of intervals in $\mathbb{R}$ will allow us to generalize the intermediate value theorem of calculus. The intermediate value theorem is the formal statement of the informal notion that the graph of a continuous function is "unbroken." The historical basis of the theorem is the concept of a function as measuring, over time, the relative position of an object moving along a straight line. Thus, if we track the position $y = f(x)$ of a moving object between time $x = a$ and some subsequent time $x = b$, we would expect the object to "visit" all of the positions y that are intermediate to $f(a)$ and $f(b)$. In short, the continuous image of the time interval $[a, b]$ should contain (at least) the full interval of positions between $f(a)$ and $f(b)$.

The secret here is the intuitively obvious fact that no interval in $\mathbb{R}$ can be split into two relatively *open* parts. Let's prove this by "brute force" for the interval $[a, b]$ (we'll do the other cases shortly).

Suppose to the contrary that $[a, b] = A \cup B$, where A and B are nonempty, disjoint, relatively open sets in $[a, b]$. We are going to find a contradiction by examining the "border" between A and B. The trouble comes from the fact that A and B are necessarily also closed in $[a, b]$, since each is the complement of an open set: $A = [a, b] \setminus B$ and $B = [a, b] \setminus A$, and so each of A and B lays claim to the "border."

To get started, we might as well assume that $b \in B$, and so $(b - \varepsilon, b] \subset B$, for some $\varepsilon > 0$, since B is open. Now let $c = \sup A$. Clearly, $a \le c \le b$, but note that, since A and B are open in $[a, b]$, we actually have $a < c < b$. (Why?) Next, it follows from the definition of c that $(c - \varepsilon, c) \cap A \neq \emptyset$ and $(c, c + \varepsilon) \cap B \neq \emptyset$ for any $\varepsilon > 0$; in fact, $(c, b] \subset B$. That is, $c \in \bar{A}$ and $c \in \bar{B}$. But then, $c \in \bar{A} \cap \bar{B} = A \cap B = \emptyset$. This contradiction shows that no such splitting of $[a, b]$ into nonempty, disjoint, open sets is available.

Based on this observation, we say that a metric space M is **disconnected** (or *not* connected) if M can be split into the union of two nontrivial open sets, that is, if there are nonempty open sets A and B in M with $A \cap B = \emptyset$ and $A \cup B = M$. The pair of open sets A and B is called a **disconnection** of M. We say that M is **connected** if no such disconnection can be found. Thus, for example, $[a, b]$ is connected.

Notice that we could just as well have used *closed* sets in our definition. If a disconnection A, B exists, then the disconnecting sets are also closed: $A = B^c$ and $B = A^c$. That is, A and B are *clopen* (simultaneously open and closed) sets. Conversely, if M contains a nontrivial clopen subset A (other than $\emptyset$ or M), then A and A^c are a disconnection for M. This gives us our first theorem:

Theorem 6.1. *M is connected if and only if M contains no nontrivial clopen sets.*

Examples 6.2

(a) $\mathbb{R}$ is connected. (This follows from Exercise 4.25, but we will give another proof shortly based on the fact that $[a, b]$ is connected.)
(b) A discrete space containing two or more points is disconnected.
(c) The empty set $\emptyset$ and any one-point space are connected (by default).
(d) The Cantor set Δ is (very!) disconnected. Indeed, it follows from Exercise 2.22 that for any $x, y \in \Delta$ with $x < y$ there is a $z \notin \Delta$ such that $x < z < y$. Thus, Δ is disconnected by the (relatively) open sets $A = [0, z) \cap \Delta$ and $B = (z, 1] \cap \Delta$.

Our terminology for connectedness is unavoidably fussy. After all, we have defined connectedness in terms of what it is *not*, namely, disconnected. To make matters worse, at least on the surface, Example 6.2 (d) and our proof that $[a, b]$ is connected both suggest the frightening prospect of "relatively connected" as an altogether separate notion. Well, fear not! Connectedness is not a relative property for metric spaces. To see why, we will need to face the relative definition head-on.

A subset E of a metric space M is *disconnected in E* if there exist disjoint, nonempty, open (*in E*) sets U and V such that $E = U \cup V$. Now, it is immediate that this gives us a pair of open sets A and B in M such that $U = A \cap E$ and $V = B \cap E$. And so "unrelating" the relative definition, by writing it in terms of A and B, yields: $A \cap E \neq \emptyset$, $B \cap E \neq \emptyset$, $(A \cap E) \cap (B \cap E) = \emptyset$, and $E = (A \cap E) \cup (B \cap E)$, or $E \subset A \cup B$. (Phew!) This mess would be greatly simplified if we could take A and B to be disjoint in M. While this need not hold true in more general settings, luck is with us in a metric space.

Lemma 6.3. *Let E be a subset of a metric space M. If U and V are disjoint open sets in E, then there are disjoint open sets A and B in M such that $U = A \cap E$ and $V = B \cap E$.*

PROOF. We will only sketch the proof, leaving the full details as an exercise. The hard work here is largely a matter of notational bookkeeping. To spare us much of this notation, let's avoid the relative metric wherever possible. We will state everything in terms of open balls in M, using the simple notation $B_\varepsilon(x)$ in place of the more cumbersome $B_\varepsilon^M(x)$.

For each $x \in U$ there is an $\varepsilon_x > 0$ such that $E \cap B_{\varepsilon_x}(x) \subset U$, because U is open in E. Likewise, for each $y \in V$ there is a $\delta_y > 0$ such that $E \cap B_{\delta_y}(y) \subset V$. Since $U \cap V = \emptyset$, we also get $E \cap B_{\varepsilon_x}(x) \cap B_{\delta_y}(y) = \emptyset$. We would like to get rid of the set E in this conclusion, and we can do so at a small price:

Claim. $B_{\varepsilon_x/2}(x) \cap B_{\delta_y/2}(y) = \emptyset$ for every $x \in U$ and $y \in V$. (Just check.)

Thus $A = \bigcup\{B_{\varepsilon_x/2}(x) : x \in U\}$ and $B = \bigcup\{B_{\delta_y/2}(y) : y \in V\}$ work. □

The conclusion to be drawn from Lemma 6.3 is that E is disconnected (in E) if and only if there exist disjoint, nonempty, open sets A and B in M such that $A \cap E \neq \emptyset$, $B \cap E \neq \emptyset$, and $E \subset A \cup B$. And it does not matter whether we take "open" to mean "open in E" or "open in M." That is, this statement reduces to the original definition in case $E = M$, and it gives the correct "relative" definition in any case (by taking $U = A \cap E$ and $V = B \cap E$). Thus, there is no harm in simply taking it as our new definition of a disconnected *set*, as opposed to a disconnected *space*. In other words, we have dodged a bullet! By adopting this harmless rewording of the definition of disconnected, and hence also a rewording of the definition of connected, we have freed the concept from any apparent dependence on the relative metric. We would be foolish to do otherwise.

Henceforth, when considering a subset E of a given metric space M, we will call a pair of disjoint open sets A and B a **disconnection** of E if $A \cap E \neq \emptyset$, $B \cap E \neq \emptyset$, and $E \subset A \cup B$. And, of course, we will say that E is a connected set if no such disconnection of E can be found.

Let's put this new definition to use by giving another characterization of the intervals in $\mathbb{R}$.

Theorem 6.4. *A subset E of $\mathbb{R}$, containing more than one point, is connected if and only if, whenever $x, y \in E$ with $x < y$, we also have $[x, y] \subset E$. That is, the connected subsets of $\mathbb{R}$ (containing more than one point) are precisely the intervals.*

PROOF. One direction is easy: If there exist points $x < z < y$ such that $x, y \in E$ but $z \notin E$, then $E \subset (-\infty, z) \cup (z, +\infty)$; that is, $A = (-\infty, z)$ and $B = (z, +\infty)$ is a disconnection of E.

For the other direction, suppose that E satisfies the condition that $[x, y] \subset E$ whenever $x, y \in E$ with $x < y$, but that E is disconnected. Then there are disjoint open sets A and B in $\mathbb{R}$ such that $A \cap E \neq \emptyset$, $B \cap E \neq \emptyset$, and $E \subset A \cup B$. Given points $a \in A \cap E$ and $b \in B \cap E$, we might as well assume that $a < b$ and hence that $[a, b] \subset E$. But now $[a, b] \subset E \subset A \cup B$; that is, A and B are a disconnection of the interval $[a, b]$. This contradicts the fact that $[a, b]$ is connected. Hence E is connected.

Finally, suppose that E satisfies $[x, y] \subset E$ whenever $x, y \in E$ with $x < y$. We want to prove that E is an interval. But it follows from this condition that E contains the open interval $(\inf E, \sup E)$, where we include the possibilities $\inf E = -\infty$ and $\sup E = +\infty$. (Why?) Thus, E must be an interval; which particular type of interval depends on the disposition of $\inf E$ and $\sup E$ as finite, or not, and as elements, or not, of E. □

We can now shed some light on the structure of open sets in $\mathbb{R}$. The proof of Theorem 4.6 shows that each nonempty open set U in $\mathbb{R}$ can be uniquely written as the union of *connected* subsets. Indeed, we wrote an open set in terms of "maximal" intervals I_x, and such intervals are actually maximal with respect to being connected subsets of U (i.e., no larger subset of U will be connected). At each $x \in U$, we took I_x to be the union of all of the open subintervals in U that contain x. Thus, I_x is both open and connected (see Exercise 6), and hence it is an open interval. The remainder of the proof of Theorem 4.6 shows that two such connected "components" of U are either identical or disjoint. There are at most countably many distinct I_x, the union of which must be all of U.

Given any set E, we call the maximal (with respect to containment) connected subsets of E the *connected components* of E. Essentially the same line of reasoning as above shows that every set can be written (uniquely) as the disjoint union of its connected components. A connected set, then, is a set with only one connected component (namely, itself).

EXERCISES

Except where noted, each of the following exercises is set in a generic metric space M with metric d.

1. Supply the missing details in the proof of Lemma 6.3.

2. Show that the only nonempty connected subsets of Δ are singletons. (We would say that Δ is *totally disconnected*.)

3. If E is a connected subset of M, and if A and B are disjoint open sets in M with $E \subset A \cup B$, prove that either $E \subset A$ or $E \subset B$.

4. Prove that E is disconnected if and only if there exist nonempty sets A and B in M satisfying $A \cap \bar{B} = \emptyset$, $B \cap \bar{A} = \emptyset$, and $E = A \cup B$.

▷ **5.** If E and F are connected subsets of M with $E \cap F \neq \emptyset$, show that $E \cup F$ is connected.

▷ **6.** More generally, if $\mathcal{C}$ is a collection of connected subsets of M, all having a point in common, prove that $\bigcup \mathcal{C}$ is connected. Use this to give another proof that $\mathbb{R}$ is connected.

▷ **7.** If every pair of points in M is contained in some connected set, show that M is itself connected.

8. If E and F are nonempty subsets of M, and if $E \cup F$ is connected, show that $\bar{E} \cap \bar{F} \neq \emptyset$.

We are more than ready to speak of continuous functions and connectedness. Our first result shows that the two-point discrete space is the canonical disconnected set.

Lemma 6.5. *M is disconnected if and only if there exists a continuous map from M* **onto** $\{0, 1\}$ *(the two-point discrete space).*

PROOF. If $f : M \to \{0, 1\}$ is onto, then $A = f^{-1}(\{0\})$ and $B = f^{-1}(\{1\})$ are disjoint, nonempty, and satisfy $A \cup B = M$. If f is also continuous, then A and B are clopen sets and so form a disconnection of M.

Conversely, if A and B are a disconnection of M, then setting $f(a) = 0$ for $a \in A$ and $f(b) = 1$ for $b \in B$ defines a continuous map f from M onto $\{0, 1\}$. (Why?) □

Lemma 6.5 is telling us that there is no *continuous* method of splitting a connected set M into two discrete "parcels." More generally, it follows that M is connected if and only if any continuous map from M into a discrete space is necessarily *constant*.

Lemma 6.5 gives a nearly perfect replacement for the definition of disconnected. All of the notational difficulties that we faced earlier are now hidden in subtleties of language. For example, we have traded the cumbersome notation of relatively open sets for the tacit understanding that continuity may mean *relative* continuity. Most convenient. All of this hard work is beginning to pay off! In fact, we can now give a very short proof of that generalized intermediate value theorem we have been looking for:

Theorem 6.6. *Let $f : (M, d) \to (N, \rho)$ be continuous, and let E be a subset of M. If E is connected, then $f(E)$ is connected.*

PROOF. Suppose that $f(E)$ is not connected. Then there exists a continuous, onto map $g : f(E) \to \{0, 1\}$. But this means that $g \circ f : E \to \{0, 1\}$ is continuous and onto. That is, E is not connected. □

To see that Theorem 6.6 is a generalization of the intermediate value theorem, we just need to bring Theorem 6.4 back into the picture: The connected subsets of $\mathbb{R}$ (containing more than one point) are precisely the intervals. Thus, the image of an interval under a nonconstant continuous function is again an interval.

Corollary 6.7. *If I is an interval in $\mathbb{R}$ and if $f : I \to \mathbb{R}$ is a nonconstant continuous function, then $f(I)$ is an interval. In particular, if $a, b \in I$ with $f(a) \neq f(b)$, then f assumes every value between $f(a)$ and $f(b)$.*

EXERCISES

Throughout, M denotes an arbitrary metric space with metric d.

▷ **9.** If $A \subset B \subset \bar{A} \subset M$, and if A is connected, show that B is connected. In particular, $\bar{A}$ is connected.

10. True or false? If $A \subset B \subset C \subset M$, where A and C are connected, then B is connected.

11. An alternate definition of connectedness for metric spaces can be phrased in terms of continuous real-valued functions: Prove that M is disconnected if and only if there is a continuous function $f : M \to \mathbb{R}$ such that $f^{-1}(\{0\}) = \emptyset$ while $f^{-1}((-\infty, 0)) \neq \emptyset$ and $f^{-1}((0, \infty)) \neq \emptyset$. [Hint: If A and B are a disconnection, consider $f(x) = d(x, A) - d(x, B)$.]

12. If M is connected and has at least two points, show that M is uncountable. [Hint: Find a nonconstant, continuous, real-valued function on M.]

▷ **13.** If $f : [a, b] \to [a, b]$ is continuous, show that f has a fixed point; that is, show that there is some point x in $[a, b]$ with $f(x) = x$.

14. Let $f : [0, 2] \to \mathbb{R}$ be continuous with $f(0) = f(2)$. Show that there is some x in $[0, 1]$ such that $f(x) = f(x + 1)$.

15. If $f : \mathbb{R} \to \mathbb{R}$ is continuous and open, show that f is strictly monotone.

16. If $f : \mathbb{R} \to \mathbb{R}$ is continuous and one-to-one, show that f is strictly monotone.

17. Prove that there does not exist a continuous function $f : \mathbb{R} \to \mathbb{R}$ satisfying $f(\mathbb{Q}) \subset \mathbb{R} \setminus \mathbb{Q}$ and $f(\mathbb{R} \setminus \mathbb{Q}) \subset \mathbb{Q}$.

18. Let A and B be closed subsets of M, and suppose that both $A \cup B$ and $A \cap B$ are connected. Prove that A and B are connected.

19. Let $I = (\mathbb{R} \setminus \mathbb{Q}) \cap [0, 1]$ and $Q = \mathbb{Q} \cap [0, 1]$, with their usual metrics. Prove that there is a continuous map from I *onto* Q, but that there does not exist a continuous map from $[0, 1]$ onto Q. [Hint: Given a sequence of rationals $0 = r_0 < r_1 < \cdots < r_n < 1$ increasing to 1, notice that I can be written as the disjoint union of the open sets $(r_{n-1}, r_n) \cap [0, 1]$, $n = 1, 2, \ldots$.]

20. Let $f : [a, b] \to \mathbb{R}$ be continuous, and suppose that f takes on no value more than twice. Show that f takes on some value exactly once. [Hint: Either the maximum or the minimum value occurs only once.] Consequently, f is piecewise monotone.

21. Suppose that $f : \mathbb{R} \to \mathbb{R}$ takes on each of its values exactly twice; that is, for each $y \in \mathbb{R}$, the set $\{x : y = f(x)\}$ has either 0 or 2 elements. Show that f is discontinuous at infinitely many points.

22. Suppose that $f : \mathbb{R} \to \mathbb{R}$ has the **intermediate value property**; that is, suppose that if $x < y$ with $f(x) \neq f(y)$, then f assumes every value intermediate to $f(x)$ and $f(y)$ on the interval (x, y). If, in addition, we assume that the graph of f is closed in $\mathbb{R}^2$, prove that f is continuous. [Hint: If f is discontinuous at b, then there is a sequence $b_n \to b$ such that $|f(b_n) - f(b)| > \varepsilon$ for some $\varepsilon > 0$ and all n. By passing to a subsequence, we may suppose that, say, $f(b_n) > f(b) + \varepsilon$ for all n. How does this help?]

23. If $f : \mathbb{R} \to \mathbb{R}$ is differentiable, prove that f' has the intermediate value property. Specifically, if $a < b$ and $f'(a) < m < f'(b)$, show that $f'(c) = m$ for some $c \in (a, b)$. [Hint: Consider $g(x) = f(x) - mx$.]

Although it follows easily from the definition (since it is given in terms of open sets), Theorem 6.6 also shows that connectedness is preserved by homeomorphisms. This observation allows us to clarify one of the harder examples from Chapter Five.

Example 6.8

Intervals that "look" different are different. Specifically, no pair of intervals from among $[a, b]$, $(a, b]$, and (a, b) can be homeomorphic. The reasons we gave in

Chapter Five can now be restated in terms of our new language. To see that $(a, b]$ and (a, b) are not homeomorphic, for example, suppose that $f : (a, b] \to (a, b)$ is one-to-one and onto, and let $c = f(b)$. (Hence, $a < c < b$.) Then, the restriction of f to $(a, b] \setminus \{b\} = (a, b)$ is still one-to-one, but now its range is the disconnected set $(a, b) \setminus \{c\} = (a, c) \cup (c, b)$. Since f maps a connected set onto a disconnected set, f cannot be continuous. The other cases are similar.

The key observation in Example 6.8 is that between two "different" intervals one can always afford to lose more points than the other before becoming disconnected. For example, $[a, b]$ can afford to give up two points and still remain connected, whereas $(a, b]$ only has one point to spare. We could stretch this same reasoning to show that the unit interval $[0, 1]$ is not homeomorphic to the unit square $[0, 1] \times [0, 1]$, for example. For this, we first need a lemma:

Lemma 6.9. *If A and B are connected, then $A \times B$ is connected.*

PROOF. Suppose that $f : A \times B \to \{0, 1\}$ is continuous. We need to show that f is constant. But, given any $a \in A$ and $b' \in B$, each of the functions $f(a, \cdot) : B \to \{0, 1\}$ and $f(\cdot, b') : A \to \{0, 1\}$ is continuous. (This follows from what we know about "the" product metric; see Exercise 3.46.) Consequently, since A and B are connected, each of these new maps must be constant.

This means that f is constant along "horizontal" and "vertical" lines in $A \times B$. Thus, $f(a, b) = f(a', b')$ because $f(a, \cdot)$ and $f(\cdot, b')$ are constant and the two functions must agree at (a, b'). (Figure 6.1 may help; f is constant along each dotted line, and these values must agree at the "intersections.") That is, f is constant. □

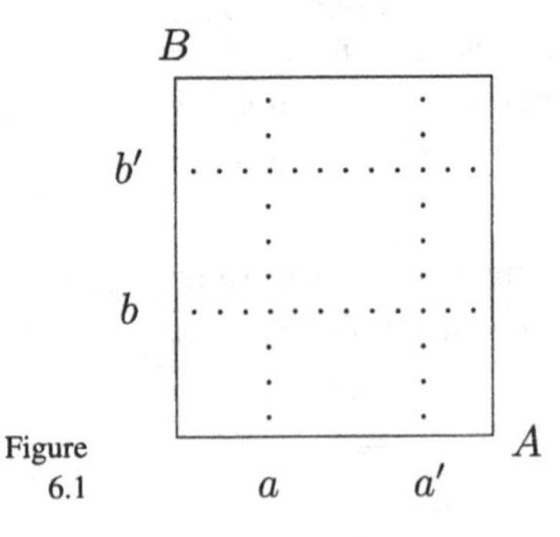

Figure 6.1

Thus $[0, 1] \times [0, 1]$ is connected, and now it is easy to see why $[0, 1] \times [0, 1]$ cannot be homeomorphic to $[0, 1]$. Indeed, $[0, 1] \setminus \{1/2\}$ is disconnected while $[0, 1] \times [0, 1]$ minus any point is still connected. (Why?) Similarly, $\mathbb{R}^2$ is connected, and essentially the same argument shows that $\mathbb{R}^2$ is not homeomorphic to $\mathbb{R}$. By induction, $\mathbb{R}^n$ is connected (we will outline a second proof in the exercises), and this line of reasoning can be used to show that $\mathbb{R}^n$ is not homeomorphic to $\mathbb{R}$ for $n > 1$. But the question of whether $\mathbb{R}^n$ is homeomorphic to $\mathbb{R}^m$ for arbitrary $n \neq m$ is *very* difficult! Nevertheless, the argument is the same in spirit (the "bigger" space is "more" connected), and it is in fact true that $\mathbb{R}^n$ is not homeomorphic to $\mathbb{R}^m$ for $n \neq m$.

EXERCISES

24. Show that $(0, 1) \times (0, 1)$, although an open set in $\mathbb{R}^2$, cannot be written as a disjoint union of open balls in $\mathbb{R}^2$. (Compare with Theorem 4.6.)

25.

(a) Give an example of a continuous function having a connected range but a disconnected domain.

(b) Let $D \subset \mathbb{R}$, and let $f : D \to \mathbb{R}$ be continuous. Prove that D is connected if $\{(x, f(x)) : x \in D\}$, the graph of f, is a connected subset of $\mathbb{R}^2$.

26. Let $f : [0, 1] \to \mathbb{R}$ be defined by $f(x) = \sin(1/x)$ for $x \neq 0$ and $f(0) = 0$. Show that although f is not continuous, the graph of f is a connected subset of $\mathbb{R}^2$. [Hint: Use Exercise 9.]

27. Let V be a normed vector space, and let $x \neq y \in V$. Show that the map $f(t) = x + t(y - x)$ is a homeomorphism from $[0, 1]$ into V. The range of f is the line segment joining x and y, and it is often written $[x, y]$ (since f is a homeomorphism, this interval notation is justified). [Hint: That f is continuous and one-to-one is easy; next show that if $f(t_n) \to z$, then (t_n) converges to some t in $[0, 1]$ with $z = f(t)$.]

28. Deduce from Exercises 7 and 27 that any normed space V is connected.

The full details will have to wait for a while, but we have enough "savvy" at this point to discuss an extremely curious and highly counterintuitive phenomenon. In spite of the fact that $[0, 1]$ and $[0, 1] \times [0, 1]$ are not homeomorphic, and in spite of the fact that the square $[0, 1] \times [0, 1]$ should, by rights, be much "bigger" than the interval $[0, 1]$, there exists a continuous *onto* map $f : [0, 1] \to [0, 1] \times [0, 1]$. (As we will see in Chapter Eight, no such map can be one-to-one. In fact, no continuous, one-to-one map from $[0, 1]$ to $[0, 1] \times [0, 1]$ can have a dense range.)

Now a map $f(t) = (x(t), y(t))$ from $[0, 1]$ to $[0, 1] \times [0, 1]$ is called a *path*, or *curve*. If the range of f "fills" the square, we say that f is a *space-filling curve*. The existence of *any* space-filling curve was considered quite shocking at one time, let alone a *continuous* space-filling curve! But, as is typical of such discoveries, once a continuous space-filling curve was shown to exist, dozens of other examples followed. We will briefly describe two such examples.

The first example is due to Peano in 1890. The idea is to define a sequence of paths that visit ever more points in the square; the "limit" path will be onto since it ultimately visits a dense set of points in the square (more on this in Chapter Eight). Figure 6.2 shows the first two paths.

Figure 6.2 shows the unit square broken into nine equal subsquares; the first path travels from $(0, 0)$ to $(1, 1)$ (i.e., from lower left to upper right) in a series of straight line paths, in the direction indicated by the circled numbers.

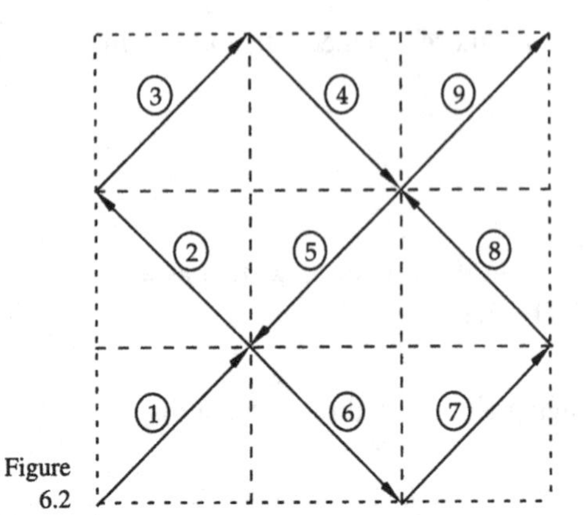

Figure 6.2

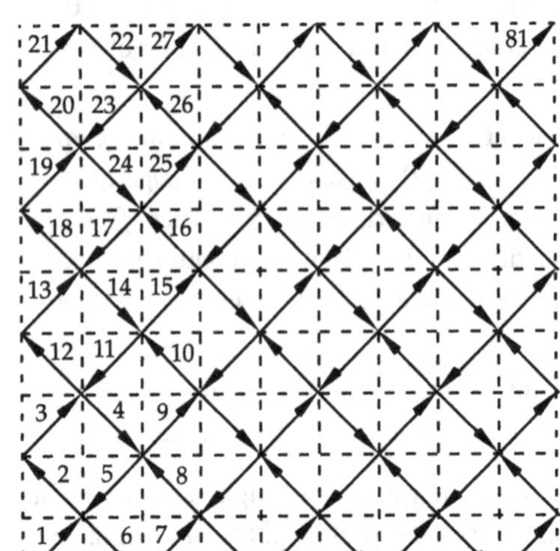

Figure 6.3

Figure 6.3 shows each of the subsquares of Figure 6.2 broken into 9 equal subsquares, giving us 81 subsquares in all. The second path travels from (0, 0) to (1, 1) by repeating the first path in "miniature" in each 3×3 block of subsquares. The new path traverses each of the nine original subsquares in the same order as before (the path wends its way up the first column of 3×3 blocks, down the center column, and up the last column). Notice, too, that the direction of each of the nine "miniature" paths is determined by the direction of the corresponding segment of the first path. That is, we enter the first 3×3 block at the lower left and exit at the upper right; we enter the second 3×3 block at the lower right and exit at the upper left; we enter the third block at the lower left and exit at the upper right, and so on.

The third path is obtained by repeating this process in each of the 81 subsquares of Figure 6.3. That is, divide each subsquare into 9 more equal subsquares, giving us 729 in all, and repeat the first path in "microminiature" in each of the new 3×3 blocks. Continue. The limit of this process (which can be made rigorous) is a continuous path mapping $[0, 1]$ *onto* the square $[0, 1] \times [0, 1]$.

Our second example of a space-filling curve is essentially due to Lebesgue in 1928. This one is even more amazing than Peano's example (if such a thing is possible). Lebesgue's idea is this: Since the Cantor function maps Δ (a tiny set) onto all of $[0, 1]$ (a big set), perhaps some variation on the Cantor function will map Δ onto the square $[0, 1] \times [0, 1]$ (an even bigger set). And it does!! *Incroyable!*

Here's the setup: Recall that each element $t \in \Delta$ can be written as $t = \sum_{n=1}^{\infty} 2a_n/3^n$, where each a_n is either 0 or 1; in other symbols, $t = 0.(2a_1)(2a_2)(2a_3)\ldots$ (base 3). Now define a map $t \mapsto (x(t), y(t))$ by $x(t) = 0.a_2a_4a_6\ldots$ (base 2) and $y(t) = 0.a_1a_3a_5\ldots$ (base 2). Each of $x(t)$ and $y(t)$ is rather like the Cantor function; each is continuous on Δ, and each maps Δ onto $[0, 1]$. (Why?) Moreover, $x(t)$ and $y(t)$ extend to continuous functions on $[0, 1]$, and the path $f(t) = (x(t), y(t))$ is actually a continuous space-filling curve (which maps Δ onto $[0, 1] \times [0, 1]$). Amazing! And now that we know the "trick," we can play this same game again to get a continuous map from Δ onto $[0, 1] \times [0, 1] \times [0, 1]$. Just take each element of Δ, written as a ternary decimal, and "spread out" the ternary decimal to make up three binary decimals, this time using every third ternary digit: $0.a_1a_4a_7\ldots$, and so on. By induction, $[0, 1]^n$ is the continuous image of Δ for every $n \geq 1$. Unbelievable! What was counterintuitive and simply out of the question moments ago has reduced to "one small step" after the fact. (And it gets even better! But we will save that story for another day.)

Notes and Remarks

For complete details of the proof that $\mathbb{R}^n$ and $\mathbb{R}^m$ are not homeomorphic for $n \neq m$, see M. H. A. Newman [1951].

For a thorough discussion of topics related to the intermediate value theorem (Corollary 6.7), including the intermediate value property for derivatives (Exercise 23), see Boas [1960].

The brand of connectedness found in Exercise 28 is called *pathwise connectedness* (or, to be precise, *arcwise connectedness*). A space is pathwise connected if there is a path (a continuous map on $[0, 1]$) joining any pair of points in the space. Exercise 7 and Theorem 6.4 show that pathwise connected spaces are connected in our sense (but not conversely – in the example given in Exercise 26, the point $(0, 0)$ cannot be connected to the rest of the graph by means of a path). Pathwise connectedness is older than connectedness; according to Willard [1970], Weierstrass used it as early as the 1880s. The modern version evolved through the efforts of several mathematicians, including Cantor, Jordan, Schoenflies, Lennes, Riesz, and Hausdorff. For a more complete history, see Wilder [1978, 1980].

For functions $f : \mathbb{R} \to \mathbb{R}$, continuity, the intermediate value property, and the connectedness of the graph of f (as a subset of $\mathbb{R}^2$) are essentially equivalent. For much more on this, see Burgess [1990]. Exercise 22 is based on the discussion in Burgess's paper, but see also Boas [1960] and Randolph [1968].

Lebesgue's simplification of Peano's space-filling curve appears in his book, *Leçons sur l'Integration* (Lebesgue [1928]), which was originally published as one of the volumes in Borel's series of monographs. Lebesgue's example was subsequently modified by I. Schoenberg in Schoenberg [1938]. For further details, see Schoenberg [1982] and Sagan [1986, 1992]. We will have more to say about the Schoenberg–Lebesgue curve later in the book.

Space-filling curves have been a constant source of fascination in the mathematical literature. New examples and simplifications of old examples continue to surface in popular journals; dozens of articles on space-filling curves have appeared in the *Monthly* over the years. Two such articles, one old and one new, are Moore [1900] and Holbrook [1991] (but see also Swift [1961], Wen [1983], and Lance and Thomas [1991]). Moore's paper is particularly interesting; he discusses Hilbert's example of a space-filling curve, Weierstrass's nondifferentiable function, and other early work. Holbrook, on the other hand, takes a novel approach: He shows that a curve $f(t) = (x(t), y(t))$ is space-filling whenever the coordinate functions $x(t)$ and $y(t)$ are stochastically independent. For a discussion of space-filling curves in general, see Boas [1960] and the articles by Whyburn [1942] and Hahn [1956b]. For a thorough treatment of related constructions, see A. N. Singh [1969].

CHAPTER SEVEN

Completeness

Totally Bounded Sets

At the end of Chapter Five we discussed the problem of defining a norm on $C(M)$, the space of continuous, real-valued functions on a metric space M. We saw that an easy solution presents itself in the case where M is finite, and the suggestion was made that it is enough for M to be "like" a finite set. In this chapter we will come one step closer to making this vague suggestion precise. To begin, we consider sets that can be written as the union of finitely many small "parcels."

A set A in a metric space (M, d) is said to be **totally bounded** if, given any $\varepsilon > 0$, there exist finitely many points $x_1, \ldots, x_n \in M$ such that $A \subset \bigcup_{i=1}^{n} B_\varepsilon(x_i)$. That is, each $x \in A$ is within ε of some x_i. For this reason, some authors would say that the set $\{x_1, \ldots, x_n\}$ is ε-**dense** in A, or that $\{x_1, \ldots, x_n\}$ is an ε-**net** for A. For our purposes, we will paraphrase the statement $A \subset \bigcup_{i=1}^{n} B_\varepsilon(x_i)$ by saying that A is **covered** by finitely many ε-balls.

In the definition of a totally bounded set A, we could easily insist that each ε-ball be centered at a point of A. Indeed, given $\varepsilon > 0$, choose $x_1, \ldots, x_n \in M$ so that $A \subset \bigcup_{i=1}^{n} B_{\varepsilon/2}(x_i)$. We may certainly assume that $A \cap B_{\varepsilon/2}(x_i) \neq \emptyset$ for each i, and so we may choose a point $y_i \in A \cap B_{\varepsilon/2}(x_i)$ for each i. By the triangle inequality, we then have $A \subset \bigcup_{i=1}^{n} B_\varepsilon(y_i)$. (Why?) That is, A can be covered by finitely many ε-balls, each centered at a point in A. More to the point, a set A is totally bounded if and only if A can be covered by finitely many *arbitrary* sets of diameter at most ε, for any $\varepsilon > 0$.

Lemma 7.1. *A is totally bounded if and only if, given $\varepsilon > 0$, there are finitely many sets $A_1, \ldots, A_n \subset A$, with* $\operatorname{diam}(A_i) < \varepsilon$ *for all i, such that $A \subset \bigcup_{i=1}^{n} A_i$.*

PROOF. First suppose that A is totally bounded. Given $\varepsilon > 0$, we may choose $x_1, \ldots, x_n \in M$ such that $A \subset \bigcup_{i=1}^{n} B_\varepsilon(x_i)$. As above, A is then covered by the sets $A_i = A \cap B_\varepsilon(x_i) \subset A$ and $\operatorname{diam}(A_i) \leq 2\varepsilon$ for each i.

Conversely, given $\varepsilon > 0$, suppose that there are finitely many sets $A_1, \ldots, A_n \subset A$, with $\operatorname{diam}(A_i) < \varepsilon$ for all i, such that $A \subset \bigcup_{i=1}^{n} A_i$. Given $x_i \in A_i$, we then have $A_i \subset B_{2\varepsilon}(x_i)$ for each i and, hence, $A \subset \bigcup_{i=1}^{n} B_{2\varepsilon}(x_i)$.

Since ε is arbitrary in either case, we are done. □

Notice that the condition in Lemma 7.1 demands that $A_1, \ldots, A_n$ be subsets of A. This is no real constraint since, after all, if A is covered by $B_1, \ldots, B_n \subset M$, then A is also covered by the sets $A_i = A \cap B_i \subset A$ and $\operatorname{diam}(A_i) \leq \operatorname{diam}(B_i)$.

Examples 7.2

(a) By the triangle inequality, a totally bounded set is necessarily bounded. (Why?) Note also that any subset of a totally bounded set is again totally bounded. (See Exercise 1.)

(b) A *finite* set is always totally bounded. In a discrete space, a set is totally bounded if and only if it is finite. (Why?)

(c) In $\mathbb{R}$, we do not get anything new: A subset of $\mathbb{R}$ is totally bounded if and only if it is bounded. (See Exercise 2.) Thus, total boundedness is apparently not a topological property; it depends intimately on the metric at hand. (See Exercise 3.)

(d) In general, not every bounded set is totally bounded. The discrete metric gives us a clue as to how we might construct such a set. Recall the sequence $e^{(n)} = (0, \ldots, 0, 1, 0, \ldots)$ in ℓ_1, where the single nonzero entry is in the nth place. Then, $\{e^{(n)} : n \geq 1\}$ is a bounded set in ℓ_1, since $\|e^{(n)}\|_1 = 1$ for all n, but *not* totally bounded. Why? Because $\|e^{(m)} - e^{(n)}\|_1 = 2$ for $m \neq n$; thus, $\{e^{(n)} : n \geq 1\}$ cannot be covered by finitely many balls of radius <2. In fact, the set $\{e^{(n)} : n \geq 1\}$ is *discrete* in its *relative* metric. (Compare with Exercise 8.)

EXERCISES

Except where noted, each of the following exercises is set in an arbitrary metric space M with metric d.

▷ **1.** If $A \subset B \subset M$, and if B is totally bounded, show that A is totally bounded.

▷ **2.** Show that a subset A of $\mathbb{R}$ is totally bounded if and only if it is bounded. In particular, if I is a closed, bounded, interval in $\mathbb{R}$ and $\varepsilon > 0$, show that I can be covered by finitely many closed subintervals $J_1, \ldots, J_n$, each of length at most ε.

3. Is total boundedness preserved by homeomorphisms? Explain. [Hint: $\mathbb{R}$ is homeomorphic to $(0, 1)$.]

4. Show that A is totally bounded if and only if A can be covered by finitely many *closed* sets of diameter at most ε for every $\varepsilon > 0$.

▷ **5.** Prove that A is totally bounded if and only if $\bar{A}$ is totally bounded.

We next give a sequential criterion for total boundedness. The key observation is isolated in:

Lemma 7.3. *Let (x_n) be a sequence in (M, d), and let $A = \{x_n : n \geq 1\}$ be its range.*

(i) *If (x_n) is Cauchy, then A is totally bounded.*

(ii) *If A is totally bounded, then (x_n) has a Cauchy subsequence.*

PROOF. (i) Let $\varepsilon > 0$. Then, since (x_n) is Cauchy, there is some index $N \geq 1$ such that $\operatorname{diam}\{x_n : n \geq N\} < \varepsilon$. Thus:

$$A = \underbrace{\{x_1\} \cup \cdots \cup \{x_{N-1}\} \cup \{x_n : n \geq N\}}_{N \text{ sets of diameter } < \varepsilon}.$$

(ii) If A is a finite set, we are done. (Why?) So, suppose that A is an infinite totally bounded set. Then A can be covered by finitely many sets of diameter <1. One of these sets, at least, must contain infinitely many points of A. Call this set A_1. But then A_1 is also totally bounded, and so it can be covered by finitely many sets of diameter $<1/2$. One of these, call it A_2, contains infinitely many points of A_1. Continuing this process, we find a decreasing sequence of sets $A \supset A_1 \supset A_2 \supset \cdots$, where each A_k contains infinitely many x_n and where $\operatorname{diam}(A_k) < 1/k$. In particular, we may choose a subsequence (x_{n_k}) with $x_{n_k} \in A_k$ for all k. (How?) That (x_{n_k}) is Cauchy is now clear since $\operatorname{diam}\{x_{n_j} : j \geq k\} \leq \operatorname{diam}(A_k) < 1/k$. □

Examples 7.4

(a) The sequence $x_n = (-1)^n$ in $\mathbb{R}$ shows that a Cauchy *subsequence* is the best that we can hope for in Lemma 7.3 (ii).

(b) Note that the sequence $(e^{(n)})$ in ℓ_1 has no Cauchy subsequence.

We are finally ready for our sequential characterization of total boundedness:

Theorem 7.5. *A set A is totally bounded if and only if every sequence in A has a Cauchy subsequence.*

PROOF. The forward implication is clear from Lemma 7.3. To prove the backward implication, suppose that A is *not* totally bounded. Then, there is some $\varepsilon > 0$ such that A cannot be covered by finitely many ε-balls. Thus, by induction, we can find a sequence (x_n) in A such that $d(x_n, x_m) \geq \varepsilon$ whenever $m \neq n$. (How?) But then, (x_n) has no Cauchy subsequence. □

All of this should remind you of the Bolzano–Weierstrass theorem – and for good reason:

Corollary 7.6. (The Bolzano–Weierstrass Theorem) *Every bounded infinite subset of $\mathbb{R}$ has a limit point in $\mathbb{R}$.*

PROOF. Let A be a bounded infinite subset of $\mathbb{R}$. Then, in particular, there is a sequence (x_n) of distinct points in A. Since A is totally bounded, there is a Cauchy subsequence (x_{n_k}) of (x_n). But Cauchy sequences in $\mathbb{R}$ converge, and so (x_{n_k}) converges to some $x \in \mathbb{R}$. Thus, x is a limit point of A. □

EXERCISES

Unless otherwise specified, each of the following exercises is set in a generic metric space (M, d).

6. Prove that A is totally bounded if and only if every sequence (x_n) in A has a subsequence (x_{n_k}) for which $d\big(x_{n_k}, x_{n_{k+1}}\big) < 2^{-k}$.

7. Show that Corollary 7.6 follows from the nested interval theorem.

8. If A is *not* totally bounded, show that A has an infinite subset B that is homeomorphic to a discrete space (where B is supplied with its relative metric). [Hint: Find

$\varepsilon > 0$ and a sequence (x_n) in A such that $d(x_n, x_m) \geq \varepsilon$ for $n \neq m$. How does this help?]

▷ **9.** Give an example of a closed bounded subset of ℓ_∞ that is not totally bounded.

▷ **10.** Prove that a totally bounded metric space M is separable. [Hint: For each n, let D_n be a finite $(1/n)$-net for M. Show that $D = \bigcup_{n=1}^{\infty} D_n$ is a countable dense set.]

11. Prove that H^∞ is totally bounded (see Exercises 3.10 and 4.48).

Complete Metric Spaces

As you can now well imagine, we want to isolate the class of metric spaces in which Cauchy sequences always converge. It follows from Theorem 7.5 that we would have an analogue of the Bolzano–Weierstrass theorem in such spaces (see Theorem 7.11). In fact, we will find that this class of metric spaces has much in common with the real line $\mathbb{R}$.

A metric space M is said to be **complete** if every Cauchy sequence in M converges – to a point in M!

Examples 7.7

(a) $\mathbb{R}$ is complete. This is a consequence of the least upper bound axiom; in fact, as we will see, the completeness of $\mathbb{R}$ is actually equivalent to the least upper bound axiom.

(b) $\mathbb{R}^n$ is complete (because $\mathbb{R}$ is).

(c) Any discrete space is complete (trivially).

(d) $(0, 1)$ is *not* complete. (Why?) Hence, completeness is not preserved by homeomorphisms. Which subsets of $\mathbb{R}$ are complete?

(e) c_0, ℓ_1, ℓ_2, and ℓ_∞ are all complete. The proofs are all very similar; we sketch the proof for ℓ_2 below and leave the rest as exercises.

(f) $C[a, b]$ is complete. The proof is not terribly difficult, but it will best serve our purposes to postpone it until Chapter Ten, where several similar proofs are collected.

The proof that ℓ_2 is complete is based on a few simple principles that will generalize to all sorts of different settings. This generality will become all the more apparent if we introduce a slight change in our notation. Since a sequence is just another name for a *function* on $\mathbb{N}$, let's agree to write an element $f \in \ell_2$ as $f = (f(k))_{k=1}^{\infty}$, in which case $\|f\|_2 = \left(\sum_{k=1}^{\infty} |f(k)|^2\right)^{1/2}$. For example, the notorious vectors $e^{(n)}$ will now be written e_n, where $e_n(k) = \delta_{n,k}$. (This is *Kronecker's delta*, defined by $\delta_{n,k} = 1$ if $n = k$ and $\delta_{n,k} = 0$ otherwise.)

Let (f_n) be a sequence in ℓ_2, where now we write $f_n = (f_n(k))_{k=1}^{\infty}$, and suppose that (f_n) is Cauchy in ℓ_2. That is, suppose that for each $\varepsilon > 0$ there exists an n_0 such that

$\|f_n - f_m\|_2 < \varepsilon$ whenever $m, n \geq n_0$. Of course, we want to show that (f_n) converges, in the metric of ℓ_2, to some $f \in \ell_2$. We will break the proof into three steps:

Step 1. $f(k) = \lim_{n\to\infty} f_n(k)$ exists in $\mathbb{R}$ for each k.

To see why, note that $|f_n(k) - f_m(k)| \leq \|f_n - f_m\|_2$ for any k, and hence $(f_n(k))_{n=1}^{\infty}$ is Cauchy in $\mathbb{R}$ for each k. Thus, f is the *obvious candidate* for the limit of (f_n), but we still have to show that the convergence takes place in the metric space ℓ_2; that is, we need to show that $f \in \ell_2$ and that $\|f_n - f\|_2 \to 0$ (as $n \to \infty$).

Step 2. $f \in \ell_2$; that is, $\|f\|_2 < \infty$.

We know that (f_n) is *bounded* in ℓ_2 (why?); say, $\|f_n\|_2 \leq B$ for all n. Thus, for any fixed $N < \infty$, we have:

$$\sum_{k=1}^{N} |f(k)|^2 = \lim_{n\to\infty} \sum_{k=1}^{N} |f_n(k)|^2 \leq B^2.$$

Since this holds for any N, we get that $\|f\|_2 \leq B$.

Step 3. Now we repeat Step 2 (more or less) to show that $f_n \to f$ in ℓ_2.

Given $\varepsilon > 0$, choose n_0 so that $\|f_n - f_m\|_2 < \varepsilon$ whenever $m, n \geq n_0$. Then, for any N and any $n \geq n_0$,

$$\sum_{k=1}^{N} |f(k) - f_n(k)|^2 = \lim_{m\to\infty} \sum_{k=1}^{N} |f_m(k) - f_n(k)|^2 \leq \varepsilon^2.$$

Since this holds for any N, we have $\|f - f_n\|_2 \leq \varepsilon$ for all $n \geq n_0$. That is, $f_n \to f$ in ℓ_2.

Examples 7.8

(a) Just having a *candidate* for a limit is not enough. Consider the sequence (f_n) in ℓ_∞ defined by $f_n = (1, \ldots, 1, 0, \ldots)$, where the first n entries are 1 and the rest are 0. The "obvious" limit is $f = (1, 1, \ldots)$ (all 1), but $\|f - f_n\|_\infty = 1$ for all n. What's wrong?

(b) Worse still, sometimes the "obvious" limit is not even in the space. Consider the same sequence as in (a) and note that each f_n is actually an element of c_0. This time, the natural candidate f is not in c_0. Again, what's wrong?

As you can see, there can be a lot of details to check in a proof of completeness, and it would be handy to have at least a few easy cases available. For example, when is a subset of a complete space complete? The answer is given as:

Theorem 7.9. *Let (M, d) be a complete metric space and let A be a subset of M. Then, (A, d) is complete if and only if A is* closed *in M.*

PROOF. First suppose that (A, d) is complete, and let (x_n) be a sequence in A that converges to some point $x \in M$. Then (x_n) is Cauchy in (A, d) and so converges to some point of A. That is, we must have $x \in A$ and, hence, A is closed.

Next suppose that (x_n) is a Cauchy sequence in (A, d). Then (x_n) is also Cauchy in (M, d). (Why?) Hence, (x_n) converges to some point $x \in M$. But A is closed and so, in fact, $x \in A$. Thus, (A, d) is complete. □

Examples 7.10

(a) $[0, 1]$, $[0, \infty)$, $\mathbb{N}$, and Δ are all complete.

(b) It follows from Theorem 7.5 that if a metric space (M, d) is both complete and totally bounded, then every sequence in M has a convergent subsequence. In particular, any closed, bounded subset of $\mathbb{R}$ is both complete and totally bounded. Thus, for example, every sequence in $[a, b]$ has a convergent subsequence. As you can easily imagine, the interval $[a, b]$ is a great place to do analysis! We will pursue the consequences of this felicitous combination of properties in the next chapter.

EXERCISES

Unless otherwise stated, (M, d) denotes an arbitrary metric space.

▷ **12.** Let A be a subset of an arbitrary metric space (M, d). If (A, d) is complete, show that A is closed in M.

13. Show that $\mathbb{R}$ endowed with the metric $\rho(x, y) = |\arctan x - \arctan y|$ is *not* complete. How about if we try $\tau(x, y) = |x^3 - y^3|$?

14. If we define

$$d(m, n) = \left| \frac{1}{m} - \frac{1}{n} \right|$$

for $m, n \in \mathbb{N}$, show that d is equivalent to the usual metric on $\mathbb{N}$ but that $(\mathbb{N}, d)$ is not complete.

15. Prove or disprove: If M is complete and $f : (M, d) \to (N, \rho)$ is continuous, then $f(M)$ is complete.

▷ **16.** Prove that $\mathbb{R}^n$ is complete under any of the norms $\|\cdot\|_1$, $\|\cdot\|_2$, or $\|\cdot\|_\infty$. [This is interesting because completeness is not usually preserved by the mere equivalence of *metrics*. Here we use the fact that all of the metrics involved are generated by *norms*. Specifically, we need the norms in question to be equivalent as functions: $\|\cdot\|_\infty \le \|\cdot\|_2 \le \|\cdot\|_1 \le n\|\cdot\|_\infty$. As we will see later, *any* two norms on $\mathbb{R}^n$ are comparable in this way.]

17. Given metric spaces M and N, show that $M \times N$ is complete if and only if both M and N are complete.

▷ **18.** Fill in the details of the proofs that ℓ_1 and ℓ_∞ are complete.

19. Prove that c_0 is complete by showing that c_0 is *closed* in ℓ_∞. [Hint: If (f_n) is a sequence in c_0 converging to $f \in \ell_\infty$, note that $|f(k)| \le |f(k) - f_n(k)| + |f_n(k)|$. Now choose n so that the $|f(k) - f_n(k)|$ is small *independent* of k.]

20. If (x_n) and (y_n) are Cauchy in (M, d), show that $\big(d(x_n, y_n)\big)_{n=1}^\infty$ is Cauchy in $\mathbb{R}$.

21. If (M, d) is complete, prove that two Cauchy sequences (x_n) and (y_n) have the same limit if (and only if) $d(x_n, y_n) \to 0$.

22. Let D be a dense subset of a metric space M, and suppose that every Cauchy sequence from D converges to some point of M. Prove that M is complete.

23. Prove that M is complete if and only if every sequence (x_n) in M satisfying $d(x_n, x_{n+1}) < 2^{-n}$, for all n, converges to a point of M.

24. Prove that the Hilbert cube H^∞ (Exercise 3.10) is complete.

25. True or false? If $f : \mathbb{R} \to \mathbb{R}$ is continuous and if (x_n) is Cauchy, then $(f(x_n))$ is Cauchy. Examples? How about if we insist that f be strictly increasing? Show that the answer is "true" if f is Lipschitz.

Our next result underlines the fact that complete spaces have a lot in common with $\mathbb{R}$.

Theorem 7.11. *For any metric space (M, d), the following statements are equivalent:*

(i) *(M, d) is complete.*

(ii) (The Nested Set Theorem) *Let $F_1 \supset F_2 \supset F_3 \supset \cdots$ be a decreasing sequence of nonempty closed sets in M with* $\operatorname{diam}(F_n) \to 0$. *Then, $\bigcap_{n=1}^\infty F_n \neq \emptyset$ (in fact, it contains exactly one point).*

(iii) (The Bolzano–Weierstrass Theorem) *Every infinite, totally bounded subset of M has a limit point in M.*

PROOF. (i) $\Longrightarrow$ (ii): (Compare this with the proof of the nested interval theorem, Theorem 1.5.) Given (F_n) as in (ii), choose $x_n \in F_n$ for each n. Then, since the F_n decrease, $\{x_k : k \geq n\} \subset F_n$ for each n, and hence $\operatorname{diam}\{x_k : k \geq n\} \to 0$ as $n \to \infty$. That is, (x_n) is Cauchy. Since M is complete, we have $x_n \to x$ for some $x \in M$. But the F_n are closed, and so we must have $x \in F_n$ for all n. Thus, $\bigcap_{n=1}^\infty F_n \neq \emptyset$.

(ii) $\Longrightarrow$ (iii): Let A be an infinite, totally bounded subset of M. Recall that we have shown that A contains a Cauchy sequence (x_n) comprised of distinct points ($x_n \neq x_m$ for $n \neq m$). Now, setting $A_n = \{x_k : k \geq n\}$, we get $A \supset A_1 \supset A_2 \supset \cdots$, each A_n is nonempty (even infinite), and $\operatorname{diam}(A_n) \to 0$. That is, (ii) *almost* applies. But, clearly, $\bar{A}_n \supset \bar{A}_{n+1} \neq \emptyset$ for each n, and $\operatorname{diam}\left(\bar{A}_n\right) = \operatorname{diam}(A_n) \to 0$ as $n \to \infty$. Thus there exists an $x \in \bigcap_{n=1}^\infty \bar{A}_n \neq \emptyset$. Now $x_n \in A_n$ implies that $d(x_n, x) \leq \operatorname{diam}\left(\bar{A}_n\right) \to 0$. That is, $x_n \to x$ and so x is a limit point of A (see Exercise 4.33).

(iii) $\Longrightarrow$ (i): Let (x_n) be Cauchy in (M, d). We just need to show that (x_n) has a convergent *subsequence*. Now, by Lemma 7.3, the set $A = \{x_n : n \geq 1\}$ is totally bounded. If A happens to be finite, we are done. (Why?) Otherwise, (iii) tells us that A has a limit point $x \in M$. It follows that some subsequence of (x_n) converges to x. (Why?) □

In particular, note that Theorem 7.11 holds for $M = \mathbb{R}$. In this case, each of the three statements in Theorem 7.11 is equivalent to the least upper bound axiom. That is, we might have instead assumed one of these three as an axiom for $\mathbb{R}$ and then deduced the existence of least upper bounds as a corollary. What's more, the fact that monotone, bounded sequences converge in $\mathbb{R}$ is also equivalent to the least upper bound axiom. (See the discussion following Theorem 1.5.) In $\mathbb{R}$, then, completeness takes on multiple personalities, with each new persona directly related to the order properties of the real numbers.

EXERCISES

Each of the following exercises is set in a metric space M with metric d.

▷ **26.** Just as with the nested interval theorem, it is essential that the sets F_n used in the nested set theorem be both closed and bounded. Why? Is the condition $\operatorname{diam}(F_n) \to 0$ really necessary? Explain.

▷ **27.** Note that the version of the Bolzano–Weierstrass theorem given in Theorem 7.11 replaces boundedness with total boundedness. Is this really necessary? Explain.

28. Suppose that every *countable, closed* subset of M is complete. Prove that M is complete.

29. Prove that M is complete if and only if, for each $r > 0$, the closed ball $\{y \in M : d(x, y) \le r\}$ is complete.

30. If (M, d) is complete, prove that every open subset G of M is homeomorphic to a complete metric space. [Hint: Let $F = M \setminus G$ and consider the metric $\rho(x, y) = d(x, y) + \left|(d(x, F))^{-1} - (d(y, F))^{-1}\right|$ on G.]

In any normed vector space, the extra algebraic structure makes completeness somewhat easier to test. That this is so can be seen through a clever observation due to Stefan Banach. In fact, Banach made so many clever observations about completeness that we now refer to a complete normed vector space as a **Banach space**.

Here's the setup: Given a sequence (x_n) in a normed vector space X, the series $\sum_{n=1}^{\infty} x_n$ is said to *converge* in X if the sequence of partial sums $\sum_{n=1}^{N} x_n$ converges to some vector $x \in X$, that is, if $\|x - \sum_{n=1}^{N} x_n\| \to 0$ as $N \to \infty$. In this case we write, as usual, $x = \sum_{n=1}^{\infty} x_n$ and we say that $\sum_{n=1}^{\infty} x_n$ is *summable* to x. In other words, $\sum_{n=1}^{\infty} x_n$ is the name that we give to the limit of the partial sums.

Now, just as in $\mathbb{R}$, sequences and series are interchangeable: Each series is really a sequence of partial sums and, conversely, each sequence is the sequence of partial sums for some series. In particular, notice that $x_n = x_1 + \sum_{i=2}^{n}(x_i - x_{i-1})$. The sequence (x_n) and the series $\sum_{i=2}^{\infty}(x_i - x_{i-1})$ live or die together; both converge or both diverge. With this tool at our disposal (and Banach's help, of course), it is not hard to see that the question of completeness for a normed space can be settled by a simple test:

Theorem 7.12. *A normed vector space X is complete if and only if every absolutely summable series in X is summable. That is, X is complete if and only if $\sum_{n=1}^{\infty} x_n$ converges in X whenever $\sum_{n=1}^{\infty} \|x_n\| < \infty$.*

PROOF. First suppose that X is complete, and let (x_n) be a sequence in X for which $\sum_{n=1}^{\infty} \|x_n\| < \infty$. If we write $s_m = \sum_{n=1}^{m} x_n$ for the sequence of partial sums, then, for $m > n$, the triangle inequality yields

$$\|s_m - s_n\| = \left\| \sum_{k=n+1}^{m} x_k \right\| \leq \sum_{k=n+1}^{m} \|x_k\|.$$

Since the partial sums of $\sum_{n=1}^{\infty} \|x_n\|$ form a convergent (and hence Cauchy) sequence, we have that $\sum_{k=n+1}^{m} \|x_k\| \to 0$ as $m, n \to \infty$. Thus, (s_n) is also a Cauchy sequence and, as such, converges in X.

Next suppose that absolutely summable series in X are summable, and let (x_n) be a Cauchy sequence in X. As always, it is enough to find a subsequence of (x_n) that converges. To this end, choose a subsequence (x_{n_k}) for which $\|x_{n_{k+1}} - x_{n_k}\| < 2^{-k}$ for all k. (How?) Then, in particular, $\sum_{k=1}^{\infty} \|x_{n_{k+1}} - x_{n_k}\|$ converges. Consequently, the series $\sum_{k=1}^{\infty} (x_{n_{k+1}} - x_{n_k})$ converges in X. As we remarked earlier, this means that the sequence $x_{n_{m+1}} = x_{n_1} + \sum_{k=1}^{m} (x_{n_{k+1}} - x_{n_k})$ converges in X. □

There is never too much of a good thing: Note that Theorem 7.12 gives us yet another characterization of completeness in $\mathbb{R}$. The familiar fact that every absolutely summable series of real numbers is summable is actually equivalent to the least upper bound axiom.

EXERCISES

31. If $\sum_{n=1}^{\infty} x_n$ is a convergent series in a normed vector space X, show that $\left\| \sum_{n=1}^{\infty} x_n \right\| \leq \sum_{n=1}^{\infty} \|x_n\|$.

32. Use Theorem 7.12 to prove that ℓ_1 is complete.

33. Let s denote the vector space of all finitely nonzero real sequences; that is, $x = (x_n) \in s$ if $x_n = 0$ for all but finitely many n. Show that s is *not* complete under the sup norm $\|x\|_\infty = \sup_n |x_n|$.

34. Prove that a normed vector space X is complete if and only if every sequence (x_n) in X satisfying $\|x_n - x_{n+1}\| < 2^{-n}$, for all n, converges to a point of X.

35. Prove that a normed vector space X is complete if and only if its closed unit ball $B = \{x \in X : \|x\| \leq 1\}$ is complete.

Fixed Points

Completeness is a useful property to have around if you are interested in solving equations. How so? Well, think about the sorts of tricks that we use in $\mathbb{R}$. How, for example, would you compute $\sqrt{2}$ "by hand"? You would most likely start by finding an *approximate* solution to the equation $x^2 = 2$ and then look for ways to improve your estimate.

Most numerical techniques give, in fact, a *sequence* of "better and better" approximate solutions, where "better and better" typically means that the error in approximation gets smaller. The completeness of $\mathbb{R}$ affords us such luxuries; we can effectively proclaim the existence of solutions without necessarily *finding* them! Once we have a Cauchy sequence of approximate solutions, completeness will finish the job.

The same holds true in any complete space. We can effectively solve certain "abstract" equations by simply displaying a Cauchy sequence of approximate solutions. One such technique, called the *method of successive approximations*, is used in the standard proof of existence for solutions to differential equations and is generally credited to Picard in 1890. (But the technique itself goes back at least to Liouville, who first published it in 1838, and it may have even been known to Cauchy.) We will see an example of this method shortly.

The modern metric space version of the method of successive approximations was explicitly stated in Banach's thesis in 1922. In this setting it is most often referred to as *Banach's contraction mapping principle*. A map $f : M \to M$ on a metric space (M, d) is called a **contraction** (or, better still, a **strict contraction**) if there is some constant α with $0 \le \alpha < 1$ such that $d(f(x), f(y)) \le \alpha\, d(x, y)$ is satisfied for all $x, y \in M$. That is, a contraction shrinks the distance between pairs of points by a factor strictly less than 1. Please note that any contraction is *automatically* continuous (since it is Lipschitz).

Banach's approach seeks to solve an "abstract" equation of the form $f(x) = x$ (this is more general than it might appear). That is, we look for a **fixed point** for f. If f is a contraction defined on a complete metric space, we can even prescribe a sequence of approximate solutions:

Theorem 7.13. *Let (M, d) be a complete metric space, and let $f : M \to M$ be a (strict) contraction. Then, f has a unique fixed point. Moreover, given any point $x_0 \in M$, the sequence of functional iterates $(f^n(x_0))$ always converges to the fixed point for f.*

[The notation f^n means the composition of f with itself n times: $f \circ f \circ \cdots \circ f$. For example, $f^2(x) = f(f(x))$, $f^3(x) = f(f^2(x))$, and so on. The sequence of functional iterates $(f^n(x))$ is called the *orbit* of x under f.]

PROOF. Let x_0 be any point in M, and consider the sequence $(f^n(x_0))$.

If $(f^n(x_0))$ converges, we are done. Indeed, if $x = \lim_{n\to\infty} f^n(x_0)$, then, since f is continuous, we have $f(x) = \lim_{n\to\infty} f\big(f^n(x_0)\big) = \lim_{n\to\infty} f^{n+1}(x_0) = \lim_{n\to\infty} f^n(x_0) = x$. And this x is unique, for if y is also a fixed point for f, then $d(x, y) = d(f(x), f(y)) \le \alpha d(x, y)$, which forces $d(x, y) = 0$.

So our goal is clear: We need to show that $(f^n(x_0))$ is a Cauchy sequence. But:

$$\begin{aligned} d\left(f^{n+1}(x_0),\, f^n(x_0)\right) &\le \alpha d\left(f^n(x_0),\, f^{n-1}(x_0)\right) \\ &\le \alpha^2 d\left(f^{n-1}(x_0),\, f^{n-2}(x_0)\right) \\ &\;\;\vdots \\ &\le \alpha^n d(f(x_0), x_0) = C\alpha^n. \end{aligned}$$

And so for $m \geq n$ the triangle inequality yields

$$d\left(f^{m+1}(x_0),\ f^n(x_0)\right) \leq \sum_{k=n}^{m} d\left(f^{k+1}(x_0),\ f^k(x_0)\right) \leq C \sum_{k=n}^{m} \alpha^k. \qquad (7.1)$$

But since $0 \leq \alpha < 1$, we have $\sum_{k=n}^{m} \alpha^k \to 0$ as $m, n \to \infty$. (Why?) Thus, $(f^n(x_0))$ is Cauchy. □

Note that the proof of Theorem 7.13 even gives us a rough estimate for the error in approximation. If we pick an initial "guess" x_0 for the fixed point x, then, by letting $m \to \infty$ in equation (7.1), we get

$$d(f^n(x_0), x) \leq d(f(x_0), x_0) \sum_{k=n}^{\infty} \alpha^k = d(f(x_0), x_0) \frac{\alpha^n}{1-\alpha}.$$

Example 7.14

Suppose that $f : [a, b] \to [a, b]$ is continuous on $[a, b]$, differentiable on (a, b), and has $|f'(x)| \leq \alpha < 1$ for all $a < x < b$. Then it follows from the mean value theorem that $|f(x) - f(y)| \leq \alpha|x - y|$ for all $x, y \in [a, b]$ and, hence, that f has a unique fixed point. See Figures 7.1 and 7.2.

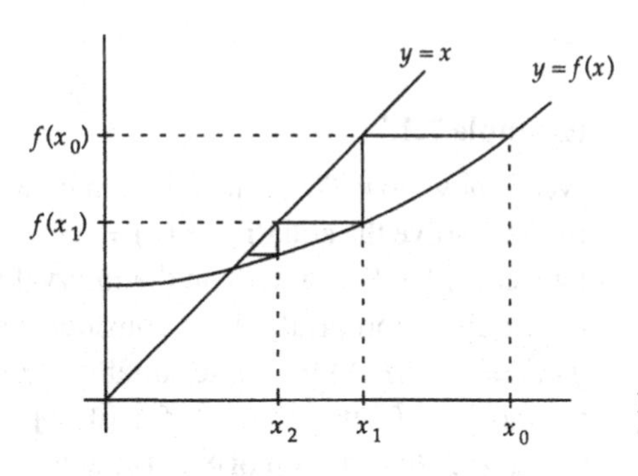

Figure 7.1

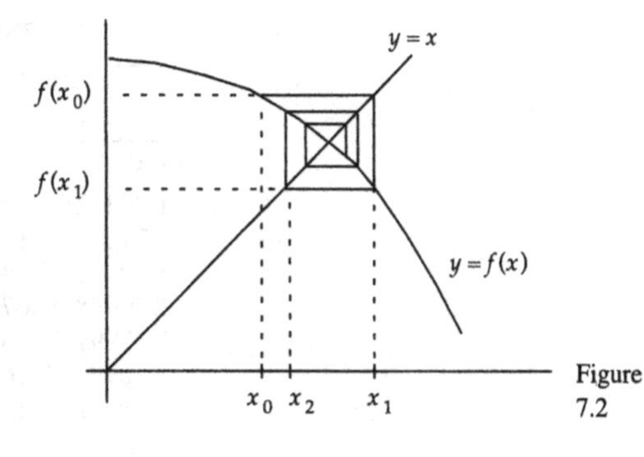

Figure 7.2

EXERCISES

36. The function $f(x) = x^2$ has two obvious fixed points: $p_0 = 0$ and $p_1 = 1$. Show that there is a $0 < \delta < 1$ such that $|f(x) - p_0| < |x - p_0|$ whenever $|x - p_0| < \delta, x \neq p_0$. Conclude that $f^n(x) \to p_0$ whenever $|x - p_0| < \delta, x \neq p_0$. This means that p_0 is an *attracting* fixed point for f; every orbit that starts out near 0 converges to 0. In contrast, find a $\delta > 0$ such that if $|x - p_1| < \delta, x \neq p_1$, then $|f(x) - p_1| > |x - p_1|$. This means that p_1 is a *repelling* fixed point for f; orbits that start out near 1 are pushed away from 1. In fact, given any $x \neq 1$, we have $f^n(x) \not\to 1$.

37. Suppose that $f : (a, b) \to (a, b)$ has a fixed point p in (a, b) and that f is differentiable at p. If $|f'(p)| < 1$, prove that p is an attracting fixed point for f. If $|f'(p)| > 1$, prove that p is a repelling fixed point for f.

38.

(a) Let $f(x) = \arctan x$. Show that $f'(0) = 1$ and that 0 is an attracting fixed point for f.

(b) Let $g(x) = x^3 + x$. Show that $g'(0) = 1$ and that 0 is a repelling fixed point for g.

(c) Let $h(x) = x^2 + 1/4$. Show that $h'(1/2) = 1$ and that $1/2$ is a fixed point for h that is neither attracting nor repelling.

39. The cubic $x^3 - x - 1$ has a unique real root x_0 with $1 < x_0 < 2$. Find it! [Hint: Iterating the function $f(x) = x^3 - 1$ won't work! Why?]

Example 7.15

We'll show how Theorem 7.13 can be used to find an estimate for, say, $\sqrt[3]{5}$. That is, we'll solve the equation $F(x) = x^3 - 5 = 0$. Now it is clear that $1 < \sqrt[3]{5} < 2$, so let's consider F as a map on the interval $[1, 2]$. And since the equation $F(x) = 0$ isn't quite appropriate, let's consider the equivalent equation $f(x) = x$, where $f(x) = x - \lambda F(x)$ for some suitably chosen $\lambda \in \mathbb{R}$. The claim is that it's possible to find $\lambda > 0$ such that (i) $f : [1, 2] \to [1, 2]$, and (ii) $|f'(x)| \leq \alpha < 1$ for $1 < x < 2$. In fact, a bit of experimentation will convince you that any $0 < \lambda < 1/6$ will do. Let's try $\lambda = 1/8$. Table 7.1 displays a few iterations of the scheme $x_{n+1} = f(x_n) = x_n - (x_n^3 - 5)/8$, starting with $x_0 = 1.5$. The last value is accurate

Table 7.1

x_n	$f(x_n)$
1.5	1.703125
1.703125	1.7106070518
1.7106070518	1.7099147854
1.7099147854	1.7099818467
1.7099818467	1.7099753773
1.7099753773	1.7099760016
1.7099760016	1.7099759414

to at least six places. Roughly speaking, each iteration increases the accuracy by one decimal place. Not bad.

EXERCISE

40. Extend the result in Example 7.15 as follows: Suppose that $F : [\,a, b\,] \to \mathbb{R}$ is continuous on $[\,a, b\,]$, differentiable in (a, b), and satisfies $F(a) < 0$, $F(b) > 0$, and $0 < K_1 \le F'(x) \le K_2$. Show that there is a unique solution to the equation $F(x) = 0$. [Hint: Consider the equation $f(x) = x$, where $f(x) = x - \lambda F(x)$ for some suitably chosen λ.]

Under suitable conditions on f, the same technique can be applied to the problem of existence and uniqueness of the solution to the initial value problem:

$$y' = f(x, y), \qquad y(0) = y_0.$$

For example, if f is continuous in some rectangle containing $(0, y_0)$ in its interior, and if f is Lipschitz in its second variable, $|f(x, y) - f(x, z)| \le K|y - z|$, for some constant K, then a unique solution exists – at least in some small neighborhood of $x = 0$. This fact was first observed by Lipschitz himself (hence the name Lipschitz condition), but Lipschitz did not have metric spaces at his disposal. Most modern proofs use some form of Banach's contraction mapping principle (often in the form of the method of successive approximations).

We will not give the full details of the proof here, but we will at least show how Banach's theorem enters the picture. For this we will want to rephrase the problem as a fixed-point problem on some complete metric space. First notice that by integrating both sides of the differential equation we get

$$y(x) = y_0 + \int_0^x f(t, y(t))\,dt \qquad (x \ge 0).$$

That is, we need a fixed point for the map $\varphi \mapsto F(\varphi)$, where

$$(F(\varphi))\,(x) = y_0 + \int_0^x f(t, \varphi(t))\,dt.$$

For simplicity, let's assume that f is defined and continuous on all of $\mathbb{R}^2$ (and still Lipschitz in its second variable). Then the integral on the right-hand side of this formula is well defined for any continuous function φ. Let's consider F as a map on $C[\,0, \delta\,]$, where $\delta > 0$ will be specified shortly. Next we'll check that F is a Lipschitz map on $C[\,0, \delta\,]$. For any $0 \le x \le \delta$, note that

$$\begin{aligned}
|(F(\varphi))\,(x) - (F(\psi))\,(x)| &= \left| \int_0^x f(t, \varphi(t))\,dt - \int_0^x f(t, \psi(t))\,dt \right| \\
&\le \int_0^x |f(t, \varphi(t)) - f(t, \psi(t))|\;dt \\
&\le K \int_0^x |\varphi(t) - \psi(t)|\;dt \\
&\le Kx \cdot \max_{0 \le t \le x} |\varphi(t) - \psi(t)| \\
&\le K\delta\; \|\varphi - \psi\|_\infty.
\end{aligned}$$

It follows that $\|F(\varphi) - F(\psi)\|_\infty \le K\delta\, \|\varphi - \psi\|_\infty$. Thus, F is a contraction on $C[0, \delta]$ provided that δ is chosen to satisfy $K\delta < 1$ and, in this case, F has a unique fixed point in $C[0, \delta]$.

Example 7.16

Consider the initial value problem $y' = 2x(1+y)$, $y(0) = 0$. By integrating both sides of the differential equation, we see that we need a function φ satisfying $\varphi(x) = \int_0^x 2t(1+\varphi(t))\,dt = \big(F(\varphi)\big)(x)$. The method of successive approximations amounts to taking an initial "guess" at the solution, say $\varphi_0 \equiv 0$, and iterating F. Thus, $\varphi_1(x) = \int_0^x 2t(1+0)\,dt = x^2$. Next, $\varphi_2(x) = \int_0^x 2t(1+t^2)\,dt = x^2 + x^4/2$. Another iteration would yield $\varphi_3(x) = x^2 + x^4/2 + x^6/6$. And so on. Finally, induction yields

$$\varphi(x) = \sum_{k=1}^{\infty} \frac{x^{2k}}{k!} = e^{x^2} - 1.$$

This solution is valid on all of $\mathbb{R}$ (and agrees, naturally, with the solution obtained by separation of variables).

EXERCISES

41. Let M be complete and let $f : M \to M$ be continuous. If f^k is a strict contraction for some integer $k > 1$, show that f has a unique fixed point.

42. Define $T : C[0, 1] \to C[0, 1]$ by $\big(T(f)\big)(x) = \int_0^x f(t)\,dt$. Show that T is *not* a strict contraction while T^2 is. What is the fixed point of T?

43. Show that each of the hypotheses of the contraction mapping principle is necessary by finding examples of a space M and a map $f : M \to M$ having no fixed point where:

(a) M is incomplete (but f is still a strict contraction).

(b) f satisfies only $d(f(x), f(y)) < d(x, y)$ for all $x \neq y$ (but M is still complete).

Completions

Completeness is a central theme in this book; it will return frequently. It may comfort you to know that every metric space can be "completed." In effect, this means that by tacking on a few "missing" limit points we can make an incomplete space complete. While the approach that we will take may not suggest anything so simple as adding a few points here and there, it is nevertheless the picture to bear in mind. In time, all will be made clear!

First, a definition. A metric space $(\hat{M}, \hat{d})$ is called a **completion** for (M, d) if

(i) $(\hat{M}, \hat{d})$ is complete, and

(ii) (M, d) is *isometric* to a *dense* subset of $(\hat{M}, \hat{d})$.

If M is already complete, then certainly $\hat{M} = M$ works. Except for this easy case, there is no obvious reason why completions should exist at all.

Formally, condition (ii) means that there is some map $i : M \to \hat{M}$ such that $d(x, y) = \hat{d}\big(i(x), i(y)\big)$ for all $x, y \in M$, and such that $i(M)$ is a dense subset of $\hat{M}$. Informally, condition (ii) says that we may regard M as an actual *subset* of $\hat{M}$ (in which case i is just the inclusion map from M into $\hat{M}$), that $\hat{d}\,|_{M\times M} = d$ (i.e., the relative metric that M inherits as a subset of $\hat{M}$ is just d), and that M is dense in $\hat{M}$.

The requirement that M is dense in $\hat{M}$ is added to insure *uniqueness* (more on this in a moment), but it is actually easy to come by. The real work comes in finding *any* complete space (N, ρ) that will accept M, isometrically, as a subset, for then we simply take $\hat{M} = \text{cl}_N M$. Notice that $\hat{M}$ is a closed subset of a complete space and hence is complete, and that M is clearly dense in $\hat{M}$.

Given a metric space M, we need to construct a complete space that is "big enough" to contain M isometrically. One way to accomplish this is to consider the collection of all bounded, real-valued functions on M. (This is roughly analogous to using the power set of M when looking for a *set* that is bigger than M.) Here's how we'll do it: Given any set M, we will define $\ell_\infty(M)$ to be the collection of all bounded, real-valued functions $f : M \to \mathbb{R}$, and we will define a norm on $\ell_\infty(M)$ in the obvious way:

$$\|f\|_\infty = \sup_{x\in M} |f(x)|.$$

This notation is consistent with that used for ℓ_∞ since, after all, a bounded sequence of real numbers is nothing other than a bounded *function* on $\mathbb{N}$. That is, $\ell_\infty = \ell_\infty(\mathbb{N})$.

The fact that $\|\cdot\|_\infty$ is a norm on $\ell_\infty(M)$ uses *the same proof* that we used for ℓ_∞. And the fact that $\ell_\infty(M)$ is complete under this norm again uses *the same proof* that we used for ℓ_∞. (See Exercises 18 and 44 and Exercise 3.21.) All of the fighting takes place in $\mathbb{R}$ and has little to do with the sets M or $\mathbb{N}$. It might help if you think of the "M" in $\ell_\infty(M)$ as simply an index set. Any index set with the same cardinality as M would suit our purposes just as well.

To find a completion for M, then, it suffices to show that (M, d) embeds isometrically into $\ell_\infty(M)$. Thus, each point $x \in M$ will have to correspond to some real-valued function on M. An obvious choice might be to associate each x with the function $t \mapsto d(x, t)$. Now this function is not necessarily bounded, but it is essentially the right choice. We just have a few details to tidy up.

Lemma 7.17. *Let (M, d) be any metric space. Then, M is isometric to a subset of $\ell_\infty(M)$.*

PROOF. Fix any point $a \in M$. To each $x \in M$ we associate an element $f_x \in \ell_\infty(M)$ by setting

$$f_x(t) = d(x, t) - d(a, t), \qquad t \in M.$$

Note that f_x is bounded since $|f_x(t)| = |d(x, t) - d(a, t)| \le d(x, a)$, a number that does not depend on t. That is, $\|f_x\|_\infty \le d(x, a)$. All that remains is to check that the correspondence $x \mapsto f_x$ is actually an isometry. But $\|f_x - f_y\|_\infty =$

$\sup_{t \in M} |d(x,t) - d(y,t)| \le d(x,y)$, from the triangle inequality, and $|d(x,t) - d(y,t)| = d(x,y)$ when $t = x$ or $t = y$. Thus, $\|f_x - f_y\|_\infty = d(x,y)$. □

Lemma 7.17 shows that M is identical to the subset $\{f_x : x \in M\}$ of $\ell_\infty(M)$. We may define a completion of M by taking $\hat{M}$ to be the *closure* of $\{f_x : x \in M\}$ in $\ell_\infty(M)$. Seem a bit complicated? Would it surprise you to learn that this completion is essentially the only one available? Well, prepare yourself!

Theorem 7.18. *If M_1 and M_2 are completions of M, then M_1 and M_2 are isometric.*

PROOF. For simplicity of notation, let's suppose that M is actually a *subset* of M_1 and M_2 (and dense in each, of course). This will make for fewer arrows to chase in the diagram below. The claim is that the identity on M "lifts" to an isometry f from M_1 onto M_2.

$$\begin{array}{ccc} M_1 & \xrightarrow{f} & M_2 \\ \cup & & \cup \\ M & \xrightarrow{I} & M \end{array}$$

Here's how. We will define $f : M_1 \to M_2$ through a series of observations. First, given $x \in M_1$, there is some sequence (x_n) in M such that $x_n \to x$ in M_1, because M is dense in M_1. In particular, (x_n) is Cauchy in M_1. But then (x_n) is also Cauchy in M_2. (Why? Recall that $(x_n) \subset M \subset M_2$.) Hence $x_n \to y$ in M_2, for some $y \in M_2$, because M_2 is complete. Now set $f(x) = y$. In other words, put $f(M_1\text{-}\lim x_n) = M_2\text{-}\lim I(x_n)$.

We first check that f is well defined. If (x_n) and (z_n) are sequences in M, and if both converge to x in M_1, then both must also converge to y in M_2 since

$$d_2(x_n, z_n) \;=\; d_1(x_n, z_n) \;=\; d(x_n, z_n) \to 0,$$

where we've written d_1 for the metric in M_1 and d_2 for the metric in M_2 (recall that both agree with d on pairs from M).

Now that we know that f is well defined, we also know that $f|_M = I$; that is, f is an extension of the identity on M. This is more or less obvious, since, if $x \in M$, we have the *constant* sequence, $x_n = x$ for all n, at our disposal.

Next let's check that f is onto. Given $y \in M_2$, there is some sequence (x_n) in M such that $x_n \to y$ in M_2 (because M is dense in M_2). But, just as before, this means that $x_n \to x$ in M_1 for some x. Clearly, $y = f(x)$.

Finally, we check that f is an isometry. Given $x, y \in M_1$, choose sequences (x_n), (y_n) in M such that $x_n \to x$ in M_1 and $y_n \to y$ in M_1. Then, $x_n \to f(x)$ in M_2 and $y_n \to f(y)$ in M_2. Consequently,

$$d_1(x,y) \;=\; \lim_{n\to\infty} d(x_n, y_n) \;=\; d_2(f(x), f(y)). \qquad \text{(Why?)} \quad \square$$

The proof of Theorem 7.18 allows us to make precise the notion of "adding on" a few points to make M complete. The points that are "added on" are limit points for entire collections of (nonconvergent) Cauchy sequences. Each point x in the completion $\hat{M}$ corresponds to the collection of *all* Cauchy sequences in M that converge to x; given one such Cauchy sequence (x_n), any other Cauchy sequence (y_n) in the same collection must be "equivalent" to (x_n) in the sense that $d(x_n, y_n) \to 0$. In fact, this is the standard construction; we define an equivalence relation on the class $\mathcal{C}$ of all Cauchy sequences in M by declaring (x_n) and (y_n) to be *equivalent* whenever $d(x_n, y_n) \to 0$. The completion of M, then, is the set of *equivalence classes* of $\mathcal{C}$ under this relation.

In the next chapter we will use a technique that is similar to the one used in the proof of Theorem 7.18 to construct extensions for maps other than isometries. The key ingredients will still be a dense domain of definition and the preservation of Cauchy sequences.

EXERCISES

Except where noted, M is an arbitrary metric space with metric d.

44. Give any set M, check that $\ell_\infty(M)$ is a complete normed vector space.

45. If M and N are equivalent sets, show that $\ell_\infty(M)$ and $\ell_\infty(N)$ are isometric. [Hint: If $g : N \to M$ is any map, then $f \mapsto f \circ g$ defines a map from $\ell_\infty(M)$ to $\ell_\infty(N)$. How does this help?]

46. If A is a dense subset of a metric space (M, d), show that (A, d) and (M, d) have the same completion (isometrically). [Hint: If $\hat{M}$ is the completion for M, then A is dense is $\hat{M}$. Why?]

47. A function $f : (M, d) \to (N, \rho)$ is said to be **uniformly continuous** if f is continuous and if, given $\varepsilon > 0$, there is always a *single* $\delta > 0$ such that $\rho(f(x), f(y)) < \varepsilon$ for any $x, y \in M$ with $d(x, y) < \delta$. That is, δ is allowed to depend on f and ε but not on x or y. Prove that any Lipschitz map is uniformly continuous.

48. Prove that a uniformly continuous map sends Cauchy sequences into Cauchy sequences.

49. Suppose that $f : \mathbb{Q} \to \mathbb{R}$ is Lipschitz. Prove that f extends uniquely to a continuous function $g : \mathbb{R} \to \mathbb{R}$. [Hint: Given $x \in \mathbb{R}$, define $g(x) = \lim_{n\to\infty} f(r_n)$, where (r_n) is a sequence of rationals converging to x.]

50. Given a point $a \in M$ and a subset $A \subset M$, show that each of the functions $x \mapsto d(x, a)$ and $x \mapsto d(x, A)$ are uniformly continuous.

51. Two metric spaces (M, d) and (N, ρ) are said to be **uniformly homeomorphic** if there is a one-to-one and onto map $f : M \to N$ such that both f and f^{-1} are *uniformly* continuous. In this case we say that f is a **uniform homeomorphism**. Prove that completeness is preserved by uniform homeomorphisms.

Just as we have solved one problem, we have raised another. We now know that every metric space has a unique completion (at least if we agree to identify isometric spaces). But suppose that the incomplete metric space that we start with carries some extra structure. Say that we need the completion of an incomplete normed *vector space*, for example. Will we have to give up the vector space structure to gain completeness? In other words, is the completion of a normed vector space still a normed vector space? In still other words, could the completion be more trouble than its worth?

Luck is with us on this question; the completion of a normed vector space is indeed a Banach space. The proof is not terribly hard, but it is rather tedious, with lots of details to verify. The key steps, though, are easy to describe.

Given a normed vector space X and its completion $\hat{X}$, we need to suitably define both addition and scalar multiplication on $\hat{X}$ (and check that $\hat{X}$ is a vector space under these), and we have to define a suitable norm on $\hat{X}$. So, suppose that we are handed x, $y \in \hat{X}$, and scalars $\alpha, \beta \in \mathbb{R}$. How do we define $\alpha x + \beta y$? Well, choose sequences (x_n), (y_n) in X such that $x_n \to x$ and $y_n \to y$ in $\hat{X}$, and define

$$\alpha x + \beta y = \lim_{n\to\infty} (\alpha x_n + \beta y_n).$$

(This makes sense because $(\alpha x_n + \beta y_n)$ is Cauchy in X.) After checking that this definition turns $\hat{X}$ into a vector space, there is only one reasonable choice for a norm on $\hat{X}$. We would set

$$\|x\| = \hat{d}(x, 0) = \lim_{n\to\infty} d(x_n, 0) = \lim_{n\to\infty} \|x_n\|$$

and check that this is actually a norm on $\hat{X}$. (If so, then it has to be complete – that is already determined by $\hat{d}$.) In this setting, X is a dense *linear subspace* of $\hat{X}$.

Notes and Remarks

Fréchet introduced complete metric spaces in his thesis, Fréchet [1906], while Hausdorff coined the term totally bounded. But much of what is in this chapter has its roots in Cantor's work: The nested set theorem for $\mathbb{R}$, a special case of Theorem 7.11(ii), is generally credited to Cantor. The metric space version is due to Fréchet.

For more on the result in Exercise 30, see Kelley [1955]. Exercise 38 is taken from Gulick [1992]. Examples 7.14 and 7.15, along with Exercise 40, are based on the presentation in Kolmogorov and Fomin [1970]. Exercise 39 is adapted from an entertaining article by Cannon and Elich [1993]. For more applications of functional iteration and its relation to chaos and fractals, see Barnsley [1988], Devaney [1992], and Edgar [1990]. For a historical survey of functional iteration, see D. F. Bailey [1989].

Picard's theorem appears in Picard [1890]. Banach's observation on completeness for normed linear spaces (Theorem 7.12) and the contraction mapping principle (Theorem 7.13) are from his thesis, Banach [1922]. You will find even more applications of Banach's contraction mapping theorem in Copson [1968], including proofs of the inverse and implicit function theorems. For an interesting application to "crinkly" curves, see Katsuura [1991].

For a brief survey of some of fixed point theory's "greatest hits," see Shaskin [1991]. Fixed point theory remains a hot research area; for a look at some of the recent developments, see Goebel and Kirk [1990].

It was Hausdorff who first showed that every metric space has a completion, and his proof is based on what he calls the Cantor–Méray theorem (the description of the irrationals in terms of Cauchy sequences of rationals). The proof given here is a hybrid; Lemma 7.17 is based on a proof given in Kuratowski [1935] (but see also Fréchet [1928] and Kaplansky [1977]) while Theorem 7.18 (and the subsequent remarks) follows the lines of Hausdorff's original proof (see, for example, Hausdorff [1937]). Note that the function f_x used in the proof of Lemma 7.17 is actually a continuous function on M – we will use this observation later to show that (under certain circumstances) M embeds isometrically into $C(M)$, the space of continuous real-valued functions on M.

We will have much more to say about uniform continuity (Exercise 47) and uniform homeomorphisms (Exercise 51) in the next chapter.

CHAPTER EIGHT

Compactness

Compact Metric Spaces

A metric space (M, d) is said to be **compact** if it is both complete and totally bounded. As you might imagine, a compact space is the best of all possible worlds.

Examples 8.1

(a) A subset K of $\mathbb{R}$ is compact if and only if K is closed and bounded. This fact is usually referred to as the Heine–Borel theorem. Hence, a closed bounded interval $[a, b]$ is compact. Also, the Cantor set Δ is compact. The interval $(0, 1)$, on the other hand, is *not* compact.

(b) A subset K of $\mathbb{R}^n$ is compact if and only if K is closed and bounded. (Why?)

(c) It is important that we not confuse the first two examples with the general case. Recall that the set $\{e_n : n \geq 1\}$ is closed and bounded in ℓ_∞ but not totally bounded – hence not compact. Taking this a step further, notice that the closed ball $\{x : \|x\|_\infty \leq 1\}$ in ℓ_∞ is not compact, whereas any closed ball in $\mathbb{R}^n$ is compact.

(d) A subset of a discrete space is compact if and only if it is *finite*. (Why?)

Just as with completeness and total boundedness, we will want to give several equivalent characterizations of compactness. In particular, since neither completeness nor total boundedness is preserved by homeomorphisms, our newest definition does not appear to be describing a topological property. Let's remedy this immediately by giving a sequential characterization of compactness that will turn out to be invariant under homeomorphisms.

Theorem 8.2. *(M, d) is compact if and only if every sequence in M has a subsequence that converges to a point in M.*

PROOF.

$$\left\{\begin{array}{c}\text{totally bounded}\\ +\\ \text{complete}\end{array}\right\} \iff \left\{\begin{array}{c}\text{every sequence in } M \text{ has}\\ \text{a Cauchy subsequence}\\ +\\ \text{Cauchy sequences converge}\end{array}\right\}. \quad \square$$

It is easy to believe that compactness is a valuable property for an analyst to have available. Convergent sequences are easy to come by in a compact space; no fussing with difficult prerequisites here! If you happen on a nonconvergent sequence, just extract a subsequence that does converge and use that one instead. You couldn't ask for more!

Given a compact space, it is easy to decide which of its subsets are compact:

Corollary 8.3. *Let A be a subset of a metric space M. If A is compact, then A is closed in M. If M is compact and A is closed, then A is compact.*

PROOF. Suppose that A is compact, and let (x_n) be a sequence in A that converges to a point $x \in M$. Then, from Theorem 8.2, (x_n) has a subsequence that converges in A, and hence we must have $x \in A$. Thus, A is closed.

Next, suppose that M is compact and that A is closed in M. Given an arbitrary sequence (x_n) in A, Theorem 8.2 supplies a subsequence of (x_n) that converges to a point $x \in M$. But since A is closed, we must have $x \in A$. Thus, A is compact. □

EXERCISES

Unless otherwise stated, (M, d) denotes a generic metric space.

▷ **1.** If K is a nonempty compact subset of $\mathbb{R}$, show that $\sup K$ and $\inf K$ are elements of K.

▷ **2.** Let $E = \{x \in \mathbb{Q} : 2 < x^2 < 3\}$, considered as a subset of $\mathbb{Q}$ (with its usual metric). Show that E is closed and bounded but *not* compact.

3. If A is compact in M, prove that $\text{diam}(A)$ is finite. Moreover, if A is nonempty, show that there exist points x and y in A such that $\text{diam}(A) = d(x, y)$.

4. If A and B are compact sets in M, show that $A \cup B$ is compact.

5. True or false? M is compact if and only if every closed ball in M is compact.

6. If A is compact in M and B is compact in N, show that $A \times B$ is compact in $M \times N$ (see Exercise 3.46).

7. If K is a compact subset of $\mathbb{R}^2$, show that $K \subset [a, b] \times [c, d]$ for some pair of compact intervals $[a, b]$ and $[c, d]$.

8. Prove that the set $\{x \in \mathbb{R}^n : \|x\|_1 = 1\}$ is compact in $\mathbb{R}^n$ under the Euclidean norm.

9. Prove that (M, d) is compact if and only if every infinite subset of M has a limit point.

10. Show that the Heine–Borel theorem (closed, bounded sets in $\mathbb{R}$ are compact) implies the Bolzano–Weierstrass theorem. Conclude that the Heine–Borel theorem is equivalent to the completeness of $\mathbb{R}$.

11. Prove that compactness is not a relative property. That is, if K is compact in M, show that K is compact in *any* metric space that contains it (isometrically).

12. Show that the set $A = \{x \in \ell_2 : |x_n| \le 1/n,\ n = 1, 2, \ldots\}$ is compact in ℓ_2. [Hint: First show that A is closed. Next, use the fact that $\sum_{n=1}^{\infty} 1/n^2 < \infty$ to show that A is "within ε" of the set $A \cap \{x \in \ell_2 : |x_n| = 0,\ n \ge N\}$.]

13. Given $c_n \ge 0$ for all n, prove that the set $\{x \in \ell_2 : |x_n| \le c_n,\ n \ge 1\}$ is compact in ℓ_2 if and only if $\sum_{n=1}^{\infty} c_n^2 < \infty$.

14. Show that the Hilbert cube H^∞ (Exercise 3.10) is compact. [Hint: First show that H^∞ is complete (Exercise 7.24). Now, given $\varepsilon > 0$, choose N so that $\sum_{n=N}^{\infty} 2^{-n} < \varepsilon$ and argue that H^∞ is "within ε" of the set $\{x \in H^\infty : |x_n| = 0$ for $n \ge N\}$.]

15. If A is a totally bounded subset of a complete metric space M, show that $\bar{A}$ is compact in M. For this reason, totally bounded sets are sometimes called *precompact* or *conditionally compact.* In fact, any set with compact closure might be labeled precompact.

16. Show that a metric space M is totally bounded if and only if its completion $\hat{M}$ is compact.

▷ **17.** If M is compact, show that M is also separable.

18. A collection (U_α) of open sets is called an *open base* for M if every open set in M can be written as a union of the U_α. For example, the collection of all open intervals in $\mathbb{R}$ with *rational* endpoints is an open base for $\mathbb{R}$ (and this is even a countable collection). (Why?) Prove that M has a countable open base if and only if M is separable. [Hint: If $\{x_n\}$ is a countable dense set in M, consider the collection of open balls with rational radii centered at the x_n.]

19. Prove that M is separable if and only if M is homeomorphic to a totally bounded metric space (specifically, a subset of the Hilbert cube). [Hint: See Exercise 4.49.]

To show that compactness is indeed a topological property, let's show that the continuous image of a compact set is again compact:

Theorem 8.4. *Let $f : (M, d) \to (N, \rho)$ be continuous. If K is compact in M, then $f(K)$ is compact in N.*

PROOF. Let (y_n) be a sequence in $f(K)$. Then, $y_n = f(x_n)$ for some sequence (x_n) in K. But, since K is compact, (x_n) has a convergent subsequence, say, $x_{n_k} \to x \in K$. Then, since f is continuous, $y_{n_k} = f(x_{n_k}) \to f(x) \in f(K)$. Thus, $f(K)$ is compact. □

Theorem 8.4 gives us a wealth of useful information. In particular, it tells us that real-valued continuous functions on compact spaces are quite well behaved:

Corollary 8.5. *Let (M, d) be compact. If $f : M \to \mathbb{R}$ is continuous, then f is* bounded. *Moreover, f* attains *its maximum and minimum values.*

PROOF. $f(M)$ is compact in $\mathbb{R}$; hence it is closed and bounded. Moreover, $\sup f(M)$ and $\inf f(M)$ are actually elements of $f(M)$. (Why?) That is, there exist $x, y \in M$ such that $f(x) \le f(t) \le f(y)$ for all $t \in M$. (In this case we would write $f(x) = \min_{t \in M} f(t)$ and $f(y) = \max_{t \in M} f(t)$.) □

Corollary 8.6. *If $f : [a, b] \to \mathbb{R}$ is continuous, then the range of f is a compact interval $[c, d]$ for some $c, d \in \mathbb{R}$.*

Corollary 8.7. *If M is a compact metric space, then $\|f\|_\infty = \max_{t \in M} |f(t)|$ defines a norm on $C(M)$, the vector space of continuous real-valued functions on M.*

EXERCISES

Throughout, M denotes a metric space with metric d.

▷ **20.** Let E be a noncompact subset of $\mathbb{R}$. Find a continuous function $f : E \to \mathbb{R}$ that is (i) not bounded; (ii) bounded but has no maximum value.

21. Prove Corollary 8.6.

22. If M is compact and $f : M \to N$ is continuous, prove that f is a closed map.

▷ **23.** Suppose that M is compact and that $f : M \to N$ is continuous, one-to-one, and onto. Prove that f is a homeomorphism.

24. Let $f : [0, 1] \to [0, 1] \times [0, 1]$ be continuous and one-to-one. Show that f cannot be onto. Moreover, show that the range of f is *nowhere dense* in $[0, 1] \times [0, 1]$. [Hint: The range of f is closed (why?); if it has nonempty interior, then it contains a closed rectangle. Argue that this rectangle is the image of some subinterval of $[0, 1]$.]

25. Let V be a normed vector space, and let $x \ne y \in V$. Show that the map $f(t) = x + t(y - x)$ is a homeomorphism from $[0, 1]$ into V. The range of f is the line segment joining x and y; it is often written $[x, y]$.

26. If $f : \mathbb{R} \to \mathbb{R}$ is both continuous and open, show that f is strictly monotone.

27. Given $f : [a, b] \to \mathbb{R}$, define $G : [a, b] \to \mathbb{R}^2$ by $G(x) = (x, f(x))$ (the range of G is the *graph* of f). Prove that the following are equivalent: (i) f is continuous; (ii) G is continuous; (iii) the graph of f is a compact subset of $\mathbb{R}^2$. [Hint: f is continuous if, whenever $x_n \to x$, there is a *subsequence* of $(f(x_n))$ that converges to $f(x)$. Why?]

28. Let $f : [a, b] \to [a, b]$ be continuous. Show that f has a fixed point. Try to prove this without appealing to the intermediate value theorem. [Hint: Consider the function $g(x) = |x - f(x)|$.]

29. Let M be a compact metric space and suppose that $f : M \to M$ satisfies $d(f(x), f(y)) < d(x, y)$ whenever $x \neq y$. Show that f has a fixed point. [Hint: First note that f is continuous; next, consider $g(x) = d(x, f(x))$.]

Corollary 8.7 would seem to suggest that compactness is the analogue of "finite" that we talked about at the end of Chapter Five. To better appreciate this, we will need a slightly more esoteric characterization of compactness. A bit of preliminary detail-checking will ease the transition.

Lemma 8.8. *In a metric space M, the following are equivalent:*

(a) *If $\mathcal{G}$ is any collection of open sets in M with $\bigcup\{G : G \in \mathcal{G}\} \supset M$, then there are finitely many sets $G_1, \ldots, G_n \in \mathcal{G}$ with $\bigcup_{i=1}^{n} G_i \supset M$.*

(b) *If $\mathcal{F}$ is any collection of closed sets in M such that $\bigcap_{i=1}^{n} F_i \neq \emptyset$ for all choices of finitely many sets $F_1, \ldots, F_n \in \mathcal{F}$, then $\bigcap\{F : F \in \mathcal{F}\} \neq \emptyset$.*

The proof of Lemma 8.8 is left as an exercise; as you might guess, De Morgan's laws do all of the work. The first condition is usually paraphrased by saying, in less than perfect English, "every *open cover* has a *finite subcover*." The second condition is abbreviated by saying "every collection of closed sets with the *finite intersection property* has nonempty intersection." These may at first seem to be unwieldy statements to work with, but each is worth the trouble. Here's why we care: Condition (a) implies that M is *totally bounded* because, for any $\varepsilon > 0$, the collection $\mathcal{G} = \{B_\varepsilon(x) : x \in M\}$ is an open cover for M. Condition (b) implies that M is *complete* because it easily implies the nested set theorem (if $F_1 \supset F_2 \supset \cdots$ are nonempty, then $\bigcap_{i=1}^{n} F_i = F_n \neq \emptyset$). Put the two together and we've got our new characterization of compactness.

Theorem 8.9. *M is compact if and only if it satisfies either* (*hence both*) 8.8 (a) *or* 8.8 (b).

PROOF. As noted above, conditions 8.8 (a) and 8.8 (b) imply that M is totally bounded and complete, hence compact. So we need to show that compactness will imply, say, 8.8 (a). To this end, suppose that M is compact, and suppose that $\mathcal{G}$ is an open cover for M that admits *no* finite subcover. We will work toward a contradiction.

Now M is totally bounded, so M can be covered by finitely many *closed* sets of diameter at most 1. It follows that at least one of these, call it A_1, cannot be covered by finitely many sets from $\mathcal{G}$. Certainly $A_1 \neq \emptyset$ (since the empty set is easy to cover!). Note that A_1 must be infinite.

Next, A_1 is totally bounded, so A_1 can be covered by finitely many closed sets of diameter at most $1/2$. At least one of these, call it A_2, cannot be covered by finitely many sets from $\mathcal{G}$.

Continuing, we get a decreasing sequence $A_1 \supset A_2 \supset \cdots \supset A_n \supset \cdots$, where A_n is closed, nonempty (infinite, actually), has $\operatorname{diam} A_n \leq 1/n$, and cannot be covered by finitely many sets from $\mathcal{G}$.

Now here's the fly in the ointment! Let $x \in \bigcap_{n=1}^{\infty} A_n$ ($\neq \emptyset$, because M is complete). Then, $x \in G \in \mathcal{G}$ for some G (since $\mathcal{G}$ is an open cover) and so, since G is open, $x \in B_\varepsilon(x) \subset G$ for some $\varepsilon > 0$. But for any n with $1/n < \varepsilon$ we would then have $x \in A_n \subset B_\varepsilon(x) \subset G$. That is, A_n is covered by a single set from $\mathcal{G}$. This is the contradiction that we were looking for. □

Just look at the tidy form that the nested set theorem takes on in a compact space:

Corollary 8.10. *M is compact if and only if every decreasing sequence of nonempty closed sets has nonempty intersection; that is, if and only if, whenever $F_1 \supset F_2 \supset \cdots$ is a sequence of nonempty closed sets in M, we have $\bigcap_{n=1}^{\infty} F_n \neq \emptyset$.*

PROOF. The forward implication is clear from Theorem 8.9. So, suppose that every nested sequence of nonempty closed sets in M has nonempty intersection, and let (x_n) be a sequence in M. Then there is some point x in the nonempty set $\bigcap_{n=1}^{\infty} \overline{\{x_k : k \geq n\}}$. (Why?) It follows that some subsequence of (x_n) must converge to x. □

Note that we no longer need to assume that the diameters of the sets F_n tend to zero; hence, $\bigcap_{n=1}^{\infty} F_n$ may contain more than one point.

Corollary 8.11. *M is compact if and only if every* countable *open cover admits a finite subcover.* (Why?)

EXERCISES

Except where noted, M is an arbitrary metric space with metric d.

▷ **30.** Prove Lemma 8.8.

31. Given an arbitrary metric space M, show that a decreasing sequence of nonempty *compact* sets in M has nonempty intersection.

32. Prove Corollary 8.11 by showing that the following two statements are equivalent.

(i) Every decreasing sequence of nonempty closed sets in M has nonempty intersection.

(ii) Every countable open cover of M admits a finite subcover; that is, if (G_n) is a sequence of open sets in M satisfying $\bigcup_{n=1}^{\infty} G_n \supset M$, then $\bigcup_{n=1}^{N} G_n \supset M$ for some (finite) N.

33. Let (M, d) be compact. Suppose that (F_n) is a decreasing sequence of nonempty closed sets in M, and that $\bigcap_{n=1}^{\infty} F_n$ is contained in some open set G. Show that $F_n \subset G$ for all but finitely many n.

34. Let A be a subset of a metric space M. Prove that A is closed in M if and only if $A \cap K$ is compact for every compact set K in M. [Hint: If (x_n) converges to x, then $\{x\} \cup \{x_n : n \geq 1\}$ is compact. (Why?)]

35. Let $\mathcal{G}$ be an open cover for M. We say that $\varepsilon > 0$ is a *Lebesgue number* for $\mathcal{G}$ if each subset of M of diameter $<\varepsilon$ is contained in some $G \in \mathcal{G}$. If M is compact, show that every open cover of M has a Lebesgue number. [Hint: If not, there exists a set E_n in M with $\text{diam}(E_n) < 1/n$ such that E_n is not contained in any $G \in \mathcal{G}$.]

36. Let F and K be disjoint, nonempty subsets of a metric space M with F closed and K compact. Show that $d(F, K) = \inf\{d(x, y) : x \in F, y \in K\} > 0$. Show that this may fail if we assume only that F and K are disjoint closed sets.

37. A real-valued function f on a metric space M is called *lower semicontinuous* if, for each real α, the set $\{x \in M : f(x) > \alpha\}$ is open in M. Prove that f is lower semicontinuous if and only if $f(x) \leq \liminf_{n\to\infty} f(x_n)$ whenever $x_n \to x$ in M.

38. If M is compact, prove that every lower semicontinuous function on M is bounded below and attains a minimum value.

39. A function $f : M \to \mathbb{R}$ is called *upper semicontinuous* if $-f$ is lower semicontinuous. Formulate the analogues of Exercises 37 and 38 for upper semicontinuous functions.

40. Let M be compact and let $f : M \to M$ satisfy $d(f(x), f(y)) = d(x, y)$ for all $x, y \in M$. Show that f is onto. [Hint: If $B_\varepsilon(x) \cap f(M) = \emptyset$, consider the sequence $(f^n(x))$.]

41. Is compactness necessary in Exercise 40? That is, is it possible for a metric space to be isometric to a proper subset of itself? Explain.

42. Let M be compact and let $f : M \to M$ satisfy $d(f(x), f(y)) \geq d(x, y)$ for all $x, y \in M$. Prove that f is an isometry of M onto itself. [Hint: First, given $x \in M$, consider $x_n = f^n(x)$. By passing to a subsequence, if necessary, we may suppose that (x_n) converges. Argue that $x_n \to x$. Next, given $x, y \in M$, show that we must have $d(f(x), f(y)) = d(x, y)$. Thus, f is an isometry into M. Finally, argue that f has dense range.]

43. Let M be compact and suppose that $f : M \to M$ is one-to-one, onto, and satisfies $d(f(x), f(y)) \leq d(x, y)$ for all $x, y \in M$. Prove that f is an isometry of M onto itself. [Hint: Exercise 42.]

Uniform Continuity

As it happens, continuous functions on compact spaces turn out to be more than simply continuous. To better appreciate this, let's first consider an easy example:

Example 8.12

The map $f : (0, 1) \to \mathbb{R}$ given by $f(x) = 1/x$ is continuous. But f does not map nearby x to nearby $f(x)$; for example, note that

$$\left|\frac{1}{n} - \frac{1}{n+1}\right| \to 0 \qquad \text{while} \qquad \left|f\left(\frac{1}{n}\right) - f\left(\frac{1}{n+1}\right)\right| = 1 \not\to 0.$$

What's going on?

We cannot overlook the fact that continuity is a pointwise phenomenon; that is, $f : M \to N$ is continuous if it is continuous *at each point* $x \in M$. And so, given $\varepsilon > 0$, the δ that "works" for one x may not work so well for another. That is, δ typically depends on x too. A shorthand reminder will help explain the situation:

$$\underbrace{\forall x \in M}_{} \quad \forall \varepsilon > 0 \quad \exists \delta(x, \varepsilon) > 0 \quad \text{such that}\ldots$$

we want to move this forward!

The question is, can we find a δ that *does not* depend on x? If so, f is called *uniformly* continuous, because a single δ "works" uniformly for all x.

Examples 8.13

(a) A Lipschitz map $f : \mathbb{R} \to \mathbb{R}$ is uniformly continuous. If f satisfies $|f(x) - f(y)| \le K|x - y|$ for all x, y, then, given any ε, the choice $\delta = \varepsilon/K$ always "works."

(b) Recall that $|\sqrt{x} - \sqrt{y}\,| \le \sqrt{|x - y|}$ holds for any $x, y \ge 0$. It follows that $f(x) = \sqrt{x}$ is uniformly continuous on $[0, \infty)$, because $\delta = \varepsilon^2$ "works" for any $\varepsilon > 0$. Note, however, that f is not Lipschitz on $[0, \infty)$, because $\sqrt{x}/x = 1/\sqrt{x} \to \infty$ as $x \to 0^+$.

It's time we gave a formal definition: We say that $f : (M, d) \to (N, \rho)$ is **uniformly continuous** if

$$\left\{\begin{array}{l}\text{for every } \varepsilon > 0 \text{ there is a } \delta > 0 \text{ (which may depend on } f \text{ and } \varepsilon) \\ \text{such that } \rho(f(x), f(y)) < \varepsilon \text{ whenever } x, y \in M \text{ satisfy } d(x, y) < \delta.\end{array}\right.$$

We can easily change this to read: f is uniformly continuous if, given $\varepsilon > 0$, there is a $\delta > 0$ such that $f\left(B_\delta^d(x)\right) \subset B_\varepsilon^\rho(f(x))$ for any $x \in M$. (Note that a uniformly continuous map is continuous – but not conversely.) Here's a picturesque rephrasing of this definition:

$$\left\{\begin{array}{l}f \text{ is uniformly continuous if (and only if), for every } \varepsilon > 0, \text{ there is a } \delta > 0 \\ \text{such that } \operatorname{diam}_N f(A) < \varepsilon \text{ whenever } A \subset M \text{ satisfies } \operatorname{diam}_M(A) < \delta. \text{ (Why?)}\end{array}\right.$$

It follows that a uniformly continuous map f sends Cauchy sequences into Cauchy sequences. (Why?)

EXERCISES

Except where noted, M is an arbitrary metric space with metric d.

▷ **44.** Show that any Lipschitz map $f : (M, d) \to (N, \rho)$ is uniformly continuous. In particular, any isometry is uniformly continuous.

45. Prove that *every* map $f : \mathbb{N} \to \mathbb{R}$ is uniformly continuous.

46. Show that $|d(x, z) - d(y, z)| \le d(x, y)$ and conclude that the map $x \mapsto d(x, z)$ is uniformly continuous on M for each fixed $z \in M$.

47. Given a nonempty subset A of M, show that $|d(x, A) - d(y, A)| \le d(x, y)$ and conclude that the map $x \mapsto d(x, A)$ is uniformly continuous on M.

▷ **48.** Prove that a uniformly continuous map sends Cauchy sequences into Cauchy sequences.

49. Show that the sum of uniformly continuous maps is uniformly continuous. Is the product of uniformly continuous maps always uniformly continuous? Explain.

50. If f is uniformly continuous on $(0, 2)$ and on $(1, 3)$, is f uniformly continuous on $(0, 3)$? If f is uniformly continuous on $[\,n, n+1\,]$ for every $n \in \mathbb{Z}$, is f necessarily uniformly continuous on $\mathbb{R}$? Explain.

51. If $f : (0, 1) \to \mathbb{R}$ is uniformly continuous, show that $\lim_{x \to 0^+} f(x)$ exists. Conclude that f is bounded on $(0, 1)$.

52. Given $f : \mathbb{R} \to \mathbb{R}$ and $a \in \mathbb{R}$, define $F(x) = [f(x) - f(a)]/(x - a)$ for $x \ne a$. Prove that f is differentiable at a if and only if F is uniformly continuous in some punctured neighborhood of a.

53. Suppose that $f : \mathbb{R} \to \mathbb{R}$ is continuous and that $f(x) \to 0$ as $x \to \pm\infty$. Prove that f is uniformly continuous.

▷ **54.** Let E be a bounded, noncompact subset of $\mathbb{R}$. Show that there is a continuous function $f : E \to \mathbb{R}$ that is not uniformly continuous.

▷ **55.** Give an example of a bounded continuous map $f : \mathbb{R} \to \mathbb{R}$ that is not uniformly continuous. Can an unbounded continuous function $f : \mathbb{R} \to \mathbb{R}$ be uniformly continuous? Explain.

56. Prove that $f : (M, d) \to (N, \rho)$ is uniformly continuous if and only if $\rho(f(x_n), f(y_n)) \to 0$ for any pair of sequences (x_n) and (y_n) in M satisfying $d(x_n, y_n) \to 0$. [Hint: For the backward implication, assume that f is *not* uniformly continuous and work toward a contradiction.]

▷ **57.** A function $f : \mathbb{R} \to \mathbb{R}$ is said to satisfy a *Lipschitz condition of order* α, where $\alpha > 0$, if there is a constant $K < \infty$ such that $|f(x) - f(y)| \le K|x - y|^\alpha$ for all x, y. Prove that such a function is uniformly continuous.

▷ **58.** Show that any function $f : \mathbb{R} \to \mathbb{R}$ having a bounded derivative is Lipschitz of order 1. [Hint: Use the mean value theorem.]

59. The Lipschitz condition is interesting only for $\alpha \le 1$; show that a function satisfying a Lipschitz condition of order $\alpha > 1$ is *constant*.

60. Show that x^α is uniformly continuous on $(0, \infty)$ if and only if $0 \le \alpha \le 1$. [Hint: For $0 < \alpha \le 1$, show that x^α is Lipschitz of order α. Next, if $\alpha = 2$, for example, notice that $\sqrt{n+1} - \sqrt{n} \to 0$ as $n \to \infty$. How does this help?]

61. Two metric spaces (M, d) and (N, ρ) are said to be **uniformly homeomorphic** if there is a one-to-one and onto map $f : M \to N$ such that both f and f^{-1} are *uniformly* continuous. In this case we say that f is a **uniform homeomorphism**. Prove that completeness is preserved by uniform homeomorphisms.

62. Two metrics d and ρ on a set M are said to be **uniformly equivalent** if the identity map between (M, d) and (M, ρ) is uniformly continuous in both directions (i.e., if the identity map is a uniform homeomorphism). If there are constants $0 < c$, $C < \infty$ such that $c\rho(x, y) \le d(x, y) \le C\rho(x, y)$ for every pair of points $x, y \in M$, prove that d and ρ are uniformly equivalent.

63. Let $d(x, y) = \|x - y\|_2$ be the usual (Euclidean) metric on $\mathbb{R}^2$, and define a second metric ρ on $\mathbb{R}^2$ by

$$\rho(x, y) = \frac{\|x - y\|_2}{\left(1 + \|x\|_2^2\right)^{1/2}\left(1 + \|y\|_2^2\right)^{1/2}}.$$

Show that d and ρ are equivalent but not uniformly equivalent.

64. Show that the metric $\rho = d/(1 + d)$ is always uniformly equivalent to d, but that there are examples in which the inequality $c\rho \le d \le C\rho$ may fail to hold (for all x, y).

It follows from our earlier observations that a uniformly continuous function maps sets of small diameter into sets of small diameter. But even more is true:

Proposition 8.14. *If $f : M \to N$ is uniformly continuous, then f maps totally bounded sets into totally bounded sets.*

PROOF. Let $A \subset M$ be totally bounded and let $\varepsilon > 0$. Since f is uniformly continuous, there is a $\delta > 0$ so that $f\left(B_\delta^d(x)\right) \subset B_\varepsilon^\rho(f(x))$ for any $x \in M$. Next, since A is totally bounded, $A \subset \bigcup_{i=1}^n B_\delta^d(x_i)$ for some $x_1, \ldots, x_n \in M$. Combining these observations yields $f(A) \subset \bigcup_{i=1}^n B_\varepsilon^\rho(f(x_i))$. Hence, $f(A)$ is totally bounded. □

We can push this further still. If the domain space M is compact, then *every* continuous function on M is actually uniformly continuous:

Theorem 8.15. *If M is a compact metric space, then every continuous map $f : M \to N$ is uniformly continuous.*

PROOF. Let $\varepsilon > 0$. For each $x \in M$, let $\delta_x > 0$ be chosen such that $\rho(f(x), f(y)) < \varepsilon$ whenever y satisfies $d(x, y) < \delta_x$. If we should be so lucky as to have $\inf_x \delta_x > 0$, then we are done. (Why?) Otherwise, we want to reduce to finitely many δ_x and take their minimum.

Now the collection $\{B_{\delta_x/2}(x) : x \in M\}$ is an open cover for M and so there are finitely many points $x_1, \ldots, x_k \in M$ such that $M \subset \bigcup_{i=1}^k B_{\eta_i}(x_i)$, where $\eta_i = \delta_{x_i}/2$. This is the reduction to finitely many δ_x that we needed. Next we take the smallest one; set $\delta = \min\{\eta_1, \ldots, \eta_k\} > 0$. We claim that this δ "works" for 2ε.

Let x and y be in M with $d(x, y) < \delta$. Now $x \in B_{\eta_i}(x_i)$ for some i, so

$$d(y, x_i) \le d(y, x) + d(x, x_i) < \delta + \eta_i \le 2\eta_i = \delta_{x_i}.$$

Thus, since we already have $d(x, x_i) < \eta_i < \delta_{x_i}$, we get

$$\rho(f(x), f(y)) \le \rho(f(x), f(x_i)) + \rho(f(x_i), f(y)) < \varepsilon + \varepsilon = 2\varepsilon. \quad \square$$

Theorem 8.15 is an important result, and so it might be enlightening to discuss two other proofs. The second (less direct) proof is based on Exercise 56. If $f : M \to N$ is *not* uniformly continuous, then it follows from Exercise 56 that there are sequences (x_n) and (y_n) in M and some $\varepsilon > 0$ such that $d(x_n, y_n) \to 0$ while $\rho(f(x_n), f(y_n)) \ge \varepsilon > 0$ for all n. (How?) If M is compact, though, we may assume that (x_n) converges to a point $x \in M$, by passing to a subsequence if necessary. The corresponding subsequence of (y_n) must also converge to x. That is, by relabeling, we may suppose that $x_n \to x$ and $y_n \to x$. But then, assuming that we started with a continuous map f, we'd have $f(x_n) \to f(x)$ and $f(y_n) \to f(x)$ and, in particular, $\rho(f(x_n), f(y_n)) \to 0$, which is a contradiction.

The third proof is "by picture." Let's first show that if $f : [a, b] \to \mathbb{R}$ is continuous, then f is uniformly continuous. To begin, let $\varepsilon > 0$. We need to find a $\delta > 0$ such that if a pair of points $x, y \in [a, b]$ satisfy $|f(x) - f(y)| \ge \varepsilon$, then x and y also satisfy $|x - y| \ge \delta$. (Why?) In other words, we want to show that the function $d(x, y) = |x - y|$ is bounded away from 0 on the set $E = \{(x, y) \in [a, b] \times [a, b] : |f(x) - f(y)| \ge \varepsilon\}$.

The square $[a, b] \times [a, b]$ is pictured in Figure 8.1. The shaded regions form the set E. Note that E cannot hit the diagonal $y = x$ because $\varepsilon > 0$. (That is, $d(x, y) = |x - y|$ is strictly positive on E.) The heart of the proof lies in the observation that E is compact, and so it must be strictly separated from the diagonal by some positive distance.

Now since f is continuous, it follows that E is a closed subset of $[a, b] \times [a, b]$ (a compact metric space), and hence is compact. This is easy enough to check by using

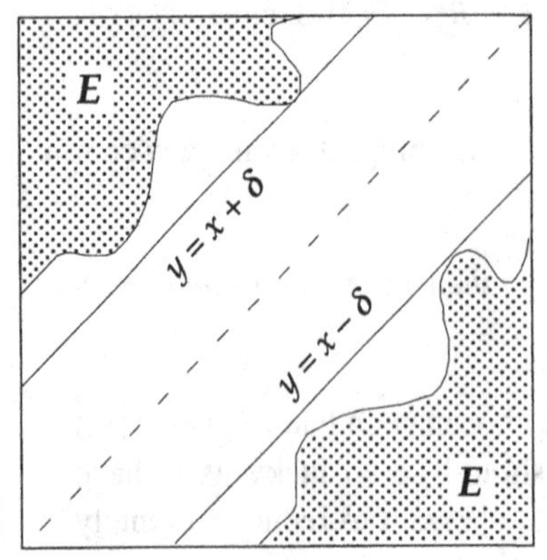

Figure 8.1

a sequential argument, but instead consider this: The function $g(x, y) = |f(x) - f(y)|$ is a continuous function on $[a, b] \times [a, b]$, and so $E = g^{-1}([\varepsilon, \infty))$ is closed. Finally, since the function $d(x, y) = |x - y|$ is continuous (and strictly positive) on E, it follows that d attains a minimum value $\delta > 0$ on E.

It is easy to modify this proof to work in the general case of a continuous function $f : (M, d) \to (N, \rho)$ on a compact space M. Essentially repeat this proof, using $d(x, y)$ in place of $|x - y|$ and $\rho(f(x), f(y))$ in place of $|f(x) - f(y)|$. The proof that the corresponding set E is a closed subset of the compact space $M \times M$ is the same. The details are left as an exercise.

Uniform continuity is often useful for finding extensions of continuous functions. Here is a variation on Theorem 7.18 that explains how this is done (you might want to recall the proof of Theorem 7.18 before reading on).

Theorem 8.16. *Let D be dense in M, let N be complete, and let $f : D \to N$ be uniformly continuous. Then, f extends uniquely to a uniformly continuous map $F : M \to N$, defined on all of M. Moreover, if f is an isometry, then so is the extension F.*

PROOF. First notice that *uniqueness is obvious*, because D is dense. That is, any two continuous functions $g, h : M \to N$ that agree on D must actually agree on all of M. *Existence* is the tough part.

We define $F : M \to N$ as follows (this is nearly the same scheme that we used in the proof of Theorem 7.18): Given $x \in M$, there is a sequence (x_n) in D such that $x_n \to x$ in M, since D is dense in M. Now (x_n) is Cauchy in D, and hence $(f(x_n))$ is Cauchy in N, because f is uniformly continuous. Thus, since N is complete, $f(x_n) \to y$ for some $y \in N$. Set $F(x) = y$. In brief, if $x = \lim_{n\to\infty} x_n$, where (x_n) is in D, then set $F(x) = \lim_{n\to\infty} f(x_n)$ in N.

First let's check that F is well defined. If (x_n) and (z_n) are two sequences in D with $x_n \to x$ and $z_n \to x$, then the sequence $x_1, z_1, x_2, z_2, \ldots$ also converges to x. Thus, $f(x_1), f(z_1), f(x_2), f(z_2), \ldots$ converges to some $y \in N$ (as above). But then we must have $f(x_n) \to y$ and $f(z_n) \to y$. (Why?)

The fact that F is an extension of f, that is, that $F|_D = f$, is obvious because f is continuous (besides, we get to use constant sequences).

Next we'll check that F is uniformly continuous. (Watch the ε's and δ's carefully here!) Let $\varepsilon > 0$, and choose $\delta > 0$ so that $\rho(f(x'), f(y')) < \varepsilon$ whenever $x', y' \in D$ with $d(x', y') < \delta$. We claim that $\delta/3$ "works" for 3ε and F. To see this it will help matters if we first make an observation: Given $x \in M$, there is an $x' \in D$ such that $d(x, x') < \delta/3$ *and* $\rho(F(x), f(x')) < \varepsilon$. (Why? Because if $x_n \to x$, where $x_n \in D$, then $f(x_n) \to F(x)$.)

The rest is easy. Given $x, y \in M$ with $d(x, y) < \delta/3$, choose $x', y' \in D$ (as above) such that $d(x, x') < \delta/3$, $d(y, y') < \delta/3$, $\rho(F(x), f(x')) < \varepsilon$, and $\rho(F(y), f(y')) < \varepsilon$. But then $d(x', y') \le d(x', x) + d(x, y) + d(y, y') < \delta$, and hence

$$\begin{aligned}\rho(F(x), F(y)) &\le \rho(F(x), f(x')) + \rho(f(x'), f(y')) + \rho(f(y'), F(y)) \\ &< \varepsilon + \varepsilon + \varepsilon = 3\varepsilon.\end{aligned}$$

Finally, note that if f is an isometry, then so is F. Given $x, y \in M$, choose (x_n) and (y_n) in D with $x_n \to x$ and $y_n \to y$. Then

$$d(x, y) = \lim_{n\to\infty} d(x_n, y_n) = \lim_{n\to\infty} \rho(f(x_n), f(y_n)) = \rho(F(x), F(y)). \quad \square$$

Corollary 8.17. *Completions are unique (up to isometry). That is, if M_1 and M_2 are completions of M, then M_1 and M_2 are isometric.*

EXERCISES

Throughout, M denotes a generic metric space with metric d.

65. If $f : (0, 1) \to \mathbb{R}$ is continuous, and if both $f(0+)$ and $f(1-)$ exist, show that the function F defined by $F(0) = f(0+)$, $F(1) = f(1-)$, and $F(x) = f(x)$ for $0 < x < 1$ is uniformly continuous on $[0, 1]$.

66. If $f : (0, 1) \to \mathbb{R}$ is uniformly continuous, show that $\lim_{x\to 0^+} f(x)$ exists. Conclude that f is bounded on $(0, 1)$.

67. Define $f : \ell_2 \to \ell_1$ by $f(x) = (x_n/n)_{n=1}^{\infty}$. Show that f is uniformly continuous.

68. Fix $y \in \ell_\infty$ and define $g : \ell_1 \to \ell_1$ by $g(x) = (x_n y_n)_{n=1}^{\infty}$. Show that g is uniformly continuous.

69. Prove Theorem 8.15 by supplying the details to the "proof by picture" in the general case.

70. Let $K = \{x \in \ell_\infty : \lim x_n = 1\}$. Prove:

(a) K is a closed (and hence complete) subset of ℓ_∞.

(b) If $T : \ell_\infty \to \ell_\infty$ is given by $T(x) = (0, x_1, x_2, \ldots)$ for $x = (x_1, x_2, \ldots)$ in ℓ_∞, that is, if T shifts the entries forward and puts 0 in the empty slot, then $T(K) \subset K$.

(c) T is an isometry on K, but T has no fixed point in K.

71. If A is dense in M, show that A and M have the same completion (isometrically).

72. Let D be dense in M. Show that M is isometric to a subset of $\ell_\infty(D)$. [Hint: First embed D into $\ell_\infty(D)$ and then apply Theorem 8.16.] In particular, every separable metric space is isometric to a subset of ℓ_∞. (But ℓ_∞ is not separable. Why?)

Equivalent Metrics

As a last topic related to both compactness and uniform continuity, we discuss several notions of equivalence for metrics (and norms). Throughout, we will suppose that d

and ρ are two metrics on the same set M. We will write $i : (M, d) \to (M, \rho)$ as the identity map and $i^{-1} : (M, \rho) \to (M, d)$ as its inverse (also the identity map, but in the other direction).

We say that d and ρ are **equivalent** if both i and i^{-1} are *continuous* (that is, if i is a homeomorphism), and we say that d and ρ are **uniformly equivalent** if i and i^{-1} are both *uniformly continuous* (that is, if i is a uniform homeomorphism). Finally, we say that d and ρ are **strongly equivalent** if both i and i^{-1} are *Lipschitz*. That is, d and ρ are strongly equivalent if there exist constants $0 < c, C < \infty$ such that $c\rho(x, y) \le d(x, y) \le C\rho(x, y)$ for all $x, y \in M$. (Some authors would state this requirement by saying that i is a *lipeomorphism*.) Actually, many authors take *strong* equivalence as their definition of simple equivalence, but, as we shall see, there are some differences between the three definitions. In any case, it is easy to see that

$$\text{strongly equivalent} \implies \text{uniformly equivalent} \implies \text{equivalent.}$$

In this section we will see that neither of these implications will reverse, in general, without some additional hypothesis.

Example 8.18

Consider $d(x, y) = |x - y|$ and $\rho(x, y) = \sqrt{|x - y|}$ on $M = [0, 1]$. Then, d and ρ are equivalent. (Recall Exercise 3.42. In fact, d and ρ are even uniformly equivalent – why?) However, $c\sqrt{|x - y|} \le |x - y|$ cannot hold for any $c > 0$ (and all x, y). That is, d and ρ are not strongly equivalent. Here's why: Replace $|x - y|$ by t and suppose that $c\sqrt{t} \le t$ for some $c > 0$ and all $0 \le t \le 1$. Then, by dividing, we would have $c \le \sqrt{t}$ for all $0 < t \le 1$, which is clearly impossible (since $\sqrt{t} \to 0$ as $t \to 0^+$).

EXERCISES

73. Given any metric space (M, d), show that the metric $\rho = d/(1 + d)$ is always uniformly equivalent to d but that there are cases in which the inequality $d \le C\rho$ may fail to hold.

74. Let $d(x, y) = \|x - y\|_2$ be the usual (Euclidean) metric on $\mathbb{R}^2$, and define a second metric ρ on $\mathbb{R}^2$ by

$$\rho(x, y) = \frac{\|x - y\|_2}{\left(1 + \|x\|_2^2\right)^{1/2}\left(1 + \|y\|_2^2\right)^{1/2}}.$$

Show that d and ρ are equivalent but not uniformly equivalent.

It is easy to imagine at least one case where equivalence and uniform equivalence should coincide. If (M, d) is *compact*, then every continuous map on M is actually uniformly continuous, and so equivalence and uniform equivalence might very well be one and the same. And so they are.

Proposition 8.19. *Suppose that (M, d) is compact and that ρ is another metric on M. Then d and ρ are equivalent if and only if d and ρ are uniformly equivalent.*

PROOF. The identity map $i : (M, d) \to (M, \rho)$ is continuous and onto; hence i is uniformly continuous and (M, ρ) is compact. Now, by applying the same reasoning to i^{-1}, it follows that i^{-1} is uniformly continuous. □

In spite of the fact that the three notions of equivalence are different, in general, we will establish the rather surprising fact that all three coincide when applied to *norms* on any vector space. To see this, we will first need to collect a few preliminary results about *linear* maps between normed vector spaces, each of which is interesting in its own right. In particular, for a linear map, we will show that continuity at a single point automatically gives us uniform continuity (and even more).

For the next several results, we suppose that $(V, \|\cdot\|)$ and $(W, |||\cdot|||)$ are normed vector spaces and that $T : V \to W$ is a linear map. That is, T is a vector space homeomorphism. This means that T "respects" vector space operations in the sense that $T(\alpha x + \beta y) = \alpha T(x) + \beta T(y)$ for any $x, y \in V$ and any scalars $\alpha, \beta \in \mathbb{R}$. In particular, a linear map always satisfies $T(0) = 0$.

Theorem 8.20. *Let $(V, \|\cdot\|)$ and $(W, |||\cdot|||)$ be normed vector spaces, and let $T : V \to W$ be a linear map. Then the following are equivalent:*

(i) *T is Lipschitz;*
(ii) *T is uniformly continuous;*
(iii) *T is continuous (everywhere);*
(iv) *T is continuous at $0 \in V$;*
(v) *there is a constant $C < \infty$ such that $|||T(x)||| \le C\|x\|$ for all $x \in V$.*

PROOF. Clearly, (i) $\Longrightarrow$ (ii) $\Longrightarrow$ (iii) $\Longrightarrow$ (iv). We need to show that (iv) $\Longrightarrow$ (v) and that (v) $\Longrightarrow$ (i) (for example). The second of these is easier, so let's start there.

(v) $\Longrightarrow$ (i): If condition (v) holds for a linear map T, then T is Lipschitz (with constant C) because $|||T(x) - T(y)||| = |||T(x-y)||| \le C\|x-y\|$ for any $x, y \in V$.

(iv) $\Longrightarrow$ (v): Suppose that T is continuous at 0. Then we may choose a $\delta > 0$ so that $|||T(x)||| = |||T(x) - T(0)||| \le 1$ whenever $\|x\| = \|x - 0\| \le \delta$.

Given $0 \ne x \in V$, we scale by the factor $\delta/\|x\|$ to get $\big\| \delta x/\|x\| \big\| = \delta$. Hence, $\big|\big|\big| T\big(\delta x/\|x\|\big)\big|\big|\big| \le 1$. But $T\big(\delta x/\|x\|\big) = (\delta/\|x\|)\, T(x)$, because T is linear, and so we get $|||T(x)||| \le (1/\delta)\|x\|$. That is, $C = 1/\delta$ works in condition (v). (Since condition (v) is trivial for $x = 0$, we only care about the case in which $x \ne 0$.) □

A linear map satisfying condition (v) of Theorem 8.20 (i.e., a continuous linear map) is often said to be **bounded**. The meaning of bounded in this context is slightly different than usual; here it means that T maps bounded sets to bounded sets. This follows from the fact that T is Lipschitz. Indeed, if $|||T(x)||| \le C\|x\|$ for all $x \in V$, then (as we saw earlier) $|||T(x) - T(y)||| \le C\|x - y\|$ for any $x, y \in V$, and hence T maps the ball about

x of radius r into the ball about $T(x)$ of radius Cr. In symbols, $T\big(B_r(x)\big) \subset B_{Cr}(T(x))$. More generally, T maps a set of diameter d into a set of diameter at most Cd. There is no danger of confusion in our using the word bounded to mean something new here; the ordinary usage of the word (as applied to functions) is uninteresting for linear maps. A nonzero linear map always has an unbounded range. (Why?)

Given normed vector spaces $(V, \|\cdot\|)$ and $(W, |||\cdot|||)$, the collection of all bounded linear maps $T : V \to W$ is itself a vector space under the usual pointwise operations on functions. That is, if $S, T : V \to W$ are continuous, linear maps, and if $\alpha, \beta \in \mathbb{R}$, then the map $\alpha S + \beta T : V \to W$, defined by

$$(\alpha S + \beta T)(x) = \alpha S(x) + \beta T(x), \qquad x \in V,$$

is again linear and continuous. The collection of all continuous, linear maps from V into W will be denoted by $B(V, W)$, where B stands for "bounded."

Theorem 8.20 provides a natural candidate for a norm on $B(V, W)$. If $T : V \to W$ is continuous and linear, we define the **norm** of T to be the smallest constant C that "works" in Theorem 8.20 (v). Thus, the norm of T is given by

$$\|T\| = \inf\{C : |||Tx||| \leq C\|x\| \text{ for all } x \in V\} = \sup_{x \neq 0} \frac{|||Tx|||}{\|x\|}.$$

That is, $\|T\|$ satisfies $|||Tx||| \leq \|T\|\,\|x\|$ for all $x \in V$, and $\|T\|$ is the smallest constant satisfying this inequality for all $x \in V$. The proof that this new expression, called the **operator norm**, actually is a norm on $B(V, W)$ is left as an exercise.

EXERCISES

75. Suppose that $f : \mathbb{R} \to \mathbb{R}$ satisfies $f(x + y) = f(x) + f(y)$ for every $x, y \in \mathbb{R}$. If f is continuous at a point $x_0 \in \mathbb{R}$, prove that there is some constant $a \in \mathbb{R}$ such that $f(x) = ax$ for all $x \in \mathbb{R}$. That is, an additive function that is continuous at even one point is linear – and hence continuous on all of $\mathbb{R}$.

76. Fix $y \in \mathbb{R}^n$ and define a linear map $L : \mathbb{R}^n \to \mathbb{R}$ by $L(x) = \langle x, y \rangle$. Show that L is continuous and compute $\|L\| = \sup_{x \neq 0} |L(x)|/\|x\|_2$. [Hint: Cauchy–Schwarz!]

77. Fix $k \geq 1$ and define $f : \ell_\infty \to \mathbb{R}$ by $f(x) = x_k$. Show that f is linear and has $\|f\| = 1$.

78. Define a linear map $f : \ell_2 \to \ell_1$ by $f(x) = (x_n/n)_{n=1}^\infty$. Is f bounded? If so, what is $\|f\|$?

79. If $S, T \in B(V, W)$, show that $S + T \in B(V, W)$ and that $\|S + T\| \leq \|S\| + \|T\|$. Using this, complete the proof that $B(V, W)$ is a normed space under the operator norm.

80. Show that the definite integral $I(f) = \int_a^b f(t)\,dt$ is continuous from $C[a, b]$ into $\mathbb{R}$. What is $\|I\|$?

81. Prove that the indefinite integral, defined by $T(f)(x) = \int_a^x f(t)\,dt$, is continuous as a map from $C[a, b]$ into $C[a, b]$. Estimate $\|T\|$.

82. For $T \in B(V, W)$, prove that $\|T\| = \sup\{|||Tx||| : \|x\| = 1\}$.

83. If V is any normed vector space, show that $B(V, \mathbb{R})$ is always complete. [Hint: Use Banach's characterization, Theorem 7.12.]

84. Prove that $B(V, W)$ is complete whenever W is complete.

Theorem 8.20, besides being merely spectacular, does even more for us: It supplies the proof that "equivalent" and "strongly equivalent" coincide for norms. (Recall that two norms are said to be equivalent if the metrics that they induce are equivalent. The same goes for strongly equivalent.)

Corollary 8.21. *Let* $\|\cdot\|$ *and* $|||\cdot|||$ *be two norms on a vector space* V. *Then,* $\|\cdot\|$ *and* $|||\cdot|||$ *are equivalent if and only if there are constants* $0 < c, C < \infty$ *such that* $c\,\|x\| \le |||x||| \le C\|x\|$ *for every* $x \in V$.

PROOF. The key here is that both the identity map $i : (V, \|\cdot\|) \to (V, |||\cdot|||)$ and its inverse i^{-1} are *linear*. Now, $\|\cdot\|$ and $|||\cdot|||$ are equivalent if and only if both i and i^{-1} are continuous. By Theorem 8.20, i and i^{-1} are continuous if and only if there exist constants $0 < c, C < \infty$ such that $|||x||| \le C\|x\|$ and $\|x\| \le c^{-1}|||x|||$ for all $x \in V$. (Why?) □

Once again, if we bring compactness into the picture, we can say even more. We will use the fact that closed balls in $\mathbb{R}^n$ are compact to prove:

Theorem 8.22. *Any two norms on a finite-dimensional vector space are equivalent.*

PROOF. Let V be an n-dimensional vector space with basis $x_1, \ldots, x_n$. We will define a specific, convenient norm on V and prove that any other norm on V is equivalent to ours. To do this, it will help if we first recall a simple fact from linear algebra.

Algebraically, V *is just* $\mathbb{R}^n$ *in disguise*. Each $x \in V$ can be uniquely written as $x = \sum_{i=1}^n \alpha_i x_i$, for some scalars $\alpha_1, \ldots, \alpha_n \in \mathbb{R}$. Thus we may think of x as the n-tuple $(\alpha_1, \ldots, \alpha_n) \in \mathbb{R}^n$. That is, the basis-to-basis map $x_i \mapsto e_i = (0, \ldots, 0, 1, 0, \ldots, 0)$ (the usual basis in $\mathbb{R}^n$) is a vector space isomorphism between V and $\mathbb{R}^n$.

Given this, we can easily define a norm on V by "borrowing" a norm from $\mathbb{R}^n$. Specifically, let

$$\left\| \sum_{i=1}^n \alpha_i x_i \right\| = \sum_{i=1}^n |\alpha_i| = \left\| \sum_{i=1}^n \alpha_i e_i \right\|_1$$

for each $x = \sum_{i=1}^n \alpha_i x_i \in V$. Since $x_1, \ldots, x_n$ is a basis, this clearly defines a norm on V:

$$\|x\| = 0 \quad \Longleftrightarrow \quad \alpha_i = 0 \quad \text{for all } i \quad \Longleftrightarrow \quad x = 0.$$

Moreover, the basis-to-basis map is a *linear isometry* between $(V, \|\cdot\|)$ and $(\mathbb{R}^n, \|\cdot\|_1)$.

Here is what we need out of all of this: The unit sphere $S = \{x \in V : \|x\| = 1\}$ is *compact* in $(V, \|\cdot\|)$ because the corresponding set in $\mathbb{R}^n$ is compact. (Why?) Now we can start the proof of the theorem!

Suppose that $|||\cdot|||$ is any other norm on V. Then, for $x = \sum_{i=1}^n \alpha_i x_i$, we have

$$\left|\left|\left|\sum_{i=1}^n \alpha_i x_i\right|\right|\right| \le \sum_{i=1}^n |\alpha_i|\, |||x_i|||$$
$$\le \left(\max_{1\le j\le n} |||x_j|||\right) \sum_{i=1}^n |\alpha_i|$$
$$= C\|x\|, \qquad \text{where } C = \max_{1\le j\le n} |||x_j|||.$$

That is, $|||x||| \le C\|x\|$ for every $x \in V$.

For the other inequality we will need to use our observation about the unit sphere S. The inequality that we have just proved tells us that $|||\cdot|||$ is a *continuous* function on $(V, \|\cdot\|)$. Indeed, $\big|\, |||x||| - |||y|||\, \big| \le |||x - y||| \le C\|x - y\|$ for any $x, y \in V$. But then, $|||\cdot|||$ is also continuous on S, and so $|||\cdot|||$ must assume a *minimum* value on S, say $c \in \mathbb{R}$. That is, $|||x||| \ge c$ whenever $\|x\| = 1$. Since this minimum is actually attained, we must also have $c > 0$. (Why?) Now we're cooking! Given $0 \ne x \in V$ we have $x/\|x\| \in S$, and hence $\big|\big|\big| x/\|x\| \big|\big|\big| \ge c$. That is, $|||x||| \ge c\,\|x\|$. □

The fact that all norms on a finite-dimensional normed space are equivalent elevates the merely spectacular to the simply phenomenal:

Corollary 8.23. *Let V and W be normed vector spaces with V finite-dimensional. Then, every linear map $T : V \to W$ is continuous.*

PROOF. Let $x_1, \ldots, x_n$ be a basis for V and let $\|\sum_{i=1}^n \alpha_i x_i\| = \sum_{i=1}^n |\alpha_i|$, as above. We may assume that this is "the" norm on V, since, by Theorem 8.22, every norm produces the same continuous functions on V.

Now if $T : (V, \|\cdot\|) \to (W, |||\cdot|||)$ is linear, we get

$$\left|\left|\left| T\left(\sum_{i=1}^n \alpha_i x_i\right)\right|\right|\right| = \left|\left|\left|\sum_{i=1}^n \alpha_i T(x_i)\right|\right|\right|$$
$$\le \sum_{i=1}^n |\alpha_i|\, |||T(x_i)|||$$
$$\le \left(\max_{1\le j\le n} |||T(x_j)|||\right) \sum_{i=1}^n |\alpha_i|.$$

That is, $|||T(x)||| \le C\|x\|$, where $C = \max_{1\le j\le n} |||T(x_j)|||$. By Theorem 8.20, T is continuous. □

Corollary 8.23 allows us to clean up a detail left over from Chapter Five:

Corollary 8.24. *Any two finite-dimensional normed vector spaces of the same dimension are uniformly homeomorphic. In fact, we can even find a linear (and hence Lipschitz) homeomorphism between them.*

Corollary 8.25. *Every finite-dimensional normed vector space is complete.* (Why?)

Corollary 8.26. *A finite-dimensional linear subspace of any normed vector space is always closed.* (Why?)

EXERCISES

85. Fill in the missing details in the proof of Theorem 8.22.

86. If $(V, \|\cdot\|)$ is an n-dimensional normed vector space, show that there is a norm $|||\cdot|||$ on $\mathbb{R}^n$ such that $(\mathbb{R}^n, |||\cdot|||)$ is linearly isometric to $(V, \|\cdot\|)$.

87. Prove Corollary 8.24.

88. Prove Corollary 8.25.

89. Corollary 8.26 is of interest because an infinite-dimensional normed space may have nonclosed subspaces. For example, show that $\{x \in \ell_1 : x_n = 0 \text{ for all but finitely many } n\}$ is a proper *dense* linear subspace of ℓ_1.

Notes and Remarks

The classical definition of compactness, due to Fréchet, is the statement of Theorem 8.2: Each sequence has a convergent subsequence. But early usages of the word "compact" often referred to what we have called precompact sets – sets whose closures are compact. In effect, then, the Bolzano–Weierstrass theorem characterizes the bounded sets as the precompact subsets of $\mathbb{R}$. Hausdorff first proved the theorem that we have taken as our starting point: A space is compact if and only if it is complete and totally bounded.

The property described in Lemma 8.8 (a) is generally taken as the formal definition of compactness for topological spaces, due to Alexandrov and Urysohn [1924] (who used the word "bicompact" in describing such spaces). It has as its basis the so-called Heine–Borel or Borel–Lebesgue theorems (a covering of a closed, bounded interval by open sets has a finite subcover). Riesz [1908] added the finite intersection property to the list for subsets of $\mathbb{R}^n$, while the general case is due to Sierpiński [1918]. For more on the early history of Theorem 8.9, see Dudley [1989], Manheim [1964], Temple [1981], Willard [1970], and the award-winning article by Hildebrandt [1926] (reprinted in Abbott [1978]). The property described in Theorem 8.2 is called *sequential compactness*, while the property described in Corollary 8.11 is called *countable compactness*. In a metric space, each of these coincides with the formal definition of compactness, but this is not always the case in more general topological spaces.

Corollary 8.11 is due to Fréchet. For more on Exercise 27, see Apostol [1975], Buck [1967], and Thurston [1989]. Exercises 29 and 40–43 are taken from Kaplansky [1977]. For more on the results stated in Exercises 28 and 29, see D. F. Bailey [1989] (and its bibliography), and Bennett and Fisher [1974]. Semicontinuity (Exercises 37–39) was introduced by Baire [1899]. See Radó [1942] for more details.

For a survey of applications of compactness in analysis, see Hewitt [1960]. For a simplified treatment of the classical theorems presented in this chapter in the case of a closed bounded interval $[a, b]$, see Botsko [1987]. Barnsley [1988] and Edgar [1990], on the other hand, illustrate certain "modern" applications of compactness.

Exercise 70 is adapted from an exercise in Hoffman [1975]. It would seem that Heine was the first to define uniform continuity for real-valued functions; he used it to prove Theorem 8.15 for real-valued functions defined on a closed bounded interval $[a, b]$. According to Dudley [1989], Heine gave a great deal of credit to unpublished lectures of Weierstrass. The metric space definition is due to Fréchet and Hausdorff. The clever "proof by picture" for Theorem 8.15 is taken from the article by D. M. Bloom [1989]. Several authors have considered the problem of characterizing those spaces for which all continuous maps are uniformly continuous; see, for example, Beer [1988], Chaves [1985], Hueber [1981], Levine [1960], and Snipes [1984].

The discussion of equivalence, strong equivalence, and uniform equivalence for metrics is based in part on the presentation in Kuller [1969]. Maddox [1989] gives an elementary computation of the norm of a linear map on $C[a, b]$ defined by an integral, as in Exercises 80 and 81.

Analysis in infinite-dimensional normed vector spaces is vastly different from the finite-dimensional case. To fully appreciate the extent of the difference is beyond our means just now, but we can at least indicate a few reasons. For one, recall that $S = \{x \in \ell_2 : \|x\|_2 = 1\}$, the unit sphere in ℓ_2, is *not* compact. (Remember the e_n?) Thus, the proofs of Theorem 8.22 and Corollary 8.23 fall apart in ℓ_2. But the same would be true of any infinite-dimensional space. In fact, it turns out that a normed linear space $(V, \|\cdot\|)$ is finite-dimensional if and only if its closed unit ball $B = \{x \in V : \|x\| \leq 1\}$ is compact. Moreover, $(V, \|\cdot\|)$ is infinite-dimensional if and only if there exists a *discontinuous* linear map $T : V \to \mathbb{R}$ if and only if V contains a proper *dense* subspace. On the other hand, Corollary 8.24 can be at least partially salvaged: Anderson [1962] has shown that all separable, infinite-dimensional Banach spaces are (mutually) homeomorphic. We cannot hope for uniformly homeomorphic here since, for example, it is known that ℓ_p and ℓ_q are not uniformly homeomorphic for any $1 \leq p < q \leq \infty$. For much more on this, see the note by Bessaga and Pełczyński [1987] in the English translation of Banach's book.

CHAPTER NINE

Category

Discontinuous Functions

We have had a lot to say so far about continuous functions, but what about discontinuous functions? Is there anything meaningful we might say about them? In order that we might ask more precise questions, let's fix our notation. Throughout this section, we will be concerned with a function $f : \mathbb{R} \to \mathbb{R}$, and we will write $D(f)$ for the set of points at which f is *discontinuous*. The questions are: What can we say about $D(f)$? What kind of set is it? Can any set be realized as the set of discontinuities of a function, or does $D(f)$ have some distinguishing characteristics? To get us started, let's recall a few examples.

Examples 9.1

(a) If f is monotone, then $D(f)$ is countable. Conversely, any countable set is the set of discontinuities for some monotone f (see Exercise 2.34).

(b) There are examples of functions f, g with $D(f) = \mathbb{Q}$ and $D(g) = \mathbb{R}$. (What are they?)

In particular, we might ask whether $D(f)$ can be a proper, uncountable subset of $\mathbb{R}$. For example, is there an f with $D(f) = \mathbb{R} \setminus \mathbb{Q}$? or with $D(f) = \Delta$? The answer to the first question is: No, and to the second: Yes, but to understand this will require a bit of machinery.

The first thing we need is a detailed description of $D(f)$. For this we will simply negate the definition of the statement "f is continuous at a":

$$a \in D(f) \iff \begin{cases} \text{there exists an } \varepsilon > 0 \text{ such that, given any } \delta > 0, \\ \text{we have } |f(x) - f(a)| \geq \varepsilon \text{ for some } x \text{ with } |x - a| < \delta. \end{cases}$$

What this means is that, given any bounded, open interval I containing a, we always have $\sup\{|f(x) - f(y)| : x, y \in I\} \geq \varepsilon$. (Why?) This supremum has a geometric description (which is why we want to use it); indeed, notice that

$$\sup_{x, y \in I} |f(x) - f(y)| = \operatorname{diam} f(I).$$

We will write our description of $D(f)$ in terms of this supremum, but first we will give it a name. Given a bounded interval I, we define $\omega(f; I)$, the **oscillation of** f **on** I, by $\omega(f; I) = \sup\{|f(x) - f(y)| : x, y \in I\}$. Note that $0 \leq \omega(f; I) \leq 2\sup_{x \in I} |f(x)|$. Of course, if f is unbounded on I, we set $\omega(f; I) = \infty$.

Also notice that $\omega(f; I)$ *decreases* as I decreases; that is, if $J \subset I$, then $\omega(f; J) \leq \omega(f; I)$. Consequently, if f is bounded in some neighborhood of a, and if we consider intervals that "shrink" to a, then the oscillations over those intervals will decrease to a fixed (finite) number. These observations allow us to define the **oscillation of f at** a, written $\omega_f(a)$, by

$$\omega_f(a) = \inf_{\substack{I \ni a \\ I \text{ open}}} \omega(f; I) = \lim_{h \to 0^+} \omega(f; (a-h, a+h)) = \lim_{h \to 0^+} \operatorname{diam} f(B_h(a)),$$

where the notation $I \ni a$ is intended as a reminder that the infimum is over bounded (open) intervals I containing a. If f is unbounded in every neighborhood of a, we set $\omega_f(a) = \infty$. We have insisted on open intervals in the definition of $\omega_f(a)$ to be consistent with the characterization of discontinuity at a that we gave earlier.

The oscillation of f at a is rather like the "jump" in the graph of f at a (if any). For example, if f is *increasing*, then $\omega_f(a) = f(a+) - f(a-)$. In any case, we always have $\omega_f(a) \geq 0$, and our earlier discussion tells us that $a \in D(f)$ if and only if $\omega_f(a) > 0$. That is, f is continuous at a if and only if $\omega_f(a) = 0$. (Why?)

Now we are ready to give a more detailed description of $D(f)$.

Theorem 9.2. *If $f : \mathbb{R} \to \mathbb{R}$, then $D(f)$ is the* **countable** *union of* **closed** *sets in $\mathbb{R}$.*

PROOF. First, let's write $D(f)$ as a countable union:

$$\begin{aligned} D(f) &= \{a : \omega_f(a) > 0\} \\ &= \{a : \omega_f(a) \geq \varepsilon \text{ for some } \varepsilon > 0\} \\ &= \bigcup_{n=1}^{\infty} \{a : \omega_f(a) \geq 1/n\} \qquad \text{(Why?)} \end{aligned}$$

Thus, we need to show that a set of the form $\{a : \omega_f(a) \geq r\}$ is closed, where $r > 0$ is fixed. Equivalently, we might show that the set $\{a : \omega_f(a) < r\}$ is open, and this is easy. If $x_0 \in \{a : \omega_f(a) < r\}$, that is, if $\omega_f(x_0) < r$, then there is some bounded open interval I containing x_0 such that $\omega(f; I) < r$. (Why?) It follows that $I \subset \{a : \omega_f(a) < r\}$, since $\omega_f(x) \leq \omega(f; I) < r$ for any $x \in I$. □

EXERCISES

1. If f is increasing, show that $\omega_f(a) = f(a+) - f(a-)$.

2. Prove that f is continuous at a if and only if $\omega_f(a) = 0$.

3. Given $f : \mathbb{R} \to \mathbb{R}$, show that $g(x) = \arctan f(x)$ satisfies $D(g) = D(f)$. Thus, in any discussion of $D(f)$, we may assume that f is bounded.

▷ 4. Let $f : [a, b] \to \mathbb{R}$ be continuous, and let $\varepsilon > 0$. Show that there is an $n \in \mathbb{N}$ such that $\omega(f; [(k-1)/n, k/n]) < \varepsilon$ for all $k = 1, \ldots, n$.

▷ 5. If A is a subset of $\mathbb{R}$ and if x is in the interior of A, show that x is a point of continuity for χ_A (the characteristic function of A). Are there any other points of continuity?

6. Compute $D(\chi_\Delta)$, where Δ is the Cantor set. If E is the set of all endpoints in Δ (see Exercise 2.23), compute $D(\chi_{\Delta\setminus E})$.

7. For which sets A is χ_A upper semicontinuous? lower semicontinuous?

8. Given any bounded function f, show that the function $\omega_f(x)$ is upper semicontinuous.

9. If E is a closed set in $\mathbb{R}$, show that $E = D(f)$ for some bounded function f. [Hint: A sum of two characteristic functions will do the trick.]

10. Is every *bounded* continuous function on $\mathbb{R}$ *uniformly* continuous?

Our earlier questions about the nature of $D(f)$ can now be rephrased: Which subsets of $\mathbb{R}$ can be written as a countable union of closed sets? In particular, is $\mathbb{R} \setminus \mathbb{Q}$ such a set? Conversely, is every countable union of closed sets the set of discontinuities for some bounded function? Before we answer these questions, it might be helpful to have a name for countable unions of closed sets (and the like).

A countable union of closed sets is called an F_σ set. Thus, the set of discontinuities $D(f)$ is an F_σ set. We might want to turn things around by taking complements, and so we also name a countable intersection of open sets; these are called G_δ sets. The letter F stands for *fermé*, or closed, while σ stands for *somme*, or sum. The letter G stands for *Gebiet*, or region – besides, it comes after F – while δ stands for *Durchschnitt*, or intersection. This is proof positive that both a Frenchman and a German had a say in our notation!

The letters δ and σ represent operations performed on the underlying class of closed sets F or on the class of open sets G. The result is often a new class of sets. For example, note that we would get nothing new by considering F_δ sets because the intersection of closed sets is again closed. In other words, $F_\delta = F$. The same goes for G_σ sets. But we do get something new by considering F_σ's and G_δ's. The set of rationals $\mathbb{Q}$, for instance, is an F_σ set, but it is obviously neither open nor closed. By taking complements, the set of irrationals $\mathbb{R} \setminus \mathbb{Q}$ is a G_δ set. We can continue this process – any combination producing something new is of interest – and consider, say, $F_{\sigma\delta}$ sets (countable intersections of F_σ sets), $G_{\delta\sigma}$ sets (countable unions of G_δ sets), and so on.

EXERCISES

11. Show that every open interval (and hence every nonempty open set) in $\mathbb{R}$ is a countable union of closed intervals, and that every closed interval in $\mathbb{R}$ is a countable intersection of open intervals.

12. More generally, in any metric space, show that every open set is an F_σ and that every closed set is a G_δ.

13. If E is an F_σ set in $\mathbb{R}$, is $E = D(f)$ for some f? (The answer is yes, but this is hard!)

The Baire Category Theorem

Recall that we have rephrased our earlier question about sets of discontinuity to read: Which subsets of $\mathbb{R}$ can be written as countable unions of closed sets? In particular, we asked whether $\mathbb{R} \setminus \mathbb{Q}$ was such a set. Obviously, we can turn things around and ask whether $\mathbb{Q}$ is a countable intersection of open sets. Now any open set containing $\mathbb{Q}$ is dense in $\mathbb{R}$, so we might first ask whether the countable intersection of dense open sets is still dense. The answer is yes:

The Baire Category Theorem for $\mathbb{R}$ 9.3. *If (G_n) is a sequence of dense, open sets in $\mathbb{R}$, then $\bigcap_{n=1}^{\infty} G_n \neq \emptyset$. In fact, $\bigcap_{n=1}^{\infty} G_n$ is* **dense** *in $\mathbb{R}$.*

PROOF. Let $x_0 \in \mathbb{R}$, and let I_0 be any open interval containing x_0. We will prove both conclusions at once by showing that $I_0 \cap \left(\bigcap_{n=1}^{\infty} G_n\right) \neq \emptyset$.

Since G_1 is dense, we know that $I_0 \cap G_1 \neq \emptyset$. But since G_1 is also open, this means that we can find some open interval $I_1 \subset I_0 \cap G_1$. By shrinking I_1 (if necessary), we may suppose that $\text{diam}(I_1) \leq 1$ and $\bar{I}_1 \subset I_0 \cap G_1$.

Now use I_1 in place of I_0 and G_2 in place of G_1. Since G_2 is dense, we have $I_1 \cap G_2 \neq \emptyset$. But G_2 is open, so there is some open interval I_2 with $\text{diam}(I_2) \leq 1/2$ such that $\bar{I}_2 \subset I_1 \cap G_2 \subset I_0 \cap G_1 \cap G_2$.

Repeat this using I_2 and G_3 in place of I_1 and G_2, and so on. What we get is a sequence of nested closed intervals, $\bar{I}_1 \supset \bar{I}_2 \supset \cdots$ with $\text{diam}(I_n) \leq 1/n$ and $\bar{I}_n \subset I_0 \cap \left(\bigcap_{k=1}^{n} G_k\right)$. Thus, by the nested interval theorem, $I_0 \cap \left(\bigcap_{k=1}^{\infty} G_k\right) \supset \bigcap_{n=1}^{\infty} \bar{I}_n \neq \emptyset$. Consequently, $\bigcap_{n=1}^{\infty} G_n$ is nonempty and dense. □

Note that Baire's theorem provides a new proof that $\mathbb{R}$ is uncountable. Indeed, if $\mathbb{R} = \{x_1, x_2, \ldots\}$, then each of the sets $G_n = \mathbb{R} \setminus \{x_n\}$ is open and dense (see Exercise 15); but they also satisfy $\bigcap_{n=1}^{\infty} G_n = \emptyset$, which contradicts Baire's theorem.

We can push this observation a bit further. A dense G_δ subset of $\mathbb{R}$ must also be an uncountable set. Here's why: If (G_n) is a sequence of open dense sets in $\mathbb{R}$ and if $\bigcap_{n=1}^{\infty} G_n = \{x_1, x_2, \ldots\}$, then the sets $\tilde{G}_n = G_n \setminus \{x_n\}$ are still open and dense, but $\bigcap_{n=1}^{\infty} \tilde{G}_n = \emptyset$, contrary to Baire's theorem. Thus, $\bigcap_{n=1}^{\infty} G_n$ is uncountable. This is the extra piece of information that we need to settle our original questions.

Corollary 9.4. *$\mathbb{Q}$ cannot be written as the countable intersection of open subsets of $\mathbb{R}$.*

Corollary 9.5. *$\mathbb{R} \setminus \mathbb{Q} \neq D(f)$ for any $f : \mathbb{R} \to \mathbb{R}$.*

By rephrasing Baire's theorem, we will be able to see another reason behind these last two corollaries.

Corollary 9.6. *If $\mathbb{R} = \bigcup_{n=1}^{\infty} E_n$, where each E_n is* **closed**, *then some E_n contains an open interval.*

PROOF. Each of the sets $G_n = \mathbb{R} \setminus E_n$ is open in $\mathbb{R}$ and $\bigcap_{n=1}^{\infty} G_n = \emptyset$. Thus, by Baire's theorem, some G_n is *not* dense. That is, some G_n misses an entire open interval. In other words, some E_n contains an interval. □

Corollary 9.7. *If* $\mathbb{R} = \bigcup_{n=1}^{\infty} E_n$, *then the* **closure** *of some* E_n *contains an interval; that is,* $\text{int}(\bar{E}_n) \neq \emptyset$ *for some* n. (Why?)

Corollary 9.8. *If* $\mathbb{R} \setminus \mathbb{Q} = \bigcup_{n=1}^{\infty} E_n$, *then the closure of some* E_n *contains an interval.*

How very different $\mathbb{R} \setminus \mathbb{Q}$ and $\mathbb{Q}$ are! The rationals are somehow very "sparse" while the irrationals are quite "thick." To appreciate this difference, and to generalize Baire's theorem to metric spaces, will require some new terminology. To begin, recall that a subset E of a metric space M is called **nowhere dense in** M if $\bar{E}$ contains no nonempty open set, that is, if the interior of $\bar{E}$ (in M) is empty. Judicious rewriting of this condition might help. Note that E is nowhere dense if and only if $\bar{E}$ is nowhere dense (obviously), and that $\bar{E}$ is nowhere dense if and only if the *complement* of $\bar{E}$ is *dense* (since every open set has to hit $(\bar{E})^c$). Consequently, E is nowhere dense in M if and only if the complement of $\bar{E}$ is an open, dense set in M.

Examples 9.9

(a) $\mathbb{N}$ and Δ are nowhere dense in $\mathbb{R}$. Also, any singleton $\{x\}$ is nowhere dense in $\mathbb{R}$. But this is not the general case; $\{x\}^\circ = \{x\}$ can, and does, happen – how?

(b) Finite unions of nowhere dense sets are again nowhere dense (see Exercise 4.56). But a countable union of nowhere dense sets may fail to be nowhere dense. For example, $\mathbb{Q}$ is not nowhere dense in $\mathbb{R}$.

(c) We have no choice but to be fussy here; note that while $\mathbb{N}$ is nowhere dense in $\mathbb{R}$, it is *not* nowhere dense relative to $\mathbb{N}$ itself. In other words, we cannot ignore the fact that we have defined the phrase "E is nowhere dense in M." The closure and the interior named in the definition refer to the closure and interior in M, not in E.

(d) In an unfortunate fluke of language, "not nowhere dense" is not the same as "dense." Indeed, $(0, 1)$ is not nowhere dense in $\mathbb{R}$, and yet it certainly is not dense in $\mathbb{R}$. It may be easier to understand the difference if we recall that some authors use the phrase *everywhere dense* in place of the single word *dense*. An everywhere dense set is one that is dense in every open set (see Exercises 4.45 and 4.46). A nowhere dense set, on the other hand, is one that is *not* dense in *any* open set (see Exercises 19 and 20, below). And so nowhere dense means "not even a little bit dense"!

Given this terminology, we next define two categories, or types, of subsets of a metric space M. A subset A of M is said to be of the **first category in** M (or, a first category set *relative* to M) if A can be written as a countable union of sets, each of which is nowhere dense in M. For example, it follows that $\mathbb{Q}$ is a first category set in $\mathbb{R}$. Some authors refer to first category sets as "meager" or "sparse" sets.

The second category consists of all those sets that fail to be in the first category. That is, a subset B of M is said to be of the **second category in** M if B is *not* of the first category. In other words, B is a second category set in M if, whenever we write $B = \bigcup_{n=1}^{\infty} E_n$, some E_n *fails* to be nowhere dense in M; that is, $\text{int}(\bar{E}_n) \neq \emptyset$ for some n. (Look familiar?)

Examples 9.10

(a) In the language of category, Corollary 9.7 says that $\mathbb{R}$ is a second category set in itself. And we could restate Corollary 9.8 by saying that $\mathbb{R} \setminus \mathbb{Q}$ is a second category set in $\mathbb{R}$. The two categories of subsets of $\mathbb{R}$ provide yet another measure of "big" versus "small" A first category set in $\mathbb{R}$, such as $\mathbb{Q}$, is "small" while a second category set in $\mathbb{R}$, such as $\mathbb{R} \setminus \mathbb{Q}$, is "big."

(b) Again we will want to be careful. The two categories of subsets of M depend on the notion of nowhere dense sets, which in turn requires that we be precise about the host space M. For example, $\mathbb{N}$ is of the *first* category in $\mathbb{R}$, but it is of the *second* category *in itself*. (Why?) In short, category is *very* relative.

Finally we can state the general theorem. The proof is exactly the same as the one we gave for $\mathbb{R}$; just repeat the proof of Theorem 9.3, using open balls instead of open intervals (and the nested set theorem in place of the nested interval theorem).

The Baire Category Theorem 9.11. *A complete metric space is of the second category in itself. That is, if M is a complete metric space, and if we write $M = \bigcup_{n=1}^{\infty} E_n$, then the closure of some E_n contains an open ball. Equivalently, if (G_n) is a sequence of dense open sets in M, then $\bigcap_{n=1}^{\infty} G_n \neq \emptyset$; in fact, $\bigcap_{n=1}^{\infty} G_n$ is dense in M.*

Note that we cannot expect a dense G_δ subset of a general metric space to be uncountable because M itself may be only countable. The fact that a dense G_δ subset of $\mathbb{R}$ is uncountable hinges on the observation that if G is open and dense in $\mathbb{R}$, then so is $G \setminus \{x\}$ (see Exercise 15).

Baire's theorem is often applied in existence proofs; after all, the conclusion is that some set is nonempty. We will see several applications of this principle later in the book. For now, let's just highlight the key fact:

Corollary 9.12. *In a complete metric space, the complement of any first category set is nonempty. In fact, it is even dense.* (Why?)

EXERCISES

Except where noted, M is an arbitrary metric space with metric d.

▷ **14.** Prove that A has an empty interior in M if and only if A^c is dense in M.

▷ **15.** If G is open and dense in $\mathbb{R}$, show that the same is true of $G \setminus \{x\}$ for any $x \in \mathbb{R}$. Is this true in any metric space? Explain.

16. Show that $\{x\}$ is nowhere dense in M if and only if x is *not* an isolated point of M.

17. Prove that a complete metric space without any isolated points is uncountable. In particular, this gives another proof that Δ is uncountable.

18. If A is either open or closed, show that $\text{bdry}(A)$ is nowhere dense in M. Is the same true of any set A?

19. Show that each of the following is equivalent to the statement that A is nowhere dense in M:

(a) $\bar{A}$ contains no nonempty open set.

(b) Each nonempty open set in M contains a nonempty open subset that is disjoint from A.

(c) Each nonempty open set in M contains an open ball that is disjoint from A.

20. If A is nowhere dense in M, and if G is a nonempty open set in M, prove that A is nowhere dense in G.

21. If $x_n \to x$ in $\mathbb{R}$, show that the set $\{x\} \cup \{x_n : n \geq 1\}$ is nowhere dense in $\mathbb{R}$. Is the same true if $\mathbb{R}$ is replaced by an arbitrary metric space M? Is every countable set nowhere dense? Explain.

22. Let (r_n) be an enumeration of $\mathbb{Q}$. For each n, let I_n be the open interval centered at r_n of radius 2^{-n}, and let $U = \bigcup_{n=1}^{\infty} I_n$. Prove that U is a proper, open, dense subset of $\mathbb{R}$ and that U^c is nowhere dense in $\mathbb{R}$.

23. Is there a dense, open set in $\mathbb{R}$ with uncountable complement? Explain.

24. Prove Corollary 9.7.

25. Prove Corollary 9.8. Deduce that the conclusion of Baire's theorem holds for $\mathbb{R} \setminus \mathbb{Q}$.

▷ **26.** Prove Theorem 9.11.

27. Let M be a complete metric space. If $M = \bigcup_{n=1}^{\infty} E_n$, where each E_n is closed, show that $D = \bigcup_{n=1}^{\infty} \text{int}(E_n)$ is dense in M. [Hint: "Estimate" $M \setminus D$.]

▷ **28.** In a metric space M, show that any subset of a first category set is still first category, and that a countable union of first category sets is again first category.

▷ **29.** In a metric space M, prove that any superset of a second category set is itself a second category set.

▷ **30.** Show that $\mathbb{N}$ is first category in $\mathbb{R}$ but second category in itself.

▷ **31.** Show that $\mathbb{Q}$ is first category in itself (thus, completeness is essential in Baire's theorem).

▷ **32.** In $\mathbb{R}$, show that any open interval (and hence any nonempty, open set) is a second category set.

33. If M is complete, is every nonempty, open set a second category set?

34. Let M be complete, and let E be an F_σ set in M. Prove that E is a first category set in M if and only if E^c is dense in M.

35. Let $f : \mathbb{R} \to \mathbb{R}$. Show that f is discontinuous on a set of the first category in $\mathbb{R}$ if and only if f is continuous at a dense set of points.

36. If M is complete, show that the complement of a first category set in M is a dense set of the second category in M. In particular, a first category set in a complete metric space must have empty interior.

37. Show that the complement of a first category set in $\mathbb{R}$ is uncountable.

38. Is the complement of a first category set necessarily a second category set? Likewise, is the complement of a second category set necessarily a first category set? Explain.

39. When is a first category set an F_σ set? Equivalently, when is a set containing a dense G_δ set itself a G_δ set?

40. Let $f : \mathbb{R} \to \mathbb{R}$ be a continuous function that is nonconstant on any interval. If A is a second category set in $\mathbb{R}$, show that $f(A)$ is also second category. [Hint: If B is closed and nowhere dense, show that $f^{-1}(B)$ is closed and nowhere dense.]

41. Let M be a complete metric space. Prove that if (E_n) is a sequence of closed sets in M, each having empty interior, then $\bigcup_{n=1}^{\infty} E_n$ has empty interior.

42. While completeness is essential in the proof of Baire's theorem, the conclusion may still hold for some incomplete spaces. Show that it holds in $\mathbb{N}$ if we use the metric $d(m,n) = |m - n|/mn$, but that $(\mathbb{N}, d)$ is not complete. [Hint: d is equivalent to the usual metric. See Exercise 7.14.]

43. If N is homeomorphic to a complete metric space M, show that the conclusion of Baire's theorem holds in N. [Hint: Homeomorphisms preserve dense open sets. Why?]

44. If M is complete, show that the conclusion of Baire's theorem holds for any *open* subset of M. [Hint: See Exercise 7.30.]

45. Fix $n > 1$, and let $f : [a,b] \to \mathbb{R}^n$ be continuous and one-to-one. Show that the range of f is nowhere dense in $\mathbb{R}^n$. [Hint: The range of f is closed (why?); if it has nonempty interior, then it contains a closed rectangle. Argue that this rectangle is the image of some subinterval of $[a,b]$.] Use this to show that $\mathbb{R}$ and $\mathbb{R}^n$ are not homeomorphic for $n > 1$.

46. Show that $\mathbb{R}^2$ cannot be written as a countable union of lines.

47. Let $\mathcal{P}$ be the vector space of all polynomials supplied with the norm $\|p\| = \max\{|a_i| : i = 0, \ldots, n\}$, where $p(x) = a_0 + a_1 x + \cdots + a_n x^n \in \mathcal{P}$. Show that $\mathcal{P}$ is not complete.

48. If W is a proper, closed, linear subspace of a normed vector space V, show that W is nowhere dense in V. [Hint: If $W \supset B_r(x)$, then $W \supset nB_1(0)$ for every n. Why?]

49. Let V be an infinite-dimensional normed vector space, and suppose that $V = \bigcup_{n=1}^{\infty} W_n$, where each W_n is a finite-dimensional subspace of V. Prove that V is *not* complete.

50. Let M be a separable metric space, and let S be a subset of M. A point $x \in S$ is said to be *a point of first category relative to* S if, for some neighborhood U of x, the set $U \cap S$ is of first category in M. If S_0 is the set of points of first category relative to S, show that S_0 is of first category in M. [Hint: M has a countable open base.]

◊

Notes and Remarks

Baire's result (for $\mathbb{R}^n$) appears in his thesis, Baire [1899]. An early (and less explicit) version of the category theorem appeared in Osgood [1897]. See Hawkins [1970] and Hobson [1927] for more details on Osgood's contribution.

Exercise 22 is adapted from Wilansky [1953b]. Diamond and Gellès [1984, 1985] discuss certain relations that exist among the various notions of "big" and "small" sets that we have encountered (and even more that we haven't!). The result stated in Exercise 50 is from Banach [1930], but see also Kuratowski [1966]. The bible for all matters categorical is Oxtoby [1971].

As mentioned earlier in this chapter, Baire's theorem has lots of applications. Here is one example (with a few details to check). The characteristic function of the rationals $\chi_{\mathbb{Q}}$ is *not* the limit of a sequence of continuous functions. Suppose, to the contrary, that there is a sequence (f_n) of continuous functions such that $\chi_{\mathbb{Q}}(x) = \lim f_n(x)$ for each $x \in \mathbb{R}$. Then, the set $A_n = \{x : f_n(x) > 1/2\}$ is open for each n and, hence, so is $G_n = \bigcup_{k=n}^{\infty} A_k = \{x : f_k(x) > 1/2 \text{ for some } k \geq n\}$. But then, $\bigcap_{n=1}^{\infty} G_n = \{x : f_n(x) > 1/2 \text{ for infinitely many } n\} = \mathbb{Q}$ (why?), and this contradicts Corollary 9.4. This example illustrates a special case of a deep result, due to both Baire and Osgood, stating that any function $f : \mathbb{R} \to \mathbb{R}$ that is the limit of a sequence of continuous functions must have a point of continuity. Various incarnations of the theorem are discussed in greater detail in Goffman [1953a], Hobson [1927], and Munroe [1965]. Myerson [1991] discusses the related problem of finding a sequence of continuous functions whose pointwise limit is finite on $\mathbb{Q}$ and infinite on $\mathbb{R} \setminus \mathbb{Q}$. We will discuss several applications of Baire's theorem in Part Two, where we will give a proof of the Baire–Osgood theorem and further details on the set of discontinuities $D(f)$ of a bounded function (especially concerning Exercises 9 and 13).

PART TWO

FUNCTION SPACES

CHAPTER TEN

Sequences of Functions

Historical Background

Unarguably, modern analysis was formed during the resolution of an important controversy (or, rather, controversies) concerning the representation of "arbitrary" functions. This controversy has unfolded slowly over the last two centuries and was put to its final rest only in our own time.

The story begins in 1746 with the famous *vibrating string problem*. Briefly, an elastic string of length L has each end fastened to one of the endpoints of the interval $[0, L]$ on the x-axis and is set into motion (as you might pluck a guitar string, for example). The problem is to determine the position $y = F(x, t)$ of the string at time t, given only its initial position $y = f(x) = F(x, 0)$ at time $t = 0$ where, for simplicity, we assume that the initial velocity $F_t(x, 0) = 0$. The function $F(x, t)$ is the solution to d'Alembert's *wave equation*: $F_{tt} = a^2 F_{xx}$, where a is a positive constant determined by certain physical properties of the string. The initial data for the problem is $F(x, 0) = f(x)$, $F_t(x, 0) = 0$, and $f(0) = 0 = f(L)$.

The controversy, initially between d'Alembert and Euler, centers around the nature of the functions f that may be permitted as initial positions. D'Alembert argued that the initial position f must be "continuous" (in the sense that f must be given by a single analytical expression or "formula"), while Euler insisted that f could be "discontinuous" (the initial position might be a series of straight line segments, as when the string is plucked in two or more places at once, in other words, a composite of two or more "formulas").

Now it is not hard to find particular solutions to the wave equation. Indeed, note that each of the functions $F(x, t) = \sin(k\pi x/L)\cos(ak\pi t/L)$, $k = 1, 2, 3, \ldots$, is a solution with corresponding initial position $F(x, 0) = \sin(k\pi x/L)$. If we assume the validity of term-by-term differentiation (that is, the "superposition" of solutions), this would suggest that any sum of the form

$$F(x, t) = \sum_{k=1}^{\infty} a_k \sin(k\pi x/L)\cos(ak\pi t/L) \tag{10.1}$$

is also a solution. In 1753, Daniel Bernoulli entered into the controversy by claiming that equation (10.1) is the most general solution to the vibrating string problem. Euler immediately took exception to Bernoulli's solution for, if we accept equation (10.1) as the general solution, it follows that the initial position f must satisfy

$$f(x) = \sum_{k=1}^{\infty} a_k \sin(k\pi x/L). \tag{10.2}$$

In other words, Bernoulli's solution suggests that the initial position f can always be represented by a sine series of the form (10.2). As Euler pointed out, the sum in equation (10.2) is odd and periodic, whereas no such assumptions can be made on f. (Since a "function" was understood to be a "formula," it was believed that the behavior of a function on an interval completely determined its behavior on the whole line.) Besides, it was inconceivable that a "discontinuous" initial position could be written as the sum of "continuous" functions. Bernoulli's arguments, which were based largely on physical principles, were unconvincing. His solution was rejected by most mathematicians of the time, including Euler and d'Alembert.

Controversy over the solution to the vibrating string problem would rage on for another 20 years and would come to involve several mathematicians, including Lagrange and Laplace.

The plot thickened in 1807, when Joseph Fourier resurrected Bernoulli's assertion. Fourier presented a paper on heat transfer in which he was able to solve for the steady-state temperature $T(x, y)$ of a rectangular metal plate with one edge placed on the interval $[-L, L]$ on the x-axis, and where the initial temperature along this edge $f(x) = T(x, 0)$ is known but is again "arbitrary." Fourier's solution is based on the premise that an arbitrary function f can be represented as a series of the form

$$f(x) = \frac{a_0}{2} + \sum_{n=1}^{\infty} \big(a_n \cos(n\pi x/L) + b_n \sin(n\pi x/L)\big).$$

Moreover, if the interval in question is instead $[0, L]$, then it suffices to use only sines (as in Bernoulli's series) or only cosines in the representation.

If, for simplicity, we take $L = \pi$, then the **Fourier series** for f over the interval $[-\pi, \pi]$ is given by

$$f(x) = \frac{a_0}{2} + \sum_{n=1}^{\infty} \big(a_n \cos nx + b_n \sin nx\big). \tag{10.3}$$

Fourier justified this equation in much the same way that Euler and Lagrange had done before him; he argued that if the **Fourier coefficients** $a_0, a_1, \ldots, b_1, b_2, \ldots$ could actually be determined, that is, if equation (10.3) could be *solved*, then it must be valid. To determine b_m, for example, we simply multiply both sides of equation (10.3) by $\sin mx$ and integrate over the interval $[-\pi, \pi]$ to obtain

$$\begin{aligned}
&\int_{-\pi}^{\pi} f(x) \sin mx \, dx \\
&\quad = \int_{-\pi}^{\pi} \left[\frac{a_0}{2} \sin mx + \sum_{n=1}^{\infty} \big(a_n \cos nx \sin mx + b_n \sin nx \sin mx\big) \right] dx
\end{aligned}$$

$$= \frac{a_0}{2} \int_{-\pi}^{\pi} \sin mx \, dx + \sum_{n=1}^{\infty} a_n \int_{-\pi}^{\pi} \cos nx \sin mx \, dx$$

$$+ \sum_{n=1}^{\infty} b_n \int_{-\pi}^{\pi} \sin nx \sin mx \, dx$$

$$= b_m \int_{-\pi}^{\pi} \sin^2 mx \, dx = b_m \pi,$$

since all of the remaining integrals are zero. A similar calculation shows that $a_m = (1/\pi) \int_{-\pi}^{\pi} f(x) \cos mx \, dx$. Thus, if we assume the existence of the various integrals in this calculation, and if we assume that term-by-term integration of the series is permitted, then equation (10.3) can be solved.

Fourier's real innovation was not in his verification of equation (10.3) – in fact, his calculations were considered to be clumsy and nonrigorous – but rather in its interpretation. Fourier argued that the Fourier coefficients of an arbitrary (but presumably bounded) function could always be determined by interpreting πb_m, for example, as the area bounded by the graph of $y = f(x) \sin mx$ and the x-axis between $x = -\pi$ and $x = \pi$. In other words, he transformed the question of existence of the series representation into the geometrically obvious "fact" that the area under a curve can always be computed.

But, as we will see later, it is not at all clear how to define the integral of an "arbitrary" function. Moreover, term-by-term integration (that is, the interchange of limits) is not so easy to justify – the question of convergence of the series enters the picture. For these reasons, Fourier's work was not well received and his ideas on trigonometric series went unpublished until the appearance of his classic book, *Théorie Analytique de la Chaleur*, in 1822.

In particular, Fourier's methods allow for a discontinuous function to be written as a sum of continuous functions (in the modern sense of the words; see Exercise 3), which was an unthinkable consequence at the time. It was so unthinkable that Cauchy was prompted to set the record straight in his famous *Cours d'Analyse* of 1821. Cauchy's refutation of Fourier's results, often called Cauchy's *wrong* theorem, states that a convergent sum of continuous functions must again be a continuous function. (The problem, as we will see, comes in the interpretation of the word "convergent.") Nevertheless, Fourier's methods seemed to work. In fact, the general consensus at the time was that *both* Cauchy and Fourier were right, although a few details would obviously have to be straightened out; this was an uncomfortable point of view in the newly born age of rigor.

As early as 1826, Abel noted that there were exceptions to Cauchy's theorem and attempted to find the "safe domain" of Cauchy's results. But the latent contradiction in Cauchy's theorem was not fully revealed until 1847, when Seidel discovered the hidden assumption in Cauchy's proof and, in so doing, introduced the concept of uniform convergence.

Although Fourier was never able to fully justify his less than rigorous arguments, the questions raised by his work would inspire mathematicians for years to come. To quote a recent article by González–Velasco:

> It was the success of Fourier's work in applications that made necessary a redefinition of the concept of function, the introduction of a definition of convergence, a reexamination of the concept of integral, and the ideas of uniform continuity and uniform convergence. It also provided motivation for the theory of sets, was in the background of ideas leading to measure theory, and contained the germs of the theory of distributions.

EXERCISES

1. Let $f(x)$ and $g(x)$ be any two distinct choices from the list 1, $\cos x$, $\sin x$, $\cos 2x$, $\sin 2x, \ldots, \cos nx$, $\sin nx$. Show that $\int_{-\pi}^{\pi} f(x)\, g(x)\, dx = 0$ while $\int_{-\pi}^{\pi} f(x)^2\, dx \neq 0$.

2. Use the result in Exercise 1 to conclude that the functions $1, \cos x, \sin x, \cos 2x$, $\sin 2x, \ldots, \cos nx$, and $\sin nx$ are linearly independent.

3. Here is one of Fourier's examples: Consider the "square wave" shown in Figure 10.1. (By including the *vertical* segments in the graph, Fourier imagined this as the graph of a continuous function.) Show that the Fourier series for this function is given by $\sum_{n=1}^{\infty} (2n)^{-1} \sin 2nx$. [Hint: Do a purely "formal" calculation of the Fourier coefficients, choosing any function values you find convenient at the points $0, \pm\pi, \ldots$ (note that the series vanishes at each of these points). This same example points up another source of controversy in Fourier's work: Does term-by-term differentiation of this series produce a series representing the derivative of the "square wave"?]

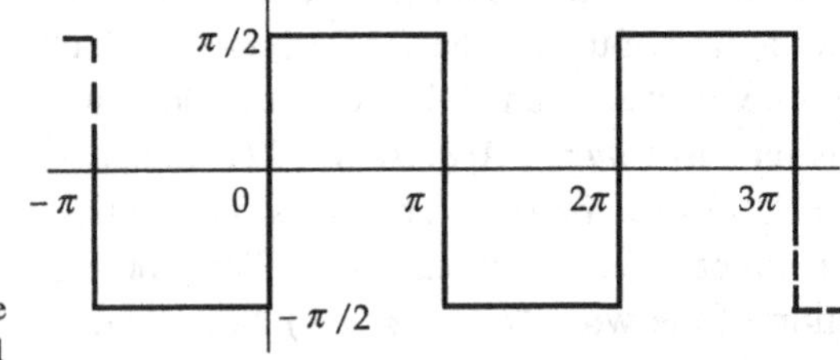

Figure 10.1

4. Let $f : \mathbb{R} \to \mathbb{R}$ be twice continuously differentiable and 2π-periodic. It follows that f' and f'' are both 2π-periodic and bounded. (Why?)

(a) Use integration by parts to show that the Fourier coefficients of f satisfy $|a_n| \le C/n$ and $|b_n| \le C/n$, for some constant C and all $n \ge 1$, and hence that $a_n \to 0$ and $b_n \to 0$.

(b) Repeat the calculation in (a) to show that $|a_n| \le C/n^2$ and $|b_n| \le C/n^2$, for some constant C and all $n \ge 1$. Use this to conclude that the Fourier series for f converges at each point of $\mathbb{R}$. (It must, in fact, converge to f, but this is somewhat harder to show.)

Pointwise and Uniform Convergence

We began our study of metric spaces in Chapter Three under the premise that such abstractions would contribute to our understanding of limits, derivatives, integrals, and sums – in other words, calculus. And while we have seen a few instances of this, we have yet to speak at any length about our very first example: The metric space $C[0,1]$. As we saw in Chapter Five, this is a space that we need to master.

In the next few chapters we will focus our attentions on $C[0,1]$ and some of its relatives. We will want to answer all of the same questions about $C[0,1]$ that we have asked of every other metric space: What are its open sets? its compact sets? Is $C[0,1]$ complete? Is it separable? And on and on. You name it, we want to know it.

The very first question we need to tackle is this: What does it mean for a sequence of functions to converge? There are many reasonable answers to this question, and we will talk about several before we are done, but only one will "do the right thing" in $C[0,1]$. For instance, given a sequence (f_n) of real-valued functions defined on $[0,1]$, we might consider the sequence of real numbers $(f_n(x))_{n=1}^{\infty}$ for each fixed x in $[0,1]$ and ask whether this sequence always converges. Or we might simply consider (f_n) as a sequence of points in the metric space $C[0,1]$ and ask whether (f_n) converges in the usual metric of $C[0,1]$. Both alternatives have their place in analysis, and both have their merits, but, for $C[0,1]$ at least, the second alternative is more appropriate.

To get a handle on this, we will want to examine both types of convergence in a variety of settings. The first type of convergence, called *pointwise convergence*, is somewhat easier to work with and, historically, is the older and more natural notion of convergence. Let's start there.

Examples 10.1

(a) Our first example takes us all the way back to Chapter One. Recall that for each fixed $x \in \mathbb{R}$, the sequence $\left((1+(x/n))^n\right)_{n=1}^{\infty}$ converges to e^x as $n \to \infty$. Said in other words, the sequence of polynomials $f_n(x) = (1+(x/n))^n$ *converge pointwise* to $f(x) = e^x$ on $\mathbb{R}$. Now this particular sequence of functions is rather well behaved; for example, recall from Exercise 1.18 that $(1+(x/n))^n$ *increases* to e^x. And, by way of bringing some calculus into the discussion, notice that for any fixed x we have

$$\frac{d}{dx}\left[\left(1+\frac{x}{n}\right)^n\right] = \left(1+\frac{x}{n}\right)^{n-1} \longrightarrow e^x = \frac{d}{dx}e^x$$

(as $n \to \infty$) and also

$$\int_0^1 \left(1+\frac{x}{n}\right)^n dx = \frac{n}{n+1}\left[\left(1+\frac{1}{n}\right)^{n+1} - 1\right] \longrightarrow e-1 = \int_0^1 e^x\,dx.$$

(b) For each n, let $g_n : [0,1] \to \mathbb{R}$ be the function whose graph is shown in Figure 10.2 (g_n is 0 outside the interval $[0,1/n]$). Then, for each $x \in [0,1]$, the sequence $g_n(x) \to 0$ as $n \to \infty$. Indeed, $g_n(0) = 0$ for any n, while if $x > 0$, then $g_n(x) = 0$ whenever $n > 1/x$. We say that $g_n \to 0$ *pointwise* on $[0,1]$.

But notice that $\int_0^1 g_n = 1 \not\to 0$. What happened? Integration is supposed to be continuous!

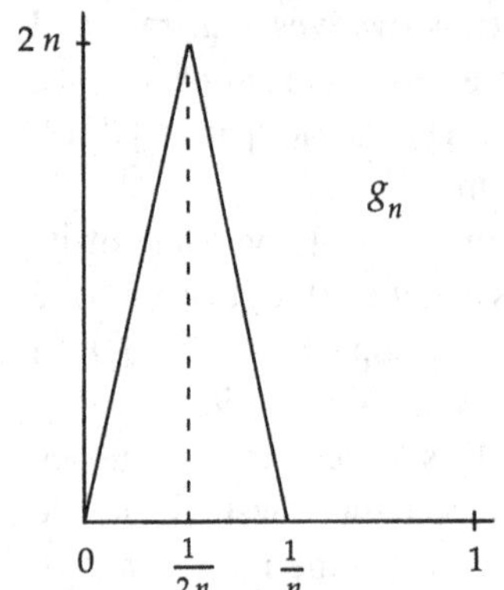

Figure 10.2

(c) Consider the sequence of functions $h_n : [0,1] \to \mathbb{R}$ given by $h_n(x) = x^{n+1}/(n+1)$. Again, $h_n \to 0$ pointwise on $[0,1]$; in fact, $|h_n(x)| \le 1/(n+1) \to 0$ as $n \to \infty$ for any x in $[0,1]$. But now what about $h_n'(x) = x^n$? Well, $h_n'(1) = 1$ for any n, and if $0 \le x < 1$, then $\lim_{n\to\infty} h_n'(x) = \lim_{n\to\infty} x^n = 0$; that is, (h_n') tends pointwise to the function k defined by $k(x) = 0$ for $0 \le x < 1$ and $k(1) = 1$. In particular,

$$\lim_{n\to\infty} h_n'(1) = 1 \ne 0 = \left(\frac{d}{dx} \lim_{n\to\infty} h_n(x)\right)\bigg|_{x=1}.$$

Isn't this annoying? To make matters worse, notice that the limit function k isn't even continuous. What's wrong?

(d) The pointwise limit of a sequence of functions has come up several times in our discussions of ℓ_1, ℓ_2, and ℓ_∞, under the alias "coordinatewise" convergence. For example, recall that in our proof that ℓ_2 is complete we found a *candidate* for the limit of a Cauchy sequence in ℓ_2 by first computing the *pointwise* limit of the sequence. That is, a sequence (f_n) in ℓ_2 is really a sequence of *functions* on $\mathbb{N}$, and so we may consider their pointwise limit $f(k) = \lim_{n\to\infty} f_n(k)$ for $k \in \mathbb{N}$. A similar device was used in Example 7.8, where we noted that the sequence $f_n = (1, \ldots, 1, 0, \ldots) \in \ell_\infty$ (where the first n entries are 1 and the rest are 0) converges pointwise on $\mathbb{N}$ to $f = (1, 1, \ldots)$ (all 1) but that this pointwise limit is not a limit in the metric of ℓ_∞. A more familiar example is provided by the ubiquitous sequence (e_n). We noted in Chapter Three that (e_n) tends pointwise to 0 on $\mathbb{N}$ but not in the metric of any of the spaces ℓ_1, ℓ_2, or ℓ_∞. Indeed, as we pointed out at the time, convergence in any of these spaces is "stronger" than pointwise convergence in the sense that convergence in the norm of ℓ_1, ℓ_2, or ℓ_∞ implies coordinatewise or pointwise convergence on $\mathbb{N}$, but not conversely. (See the discussion immediately preceding Exercise 3.40 and Exercise 3.40 itself for a positive result in this vein.)

(e) A similar line of reasoning applies to $\mathbb{R}^n$ as well. In this case we might consider an element of $\mathbb{R}^n$ as a function on the set $\{1, \ldots, n\}$ (as we did in our discussion of $C(M)$, where M is a finite set, at the end of Chapter Five). In $\mathbb{R}^n$, of course, coordinatewise convergence of sequences coincides with convergence in any norm. (Why?)

Our first three examples concerned the *interchange* of limits, as in $\lim_{n\to\infty} \int f_n = \int \lim_{n\to\infty} f_n$. While the interchange of pointwise limits worked just fine in Example 10.1 (a), it failed miserably in the next two examples. The interchange of limits typically requires something more than just pointwise convergence. In any case, pointwise convergence is evidently not the "right" mode of convergence for $C[0,1]$ because we already know that integration acts continuously on $C[0,1]$ and so should commute with a limit *in the metric of* $C[0,1]$. Before we say more, let's examine the formal definition of pointwise convergence.

Let X be any set, let (Y, ρ) be a metric space, and let f and (f_n) be functions mapping X into Y. We say that the sequence (f_n) converges **pointwise** to f on X if, for each $x \in X$, the sequence $(f_n(x))$ converges to $f(x)$ in Y. That is,

> (f_n) converges pointwise to f on X if, for each point $x \in X$ and for each $\varepsilon > 0$, there is an integer $N \geq 1$ (which depends on both x and ε) such that $\rho(f_n(x), f(x)) < \varepsilon$ whenever $n \geq N$.

Please note that since we are interested only in the distance between function values, pointwise convergence has very little to do with the domain space X; all we need is a distance function on (and, hence, a notion of convergence in) the target space Y. In discussing pointwise convergence, you may find it helpful to think of a sequence of functions (f_n) as simply a "table" of values, with n determining the "rows" and each $x \in X$ determining a "column." The values $f_1(x)$, as x ranges over X, are put in the first row; the values $f_2(x)$, for $x \in X$, are put in the second row; and so on. To say that (f_n) converges pointwise means that each "column" of values, taken one at a time, converges (as $n \to \infty$).

Also notice that since the convergence of a sequence $(f_n(x))$ is tested at each fixed x, one x at a time, the rate of convergence $N = N(x, \varepsilon)$ at one x may be vastly different than at another x. In our "tabular" framework this means that nearby *rows* in the table formed by a pointwise convergent sequence of functions might be very different when compared over all x. All we can say with certainty is that the entries in a single *column* eventually begin to look alike, provided that we read beyond some Nth row – and just how far down the column we have to read before this happens may vary with each column or x value. This point is well illustrated by several of our earlier examples; let's take another look:

Examples 10.2

(a) While the sequence $f_n(x) = (1 + (x/n))^n$ converges pointwise on $\mathbb{R}$ to $f(x) = e^x$, note that since each f_n is a *polynomial* in x, each is necessarily *unbounded* for large x. In particular, for n fixed, $|(1 + (x/n))^n| \to \infty$ as $x \to -\infty$, while $e^x \to 0$ as $x \to -\infty$. Thus, for any fixed n, we have $|f_n(x) - f(x)| \to \infty$ as $x \to -\infty$. A more delicate calculation (with n still fixed) will also show that $|f_n(x) - f(x)| \to \infty$ as $x \to \infty$. Just how large to take x before, say, $|f_n(x) - f(x)| \geq 1$, will vary with each n.

(b) Consider the sequence (g_n) of Example 10.1 (b). Although $g_n(x) \to 0$ as $n \to \infty$ for each fixed x, there are plenty of x for which an individual $g_n(x)$ is far from 0. In particular, $g_n(1/2n) = 2n \to \infty$. (At $x = (1/2n)$, we would need $N \geq 1/x = 2n$ to have $g_N(x) = 0$.)

(c) Next consider the sequence $k_n(x) = x^n$ on $[0, 1]$. Pictured in Figure 10.3 are the graphs of k_n for $n = 1, 2, 4, 6$, and 16. As noted earlier, $k_n(1) = 1$ for every n, while $k_n(x) \to 0$ for $x < 1$. That is, (k_n) converges pointwise to the function k in Example 10.1 (c). But notice, too, that near $x = 1$ each $k_n(x)$ is necessarily far from 0. In fact, $k_n\big(1/\sqrt[n]{2}\big) = 1/2$ for every n while $\sqrt[n]{2} \to 1$ as $n \to \infty$.

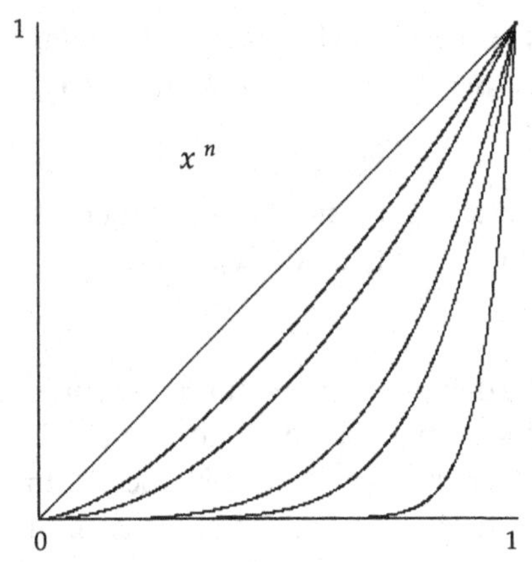

Figure 10.3

Now that we have had a chance to play around with an inappropriate mode of convergence in $C[0, 1]$, let's see if we can do better. We already know a metric on $C[0, 1]$, and so we know what it means for a sequence (f_n) in $C[0, 1]$ to converge to a function f in the metric of $C[0, 1]$; it means that $\| f_n - f \|_\infty \to 0$ as $n \to \infty$. That is, $\sup_{0\leq x\leq 1} |f_n(x) - f(x)| \to 0$ as $n \to \infty$. If we expand this into an "ε, N" statement, we will be able to compare it with the definition of pointwise convergence:

> $f_n \to f$ in the norm of $C[0, 1]$ if, for every $\varepsilon > 0$, there is some N (which may depend on ε) such that $\sup_{0\leq x\leq 1} |f_n(x) - f(x)| < \varepsilon$ for all $n \geq N$.

And now let's remove that supremum:

> $f_n \to f$ in the norm of $C[0, 1]$ if, for every $\varepsilon > 0$, there is some N (which may depend on ε) such that $|f_n(x) - f(x)| < \varepsilon$ for all $0 \leq x \leq 1$ and all $n \geq N$. In other words, the inequality $|f_n(x) - f(x)| < \varepsilon$ is to hold *uniformly* in x (for large n).

Again appealing to our "tabular" analogy, the table for a sequence (f_n) that converges in the norm of $C[0, 1]$ has the property that all of the rows, beyond some Nth row, are *uniformly* similar, independent of the columns. The key, of course, is the sup-norm; we have insisted that the *maximum* pointwise difference between f_n and f be made small. To put this in more familiar terms, recall that (f_n) converges to f in the metric of $C[0,1]$ if (f_n) is eventually in $B_\varepsilon(f) = \{g \in C[0,1] : \|f-g\|_\infty < \varepsilon\}$, and that $B_\varepsilon(f)$ is the set of functions in $C[0,1]$ whose graphs are at a *maximum* vertical distance of ε from the graph of f. Another picture might help; see Figure 10.4.

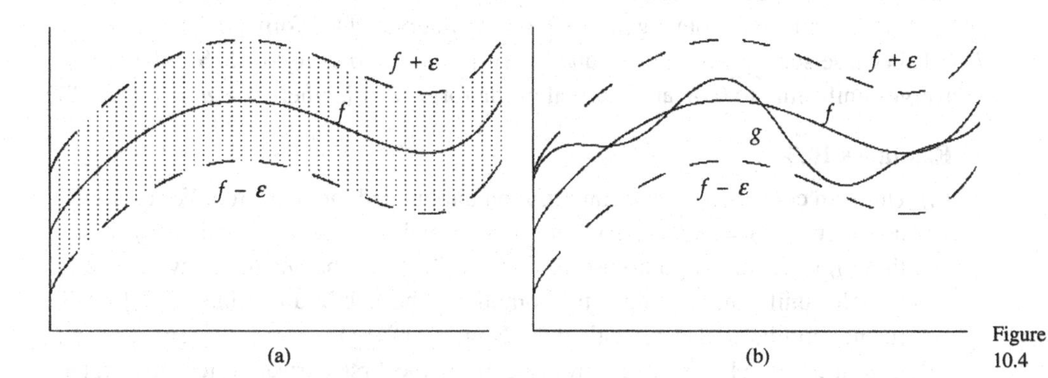

Figure 10.4

The shaded region in Figure 10.4 (a) is the set $\{(x,y) : |y - f(x)| < \varepsilon\}$. A function $g \in C[0,1]$ is in $B_\varepsilon(f)$ precisely when its graph lies within this region, as depicted in Figure 10.4 (b).

Let's recall our first few examples. For the sequence (g_n) in Example 10.1 (b) we have $\|g_n\|_\infty = \|g_n - 0\|_\infty = 2n \not\to 0$. Thus, while (g_n) does converge pointwise to 0 on $[0,1]$, it does not converge to 0 in the metric of $C[0,1]$. In fact, (g_n) cannot converge to *any* function in the metric of $C[0,1]$ since it is not a bounded sequence in $C[0,1]$. For the sequence (h_n) in Example 10.1 (c) we have $\|h_n\|_\infty = 1/(n+1) \to 0$, and hence (h_n) converges to 0 in the metric of $C[0,1]$. Finally, the sequence (k_n) of Example 10.2 (c) does not converge to any function in $C[0,1]$ (the function k certainly is not a candidate since it is not continuous). Why? Because (k_n) is not a Cauchy sequence in $C[0,1]$: Indeed, $\|k_n - k_{2n}\|_\infty \ge \left|k_n\left(1/\sqrt[n]{2}\right) - k_{2n}\left(1/\sqrt[n]{2}\right)\right| = (1/2) - (1/4) = 1/4$.

Convergence in the metric of $C[0,1]$ is called **uniform convergence**. It has little to do with continuous functions and a lot to do with the sup-norm (which, for this reason, is sometimes called the *uniform* norm). The formal definition should explain everything.

Let X be any set, let (Y,ρ) be a metric space, and let f and (f_n) be functions mapping X into Y. We say that the sequence (f_n) converges **uniformly** to f on X if, for each $\varepsilon > 0$, there is some $N \ge 1$ (which may depend on ε) such that $\rho(f_n(x), f(x)) < \varepsilon$ for all $x \in X$ and all $n \ge N$.

To highlight the fact that $\rho(f_n(x), f(x))$ is *uniformly* small for all $x \in X$, we might replace it by $\sup_{x\in X} \rho(f_n(x), f(x))$; that is, note that (f_n) converges uniformly to f if and only if, for each $\varepsilon > 0$, there is some N such that $\sup_{x\in X} \rho(f_n(x), f(x)) < \varepsilon$ for all $n \ge N$. (Why?) Said in still other words, (f_n) converges uniformly to f on X if and only if $\sup_{x\in X} \rho(f_n(x), f(x)) \to 0$ as $n \to \infty$. (Look familiar?)

Notice that a uniformly convergent sequence is also pointwise convergent (to the same limit). In other words, uniform convergence is "stronger" than pointwise convergence. (Why?)

In this notation we would say that the sequence (g_n) of Example 10.1 (b) converges pointwise to 0 on $[0, 1]$, but not uniformly; the sequence (h_n) of Example 10.1 (c) converges uniformly to 0 on $[0, 1]$; and the sequence (k_n) of Example 10.2 (c) converges pointwise to k on $[0, 1]$, but not uniformly. Notice, too, that uniform convergence depends on the underlying domain. Indeed, although (k_n) is not uniformly convergent on all of $[0, 1]$, it is uniformly convergent (to 0) on any interval of the form $[0, b]$, where $0 < b < 1$, because $\sup_{0 \le x \le b} |k_n(x)| = \sup_{0 \le x \le b} |x^n| = b^n \to 0$ as $n \to \infty$. Similarly, (g_n) converges uniformly to 0 on any interval of the form $[a, 1]$, where $0 < a < 1$. (Why?)

Examples 10.3

(a) Uniform convergence is meaningful on unbounded intervals, too. For example, consider $f_n(x) = x/(1 + nx^2)$ for $x \in \mathbb{R}$ and $n = 1, 2, \ldots$. It is easy to see that (f_n) converges pointwise to 0 on $\mathbb{R}$. To test whether the convergence is actually uniform, we might try computing the maximum value of $|f_n|$ on $\mathbb{R}$ (using familiar tools from calculus). Now $f_n'(x) = (1 - nx^2)/(1 + nx^2)^2$, which is 0 at $x = \pm 1/\sqrt{n}$, and it follows from the first derivative test that $f_n(\pm 1/\sqrt{n}) = \pm 1/(2\sqrt{n})$ are the maximum and minimum values of f_n. That is, $\sup_{x \in \mathbb{R}} |f_n(x)| = 1/(2\sqrt{n}) \to 0$ as $n \to \infty$, and so (f_n) converges uniformly to 0 on $\mathbb{R}$.

(b) Uniform convergence is also meaningful for unbounded functions. A somewhat contrived example should be sufficient to see what is going on. If we set $g_n(x) = x^3 + (1/n)$ for $x \in \mathbb{R}$ and $n = 1, 2, \ldots$, then, clearly, (g_n) converges uniformly to $g(x) = x^3$ on $\mathbb{R}$. (Why?) In other words, the functions g_n need not be bounded; the important thing is that the difference $g_n - g$ must be bounded (and tend uniformly to 0 of course).

(c) For bounded, real-valued functions on $\mathbb{N}$, uniform convergence is the same as convergence in the metric of ℓ_∞. That is, if $f, f_n \in \ell_\infty$, then (f_n) converges uniformly to f on $\mathbb{N}$ if and only if $\| f_n - f \|_\infty \to 0$ as $n \to \infty$.

(d) If we identify $\mathbb{R}^n$ with the real-valued functions on the set $\{1, \ldots, n\}$, then uniform convergence on $\{1, \ldots, n\}$ coincides with convergence in *any* norm on $\mathbb{R}^n$. (Why?)

By way of shorthand, we will occasionally (and sparingly) use the following notation. We write $f_n \to f$ on X, or $f_n \overset{X}{\to} f$ (with no additional quantifiers), to mean that (f_n) converges *pointwise* to f on X. We write $f_n \rightrightarrows f$ on X, or $f_n \overset{X}{\rightrightarrows} f$, to mean that (f_n) converges *uniformly* to f on X. This notation is intended as a visual reminder that uniform convergence is "stronger" than pointwise convergence. But, just to be on the safe side, any additional quantifiers always take precedence; for example, the statements "$f_n \to f$ uniformly on X" and "$f_n \to f$ in (the metric of) $C[0, 1]$" should be interpreted to mean that (f_n) converges uniformly to f. Obviously, we will have to be careful to avoid any confusion caused by this variety of notations. A comparison of the "abbreviated" definitions of pointwise versus uniform convergence pinpoints their differences: $f_n \overset{X}{\to} f$ means

$$\forall x \in X,\ \forall \varepsilon > 0,\ \exists N \ge 1 \text{ such that } \rho(f_n(x), f(x)) < \varepsilon,\ \forall n \ge N,$$

while $f_n \overset{X}{\rightrightarrows} f$ means

$$\forall \varepsilon > 0, \ \exists N \geq 1 \text{ such that } \rho(f_n(x), f(x)) < \varepsilon, \ \forall x \in X, \ \forall n \geq N.$$

In other words, just as in the case of uniform continuity, the quantifier "$\forall x$" has moved forward (and so ε and N no longer depend on x).

EXERCISES

5. Suppose that $f_n : [a, b] \to \mathbb{R}$ is an increasing function for each n, and that $f(x) = \lim_{n\to\infty} f_n(x)$ exists for each x in $[a, b]$. Is f increasing?

6. Let $f_n : [a, b] \to \mathbb{R}$ satisfy $|f_n(x)| \leq 1$ for all x and n. Show that there is a subsequence (f_{n_k}) such that $\lim_{k\to\infty} f_{n_k}(x)$ exists for each *rational* x in $[a, b]$. [Hint: This is a "diagonalization" argument.]

▷ **7.** Let (f_n) and (g_n) be real-valued functions on a set X, and suppose that (f_n) and (g_n) converge uniformly on X. Show that $(f_n + g_n)$ converges uniformly on X. Give an example showing that $(f_n g_n)$ need not converge uniformly on X (although it will converge pointwise, of course).

8. Let $f_n : \mathbb{R} \to \mathbb{R}$, and suppose that $f_n \rightrightarrows 0$ on every closed, bounded interval $[a, b]$. Does it follow that $f_n \rightrightarrows 0$ on $\mathbb{R}$? Explain.

▷ **9.** For each of the following sequences, determine the pointwise limit on the given interval (if it exists) and the intervals on which the convergence is uniform (if any):

(a) $f_n(x) = x^n$ on $(-1, 1]$;
(b) $f_n(x) = n^2 x(1 - x^2)^n$ on $[0, 1]$;
(c) $f_n(x) = nx/(1 + nx)$ on $[0, \infty)$;
(d) $f_n(x) = nx/(1 + n^2x^2)$ on $[0, \infty)$;
(e) $f_n(x) = xe^{-nx}$ on $[0, \infty)$;
(f) $f_n(x) = nxe^{-nx}$ on $[0, \infty)$.

In each of the above examples, will term-by-term integration or differentiation lead to a correct result?

10. Let $f : \mathbb{R} \to \mathbb{R}$ be uniformly continuous, and define $f_n(x) = f(x + (1/n))$. Show that $f_n \rightrightarrows f$ on $\mathbb{R}$.

11. Suppose that $f_n \rightrightarrows f$ on $\mathbb{R}$, and that $f : \mathbb{R} \to \mathbb{R}$ is continuous. Show that $f_n(x + (1/n)) \to f(x)$ (pointwise) on $\mathbb{R}$.

12. Prove that a sequence of functions $f_n : X \to \mathbb{R}$, where X is any set, is uniformly convergent if and only if it is *uniformly Cauchy*. That is, prove that there exists some $f : X \to \mathbb{R}$ such that $f_n \rightrightarrows f$ on X if and only if, for each $\varepsilon > 0$, there exists an $N \geq 1$ such that $\sup_{x\in X} |f_n(x) - f_m(x)| < \varepsilon$ whenever $m, n \geq N$. [Hint: Notice that if (f_n) is uniformly Cauchy, then it is also *pointwise Cauchy*. That is, if $\sup_{x\in X} |f_n(x) - f_m(x)| \to 0$ as $m, n \to \infty$, then $(f_n(x))$ is Cauchy in $\mathbb{R}$ for each $x \in X$.]

13. Here is a "negative" test for uniform convergence: Suppose that (X, d) and (Y, ρ) are metric spaces, that $f_n : X \to Y$ is continuous for each n, and that (f_n) converges pointwise to f on X. If there exists a sequence (x_n) in X such that $x_n \to x$ in X but $f_n(x_n) \not\to f(x)$, show that (f_n) *does not* converge uniformly to f on X.

Interchanging Limits

As we have seen, pointwise convergence is not always enough to guarantee the interchange of limits. In this section we will see that uniform convergence, on the other hand, does often allow for an interchange of limits.

As a first result along these lines, we will prove that the uniform limit of a sequence of continuous functions is again continuous. (Compare this with Cauchy's "wrong" theorem.)

Theorem 10.4. *Let (X, d) and (Y, ρ) be metric spaces, and let f and (f_n) be functions mapping X into Y. If (f_n) converges uniformly to f on X, and if each f_n is continuous at $x \in X$, then f is also continuous at x.*

PROOF. Let $\varepsilon > 0$. Since (f_n) converges uniformly to f, we can find an m such that $\rho(f(y), f_m(y)) < \varepsilon/3$ for all $y \in X$ (we only need one such m). Next, since f_m is continuous at x, there is a $\delta > 0$ such that $\rho(f_m(x), f_m(y)) < \varepsilon/3$ whenever $d(x, y) < \delta$. Thus, if $d(x, y) < \delta$, then

$$\begin{aligned}\rho(f(x), f(y)) &\le \rho(f(x), f_m(x)) + \rho(f_m(x), f_m(y)) + \rho(f_m(y), f(y)) \\ &< \varepsilon/3 + \varepsilon/3 + \varepsilon/3 = \varepsilon. \quad \square\end{aligned}$$

To see that Theorem 10.4 is indeed a statement about the interchange of limits, let's rewrite its conclusion. If $x_m \to x$ in X, then

$$f(x) = \lim_{n\to\infty} f_n(x) = \lim_{n\to\infty} \lim_{m\to\infty} f_n(x_m),$$

since (f_n) converges pointwise to f and each f_n is continuous at x. To say that f is also continuous at x would mean that

$$f(x) = \lim_{m\to\infty} f(x_m) = \lim_{m\to\infty} \lim_{n\to\infty} f_n(x_m).$$

Thus, in the presence of uniform convergence, we must have

$$\lim_{n\to\infty} \lim_{m\to\infty} f_n(x_m) = \lim_{m\to\infty} \lim_{n\to\infty} f_n(x_m).$$

In particular, Theorem 10.4 tells us that the space $C[a, b]$ is closed under the taking of uniform limits. That is, if (f_n) is a sequence in $C[a, b]$, and if (f_n) converges uniformly to f on $[a, b]$, then $f \in C[a, b]$. This is very comforting since, as we have seen, convergence in the metric of $C[a,b]$ coincides with uniform convergence. Specifically,

$$f_n \to f \text{ in } C[a,b] \iff \|f_n - f\|_\infty \to 0 \iff f_n \rightrightarrows f \text{ on } [a,b].$$

EXERCISES

▷ **14.** Let $f_n : \mathbb{R} \to \mathbb{R}$ be continuous for each n, and suppose that $f_n \rightrightarrows f$ on each closed, bounded interval $[a,b]$. Show that f is continuous on $\mathbb{R}$.

15. Let (X,d) and (Y,ρ) be metric spaces, and let $f, f_n : X \to Y$ with $f_n \rightrightarrows f$ on X. If each f_n is continuous at $x \in X$, and if $x_n \to x$ in X, prove that $\lim_{n\to\infty} f_n(x_n) = f(x)$.

16. Let (X,d) and (Y,ρ) be metric spaces, and let $f, f_n : X \to Y$ with $f_n \rightrightarrows f$ on X. Show that $D(f) \subset \bigcup_{n=1}^\infty D(f_n)$, where $D(f)$ is the set of discontinuities of f.

17. Suppose that $f, f_n : X \to \mathbb{R}$.

(a) Show that the set on which (f_n) converges pointwise to f is given by $\bigcap_{k=1}^\infty \bigcup_{m=1}^\infty \bigcap_{n=m}^\infty \{x : |f_n(x) - f(x)| \le (1/k)\}$.

(b) What is the set on which $(f_n(x))$ is Cauchy? If X is a metric space, and if each f_n is continuous on X, what type of set is this?

▷ **18.** Here is a partial converse to Theorem 10.4, called *Dini's theorem*. Let X be a compact metric space, and suppose that the sequence (f_n) in $C(X)$ increases pointwise to a *continuous* function $f \in C(X)$; that is, $f_n(x) \le f_{n+1}(x)$ for each n and x, and $f_n(x) \to f(x)$ for each x. Prove that the convergence is actually uniform. The same is true if (f_n) decreases pointwise to f. [Hint: First reduce to the case where (f_n) decreases pointwise to 0. Now, given $\varepsilon > 0$, consider the (open) sets $U_n = \{x \in X : f_n(x) < \varepsilon\}$.] Give an example showing that $f \in C(X)$ is necessary.

Our next two results supply an interchange of limits for integrals and derivatives.

Theorem 10.5. *Suppose that $f_n : [a,b] \to \mathbb{R}$ is continuous for each n, and that (f_n) converges uniformly to f on $[a,b]$. Then $\int_a^b f_n(x)\,dx \to \int_a^b f(x)\,dx$.*

PROOF. Note that since $f \in C[a,b]$, the integral of f is defined! Next,

$$\left| \int_a^b f_n(x)\,dx - \int_a^b f(x)\,dx \right| \le \int_a^b |f_n(x) - f(x)|\,dx$$
$$\le (b-a)\|f_n - f\|_\infty \to 0. \quad \square$$

Example 10.6

Suppose that the trigonometric series $(a_0/2) + \sum_{n=1}^\infty (a_n \cos nx + b_n \sin nx)$ is uniformly convergent on the interval $[-\pi, \pi]$. Then, according to Theorem 10.4, its sum $g(x)$ is a continuous function on $[-\pi, \pi]$. It now follows from Theorem 10.5 that this series must, in fact, be the Fourier series for $g(x)$. Indeed, for any $k = 1, 2, 3, \ldots$, we have

$$\begin{aligned}\int_{-\pi}^{\pi} g(x)\,\sin kx\,dx &= \int_{-\pi}^{\pi}\left[\frac{a_0}{2} + \sum_{n=1}^{\infty}(a_n\cos nx + b_n\sin nx)\right]\sin kx\,dx \\ &= \frac{a_0}{2}\int_{-\pi}^{\pi}\sin kx\,dx + \sum_{n=1}^{\infty} a_n \int_{-\pi}^{\pi}\cos nx\,\sin kx\,dx \\ &\quad + \sum_{n=1}^{\infty} b_n \int_{-\pi}^{\pi}\sin nx\,\sin kx\,dx \\ &= \pi b_k,\end{aligned}$$

since Theorem 10.5 grants term-by-term integration. (Why?) A similar calculation shows that $\pi a_k = \int_{-\pi}^{\pi} g(x)\cos kx\,dx$. We will return to this issue in subsequent chapters.

Now that we know how to exchange limits and integrals, the Fundamental Theorem of Calculus will tell us how to exchange limits and derivatives. While our next result may look "overspecified," it's really very useful.

Theorem 10.7. *Suppose that (f_n) is a sequence of real-valued functions, each having a continuous derivative on $[a, b]$, and suppose that the sequence of derivatives (f_n') converges uniformly to a function g on $[a, b]$. If $(f_n(x_0))$ converges at any point x_0 in $[a, b]$, then, in fact, (f_n) converges uniformly to a differentiable function f on $[a, b]$. Moreover, $f' = g$. That is, (f_n') converges uniformly to f' on $[a, b]$.*

PROOF. Let's first check that (f_n) converges pointwise to some function f on $[a, b]$. Let $C = \lim_{n\to\infty} f_n(x_0)$. Then, for any $x \in [a, b]$ we have

$$f_n(x) = f_n(x_0) + \int_{x_0}^{x} f_n' \to C + \int_{x_0}^{x} g,$$

since $f_n' \rightrightarrows g$. Thus, $f_n \to f$, where $f(x) = C + \int_{x_0}^{x} g$. It follows that $f(x) = f(a) + \int_a^x g$ for any x in $[a, b]$. The right-hand side of this expression is (continuously) differentiable and, hence, so is f. Moreover, $f' = g$. That is, $f_n' \rightrightarrows f'$ on $[a, b]$.

Finally, to show that (f_n) converges uniformly to f, we just repeat our first calculation:

$$\begin{aligned}|f_n(x) - f(x)| &= \left| f_n(a) - f(a) + \int_a^x (f_n' - f')\right| \\ &\le |f_n(a) - f(a)| + \int_a^x |f_n' - f'| \\ &\le |f_n(a) - f(a)| + (b-a)\|f_n' - f'\|_\infty \to 0.\end{aligned}$$

The right-hand side tends to 0 independent of x; hence, $|f_n(x) - f(x)| \to 0$ uniformly in x. □

EXERCISES

19. Suppose that (f_n) is a sequence of functions in $C[0,1]$ and that $f_n \rightrightarrows f$ on $[0,1]$. True or false? $\int_0^{1-(1/n)} f_n \to \int_0^1 f$.

20. $C^{(1)}[a,b]$ is the vector space of all functions $f : [a,b] \to \mathbb{R}$ having a continuous first derivative on $[a,b]$. Show that $C^{(1)}[a,b]$ is complete under the norm $\|f\|_{C^{(1)}} = \max_{a \le x \le b} |f(x)| + \max_{a \le x \le b} |f'(x)|$.

21. Use Dini's theorem to conclude that the sequence $(1 + (x/n))^n$ converges uniformly to e^x on every compact interval in $\mathbb{R}$. How does this explain the findings in Example 10.1 (a)?

22. Recall that we have defined a metric on $C(\mathbb{R})$ by setting $d(f,g) = \sum_{n=1}^{\infty} 2^{-n} d_n(f,g)/(1 + d_n(f,g))$, where $d_n(f,g) = \max_{|t| \le n} |f(t) - g(t)|$ (see Exercise 5.64). Prove that (f_n) converges to f in the metric of $C(\mathbb{R})$ if and only if (f_n) converges uniformly to f on every *compact* subset of $\mathbb{R}$. For this reason, convergence in $C(\mathbb{R})$ is sometimes called *uniform convergence on compacta.*

The Space of Bounded Functions

Given a set X, we write $B(X)$ for the vector space of all bounded, real-valued functions $f : X \to \mathbb{R}$, and we supply $B(X)$ with the sup-norm $\|f\|_\infty = \sup_{x \in X} |f(x)|$. That is, $B(X)$ is just $\ell_\infty(X)$ with a new name. (The notation $B(X)$ is somewhat more commonplace than $\ell_\infty(X)$.) Thus, convergence in $B(X)$ is the *same* as uniform convergence. Specifically,

$$f_n \to f \text{ in } B(X) \iff \|f_n - f\|_\infty \to 0 \iff f_n \rightrightarrows f \text{ on } X.$$

Moreover, $B(X)$ is *complete* under the sup-norm. The proof is exactly the same as that for $\ell_\infty(X)$, of course, which means that it is essentially the same as that for ℓ_∞. (Compare the proof of the following lemma with the "three-step" procedure outlined in Chapter Seven.)

Lemma 10.8. *If (f_n) is a Cauchy sequence in $B(X)$, then (f_n) converges uniformly to some $f \in B(X)$. Moreover,* $\sup_n \|f_n\|_\infty < \infty$ *and* $\|f_n\|_\infty \to \|f\|_\infty$ *as* $n \to \infty$.

PROOF. The last two assertions follow from general principles: If (f_n) is a Cauchy sequence in $B(X)$, then (f_n) is also a bounded sequence in $B(X)$; that is, $\sup_n \|f_n\|_\infty < \infty$. And if (f_n) converges to f in the norm of $B(X)$, then $\|f_n\|_\infty \to \|f\|_\infty$ as $n \to \infty$. (Why?)

Now, if (f_n) is Cauchy in $B(X)$, then (f_n) is also pointwise Cauchy; that is, for each $x \in X$ we have $|f_m(x) - f_n(x)| \le \|f_m - f_n\|_\infty \to 0$ as $m, n \to \infty$, and so $(f_n(x))$ is a Cauchy sequence in $\mathbb{R}$ for each $x \in X$. Consequently, $f(x) = \lim_{n\to\infty} f_n(x)$ exists for each $x \in X$. But, as we have already noted, (f_n) is a bounded sequence in $B(X)$; thus, $|f(x)| = \lim_{n\to\infty} |f_n(x)| \le \sup_n \|f_n\|_\infty = C$, and hence $\|f\|_\infty \le C$, too. That is, $f \in B(X)$.

Finally, to see that (f_n) converges uniformly to f, let $\varepsilon > 0$ and $x \in X$. Then

$$|f(x) - f_n(x)| = \lim_{m\to\infty} |f_m(x) - f_n(x)| \le \varepsilon,$$

for all n sufficiently large, since $|f_m(x) - f_n(x)| \le \|f_m - f_n\|_\infty < \varepsilon$ for all m, n sufficiently large. And since this estimate is independent of x, we get $\|f - f_n\|_\infty \le \varepsilon$ for all n sufficiently large. □

A Cauchy sequence in $B(X)$ is often said to be **uniformly Cauchy**, while a bounded sequence in $B(X)$ is often said to be **uniformly bounded** to emphasize the presence of the uniform, or sup-norm.

The fact that $B(X)$ is complete is even more meaningful in the case where X is a metric space, for then we may also consider the space $C(X)$ of continuous, real-valued functions on X. Now continuous functions on X are not necessarily bounded; in other words, $C(X)$ is not, in general, a subspace of $B(X)$. Thus we are led to consider the vector space $C_b(X) = C(X) \cap B(X)$, of all *bounded*, continuous, real-valued functions on X. It follows from Theorem 10.4 that $C_b(X)$ is a *closed* subspace of $B(X)$; hence $C_b(X)$ is *complete* under the sup-norm. (Why?)

If X is a *compact* metric space, then $C_b(X) = C(X)$ and, what's more, we may use the simpler expression $\|f\|_\infty = \max_{x\in X} |f(x)|$ in place of the sup-norm on $C(X)$. (Why?) In particular, $C[\,a, b\,]$ is a complete normed vector space under the sup-norm (i.e., under uniform convergence).

Now that we know that $B(X)$ is a complete normed vector space, we may take advantage of yet another observation from Chapter Seven, namely, Banach's characterization of completeness for normed spaces. The following special case of Theorem 7.12 is often called the **Weierstrass M-test.**

Lemma 10.9. *Let (g_n) be a sequence in $B(X)$ satisfying $\sum_{n=1}^\infty \|g_n\|_\infty < \infty$. Then $\sum_{n=1}^\infty g_n$ converges in $B(X)$; that is, $\sum_{n=1}^\infty g_n$ converges uniformly on X. Moreover, $\|\sum_{n=1}^\infty g_n\|_\infty \le \sum_{n=1}^\infty \|g_n\|_\infty$.*

The usual notation in most advanced calculus books is to set $M_n = \|g_n\|_\infty = \sup_{x\in X} |g_n(x)|$ (for the Max of the nth term), and consequently to require that $\sum_{n=1}^\infty M_n < \infty$. Hence the name "$M$-test."

Application 10.10. (Power Series) *If the power series $\sum_{n=0}^\infty a_n x^n$ converges for some $x_0 \ne 0$, then it converges uniformly (and absolutely) on every interval $|x| \le R$, where $0 < R < |x_0|$. Hence, the sum represents a continuous function for $|x| < |x_0|$. Moreover, term-by-term differentiation (in $|x| < |x_0|$) or integration (over $[\,a, b\,] \subset (-|x_0|, |x_0|)$) leads to a correct result.*

PROOF. First notice that if $\sum_{n=0}^\infty a_n x_0^n$ converges, then the terms in the series are bounded, say, $|a_n||x_0|^n \le C$ for all n. Next fix $0 < R < |x_0|$, and let $r = R/|x_0| < 1$.

Then, for $|x| \le R$, we get

$$|a_n x^n| = \left| a_n x_0^n \left(\frac{x}{x_0}\right)^n \right| \le Cr^n.$$

Thus, since $\sum_{n=0}^{\infty} Cr^n < \infty$, we have $\sum_{n=0}^{\infty} a_n x^n$ converging uniformly (and absolutely) on $|x| \le R$ by the M-test. The sum $\sum_{n=0}^{\infty} a_n x^n$ is then continuous on $(-|x_0|, |x_0|)$ because it is continuous on each interval $[-R, R]$ for $R < |x_0|$.

Term-by-term integration over $[a, b] \subset (-|x_0|, |x_0|)$ follows from Theorem 10.5 (applied to $f_n(x) = \sum_{k=0}^{n} a_k x^k$).

Finally, to show that the sum is differentiable, we appeal to Theorem 10.7 (again applied to $f_n(x) = \sum_{k=0}^{n} a_k x^k$). The proof relies on the same technique used above, but now we make use of the fact that $\sum_{n=1}^{\infty} nr^{n-1}$ converges for $0 \le r < 1$. It follows that the series $\sum_{n=1}^{\infty} na_n x^{n-1}$ converges uniformly (and absolutely) for $|x| \le R$, where $R < |x_0|$, and so it must converge to $(d/dx)\left(\sum_{n=0}^{\infty} a_n x^n\right)$. □

Application 10.11. (A Space-Filling Curve) *We next construct a pair of continuous functions* $x(t)$ *and* $y(t)$ *on* $[0, 1]$ *such that the curve* $t \mapsto (x(t), y(t))$ *fills the unit square* $[0, 1] \times [0, 1]$. *In fact, our construction will show that the curve maps* Δ *onto* $[0, 1] \times [0, 1]$.

PROOF. To begin, we define a map $f : \mathbb{R} \to [0, 1]$ as follows: Let $f(t) = 0$ for $0 \le t \le 1/3$, let $f(t) = 3t - 1$ for $1/3 < t < 2/3$, and let $f(t) = 1$ for $2/3 \le t \le 1$. Note that if $t \in \Delta$, then $f(t)$ is the first digit in the ternary decimal expansion of t. We next extend f to all of $\mathbb{R}$ by taking f to be *even* and *periodic*, of period 2, as shown in Figure 10.5.

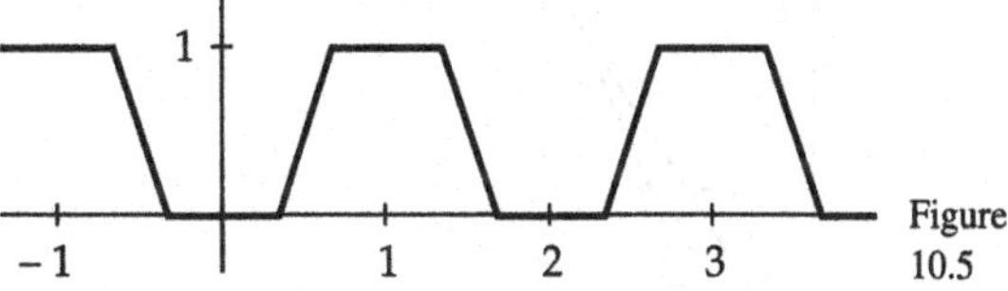

Figure 10.5

The basis of our construction lies in the observation that the function $g(t) = \sum_{k=0}^{\infty} 2^{-k-1} f\left(3^k t\right)$ agrees with the Cantor function for $t \in \Delta$. That is, $g(t)$ is another extension of the Cantor function to $[0, 1]$ (indeed, to all of $\mathbb{R}$). To see that this is so, let $t = 0.(2a_0)(2a_1)(2a_2)\cdots$ (base 3), where each a_k is 0 or 1, be a point in Δ. Then, since f is periodic with period 2, we have

$$\begin{aligned} f(3^k t) &= f\big(0.(2a_k)(2a_{k+1})(2a_{k+2})\cdots \text{ (base 3)}\big) \qquad \text{(Why?)} \\ &= 0 \quad \text{if } a_k = 0, \qquad \text{since } 0.0b_2b_3\cdots \text{ (base 3)} \in [0, 1/3] \\ &= 1 \quad \text{if } a_k = 1, \qquad \text{since } 0.2b_2b_3\cdots \text{ (base 3)} \in [2/3, 1]. \end{aligned}$$

That is, $f(3^k t) = a_k$ for $t \in \Delta$ and hence

$$g(t) = \sum_{k=0}^{\infty} 2^{-k-1} a_k = 0.a_0a_1a_2\cdots \text{ (base 2)}.$$

Now we are ready to define our curve; set $x(t) = \sum_{k=0}^{\infty} 2^{-k-1} f\left(3^{2k}t\right)$ and $y(t) = \sum_{k=0}^{\infty} 2^{-k-1} f\left(3^{2k+1}t\right)$. By the M-test, x and y are continuous on all of $\mathbb{R}$ and, clearly, each maps $\mathbb{R}$ *into* $[0, 1]$. (Why?)

To see that $\big(x(t), y(t)\big)$ fills the square, let $x_0, y_0 \in [0, 1]$ and write their base 2 decimal expansions just so:

$$x_0 = 0.a_0a_2a_4\cdots \text{ (base 2)} \qquad \text{and} \qquad y_0 = 0.a_1a_3a_5\cdots \text{ (base 2)}.$$

Now set $t_0 = 0.(2a_0)(2a_1)(2a_2)(2a_3)\cdots$ (base 3) $\in \Delta$. Then $x(t_0) = x_0$ and $y(t_0) = y_0$ since $f(3^k t_0) = a_k$ for each k. Thus the curve maps Δ onto $[0, 1] \times [0, 1]$. □

The M-test can be used to give yet another description of the Cantor function; this one more in the spirit of our "middle thirds" construction (see Chapter Two). Specifically, we will simultaneously build the nth level Cantor set (we called this set I_n in Chapter Two) and an nth level polygonal approximation $f_n : [0, 1] \to [0, 1]$ to the Cantor function f. (A *polygonal* function is a continuous function whose graph consists of finitely many straight line segments. Thus, a polygonal function is completely determined by its values at the finitely many "nodes" $x_1, \ldots, x_k$ corresponding to the finitely many "vertices" of its graph.)

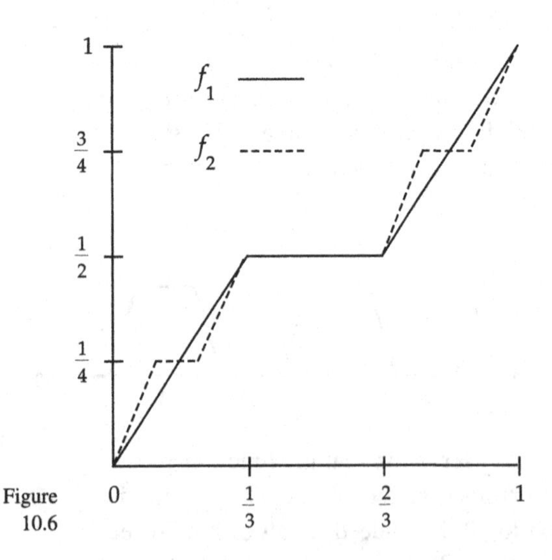

Figure 10.6

To define the first approximation f_1, set $f_1(0) = 0$, $f_1(1/3) = 1/2 = f_1(2/3)$, and $f_1(1) = 1$, and then extend f_1 to all of $[0, 1]$ by "connecting the dots." That is, f_1 is a polygonal function on $[0, 1]$ with "nodes" at the endpoints 0, 1/3, 2/3, and 1 of $I_1 = [0, 1/3] \cup [2/3, 1]$, as shown in Figure 10.6. Note that f_1 is constant on the first "discarded" interval $J_1 = (1/3, 2/3)$.

The second polygonal approximation f_2 is obtained by adding a few more nodes to the definition of f_1; namely, let f_2 agree with f_1 at each of the points 0, 1/3, 2/3, and 1, and now include $f_2(1/9) = 1/4 = f_2(2/9)$, and $f_2(7/9) = 3/4 = f_2(8/9)$, as shown in Figure 10.6. Again, f_2 has nodes at the endpoints of $I_2 = [0, 1/9] \cup [2/9, 1/3] \cup [2/3, 7/9] \cup [8/9, 1]$, and f_2 is constant on each subinterval of $J_2 =$

$(1/9, 2/9) \cup (1/3, 2/3) \cup (7/9, 8/9)$. Note that the graph of f_2 contains two "scaled-down" copies of the graph of f_1.

Can you see how we will define f_3? We will add eight more nodes to the definition of f_2, corresponding to the eight new endpoints introduced in I_3, and we will take f_3 to be constant on each of the subintervals of J_3 (using $1/8$, $3/8$, $5/8$, and $7/8$ as the four new values), so that f_3 agrees with f_2 on J_2 and agrees with f_1 on J_1. If you draw the graph of f_3, you will see four "miniature" copies of the graph of f_1 (or two copies of the graph of f_2).

If we continue this process, we will get a sequence of increasing, continuous, polygonal functions (f_n) on $[0, 1]$ such that f_n is constant on each subinterval of J_n and linear on each subinterval of I_n. In particular, each f_n is designed to agree with the Cantor function f on J_n. Using induction (based on the graphs on f_1 and f_2 and "scaling"), it is not hard to see that $\| f_{n+1} - f_n \|_\infty \le 2^{-n-1}$ for any n. Thus, the series $f_1 + \sum_{n=1}^{\infty}(f_{n+1} - f_n)$ converges uniformly to an increasing continuous function g on $[0, 1]$ (in other words, $f_n \rightrightarrows g$). But then g must agree with the Cantor function f on $\bigcup_{n=1}^{\infty} J_n = [0, 1] \setminus \Delta$, a dense subset of $[0, 1]$. Consequently, $g = f$.

Next, let's resolve an issue left over from Chapter Nine, namely, the converse to Theorem 9.2: Every F_σ subset of $\mathbb{R}$ can be realized as the set of discontinuities of some (bounded) function $f : \mathbb{R} \to \mathbb{R}$.

Application 10.12. (Discontinuous Functions) *Let F be a nonempty F_σ subset of $\mathbb{R}$. Then, $F = D(f)$ for some bounded function $f : \mathbb{R} \to \mathbb{R}$.*

PROOF. Write $F = \bigcup_{n=1}^{\infty} F_n$, where each F_n is a closed set in $\mathbb{R}$. Since finite unions of closed sets are again closed, we may assume that $F_n \subset F_{n+1}$ for each n. Now, for each n, let $G_n = \mathbb{Q} \cap F_n^\circ$, the rationals in the interior of F_n, and let $f_n = \chi_{F_n} - \chi_{G_n} = \chi_{F_n \setminus G_n}$. Then, f_n is clearly continuous at each point in the complement of F_n, and f_n is discontinuous on F_n since the oscillation of f_n is 1 at each point of F_n. (Why?) Thus, $D(f_n) = F_n$.

Next, let $f = \sum_{n=1}^{\infty} 4^{-n} f_n$. It follows from the M-test (and Theorem 10.4) that f is a bounded function on $\mathbb{R}$ that is continuous on the complement of F. To see that f is discontinuous at each point of F, let $x \in F$ and choose n such that $x \in F_n \setminus F_{n-1}$. Then $x \in F_k$ for all $k > n$ and, hence, the oscillation of f at x is at least $4^{-n} - \sum_{k>n} 4^{-k} = 4^{-n}\,(2/3) > 0$. □

As a final application of the M-test, we construct a continuous nondifferentiable function. The first published example of such a function was given by Weierstrass, who showed that the function $f(x) = \sum_{n=1}^{\infty} a^n \cos(b^n x)$, where $0 < a < 1$ and b is an odd integer satisfying $ab > 1 + 3\pi/2$, fails to have a finite derivative at any point. The following is a simplified version of Weierstrass's example.

Application 10.13. (Nowhere Differentiable Functions) *Given $x \in \mathbb{R}$, let $g(x)$ denote the distance from x to the nearest integer, and define $f(x) = \sum_{n=0}^{\infty} 2^{-n} g(2^n x)$. Then, f is a bounded (uniformly) continuous function on $\mathbb{R}$ that fails to have a finite derivative at any point of $\mathbb{R}$.*

PROOF. The graph of $g(x)$ is pictured in Figure 10.7. Note that g has period 1 while $g(2^n x)$ has period 2^{-n}. In particular, if x is a dyadic rational, $x = i2^{-n}$, for some integers i and $n \geq 1$, then $2^k x$ is an integer for all $k \geq n$, and so $g(2^k x) = 0$ for all $k \geq n$.

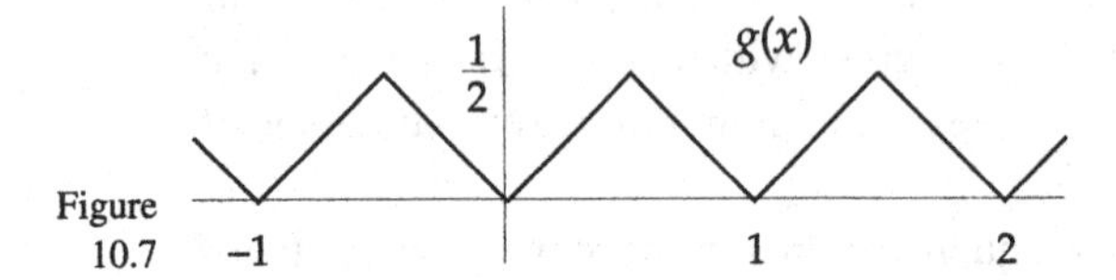

Figure 10.7

By the M-test, f is a bounded continuous function on $\mathbb{R}$. (Since f is periodic with period 1, note that f is actually uniformly continuous.) Now, if f has a finite (two-sided) derivative at some (fixed) $x \in \mathbb{R}$, then

$$\frac{f(v_n) - f(u_n)}{v_n - u_n} \to f'(x)$$

for any (u_n) and (v_n) with $u_n \leq x \leq v_n$, $u_n < v_n$, and $v_n - u_n \to 0$. (Why?) To show that f is nondifferentiable, then, we will show that this limit fails to exist for a suitable choice of (u_n) and (v_n).

Given $n \geq 1$, let u_n and v_n be the pair of successive dyadic rationals satisfying $u_n \leq x < v_n$ and $v_n - u_n = 2^{-n}$. Then

$$\frac{f(v_n) - f(u_n)}{v_n - u_n} = \sum_{k=0}^{n-1} \frac{g(2^k v_n) - g(2^k u_n)}{2^k v_n - 2^k u_n}.$$

But $2^k u_n = 2^{k-n} 2^n u_n = 2^{k-n} i$ and $2^k v_n = 2^{k-n}(i+1)$, for some integer i. Since $2^{k-n} \leq 1/2$ for $k < n$, this means that $2^k u_n$ and $2^k v_n$ both lie in the same "half-period" for g and hence that g is linear on the interval $[2^k u_n, 2^k v_n]$. Thus each of the difference quotients in the sum on the right is ± 1; that is,

$$d_n = \frac{f(v_n) - f(u_n)}{v_n - u_n} = \sum_{k=0}^{n-1} \pm 1.$$

Hence, the sequence of difference quotients (d_n) cannot converge to a finite limit because successive terms always differ by at least 1. □

EXERCISES

▷ **23.** Show that $B(X)$ is an algebra of functions; that is, if $f, g \in B(X)$, then so is fg and $\|fg\|_\infty \leq \|f\|_\infty \|g\|_\infty$. Moreover, if $f_n \to f$ and $g_n \to g$ in $B(X)$, show that $f_n g_n \to fg$ in $B(X)$. (Thus, multiplication is continuous in $B(X)$. Compare this with Exercise 7.)

24. $B(X)$ is also a lattice: If $f, g \in B(X)$, show that the functions $f \vee g = \max\{f, g\}$ and $f \wedge g = \min\{f, g\}$ (defined pointwise, just as in Chapter Five) are also in $B(X)$ and satisfy $\|f \vee g\|_\infty \le \max\{\|f\|_\infty, \|g\|_\infty\}$ and $\|f \wedge g\|_\infty \le \max\{\|f\|_\infty, \|g\|_\infty\}$.

25. Show that $B[0,1]$ is *not* separable. [Hint: This is analogous to the proof that ℓ_∞ is not separable. Consider the collection of characteristic functions of the intervals $[0,x]$ for $0 < x < 1$.]

26. If $\sum_{n=1}^\infty |a_n| < \infty$, prove that $\sum_{n=1}^\infty a_n \sin nx$ and $\sum_{n=1}^\infty a_n \cos nx$ are uniformly convergent on $\mathbb{R}$.

27. Show that $\sum_{n=1}^\infty x^2/(1+x^2)^n$ converges for all $|x| \le 1$, but that the convergence is *not* uniform. [Hint: Find the sum!]

28. Let $f_n : \mathbb{R} \to \mathbb{R}$ be continuous, and suppose that (f_n) converges uniformly on $\mathbb{Q}$. Show that (f_n) actually converges uniformly on all of $\mathbb{R}$. [Hint: Show that (f_n) is uniformly Cauchy.]

29.

(a) For which values of x does $\sum_{n=1}^\infty ne^{-nx}$ converge? On which intervals is the convergence uniform?

(b) Conclude that $\int_1^2 \sum_{n=1}^\infty ne^{-nx}\,dx = e/(e^2-1)$.

30. Prove that $\sum_{n=1}^\infty x/[n^\alpha(1+nx^2)]$ converges uniformly on every bounded interval in $\mathbb{R}$ provided that $\alpha > 1/2$. Is the convergence uniform on all of $\mathbb{R}$?

31. Show that $\lim_{x\to 1} \sum_{n=1}^\infty nx^2/(n^3+x^2) = \sum_{n=1}^\infty n/(n^3+1)$.

32.

(a) If $\sum_{n=1}^\infty |a_n| < \infty$, show that $\sum_{n=1}^\infty a_n e^{-nx}$ is uniformly convergent on $[0,\infty)$.

(b) If we assume only that (a_n) is bounded, show that $\sum_{n=1}^\infty a_n e^{-nx}$ is uniformly convergent on $[\delta,\infty)$ for every $\delta > 0$.

▷ **33.** Define $I(x) = 0$ for $x \le 0$ and $I(x) = 1$ for $x > 0$. Given sequences (x_n) and (c_n) in $\mathbb{R}$, with $\sum_{n=1}^\infty |c_n| < \infty$, show that $f(x) = \sum_{n=1}^\infty c_n I(x - x_n)$ defines a bounded function on $\mathbb{R}$ that is continuous except, possibly, at the x_n.

34. Let $0 \le g_n \in C[a,b]$. If $\sum_{n=1}^\infty g_n$ converges pointwise to a continuous function on $[a,b]$, show that $\sum_{n=1}^\infty g_n$ converges uniformly on $[a,b]$.

35. For which $\alpha \in \mathbb{R}$ is $\sum_{n=1}^\infty xn^\alpha e^{-nx}$ a continuous function on $(0,\infty)$? on $[0,\infty)$?

36. Show that both $\sum_{n=1}^\infty x^n(1-x)$ and $\sum_{n=1}^\infty (-1)^n x^n (1-x)$ are convergent on $[0,1]$, but only one converges uniformly. Which one? Why?

37. Where does $\sum_{n=1}^\infty x^n/(1+x^n)$ converge? On which intervals does it converge uniformly?

38. Let (f_n) be a sequence of continuous functions on $(0,\infty)$ with $|f_n(x)| \le n$ for every $x > 0$ and $n \ge 1$, and such that $\lim_{x\to\infty} f_n(x) = 0$ for each n. Show that $f(x) = \sum_{n=1}^\infty 2^{-n} f_n(x)$ defines a continuous function on $(0,\infty)$ that also satisfies $\lim_{x\to\infty} f(x) = 0$.

39. Show that $C(\mathbb{R})$ is complete. [Hint: Use the fact that $C[-n,n]$ is complete for each n. See Exercise 22.]

40. For any metric space X, show that X is isometric to a subset of $C_b(X)$. [Hint: Mimic the proof of Lemma 7.17, showing that X embeds into $\ell_\infty(X) = B(X)$.] Conclude that X has a completion.

◊

Notes and Remarks

For more on the history of the vibrating string problem, see Carslaw [1930], Hobson [1927, Volume II], Kline [1972], Langer [1947], Rogosinki [1950], Van Vleck [1914], and the excerpt "Riemann on Fourier series and the Riemann integral" in Birkhoff [1973] (wherein you will also find three excerpts from Fourier's work); the excerpt is from Riemann [1902], in which Riemann develops his concept of the integral to address the problem of representing continuous functions by trigonometric series. For a detailed solution of the vibrating string problem see Folland [1992] or Tolstov [1962].

For a brief history of Fourier analysis, see the articles by Coppel [1969], Gibson [1893], Jackson [1920], Jeffery [1956], and Langer [1947]. For more recent commentary see Grattan-Guinness [1970], Halmos's "Progress Report" on Fourier series, Halmos [1978], the follow-up article by Bochner [1979], and Zygmund [1976]. For more information on Fourier himself see the biographies by Grattan-Guinness [1972] and Herivel [1975], the article by González-Velasco [1992] (which is the source of the quote at the beginning of the chapter), and Körner [1988]. In addition to containing entertaining historical tidbits, Körner's book is an excellent introduction to Fourier analysis. For an enlightening discussion of the impact of Cauchy's famous "wrong" theorem and its connection with Fourier's work, see Lakatos [1976].

For more details on Exercise 4 (and related issues), see Jackson [1926, 1934a, 1941], Rogosinski [1950], and Simon [1969]. The first general convergence result for Fourier series is generally attributed to Jordan [1881].

Pointwise convergence is "as old as the hills," and it is at least as old as calculus itself. Uniform convergence was first introduced by Seidel [1847], and in the same year by George Stokes [1848]; see Hardy [1918], Hawkins [1970], and Lakatos [1976]. Once the notion of uniform convergence was recognized as the proper tool for the preservation of continuity in the limit, Weierstrass and his students began a "witch hunt" for the uses of Cauchy's theorem during the previous 50 years, in an attempt to set the record straight. The age of rigor would come to full maturity under Weierstrass's guidance. For more about Weierstrass himself, see Polubarinova-Kochina [1966].

The example of a space-filling curve given in Application 10.11 is due to Schoenberg, by way of Lebesgue, and first appeared in Schoenberg [1938]. Curiously, Schoenberg's curve turns out to be nowhere differentiable, whereas Lebesgue's (the one that we discussed in Chapter Six) is differentiable almost everywhere. For more on this see Schoenberg [1982] or Sagan [1986, 1992]. The Schoenberg–Lebesgue example is typical of a wider class of space-filling curves; in particular, the curve $(x(t), y(t))$ is space-filling whenever x and y are stochastically independent. See Holbrook [1991].

The construction in Application 10.12 is based on the presentation in Oxtoby [1971].

Both Weierstrass and Riemann spoke of continuous, nowhere differentiable functions in their lectures as early as 1861 (and other examples of such functions are now known to have existed prior to 1861), but the first published example is due to Weierstrass, an example that finally appeared in du Bois–Reymond [1875]. See also Weierstrass [1895, Vol. 2, pp. 71–76]. For more about Riemann's examples, see Hardy [1916], Hawkins [1970], Neuenschwander [1978], Segal [1978], and A. Smith [1972].

The example of a continuous, nowhere differentiable function constructed in Application 10.13 is generally credited to van der Waerden [1930]. The particulars of the present construction are taken from Billingsley [1982], but see also Boas [1960].

A great deal has been written about nondifferentiable functions in general and Weierstrass's example in particular. A short but thorough historical account is given in Hobson [1927, Volume II], but see also Hardy [1916]. A longer account, which includes some discussion of space-filling curves, is given in Singh [1969].

Exercise 40 pinpoints our interest in $C_b(X)$ and $B(X)$: They are "universal" metric spaces. In order to "know" all metric spaces, it is enough to know just the spaces $C_b(X)$. We will have more to say about this point of view in the next chapter. For now, simply notice that $C_b(X)$ *determines* X in the sense that the bounded, continuous, real-valued functions on X determine the closed sets in X (see Chapter Five). For detailed proofs of the results in Exercises 40, see Kaplansky [1977].

CHAPTER ELEVEN

The Space of Continuous Functions

The Weierstrass Theorem

While we now know something about convergence in $C(X)$, there are many more things that we would like to know about $C(X)$. We will find the task unmanageable, however, unless we place some restrictions on the metric space X. If we focus our attention on the case when X is *compact*, for example, we will be afforded plenty of extra machinery: In this case, $C(X)$ is not only a vector space, an algebra, and a lattice (where algebraic operations are defined pointwise), but also a *complete* normed space under the sup-norm. With all of these tools to work with, we will be able to accomplish quite a bit. And at least a few of our results will apply equally well to the space $C_b(X)$ of bounded continuous functions on a general metric space X. For the remainder of this chapter, then, *unless otherwise specified, X will denote a compact metric space.*

We will concentrate on two questions in particular, and each of these will lead to some interesting applications:

- Is $C(X)$ separable? More importantly, are there any "useful" dense *subspaces*, or even dense *subalgebras*, or *sublattices* of $C(X)$?
- What are the *compact* subsets of $C(X)$? And are such sets "useful"?

Either question is tough to answer in full generality, but the first one has a very satisfactory and easy to understand answer for $C[a, b]$. Since $C[a, b]$ is such an important space for our purposes, besides being the obvious place to start, we will spend much of our efforts on just this case. An initial simplification will help (see Exercise 5.63).

Lemma 11.1. *There is a linear isometry from $C[0, 1]$ onto $C[a, b]$ that maps polynomials to polynomials.*

PROOF. Define $\sigma : [a, b] \to [0, 1]$ by $\sigma(x) = (x - a)/(b - a)$ for $a \le x \le b$. Then σ is a homeomorphism, and the map $T_\sigma(f) = f \circ \sigma$ defines a *linear isometry* from $C[0, 1]$ onto $C[a, b]$. Indeed, T_σ is clearly linear. It is one-to-one and onto because it has an obvious inverse, namely, $T_{\sigma^{-1}}(h) = h \circ \sigma^{-1}$, where $\sigma^{-1}(t) = a + t(b - a)$ for $0 \le t \le 1$. Finally, it is an isometry because σ is onto: $\max_{a \le x \le b} |f(\sigma(x))| = \max_{t \in \sigma[a,b]} |f(t)| = \max_{0 \le t \le 1} |f(t)|$.

Moreover, T_σ is both a *lattice isomorphism* and an *algebra isomorphism*. That is, $T_\sigma(f) \le T_\sigma(g)$ if and only if $f \le g$, and $T_\sigma(fg) = T_\sigma(f)\, T_\sigma(g)$. In particular, note that T_σ maps polynomials to polynomials: If $p(t) = \sum_{k=0}^{n} a_k t^k$ is a polynomial in t, then $p(\sigma(x)) = \sum_{k=0}^{n} a_k [(x-a)/(b-a)]^k$ is a polynomial in x. □

The proof of Lemma 11.1 tells us that $C[a,b]$ and $C[0,1]$ are, for our purposes, *identical*. The point here is that we need only concern ourselves with a single choice of the interval $[a,b]$, and $[0,1]$ is often most convenient. Virtually any result that we might obtain about $C[0,1]$ will readily transfer to $C[a,b]$. To begin, we will show that $C[a,b]$ is separable by showing that $C[0,1]$ is separable. We will give two proofs of this result, the first of which is a "proof by picture" while the second is more analytical.

Theorem 11.2. *$C[0,1]$ is separable.*

PROOF. Let $f \in C[0,1]$, and let $\varepsilon > 0$. We first approximate f by a polygonal function, as shown in Figure 11.1. Since f is uniformly continuous, we can find a sufficiently large n so that $|f(x) - f(y)| < \varepsilon$ whenever $|x - y| < 1/n$. This means that the polygonal function g defined by $g(k/n) = f(k/n)$, for $k = 0, \ldots, n$, and g linear on each interval $(k/n, (k+1)/n)$ satisfies $\|f - g\|_\infty \leq \varepsilon$. (Why?)

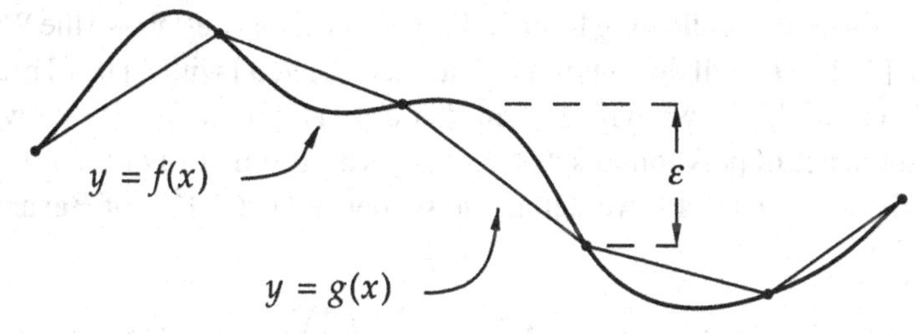

Figure 11.1

Next we modify our approximating function: Let h be another polygonal function that also has nodes at k/n for $k = 0, \ldots, n$, but with $h(k/n)$ *rational* and satisfying $|h(k/n) - g(k/n)| < \varepsilon$ for each k. Then, $\|g - h\|_\infty < \varepsilon$ and, consequently, $\|f - h\|_\infty < 2\varepsilon$.

We're done! The set of all polygonal functions taking only rational values at the nodes $(k/n)_{k=0}^{n}$, for some n, is countable. (See Exercise 1) □

EXERCISES

▷ **1.** For each n, let Q_n be the set of all polygonal functions that have nodes at $x = k/n$, $k = 0, \ldots, n$, and that take on only rational values at these points. Check that Q_n is a countable set, and hence that the union of the Q_n's is a countable dense set in $C[0,1]$.

2. Let $a = x_1 < x_2 < \cdots < x_n = b$ be distinct points in $[a,b]$, and let S_n be the set of all polygonal functions having nodes at the x_k. Show that S_n is an n-dimensional subspace of $C[a,b]$ spanned by the "angles" $\varphi_k(x) = |x - x_k| +$

$(x - x_k)$, for $k = 1, \ldots, n-1$, and the constant function $\varphi_0(x) = 1$. Specifically, show that each $h \in S_n$ can be uniquely written as $h(x) = \sum_{k=0}^{n-1} c_k \varphi_k(x)$. [Hint: The system of equations $h(x_k) = c_0 + 2\sum_{i=1}^{k-1} c_i(x_k - x_i)$, $k = 1, \ldots, n$, can be solved for the c_i. Why? How does this help?]

3. Prove that every polygonal function is Lipschitz. Thus, the Lipschitz functions are dense in $C[a, b]$.

Our second proof that $C[a, b]$ is separable uses a much more convenient dense set (at least for our purposes).

The Weierstrass Approximation Theorem 11.3. *Given $f \in C[a, b]$ and $\varepsilon > 0$, there is a polynomial p such that $\|f - p\|_\infty < \varepsilon$. Hence, there is a sequence of polynomials (p_n) such that $p_n \rightrightarrows f$ on $[a, b]$.*

The Weierstrass theorem leads to a second proof that $C[a, b]$ is separable. Indeed, given a polynomial p and any $\varepsilon > 0$, we can find another polynomial q with *rational* coefficients such that $\|p - q\|_\infty < \varepsilon$ on $[a, b]$. (How?) Since the set of polynomials with rational coefficients is a countable set, this implies that $C[a, b]$ is separable.

Of course, following Lemma 11.1, we need only establish the Weierstrass theorem for $C[0, 1]$. (Recall that our identification of $C[a, b]$ with $C[0, 1]$ preserves polynomials.) The proof that we will give in this case is quite explicit; we will actually display a sequence of polynomials that converges uniformly to a given $f \in C[0, 1]$. Specifically, given $f \in C[0, 1]$, we define the sequence $\big(B_n(f)\big)_{n=1}^{\infty}$ of **Bernstein polynomials** for f by

$$\big(B_n(f)\big)(x) = \sum_{k=0}^{n} f\left(\frac{k}{n}\right) \cdot \binom{n}{k} x^k (1-x)^{n-k}, \qquad 0 \le x \le 1.$$

Please note that $B_n(f)$ is a polynomial of degree at most n. Also, it is easy to see that $\big(B_n(f)\big)(0) = f(0)$ and $\big(B_n(f)\big)(1) = f(1)$. In general, $\big(B_n(f)\big)(x)$ is an average of the numbers $f(k/n)$, $k = 0, \ldots, n$ (more on this later).

We will prove Weierstrass's theorem by proving:

S. N. Bernstein's Theorem 11.4. *$B_n(f) \rightrightarrows f$ on $[0, 1]$ for each f in $C[0, 1]$.*

The proof of Bernstein's theorem is easy once we catalogue a few facts about the polynomials $B_n(f)$. For later reference, let's agree to write

$$f_0(x) = 1, \qquad f_1(x) = x, \qquad \text{and} \qquad f_2(x) = x^2.$$

Among other things, the following lemma establishes Bernstein's theorem for these three polynomials. Curiously, these few special cases will imply the general result.

Lemma 11.5.

(i) $B_n(f_0) = f_0$ *and* $B_n(f_1) = f_1$.

(ii) $B_n(f_2) = \left(1 - \frac{1}{n}\right) f_2 + \frac{1}{n} f_1$, *and hence* $B_n(f_2) \rightrightarrows f_2$.

(iii) $\displaystyle\sum_{k=0}^{n}\left(\frac{k}{n}-x\right)^2\binom{n}{k}x^k(1-x)^{n-k}=\frac{x(1-x)}{n}\le\frac{1}{4n},\qquad \text{if } 0\le x\le 1.$

(iv) *Given $\delta>0$ and $0\le x\le 1$, let F denote the set of k in $\{0,\ldots,n\}$ for which $\left|(k/n)-x\right|\ge\delta$. Then*

$$\sum_{k\in F}\binom{n}{k}x^k(1-x)^{n-k}\le\frac{1}{4n\delta^2}.$$

PROOF. The fact that $B_n(f_0)=f_0$ follows from the binomial formula:

$$\sum_{k=0}^{n}\binom{n}{k}x^k(1-x)^{n-k}=[x+(1-x)]^n=1.$$

To see that $B_n(f_1)=f_1$, first notice that for $k\ge 1$ we have

$$\frac{k}{n}\binom{n}{k}=\frac{(n-1)\,!}{(k-1)\,!\,(n-k)\,!}=\binom{n-1}{k-1}.$$

Consequently,

$$\begin{aligned}\sum_{k=0}^{n}\frac{k}{n}\binom{n}{k}x^k(1-x)^{n-k}&=x\sum_{k=1}^{n}\binom{n-1}{k-1}x^{k-1}(1-x)^{n-k}\\&=x\sum_{j=0}^{n-1}\binom{n-1}{j}x^j(1-x)^{(n-1)-j}=x.\end{aligned}$$

Next, to compute $B_n(f_2)$, we rewrite twice:

$$\begin{aligned}\left(\frac{k}{n}\right)^2\binom{n}{k}=\frac{k}{n}\binom{n-1}{k-1}&=\frac{n-1}{n}\cdot\frac{k-1}{n-1}\binom{n-1}{k-1}+\frac{1}{n}\binom{n-1}{k-1}, &&\text{if } k\ge 1\\&=\left(1-\frac{1}{n}\right)\binom{n-2}{k-2}+\frac{1}{n}\binom{n-1}{k-1}, &&\text{if } k\ge 2.\end{aligned}$$

Thus,

$$\begin{aligned}&\sum_{k=0}^{n}\left(\frac{k}{n}\right)^2\binom{n}{k}x^k(1-x)^{n-k}\\&\quad=\left(1-\frac{1}{n}\right)\sum_{k=2}^{n}\binom{n-2}{k-2}x^k(1-x)^{n-k}+\frac{1}{n}\sum_{k=1}^{n}\binom{n-1}{k-1}x^k(1-x)^{n-k}\\&\quad=\left(1-\frac{1}{n}\right)x^2+\frac{1}{n}x,\end{aligned}$$

which establishes (ii) since $\|B_n(f_2)-f_2\|_\infty=(1/n)\|f_1-f_2\|_\infty\to 0$ as $n\to\infty$.

To prove (iii) we combine the observations in (i) and (ii) and simplify. Since $((k/n)-x)^2=(k/n)^2-2x(k/n)+x^2$, we get

$$\begin{aligned}\sum_{k=0}^{n}\left(\frac{k}{n}-x\right)^2\binom{n}{k}x^k(1-x)^{n-k}&=\left(1-\frac{1}{n}\right)x^2+\frac{1}{n}x-2x^2+x^2\\&=\frac{1}{n}x(1-x)\le\frac{1}{4n},\end{aligned}$$

for $0\le x\le 1$.

Finally, to prove (iv), let $0 \le x \le 1$ and note that $1 \le ((k/n)-x)^2/\delta^2$ for $k \in F$. Hence,

$$\begin{aligned}\sum_{k\in F}\binom{n}{k}x^k(1-x)^{n-k} &\le \frac{1}{\delta^2}\sum_{k\in F}\left(\frac{k}{n}-x\right)^2\binom{n}{k}x^k(1-x)^{n-k}\\ &\le \frac{1}{\delta^2}\sum_{k=0}^{n}\left(\frac{k}{n}-x\right)^2\binom{n}{k}x^k(1-x)^{n-k}\\ &\le \frac{1}{4n\delta^2}, \qquad \text{from (iii).} \quad \square\end{aligned}$$

Now we are ready for *the proof of Bernstein's theorem*:

PROOF. Let $f \in C[0,1]$ and let $\varepsilon > 0$. Then, since f is uniformly continuous, there is a $\delta > 0$ such that $|f(x) - f(y)| < \varepsilon/2$ whenever $|x-y| < \delta$. Now we use Lemma 11.5 to estimate $\|f - B_n(f)\|_\infty$. First notice that since the numbers $\binom{n}{k}x^k(1-x)^{n-k}$ are nonnegative and sum to 1, we have

$$\begin{aligned}&\left|f(x) - \sum_{k=0}^{n}\binom{n}{k}f\left(\frac{k}{n}\right)x^k(1-x)^{n-k}\right|\\ &\quad = \left|\sum_{k=0}^{n}\left(f(x) - f\left(\frac{k}{n}\right)\right)\binom{n}{k}x^k(1-x)^{n-k}\right|\\ &\quad \le \sum_{k=0}^{n}\left|f(x) - f\left(\frac{k}{n}\right)\right|\binom{n}{k}x^k(1-x)^{n-k}.\end{aligned}$$

Now fix n (to be specified in a moment). Given $0 \le x \le 1$, let F denote the set of k in $\{0,\ldots,n\}$ for which $|(k/n) - x| \ge \delta$. Then $|f(x) - f(k/n)| < \varepsilon/2$ for $k \notin F$, while $|f(x) - f(k/n)| \le 2\|f\|_\infty$ for $k \in F$. Thus,

$$\begin{aligned}&\left|f(x) - (B_n(f))(x)\right|\\ &\quad \le \frac{\varepsilon}{2}\sum_{k\notin F}\binom{n}{k}x^k(1-x)^{n-k} + 2\|f\|_\infty\sum_{k\in F}\binom{n}{k}x^k(1-x)^{n-k}\\ &\quad < \frac{\varepsilon}{2}\cdot 1 + 2\|f\|_\infty \cdot \frac{1}{4n\delta^2}, \qquad \text{from Lemma 11.5 (iv),}\\ &\quad < \varepsilon, \qquad \text{provided that } n > \|f\|_\infty/\varepsilon\delta^2.\end{aligned}$$

Since this choice of n does not depend on x, we get that $\|B_n(f) - f\|_\infty < \varepsilon$ whenever $n > \|f\|_\infty/\varepsilon\delta^2$. $\square$

There is a probabilistic interpretation of Bernstein's result. To see this, fix an x in $[0,1]$, and consider a "game" with probability of success equal to x and, hence, probability of failure equal to $1-x$. For instance, a coin might be weighted so as to come up heads with probability x and tails with probability $1-x$. Then, the probability of exactly k successes in n independent trials of the game is given by $\binom{n}{k}x^k(1-x)^{n-k}$. This is one of the terms in the so-called *binomial distribution*. The first part of Lemma 11.5

says that this is, indeed, a probability distribution since $\sum_{k=0}^{n}\binom{n}{k}x^k(1-x)^{n-k} = 1$, and that the *mean* of this distribution is $\sum_{k=0}^{n} k\binom{n}{k}x^k(1-x)^{n-k} = nx$. The second part of Lemma 11.5 computes its *variance* as $\sum_{k=0}^{n}(k-nx)^2\binom{n}{k}x^k(1-x)^{n-k} = nx(1-x)$. The last part of Lemma 11.5 is (Chebyshev's version of) *Bernoulli's law of large numbers*; it says that, for large n, most of the weight of the distribution is concentrated near the mean. Those terms for which $|k-nx|$ is large do not contribute much to the distribution. In other words, if n is large, the most likely outcome of n trials is to have roughly nx successes. Thus, the average number of successes in a large number of trials is a good estimate for the actual probability of success.

The binomial distribution in the case $n = 12$ and $x = 1/3$ is depicted in Figure 11.2. Note that the most likely outcome is $k = nx = 4$ successes; the probabilities of $k = 10$, 11, or 12 successes are so small that they do not even register on the graph.

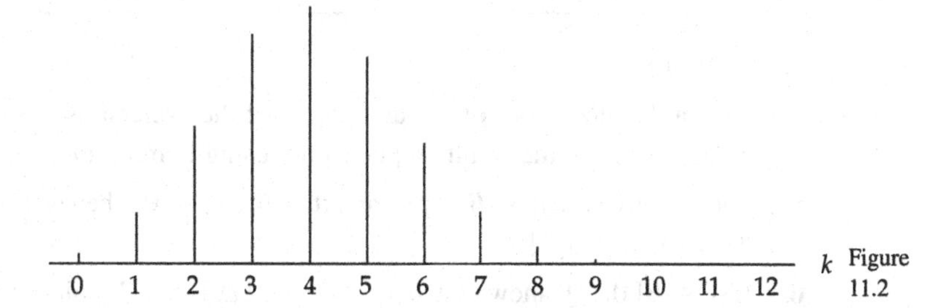

Figure 11.2

To phrase Bernstein's theorem in this language, consider $f \in C[0,1]$ as the "payoff" for the game; if there are k successes in n trials, we win (or lose) an amount equal to $f(k/n)$. What are our expected winnings, given that the probability of success on any one trial is x? It is exactly $(B_n(f))(x)$! For n large, then, we would expect our winnings to be approximately $f(x)$. The law of large numbers and the uniform continuity of f are responsible for the fact that this approximation is uniform (it depends on f and n, but not on x).

We will see in the next chapter that Theorem 11.3 will generalize to $C(X)$, where X is compact. On the other hand, the "easy" proof given for Theorem 11.2 would be hard to mimic in a more general setting. The major difference between the two results is that the polynomials form a *subalgebra* of $C[a,b]$ while the polygonal functions form only a subspace. The fact that the Weierstrass theorem admits an algebraic interpretation along these lines will prove very useful in the next chapter.

The Weierstrass theorem affords us some small insight into *the moment problem*. The problem, loosely stated, is this: Consider a thin metal rod placed along the interval $[a,b]$ on the x-axis, and suppose that we know the density of the rod at each point x as a function $f(x)$ in $C[a,b]$. The question is: Does the sequence of moments $\mu_n = \int_a^b x^n f(x)\,dx$ (about the y-axis) uniquely determine f? If we knew the sequence of numbers (μ_n), could we actually reconstruct f? The answer, as it happens, is yes, but it is a bit beyond our means at this point. We can, however, say this much: The solution, if it exists, has to be unique. That is, if two functions f and g in $C[a,b]$ have the same moment sequence, then f and g must be identical. Thanks to the linearity of the integral, it is enough to establish the following:

Application 11.6. *If $f \in C[a,b]$, and if $\int_a^b x^n f(x)\,dx = 0$ for each $n = 0, 1, 2, \ldots$, then $f = 0$.*

PROOF. From the Weierstrass theorem, there is a sequence of polynomials (p_n) such that $p_n \rightrightarrows f$ on $[a,b]$. Hence, $f \cdot p_n \rightrightarrows f^2$ on $[a,b]$, and so

$$\int_a^b f^2(x)\,dx = \lim_{n\to\infty} \int_a^b f(x)\,p_n(x)\,dx.$$

But since $\int_a^b x^n f(x)\,dx = 0$ for each n (and since the integral is linear), it follows that $\int_a^b f(x)\,p_n(x)\,dx = 0$ for each n. That is, $\int_a^b f^2(x)\,dx = 0$. Since f is continuous, this means that $f = 0$. (Why?) □

EXERCISES

4. Give a detailed proof of the assertion that the Weierstrass theorem for general $[a,b]$ follows from the result on $[0,1]$ (by using Lemma 11.1).

5. Show that $|B_n(f)| \le B_n(|f|)$, and that $B_n(f) \ge 0$ whenever $f \ge 0$. Conclude that $\|B_n(f)\|_\infty \le \|f\|_\infty$.

6. If $f \in B[0,1]$, show that $B_n(f)(x) \to f(x)$ at each point of continuity of f.

▷ **7.** If p is a polynomial and $\varepsilon > 0$, prove that there is a polynomial q with rational coefficients such that $\|p - q\|_\infty < \varepsilon$ on $[0,1]$.

8. Prove that $C(\mathbb{R})$ is separable.

▷ **9.** Let $\mathcal{P}_n$ denote the set of polynomials of degree at most n, considered as a subset of $C[a,b]$. Clearly, $\mathcal{P}_n$ is a subspace of $C[a,b]$ of dimension $n+1$. Also, $\mathcal{P}_n$ is closed in $C[a,b]$. (Why?) How do you know that $\mathcal{P}$, the union of all of the $\mathcal{P}_n$, is not all of $C[a,b]$? That is, why are there necessarily nonpolynomial elements in $C[a,b]$?

10. Let (x_i) be a sequence of numbers in $(0,1)$ such that $\lim_{n\to\infty}(1/n)\sum_{i=1}^n x_i^k$ exists for every $k = 0, 1, 2, \ldots$. Show that $\lim_{n\to\infty}(1/n)\sum_{i=1}^n f(x_i)$ exists for every $f \in C[0,1]$.

11. Several proofs of the Weierstrass theorem are based on a special case that can be checked independently: There is a sequence of polynomials (P_n) that converges uniformly to $|x|$ on $[-1,1]$. Here is an outline of an elementary proof:

(a) Define (P_n) recursively by $P_{n+1}(x) = P_n(x) + [x - P_n(x)^2]/2$, where $P_0(x) = 0$. Clearly, each P_n is a polynomial.

(b) Check that $0 \le P_n(x) \le P_{n+1}(x) \le \sqrt{x}$ for $0 \le x \le 1$. Use Dini's theorem (Exercise 10.18) to conclude that $P_n(x) \rightrightarrows \sqrt{x}$ on $[0,1]$.

(c) $P_n(x^2)$ is also a polynomial, and $P_n(x^2) \rightrightarrows |x|$ on $[-1,1]$.

Since a polygonal function can be written in the form $\sum_{i=1}^k a_i|x - x_i| + bx + d$, it follows that every polygonal function can be uniformly approximated by polynomials. The Weierstrass theorem now follows from the proof of Theorem 11.2.

▷ **12.** Let p_n be a polynomial of degree m_n, and suppose that $p_n \rightrightarrows f$ on $[a, b]$, where f is *not* a polynomial. Show that $m_n \to \infty$.

13. Show that the set of all polynomials $\mathcal{P}$ is a first category set in $C[a, b]$.

14. Let $f \in C[a, b]$ be continuously differentiable, and let $\varepsilon > 0$. Show that there is a polynomial p such that $\|f - p\|_\infty < \varepsilon$ *and* $\|f' - p'\|_\infty < \varepsilon$. Conclude that $C^{(1)}[a, b]$ is separable.

15. Construct a sequence of polynomials that converge uniformly on $[0, 1]$ but whose derivatives fail to converge uniformly.

16. Prove that there is a sequence of polynomials (p_n) such that $p_n \to 0$ pointwise on $[0, 1]$, but such that $\int_0^1 p_n(x)\,dx \to 3$.

17. Suppose that $f : [1, \infty) \to \mathbb{R}$ is continuous and that $\lim_{x\to\infty} f(x)$ exists. For $\varepsilon > 0$, show that there is a polynomial p such that $|f(x) - p(1/x)| < \varepsilon$ for all $x \geq 1$.

18. Find $B_n(f)$ for $f(x) = x^3$. [Hint: $k^2 = (k-1)(k-2) + 3(k-1) + 1$.] Note that the same calculation can be used to show that if $f \in \mathcal{P}_m$, then $B_n(f) \in \mathcal{P}_m$ for any $n > m$.

19. Here is an alternate approach to Exercise 14: If f is continuously differentiable on $[0, 1]$, show that $B_{n+1}(f)' \rightrightarrows f'$ on $[0, 1]$. [Hint: The mean value theorem and a bit of rewriting allow for the comparison of $B_{n+1}(f)'$ and $B_n(f')$. If we set $p_{n,k}(x) = \binom{n}{k} x^k (1-x)^{n-k}$, show that $p'_{n+1,k} = (n+1)(p_{n,k-1} - p_{n,k})$.]

$\mathrm{Lip}_K\alpha$ denotes the set of functions $f \in C[0, 1]$ that are Lipschitz of order α with constant K on $[0, 1]$, where $0 < \alpha \leq 1$ and $0 < K < \infty$. That is, $f \in \mathrm{Lip}_K\alpha$ if $|f(x) - f(y)| \leq K|x - y|^\alpha$ for all $x, y \in [0, 1]$. (See Exercises 8.57–8.60 for more details.) We write $\mathrm{Lip}\,\alpha$ for the set of f that are in $\mathrm{Lip}_K\alpha$ for some K; that is, $\mathrm{Lip}\,\alpha = \bigcup_{K=1}^\infty \mathrm{Lip}_K\alpha$.

20. Show that $\mathrm{Lip}_K\alpha$ is closed in $C[0, 1]$. In fact, if a sequence (f_n) in $\mathrm{Lip}_K\alpha$ converges *pointwise* to f on $[0, 1]$, show that $f \in \mathrm{Lip}_K\alpha$. Is $\mathrm{Lip}_K\alpha$ a subspace of $C[0, 1]$?

21. Show that $\mathrm{Lip}\,\alpha$ is a subspace of $C[0, 1]$. Is $\mathrm{Lip}\,\alpha$ a subalgebra of $C[0, 1]$?

22. Show that every polynomial is in $\mathrm{Lip}\,1$, but that $\sqrt{x}$, for example, is not.

23. Show that $x^\alpha \in \mathrm{Lip}\,\alpha$. For which $\beta \geq 0$ is $x^\beta \in \mathrm{Lip}\,\alpha$?

24. Prove that $\mathrm{Lip}\,1$ is *not* closed in $C[0, 1]$. In fact, $\mathrm{Lip}\,1$ is both dense and of first category in $C[0, 1]$. [Hint: For $\varepsilon > 0$, find $f \notin \mathrm{Lip}_K 1$ with $\|f\|_\infty < \varepsilon$. That is, show that $\mathrm{Lip}_K 1$ is nowhere dense.]

25. Prove that the set $\mathcal{P}$ of all polynomials is both dense and of first category in $C^{(1)}[0, 1]$.

26. For each $f \in \mathrm{Lip}\,\alpha$, define $N_\alpha(f) = \sup_{x \neq y} \big[\, |f(x) - f(y)| \,/\, |x - y|^\alpha \,\big]$.
(a) Show that N_α defines a seminorm on $\mathrm{Lip}\,\alpha$.
(b) Show that $\|f\|_{\mathrm{Lip}\,\alpha} = \|f\|_\infty + N_\alpha(f)$ defines a complete norm on $\mathrm{Lip}\,\alpha$.

Trigonometric Polynomials

In a follow-up to the paper in which Weierstrass established his famous theorem on approximation by algebraic polynomials, he proved an analogous result on approximation by trigonometric polynomials. In this section we will outline Lebesgue's elementary proof of Weierstrass's result.

To begin, a **trigonometric polynomial** (or, briefly, a trig polynomial) is a finite linear combination of the functions $\cos kx$ and $\sin kx$ for $k = 0, \ldots, n$, that is, a function of the form

$$T(x) = a_0 + \sum_{k=1}^{n} (a_k \cos kx + b_k \sin kx), \tag{11.1}$$

where, for our purposes, the a_k and b_k are real numbers. The *degree* of a trig polynomial, as you might expect, is the order of its highest nonzero coefficient; thus, the trig polynomial T displayed above has degree exactly n if at least one of a_n or b_n is different from 0.

Our first project is to justify the use of the word "polynomial" here by showing that a trigonometric polynomial is actually an algebraic polynomial (of the same degree) in $\cos x$ and $\sin x$.

Lemma 11.7. $\cos nx$ *and* $\sin(n+1)x / \sin x$ *can be written as polynomials of degree exactly n in* $\cos x$ *for any integer* $n \geq 1$.

PROOF. By using the recurrence formula,

$$\cos kx + \cos(k-2)x = 2\cos(k-1)x \cos x,$$

it is easy to check that $\cos 2x = 2\cos^2 x - 1$, $\cos 3x = 4\cos^3 x - 3\cos x$, and $\cos 4x = 8\cos^4 x - 8\cos^2 x + 1$. More generally, it follows by induction that $\cos nx$ is a polynomial of degree n in $\cos x$ with leading coefficient 2^{n-1}. Using this fact and the identity

$$\sin(k+1)x - \sin(k-1)x = 2\cos kx \sin x,$$

it follows (again by induction) that $\sin(n+1)x$ can be written as $\sin x$ times a polynomial of degree n in $\cos x$ with leading coefficient 2^n. □

EXERCISES

▷ **27.** Let T be a trig polynomial. Prove:
(a) If T is an odd function, then T can be written using only cosines.
(b) If T is an even function, then T can be written using only sines.

28. Show that there is an algebraic polynomial $p(t)$ of degree exactly $2k$ such that $\sin^{2k} x = p(\cos x)$.

▷ **29.** Given a trig polynomial $T(x)$ of degree n, show that there is an algebraic polynomial $p(t, s)$ of degree exactly n (in two variables) such that $T(x) =$

$p(\cos x, \sin x)$. [Hint: $p(t, s)$ can be chosen to be of the form $q(t) + r(t)s$ for some polynomials q and r.] If T is an even function, then there is an algebraic polynomial $p(t)$ of degree exactly n such that $T(x) = p(\cos x)$.

Conversely, every algebraic polynomial in $\cos x$ and $\sin x$ is also a trig polynomial (of the same degree). One way to see this is by induction:

30.

(a) Show that an algebraic polynomial in $\cos x$ and $\sin x$ can always be written using only functions of the form $\cos^n x$ and $\cos^m x \sin x$.

(b) Use induction to show that $\cos^n x$ is a trig polynomial of degree exactly n; in particular, $\cos^n x$ can be written as $\sum_{k=0}^{n} b_k \cos kx$, where $b_n = 2^{-n+1}$. [Hint: $2\cos\alpha\cos\beta = \cos(\alpha+\beta) + \cos(\alpha-\beta)$.]

(c) Show that $\cos^m x \sin x$ is a trig polynomial of degree exactly $m + 1$.

Our insights on trig polynomials will shed some light on the Fourier series representation of a continuous function.

31. Let $f : \mathbb{R} \to \mathbb{R}$ be continuous and 2π-periodic, and suppose that all of the Fourier coefficients for f vanish; that is, $\int_{-\pi}^{\pi} f(x) \cos nx\, dx = 0$ and $\int_{-\pi}^{\pi} f(x) \sin nx\, dx = 0$ for all $n = 0, 1, 2, \ldots$. This exercise outlines a proof, due to Lebesgue, that $f = 0$.

(a) If $f(x_0) = c > 0$ for some point x_0, then there exists $0 < \delta < \pi$ such that $f(x) \ge c/2$ for all x with $|x - x_0| \le \delta$.

(b) The functions $T_m(x) = [\,1 + \cos(x - x_0) - \cos\delta\,]^m$, $m = 1, 2, 3, \ldots$, satisfy $T_m(x) \ge 1$ for $|x - x_0| \le \delta$ and $|T_m(x)| < 1$ elsewhere in the interval $[\,x_0 - \pi, x_0 + \pi\,]$. In fact, the sequence (T_m) converges uniformly to 0 on the intervals $[\,x_0 - \pi, x_0 - \delta'\,]$ and $[\,x_0 + \delta', x_0 + \pi\,]$ for any $\delta < \delta' < \pi$.

(c) By first taking δ' sufficiently close to δ and then choosing m sufficiently large, show that $\int_{x_0-\pi}^{x_0+\pi} f(x)\, T_m(x)\, dx \ge \delta c/2 > 0$.

(d) By showing that T_m is a trig polynomial of degree m, conclude from our assumptions on f that $\int_{-\pi}^{\pi} f(x)\, T_m(x)\, dx = 0$, a contradiction.

The trig polynomials belong to the set of all 2π-periodic continuous functions $f : \mathbb{R} \to \mathbb{R}$, a space that we will denote by $C^{2\pi}$. If we write $\mathcal{T}_n$ to denote the collection of trig polynomials of degree at most n, then $\mathcal{T}_n$ is a subspace (and even a subalgebra) of $C^{2\pi}$.

A bit of linear algebra will now permit us to summarize our results quite succinctly (giving an alternate proof to Exercise 29 while we're at it). First, the $2n + 1$ functions in the set

$$\mathcal{A} = \{\,1,\ \cos x,\ \cos 2x,\ \ldots,\ \cos nx,\ \sin x,\ \sin 2x,\ \ldots,\ \sin nx\,\}$$

are *linearly independent*; the easiest way to see this is to notice that we may define an *inner product* on $C^{2\pi}$ under which these functions are *orthogonal*. Specifically,

$$\langle f, g\rangle = \int_{-\pi}^{\pi} f(x)\, g(x)\, dx = 0, \qquad \langle f, f\rangle = \int_{-\pi}^{\pi} f(x)^2\, dx \ne 0$$

for any pair of functions $f \neq g \in \mathcal{A}$. (See Exercises 10.2 and 10.3 or Exercise 33, below. We will pursue this observation in greater detail later in the book.) Second, we have shown that each element of $\mathcal{A}$ lives in the space spanned by the $2n+1$ functions in the set

$$\mathcal{B} = \{\, 1,\ \cos x,\ \cos^2 x,\ \ldots,\ \cos^n x,\ \sin x,\ \cos x \sin x,\ \ldots,\ \cos^{n-1} x \sin x \,\}.$$

That is,

$$\mathcal{T}_n \;=\; \operatorname{span} \mathcal{A} \;\subset\; \operatorname{span} \mathcal{B}.$$

By comparing dimensions, we have

$$2n+1 = \dim \mathcal{T}_n = \dim(\operatorname{span} \mathcal{A}) \leq \dim(\operatorname{span} \mathcal{B}) \leq 2n+1,$$

and hence we must have $\operatorname{span} \mathcal{A} = \operatorname{span} \mathcal{B}$. The point here is that $\mathcal{T}_n$ is a finite-dimensional subspace of $C^{2\pi}$ of dimension $2n+1$, and we may use either one of these sets of functions as a basis for $\mathcal{T}_n$.

EXERCISES

32. Show that the product of two trig polynomials is again a trig polynomial. Consequently, the collection of all trig polynomials is both a subspace and a *subalgebra* of $C^{2\pi}$.

33.

(a) Check that the functions $1, \cos x, \sin x, \ldots, \cos nx, \sin nx$ are orthogonal. That is, show that $\int_{-\pi}^{\pi} fg = 0$ for any pair of functions $f \neq g$ from this list, and that $\int_{-\pi}^{\pi} f^2 \neq 0$ for any f from the list.

(b) Conclude that the functions $1, \cos x, \sin x, \ldots, \cos nx, \sin nx$ are linearly independent (over either $\mathbb{R}$ or $\mathbb{C}$). [Hint: Show that the coefficients in equation (11.1) can be uniquely determined.]

34. Show that the functions $e^{ikx} = \cos kx + i \sin kx$, $k = -n, \ldots, n$, are linearly independent (again, over either $\mathbb{R}$ or $\mathbb{C}$). [Hint: The integral of a complex-valued function $f = u + iv$, where u and v are real-valued, is defined as $\int f = \int u + i \int v$.]

An alternate approach here is to note that every trig polynomial is actually an algebraic polynomial with *complex* coefficients in $z = e^{ix} = \cos x + i \sin x$ and $\bar{z} = e^{-ix} = \cos x - i \sin x$, that is, a linear combination of complex exponentials of the form

$$\sum_{k=-n}^{n} c_k e^{ikx}, \tag{11.2}$$

where the c_k are allowed to be complex numbers. We will call this form a *complex* trig polynomial (of degree n) and distinguish it from our original form by referring to that as a *real* trig polynomial.

Using DeMoivre's formula $(\cos x + i\sin x)^n = \cos nx + i\sin nx$, we can give an alternate proof of Lemma 11.7. Indeed, notice that

$$\begin{aligned}\cos nx &= \mathrm{Re}\left[(\cos x + i\sin x)^n\right]\\ &= \mathrm{Re}\left[\sum_{k=0}^{n}\binom{n}{k} i^k \sin^k x \cos^{n-k} x\right]\\ &= \sum_{k=0}^{[n/2]}\binom{n}{2k}(\cos^2 x - 1)^k \cos^{n-2k} x,\end{aligned}$$

where we have written $i^2\sin^2 x = \cos^2 x - 1$. The coefficient of $\cos^n x$ on the right-hand side is then

$$\sum_{k=0}^{[n/2]}\binom{n}{2k} = \frac{1}{2}\sum_{k=0}^{n}\binom{n}{k} = 2^{n-1}.$$

(All of the binomial coefficients together sum to $(1+1)^n = 2^n$, but the even or odd terms, taken separately, sum to exactly half this amount since $(1+(-1))^n = 0$.)

Similarly,

$$\begin{aligned}\sin(n+1)x &= \mathrm{Im}\left[(\cos x + i\sin x)^{n+1}\right]\\ &= \mathrm{Im}\left[\sum_{k=0}^{n+1}\binom{n+1}{k}(i\sin x)^k \cos^{n+1-k} x\right]\\ &= \sum_{k=0}^{[(n+1)/2]-1}\binom{n+1}{2k+1}(\cos^2 x - 1)^k \cos^{n-2k} x \sin x,\end{aligned}$$

where we have written $(i\sin x)^{2k+1} = i(\cos^2 x - 1)^k \sin x$. The coefficient of $\cos^n x \sin x$ on the right-hand side is

$$\sum_{k=0}^{[(n+1)/2]-1}\binom{n+1}{2k+1} = \frac{1}{2}\sum_{k=0}^{n+1}\binom{n+1}{k} = 2^n.$$

Obviously, every real trig polynomial can be written as a complex trig polynomial, since $\cos nx = (1/2)(e^{inx} + e^{-inx})$ and $\sin nx = (1/2i)(e^{inx} - e^{-inx})$, but notice that, in general, we must use *complex* coefficients c_k to represent real trig polynomials. Conversely, every complex trig polynomial can be written as a linear combination of sines and cosines but, again, typically with complex coefficients.

The point here is that only *certain* complex trig polynomials represent real-valued functions. Indeed, the real trig polynomials correspond to the *real parts* of the complex trig polynomials. To see this, notice that equation (11.2) represents a real-valued function if and only if

$$\sum_{k=-n}^{n} c_k e^{ikx} = \overline{\sum_{k=-n}^{n} c_k e^{ikx}} = \sum_{k=-n}^{n} \bar{c}_{-k} e^{ikx};$$

that is, $c_k = \bar{c}_{-k}$ for each k. In particular, c_0 must be real, and hence

$$\begin{aligned}
\sum_{k=-n}^{n} c_k e^{ikx} &= c_0 + \sum_{k=1}^{n} (c_k e^{ikx} + c_{-k} e^{-ikx}) \\
&= c_0 + \sum_{k=1}^{n} (c_k e^{ikx} + \bar{c}_k e^{-ikx}) \\
&= c_0 + \sum_{k=1}^{n} \big[\, (c_k + \bar{c}_k) \cos kx + i(c_k - \bar{c}_k) \sin kx \,\big] \\
&= c_0 + \sum_{k=1}^{n} \big[\, 2\mathrm{Re}(c_k) \cos kx - 2\mathrm{Im}(c_k) \sin kx \,\big],
\end{aligned}$$

which is of the form (11.1) with a_k and b_k real.

Conversely, given any real trig polynomial (11.1), we have

$$\begin{aligned}
a_0 &+ \sum_{k=1}^{n} \left(a_k \cos kx + b_k \sin kx \right) \\
&= a_0 + \sum_{k=1}^{n} \left[\left(\frac{a_k - i b_k}{2} \right) e^{ikx} + \left(\frac{a_k + i b_k}{2} \right) e^{-ikx} \right],
\end{aligned}$$

which is of the form (11.2) with $c_k = \bar{c}_{-k}$ for each k.

The real trig polynomials of degree n are the real linear span of the functions $1, \cos x, \sin x, \ldots, \cos nx, \sin nx$, and hence form a vector space of dimension $2n + 1$ over $\mathbb{R}$. The complex trig polynomials of degree n are the complex linear span of $1, \cos x, \sin x, \ldots, \cos nx, \sin nx$, and so form a vector space of dimension $2n + 1$ over $\mathbb{C}$, or of dimension $2(2n + 1)$ over $\mathbb{R}$. Obviously, if we want to restrict our attention to real-valued functions, we want only "half" of the complex trig polynomials.

Now we are ready to talk about approximating a continuous function by a trig polynomial. (Henceforth, "trig polynomial" means "real trig polynomial.") Since each trig polynomial is periodic with period 2π, though, we would only expect to approximate functions that were likewise periodic with period 2π. In fact, it is easy to see that even the pointwise limit of a sequence of periodic functions is again periodic, and so the same will be true for uniform limits.

Each $f \in C^{2\pi}$ is completely determined by its values on, say, $[-\pi, \pi\,]$, and so we can norm $C^{2\pi}$ by setting $\|f\|_\infty = \max_{|x| \le \pi} |f(x)|$. Please note that each element of $C^{2\pi}$ is necessarily uniformly continuous on $\mathbb{R}$. (Why?)

Weierstrass's Second Theorem 11.8. *Given $f \in C^{2\pi}$ and $\varepsilon > 0$, there is a trig polynomial T such that $\|f - T\|_\infty < \varepsilon$. Hence, there is a sequence of trig polynomials (T_n) such that $T_n \rightrightarrows f$ on $\mathbb{R}$.*

We will show that Weierstrass's second theorem follows from his first (Theorem 11.3). To begin, we need a simple lemma.

Lemma 11.9. *Given an even function $f \in C^{2\pi}$ and $\varepsilon > 0$, there is an even trig polynomial T such that $\|f - T\|_\infty < \varepsilon$.*

PROOF. The simple trick here is to note that $g(y) = f(\arccos y)$ defines a continuous function for $-1 \le y \le 1$. Thus, by Theorem 11.3, there is an algebraic polynomial p such that $\max_{|y|\le 1} |f(\arccos y) - p(y)| < \varepsilon$. But then, $T(x) = p(\cos x)$ is an even trig polynomial, and, clearly, $\max_{0\le x\le \pi} |f(x) - p(\cos x)| < \varepsilon$. Since f is even, it follows that $\|f - T\|_\infty < \varepsilon$. □

The rest of *the proof of Weierstrass's second theorem* consists of several clever applications of Lemma 11.9.

PROOF. Given $f \in C^{2\pi}$, note that both of the functions

$$f(x) + f(-x), \qquad \text{and} \qquad [f(x) - f(-x)]\sin x$$

are *even*. Thus, from Lemma 11.9, there are even trig polynomials T_1 and T_2 such that

$$f(x) + f(-x) = T_1(x) + d_1(x) \qquad \text{and} \qquad [f(x) - f(-x)]\sin x = T_2(x) + d_2(x),$$

where $\|d_1\|_\infty < \varepsilon/4$ and $\|d_2\|_\infty < \varepsilon/4$. By multiplying the first equation by $\sin^2 x$, the second by $\sin x$, and adding the results, we get

$$f(x)\sin^2 x = T_3(x) + d_3(x), \tag{11.3}$$

where T_3 is a trig polynomial and $\|d_3\|_\infty < \varepsilon/2$. But since this is true for any $f \in C^{2\pi}$, it must also hold for the function $f(x - \pi/2)$; in other words, we also have $f(x - \pi/2)\sin^2 x = T_4(x) + d_4(x)$, where T_4 is a trig polynomial and $\|d_4\|_\infty < \varepsilon/2$. Thus, after replacing x by $x + \pi/2$, we have

$$f(x)\cos^2 x = T_5(x) + d_5(x), \tag{11.4}$$

where T_5 is a trig polynomial and $\|d_5\|_\infty < \varepsilon/2$. Finally, adding equations (11.3) and (11.4),

$$f(x) = T_6(x) + d_6(x),$$

where T_6 is a trig polynomial and $\|d_6\|_\infty < \varepsilon$. That is, $\|f - T_6\|_\infty < \varepsilon$. □

To round off our discussion of Weierstrass's second theorem, we next show that Theorem 11.8 implies Theorem 11.3. By Lemma 11.1, it is enough to show that Theorem 11.3 holds in, say, $C[-1, 1]$. But, given $f \in C[-1, 1]$, note that $f(\cos x) \in C[0, \pi]$. In fact, $f(\cos x)$ defines an *even* function in $C^{2\pi}$. Thus, by Theorem 11.8, there is a trig polynomial T such that $|f(\cos x) - T(x)| < \varepsilon$ for all $x \in \mathbb{R}$. Then, since $f(\cos x)$ is even, it follows that $|f(\cos x) - T(-x)| < \varepsilon$ for all $x \in \mathbb{R}$, too. Hence, the *even* trig polynomial $g(x) = [T(x) + T(-x)]/2$ likewise satisfies $|f(\cos x) - g(x)| < \varepsilon$ for all $x \in \mathbb{R}$. (Why?) Finally, from Exercise 29, there is an algebraic polynomial p such that $g(x) = p(\cos x)$. That is, $|f(\cos x) - p(\cos x)| < \varepsilon$ for all $x \in \mathbb{R}$, and hence $|f(t) - p(t)| < \varepsilon$ for all $t \in [-1, 1]$.

The conclusion here is that Weierstrass's two theorems are logically equivalent. This observation may seem pointless; after all, we used Theorem 11.3 to prove Theorem 11.8. But there are many independent proofs of Weierstrass's two theorems. The real point here is that it is necessary only to prove one of the two; the other will follow from elementary arguments. We will find plenty of applications of Weierstrass's approximation theorems in Part Three.

EXERCISES

35. Prove that $C^{2\pi}$ is complete.

36. Prove that $C^{2\pi}$ is separable.

37. Let f be Riemann integrable on $[-\pi, \pi]$, and let $\varepsilon > 0$. Prove:
(a) There is a function $g \in C[-\pi, \pi]$ satisfying $\int_{-\pi}^{\pi} |f(x) - g(x)|\, dx < \varepsilon$.
(b) There is a continuous, 2π-periodic function $h \in C^{2\pi}$ satisfying $\int_{-\pi}^{\pi} |f(x) - h(x)|\, dx < \varepsilon$.
(c) There is a trig polynomial T with $\int_{-\pi}^{\pi} |f(x) - T(x)|\, dx < \varepsilon$.

38. Show that each element of $C^{2\pi}$ is uniquely determined by its Fourier series. That is, show that if $f \in C^{2\pi}$, and if $\int_{-\pi}^{\pi} f(x) \cos nx\, dx = 0$, and $\int_{-\pi}^{\pi} f(x) \sin nx\, dx = 0$ for all $n = 0, 1, 2, \ldots$, then $f = 0$. [Hint: For an easy proof, modify the argument used in Application 11.6.]

39. Let $f \in C^{2\pi}$. If the Fourier series for f is uniformly convergent on $\mathbb{R}$, prove that it must, in fact, converge to f. [Hint: Combine the arguments of Example 10.6 and the previous exercise.]

40. If $f : \mathbb{R} \to \mathbb{R}$ is twice continuously differentiable and 2π-periodic, prove that the Fourier series for f converges uniformly to f. [Hint: See Exercise 10.4.]

Infinitely Differentiable Functions

The value in approximating by algebraic or trigonometric polynomials should be obvious: Polynomials are well behaved. Either type of polynomial is not only continuous, but differentiable. In fact, either sort of polynomial has continuous derivatives of all orders; in other words, they are *infinitely differentiable*. Thus, while the typical function in $C[0, 1]$ or $C^{2\pi}$ may not be differentiable at *any* point, it is nevertheless close to one that is infinitely differentiable. Our goal in this section is to show how this result extends to $C(\mathbb{R})$. The Weierstrass theorem will do most of the work for us; all that is lacking is a method for constructing infinitely differentiable functions with certain prescribed properties.

The class of infinitely differentiable functions $f : \mathbb{R} \to \mathbb{R}$ is denoted by $C^\infty(\mathbb{R})$. That is, $f \in C^\infty(\mathbb{R})$ if and only if f has continuous derivatives of all orders on $\mathbb{R}$. Obviously, $C^\infty(\mathbb{R})$ is both a subspace and a subalgebra of $C(\mathbb{R})$.

Lemma 11.10. *There is an $f \in C^\infty(\mathbb{R})$ such that $f(x) = 0$ for $x \leq 0$ and $f(x) > 0$ for $x > 0$.*

PROOF. Define f by $f(x) = 0$ for $x \le 0$ and $f(x) = e^{-1/x}$ for $x > 0$. It is clear that f is infinitely differentiable everywhere except, possibly, at $x = 0$. Notice that $f'(x) = x^{-2}e^{-1/x}$ and $f''(x) = (x^{-4} - 2x^{-3})e^{-1/x}$ for $x > 0$. Using induction, it is easy to see that $f^{(k)}(x) = p_k(x^{-1})f(x)$ for $x > 0$, where p_k is a polynomial of degree at most $2k$. Of course, $f^{(k)}(x) = 0$ for $x < 0$ and any k.

To see that f is continuous at 0, first note that if $y > 0$, then $e^y = \sum_{n=0}^{\infty} y^n/n! > y^m/m!$ for any $m = 0, 1, 2, \ldots$. Thus, for $x > 0$,

$$0 < f(x) = e^{-1/x} = \left(e^{1/x}\right)^{-1} < m!\,x^m,$$

for $m = 0, 1, 2, \ldots$. In particular, $f(x) \to 0$ as $x \to 0$. Likewise, $f(x)/x \to 0$ as $x \to 0$. That is, f' exists and is continuous at $x = 0$, and $f'(0) = 0$.

Suppose that we have shown that $f^{(k)}$ exists and is continuous at 0. Then, of course, $f^{(k)}(0) = 0$. Thus, $f^{(k)}(x)/x = x^{-1}p_k(x^{-1})f(x)$. And since p_k has degree at most $2k$, and since $|f(x)| \le (2k+2)!\,x^{2k+2}$, it follows that $f^{(k)}(x)/x \to 0$ as $x \to 0$. That is, $f^{(k+1)}(0)$ exists and equals 0. A similar argument shows that $f^{(k+1)}(x) = p_{k+1}(x)f(x) \to 0$ as $x \to 0$; that is, $f^{(k+1)}$ is continuous at 0. By induction, $f^{(k)}$ exists and is continuous at 0 for all k. □

The function f constructed in Lemma 11.10 is an important example. All of the derivatives of f vanish at 0, but f is not identically 0. Thus, the Taylor series expansion for f about 0 does not converge to f. In fact, no convergent power series $\sum_{n=0}^{\infty} a_n x^n$ can represent f in any neighborhood of 0.

Given f, it is easy to construct all sorts of C^∞ functions:

Lemma 11.11. *There is a $g \in C^\infty(\mathbb{R})$ such that $g(x) = 0$ for $|x| \ge 1$ and $g(x) > 0$ for $|x| < 1$.*

PROOF. Let $g(x) = f(x+1)f(1-x)$, where f is the function constructed in Lemma 11.10. □

Lemma 11.12. *There is an $h \in C^\infty(\mathbb{R})$ such that*

(i) $h(x) = 0$ *for* $|x| \ge 1$, $0 < h(x) \le 1$ *for* $|x| < 1$, *and* $h(0) = 1$;

(ii) *Given* $n \in \mathbb{Z}$ *and* $n \le x < n+1$, *we have* $h(x-n) + h(x-n-1) = 1$, *while* $h(x-k) = 0$ *for any integer* $k < n$ *or* $k > n+1$.

PROOF. Let g be the function constructed in Lemma 11.11, and consider the function $G(x) = \sum_{n \in \mathbb{Z}} g(x-n)$. This series is actually a finite sum in a small neighborhood about any point $x \in \mathbb{R}$. Indeed, if $n \in \mathbb{Z}$ is chosen so that $n - 1 < x < n + 1$, then at most three terms, namely $g(x-n+1)$, $g(x-n)$, and $g(x-n-1)$, are nonzero (and at least one is strictly positive). That is, $G(x) = g(x-n+1) + g(x-n) + g(x-n-1)$ on $n - 1 < x < n + 1$ (and $G(x) = g(x-n) + g(x-n-1)$ if $n \le x < n+1$). Consequently, the series converges to an infinitely differentiable function $G(x)$ and, moreover, $G(x) > 0$ for any x. Finally, if we set $h(x) = g(x)/G(x)$, then it is easy to check that h has the properties stated in the lemma. □

Now let's bring the Weierstrass theorem back into the picture.

Theorem 11.13. *Given $f \in C(\mathbb{R})$ and $\varepsilon > 0$, there is a function $\varphi \in C^\infty(\mathbb{R})$ such that $|f(x) - \varphi(x)| < \varepsilon$ for all $x \in \mathbb{R}$. Hence, there is a sequence (φ_n) in $C^\infty(\mathbb{R})$ such that $\varphi_n \rightrightarrows f$ on $\mathbb{R}$.*

PROOF. For each $n \in \mathbb{Z}$, Theorem 11.3 supplies a polynomial p_n such that $|f(x) - p_n(x)| < \varepsilon$ for all $n-1 \le x \le n+1$. Now define φ by $\varphi(x) = \sum_{n \in \mathbb{Z}} p_n(x)h(x-n)$, where h is the function constructed in Lemma 11.12. This series is actually a finite sum over any bounded interval, so $\varphi \in C^\infty(\mathbb{R})$. And, from Lemma 11.12 (ii), if $n \le x < n+1$, then

$$\varphi(x) = p_n(x)h(x-n) + p_{n+1}(x)h(x-n-1).$$

Thus, for $n \le x < n+1$, we get

$$\begin{aligned} |f(x) - \varphi(x)| &= \big|h(x-n)[f(x) - p_n(x)] + h(x-n-1)[f(x) - p_{n+1}(x)]\big| \\ &\le h(x-n)|f(x) - p_n(x)| + h(x-n-1)|f(x) - p_{n+1}(x)| \\ &< \varepsilon, \end{aligned}$$

since $h \ge 0$ and $h(x-n) + h(x-n-1) = 1$. □

EXERCISES

41. Given $a < b$, modify the construction in Lemma 11.11 to find a function $\varphi \in C^\infty(\mathbb{R})$ with $\varphi(x) = 0$ for $x \notin (a, b)$ and $\varphi(x) > 0$ for $x \in (a, b)$.

42. Given $a < b$, show that there is an $\psi \in C^\infty(\mathbb{R})$ such that $\psi(x) = 0$ for $x \le a$, $0 < \psi(x) < 1$ for $a < x < b$, and $\psi(x) = 1$ for $x \ge b$. [Hint: Consider $\psi(x) = c\int_{-\infty}^{x} \varphi$, where φ is as in Exercise 41.]

43. Given $a < b$ and $\varepsilon > 0$, show that there is a function $\varphi \in C^\infty(\mathbb{R})$ such that $\varphi(x) = 0$ for $x \notin [a-\varepsilon, b+\varepsilon]$, $\varphi(x) = 1$ for $x \in [a, b]$, and $0 < \varphi(x) < 1$ otherwise.

▷ **44.** Let h be the function constructed in Lemma 11.12. Given any integer $n \in \mathbb{Z}$ and any positive integer $k \in \mathbb{N}$, show that $\sum_{i=n}^{n+k} h(x-i) = 1$ for $n \le x \le n+k$.

Equicontinuity

We next turn our attention to the second question raised at the beginning of the chapter: Given a compact metric space X, what are the *compact* subsets of $C(X)$? Since $C(X)$ is complete, we know that this is the same as asking: What are the *totally bounded* subsets of $C(X)$? (Because the compact sets in $C(X)$ are just the closures of the totally bounded sets.) If we recall the Bolzano–Weierstrass characterization of total boundedness, we can rephrase the question yet again to read: When does a (uniformly) bounded sequence in $C(X)$ have a (uniformly) convergent subsequence? We will see in this section that this last question is asking for the missing ingredient in the formula

$$\text{pointwise convergence} + \boxed{???} \Longrightarrow \text{uniform convergence.}$$

To begin, let's make a few easy observations. Recall that, *throughout this chapter, unless otherwise specified, X denotes a compact metric space.*

Examples 11.14

(a) If (f_n) is a uniformly convergent sequence in $C(X)$, and if $f_n \rightrightarrows f$ on X, then the set $\{f\} \cup \{f_n : n \geq 1\}$ is compact in $C(X)$. (Why?)

(b) A collection of real-valued functions $\mathcal{F}$ on (a set) X is said to be **uniformly bounded** if the set $\{f(x) : x \in X, f \in \mathcal{F}\}$ is bounded (in $\mathbb{R}$), that is, if $\sup_{f \in \mathcal{F}} \sup_{x \in X} |f(x)| = \sup_{f \in \mathcal{F}} \|f\|_\infty < \infty$. In other words, uniformly bounded means bounded in the metric of $B(X)$ (or $C(X)$). Clearly, any uniformly convergent sequence in $B(X)$ is uniformly bounded.

The point to Example 11.14 (a) is that we already know some easy compact subsets of $C(X)$, and Example 11.14 (b) is reminding us that boundedness is a necessary condition for compactness (or total boundedness). But, as you might suspect, a totally bounded set should be something more than merely bounded. The extra ingredient here is called *equicontinuity*.

Let $\mathcal{F}$ be a collection of real-valued continuous functions on a metric space X. If, given $\varepsilon > 0$, a single δ can always be chosen to "work" (in the ε-δ definition of continuity) simultaneously for every $f \in \mathcal{F}$ *and* every $x \in X$, then $\mathcal{F}$ is called **equicontinuous** (or, sometimes, *uniformly equicontinuous*). That is, $\mathcal{F}$ is equicontinuous if, given $\varepsilon > 0$, there is a $\delta > 0$ such that whenever $x, y \in X$ satisfy $d(x, y) < \delta$, we then have $|f(x) - f(y)| < \varepsilon$ for all $f \in \mathcal{F}$. In short, an equicontinuous collection of functions is "uniformly uniformly continuous."

Examples 11.15

(a) Clearly, any *finite* subset of $C(X)$ is equicontinuous. (Why?) Also note that any subset of an equicontinuous set of functions is again equicontinuous.

(b) Given $0 < K < \infty$ and $0 < \alpha \leq 1$, recall that $\text{Lip}_K\alpha$ is the collection of all $f \in C[0, 1]$ that satisfy $|f(x) - f(y)| \leq K|x - y|^\alpha$ for $x, y \in [0, 1]$. It is easy to see that $\text{Lip}_K\alpha$ is equicontinuous. (Why?) But $\text{Lip}_K\alpha$ is not totally bounded, since it is not bounded in $C[0, 1]$ (it always contains the constant functions).

EXERCISES

45. A collection of real-valued functions $\mathcal{F}$ on (a set) X is said to be **pointwise bounded** if, for each $x \in X$, the set $\{f(x) : f \in \mathcal{F}\}$ is bounded (in $\mathbb{R}$), that is, if $\sup_{f \in \mathcal{F}} |f(x)| < \infty$ for each $x \in X$. If (f_n) is a pointwise convergent sequence of real-valued functions, show that (f_n) is also pointwise bounded.

46. Prove that a uniformly bounded collection of functions is also pointwise bounded. Give an example of a collection of functions that is pointwise bounded but not uniformly bounded.

47. If a sequence (f_n) in $B[a,b]$ is pointwise bounded, show that some subsequence of (f_n) converges pointwise on the set of rationals in $[a,b]$. [Hint: Diagonalize!]

48. Let X be a compact metric space. Prove that an equicontinuous subset of $C(X)$ is pointwise bounded if and only if it is uniformly bounded.

49. A collection $\mathcal{F}$ of real-valued continuous functions on a metric space X is said to be **equicontinuous at a point** $x \in X$ if, for each $\varepsilon > 0$, there is a single $\delta > 0$ that "works" at x for every $f \in \mathcal{F}$. That is, $\mathcal{F}$ is equicontinuous at x if, given $\varepsilon > 0$, there is a $\delta > 0$, which may depend on x, such that whenever $y \in X$ satisfies $d(x,y) < \delta$ then $|f(x) - f(y)| < \varepsilon$ for all $f \in \mathcal{F}$. If X is a compact metric space, prove that a subset of $C(X)$ is equicontinuous if and only if it is equicontinuous at each point of X.

50. Show that a bounded subset of $C^{(1)}[a,b]$ is equicontinuous.

▷ **51.** Let X be a compact metric space, and let (f_n) be a sequence in $C(X)$. If (f_n) is uniformly convergent, show that (f_n) is both uniformly bounded and equicontinuous.

▷ **52.** Let X be a compact metric space, and let (f_n) be an equicontinuous sequence in $C(X)$. If (f_n) is pointwise convergent, prove that, in fact, (f_n) is uniformly convergent.

53. Let X be a compact metric space, and let (f_n) be a sequence in $C(X)$. If (f_n) decreases pointwise to 0, show that (f_n) is equicontinuous. [Hint: Exercise 49.] Combine this observation with the result in Exercise 52 to give another proof of Dini's theorem (Exercise 10.18).

54. Let X be a compact metric space, and let (f_n) be an equicontinuous sequence in $C(X)$. Show that $C = \{x \in X : (f_n(x)) \text{ converges}\}$ is a closed set in X.

55. If (f_n) is an equicontinuous sequence in $C[a,b]$, and if $(f_n(x))$ converges at each rational in $[a,b]$, prove that (f_n) is uniformly convergent on $[a,b]$. [Hint: Exercises 54 and 52.]

56. (Arzelà–Ascoli, utility grade): If (f_n) is an equicontinuous, pointwise bounded sequence in $C[a,b]$, then some subsequence of (f_n) converges uniformly on $[a,b]$. [Hint: Exercises 47 and 55.]

Lemma 11.16. *If $\mathcal{F}$ is a totally bounded subset of $C(X)$, then $\mathcal{F}$ is uniformly bounded and equicontinuous.*

PROOF. Since a totally bounded set is necessarily also (uniformly) bounded, we only have to prove that $\mathcal{F}$ is equicontinuous. So, let $\varepsilon > 0$.

Since $\mathcal{F}$ is totally bounded, it has a finite $\varepsilon/3$-net; that is, there exist $f_1, \ldots, f_n \in \mathcal{F}$ such that each $f \in \mathcal{F}$ satisfies $\|f - f_i\|_\infty < \varepsilon/3$ for some i. Since the set $\{f_1, \ldots, f_n\}$ is equicontinuous, there is a $\delta > 0$ such that $|f_i(x) - f_i(y)| < \varepsilon/3$ whenever $d(x, y) < \delta$. We now claim that this same δ "works" for every $f \in \mathcal{F}$. Indeed, given $f \in \mathcal{F}$, first choose i such that $\|f - f_i\|_\infty < \varepsilon/3$. Then, given x and y with $d(x, y) < \delta$, we have

$$\begin{aligned} |f(x) - f(y)| &\le |f(x) - f_i(x)| + |f_i(x) - f_i(y)| + |f_i(y) - f(y)| \\ &< \varepsilon/3 + \varepsilon/3 + \varepsilon/3 = \varepsilon. \end{aligned}$$

Thus, $\mathcal{F}$ is equicontinuous. □

Corollary 11.17. *If (f_n) is a uniformly convergent sequence in $C(X)$, then (f_n) is uniformly bounded and equicontinuous.*

Lemma 11.16 essentially characterizes the compact subsets of $C(X)$.

The Arzelà–Ascoli Theorem 11.18. *Let X be a compact metric space, and let $\mathcal{F}$ be a subset of $C(X)$. Then $\mathcal{F}$ is compact if and only if $\mathcal{F}$ is closed, uniformly bounded, and equicontinuous.*

PROOF. The forward implication follows from Lemma 11.16; that is, a compact subset of $C(X)$ is necessarily closed, uniformly bounded, and equicontinuous. We need to prove the backward implication. So, suppose that $\mathcal{F}$ is closed, uniformly bounded, and equicontinuous, and let (f_n) be a sequence in $\mathcal{F}$. We need to show that (f_n) has a uniformly convergent subsequence.

First note that (f_n) is equicontinuous. (Why?) Thus, given $\varepsilon > 0$, there is a $\delta > 0$ such that if $d(x, y) < \delta$, then $|f_n(x) - f_n(y)| < \varepsilon/3$ for all n.

Next, since X is totally bounded, X has a finite δ-net; there exist $x_1, \ldots, x_k \in X$ such that each $x \in X$ satisfies $d(x, x_i) < \delta$ for some i. Now, since (f_n) is also uniformly bounded (why?), each of the sequences $\left(f_n(x_i)\right)_{n=1}^\infty$ is bounded (in $\mathbb{R}$) for $i = 1, \ldots, k$. Thus, by passing to a subsequence of the f_n (and relabeling), we may suppose that $\left(f_n(x_i)\right)_{n=1}^\infty$ converges for each $i = 1, \ldots, k$. (How?) In particular, we can find some N such that $|f_m(x_i) - f_n(x_i)| < \varepsilon/3$ for any $m, n \ge N$ and any $i = 1, \ldots, k$.

And now we are done! Given $x \in X$, first find i such that $d(x, x_i) < \delta$, and then, whenever $m, n \ge N$, we will have

$$\begin{aligned} |f_m(x) - f_n(x)| &\le |f_m(x) - f_m(x_i)| + |f_m(x_i) - f_n(x_i)| + |f_n(x_i) - f_n(x)| \\ &< \varepsilon/3 + \varepsilon/3 + \varepsilon/3 = \varepsilon. \end{aligned}$$

That is, (f_n) is uniformly Cauchy, since our choice of N does not depend on x. Since $\mathcal{F}$ is closed in $C(X)$, it follows that (f_n) converges uniformly to some $f \in \mathcal{F}$. □

Compare the following result to Exercise 56.

Corollary 11.19. *Let X be a compact metric space. If (f_n) is a uniformly bounded, equicontinuous sequence in $C(X)$, then some subsequence of (f_n) converges uniformly on X.*

EXERCISES

57. Suppose that $f_n : [\,a, b\,] \to \mathbb{R}$ is a sequence of differentiable functions satisfying $|f_n'(x)| \le 1$ for all n and x. Prove that some subsequence of (f_n) is uniformly convergent.

58. For K and α fixed, show that $\{f \in \text{Lip}_K\alpha : f(0) = 0\}$ is a compact subset of $C[\,0, 1\,]$.

59. For each n, show that $\{f \in \text{Lip}1 : \|f\|_{\text{Lip}1} \le n\}$ is a compact subset of $C[\,0, 1\,]$. Use this to give another proof that $C[\,0, 1\,]$ is separable. [Hint: See Exercises 24 and 26.]

60. If (f_n) is an equicontinuous sequence in $C^{(1)}[\,a, b\,]$, is it necessarily true that the sequence of derivatives (f_n') is uniformly bounded? Explain.

61. For the sake of a characterization that is easier to test, it is convenient to weaken one of the hypotheses in the Arzelà–Ascoli theorem. Given a compact metric space X and a subset $\mathcal{F}$ of $C(X)$, prove that $\mathcal{F}$ is compact if and only if $\mathcal{F}$ is closed, *pointwise bounded*, and equicontinuous. [Hint: Just repeat the proof of Theorem 11.18!]

62. Let X be a compact metric space, and let $\mathcal{F}$ be a subset of $C(X)$.

(a) If $\mathcal{F}$ is pointwise bounded, prove that the closure of $\mathcal{F}$ in $C(X)$ is also pointwise bounded.

(b) If $\mathcal{F}$ is uniformly bounded, prove that the closure of $\mathcal{F}$ in $C(X)$ is also uniformly bounded.

(c) True or false? If $\mathcal{F}$ is equicontinuous, then the closure of $\mathcal{F}$ in $C(X)$ is also equicontinuous.

63. Define $T : C[\,a, b\,] \to C[\,a, b\,]$ by $(Tf)(x) = \int_a^x f$. Show that T maps bounded sets into equicontinuous (and hence compact) sets. [Hint: Tf is Lipschitz with constant $\|f\|_\infty$.]

64. Let (f_n) be a sequence in $C[\,a, b\,]$ with $\|f_n\|_\infty \le 1$ for all n and define $F_n(x) = \int_a^x f_n(t)\,dt$. Show that some subsequence of (F_n) is uniformly convergent.

65. Let $K(x, t)$ be a continuous function on the square $[\,a, b\,] \times [\,a, b\,]$.

(a) Given $f \in C[\,a, b\,]$, show that $g(x) = \int_a^b f(t)\,K(x, t)\,dt$ defines a continuous function $g \in C[\,a, b\,]$.

(b) Define $T : C[\,a, b\,] \to C[\,a, b\,]$ by $(Tf)(x) = \int_a^b f(t)\,K(x, t)\,dt$. Show that T maps bounded sets into equicontinuous sets. In particular, T is continuous.

66. Suppose that $F : \mathbb{R}^2 \to \mathbb{R}$ is continuous and Lipschitz in its second variable: $|F(r, s) - F(r, t)| \le K|s - t|$.

(a) If $f \in C[a, b]$, show that $g(x) = \int_a^x F(t, f(t))\,dt$ defines a continuous function $g \in C[a, b]$. [Hint: F is bounded on rectangles.]

(b) Define $T : C[a, b] \to C[a, b]$ by $(Tf)(x) = \int_a^x F(t, f(t))\,dt$. Show that T is continuous. [Hint: T is not linear, but it is Lipschitz.] Consequently, T achieves a minimum on any compact set in $C[a, b]$.

(c) Show that T maps bounded sets into equicontinuous sets. [Hint: Estimate the Lipschitz constant of Tf.]

Continuity and Category

In Chapter Ten we gave examples showing that the pointwise limit of a sequence of continuous functions need not be everywhere continuous. And, in general, we know that some extra ingredient is needed to ensure such a strong conclusion. But is it possible that the pointwise limit of a sequence of continuous functions could be everywhere discontinuous? For example, is it possible to express $\chi_{\mathbb{Q}}$ as the pointwise limit of a sequence of continuous functions on $\mathbb{R}$?

As it happens, the pointwise limit of a sequence of continuous functions on $\mathbb{R}$ must have *lots* of points of continuity.

The Baire–Osgood Theorem 11.20. *Let $f_n : \mathbb{R} \to \mathbb{R}$ be continuous for each n, and suppose that $f(x) = \lim_{n\to\infty} f_n(x)$ exists (as a real number) for each $x \in \mathbb{R}$. Then $D(f)$ is a first category set in $\mathbb{R}$. In particular, f is continuous at a dense set of points in $\mathbb{R}$.*

PROOF. From Theorem 9.2 we know that $D(f) = \bigcup_{n=1}^{\infty}\{x : \omega_f(x) \ge 1/n\}$ is the countable union of closed sets. Thus, it suffices to show, for any $\varepsilon > 0$, that the closed set $F = \{x : \omega_f(x) \ge 5\varepsilon\}$ is nowhere dense. The proof of this fact may seem rather indirect, but have patience!

Consider the sets $E_n = \bigcap_{i,j\ge n}\{x : |f_i(x) - f_j(x)| \le \varepsilon\}$. Since (f_n) is pointwise convergent, we know that $\bigcup_{n=1}^{\infty} E_n = \mathbb{R}$. Notice, too, that each E_n is a closed set (because the f_i are continuous).

Given any closed interval I, we want to show that $I \not\subset F$, for then it will follow that F contains no open intervals either (that is, F has an empty interior). We will take a first step in this direction by applying the Baire category theorem to $I = \bigcup_{n=1}^{\infty}(E_n \cap I)$. Since I is complete, and since each E_n is closed, it follows that, for some n, the set $E_n \cap I$ contains an entire open interval J. We are going to show that $J \subset F^c = \{x : \omega_f(x) < 5\varepsilon\}$, and hence that $I \not\subset F$.

Since $J \subset E_n$, we have $|f_i(x) - f_j(x)| \le \varepsilon$ for all $x \in J$ and all $i, j \ge n$. Thus, $|f(x) - f_n(x)| \le \varepsilon$ for all $x \in J$. (Why?) Next we use the fact that f_n is continuous: For each $x_0 \in J$ there is an open interval $I_{x_0} \subset J$, containing x_0, such that $|f_n(x) - f_n(x_0)| \le \varepsilon$ for all $x \in I_{x_0}$. But then it follows from the triangle inequality that $|f(x) - f_n(x_0)| \le 2\varepsilon$ for all $x \in I_{x_0}$ and, finally, that $|f(x) - f(y)| \le 4\varepsilon$ for all $x, y \in I_{x_0}$. That is, we have shown that $\omega_f(x_0) \le \omega(f; I_{x_0}) \le 4\varepsilon$, and hence that $x_0 \notin F$. □

Corollary 11.21. *Let $f : \mathbb{R} \to \mathbb{R}$. Then, $D(f)$ is a first category set in $\mathbb{R}$ if and only if f is continuous at a dense set of points.*

PROOF. An F_σ subset of $\mathbb{R}$ is a first category set if and only if its complement is dense. □

Examples 11.22

(a) $\chi_{\mathbb{Q}}$ cannot be written as the limit of a sequence of continuous functions. (Why?) However, we do have $\chi_{\mathbb{Q}}(x) = \lim_{m\to\infty} \lim_{n\to\infty} \left(\cos m!\,\pi x\right)^{2n}$.

(b) If $f : \mathbb{R} \to \mathbb{R}$ is everywhere differentiable, then f' must have a point of continuity, since f' is then the limit of a sequence of continuous functions: $f'(x) = \lim_{n\to\infty} n\,[f(x + (1/n)) - f(x)]$.

Since the subject of derivatives has come up in conjunction with the Baire category theorem, now is probably a good time to discuss Banach's proof of the *existence* of continuous nowhere differentiable functions. Rather than pursue the "hard" technicalities that we saw in Chapter Ten, we will take this as an excuse to demonstrate some of the advantages of the "soft" approach.

To begin, let F denote the set of all functions in $C[0, 1]$ having a finite derivative at some point of $[0, 1]$. Banach's wonderfully clever observation is that F is a first category set in (the complete space) $C[0, 1]$. Since this means that the complement of F is dense in $C[0, 1]$, it would be fair to say that "most" continuous functions on $[0, 1]$ fail to have a finite derivative at even a single point. Isn't this curious? Without displaying a single concrete example, Banach's observation shows that nondifferentiability is the rule, rather than the exception, for elements of $C[0, 1]$.

For each $n \geq 2$, consider the set E_n consisting of those $f \in C[0, 1]$ such that, for some $0 \leq x \leq 1 - (1/n)$, we have $|f(x + h) - f(x)| \leq nh$ for all $0 < h < 1 - x$. In particular, any $f \in C[0, 1]$ having a right-hand derivative at most n in magnitude at even one point in $[0, 1 - (1/n)]$ is in E_n. The set $E = \bigcup_{n=2}^{\infty} E_n$ consists of all of those $f \in C[0, 1]$ that have bounded right-hand difference quotients at some x in $[0, 1)$. In particular, any $f \in C[0, 1]$ having a finite right-hand derivative at even one point in $[0, 1)$ is in E. We will show that E is a first category set in $C[0, 1]$ by showing that each E_n is closed and nowhere dense in $C[0, 1]$.

First, let's show that the complement of E_n is dense in $C[0, 1]$. Once we have established that E_n is closed, this will prove that E_n is nowhere dense. Given $\varepsilon > 0$, we need to show that an arbitrary $g \in C[0, 1]$ is within ε of some $f \notin E_n$. Since the polygonal functions are dense in $C[0, 1]$, it is enough to consider the case where g is polygonal. But now our job is easy: We just argue that we can find a "sawtooth" function f, having right-hand derivatives bigger than n in magnitude, that is within ε of g, as shown in Figure 11.3.

Next, let's check that E_n is closed. Suppose that (f_k) is a sequence from E_n, and that (f_k) converges uniformly to some f in $C[0, 1]$. We need to show that $f \in E_n$. Now there is a corresponding sequence (x_k) with $0 \leq x_k \leq 1-(1/n)$ such that $|f_k(x_k+h)-f(x_k)| \leq nh$ for all $0 < h < 1 - x_k$. By passing to a subsequence, if necessary (and relabeling), we may suppose that $x_k \to x$, where $0 \leq x \leq 1-(1/n)$. We will take the corresponding subsequence of (f_k), too (likewise relabeled). Thus, $f_k \rightrightarrows f$ and $x_k \to x$.

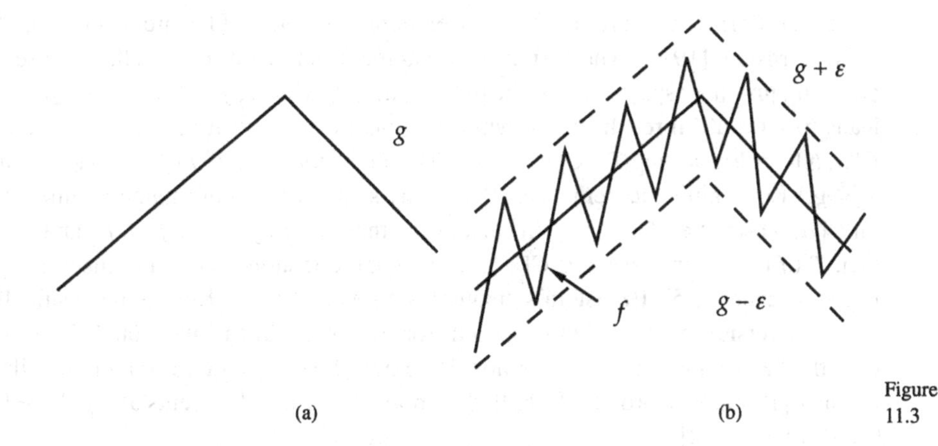

Figure 11.3

If $0 < h < 1-x$, then $0 < h < 1-x_k$ for all k sufficiently large. Thus, if k sufficiently large, we have

$$\begin{aligned}|f(x+h) - f(x)| &\le |f(x+h) - f(x_k+h)| + |f(x_k+h) - f_k(x_k+h)| \\ &\quad + |f_k(x_k+h) - f_k(x_k)| + |f_k(x_k) - f(x_k)| + |f(x_k) - f(x)| \\ &\le |f(x+h) - f(x_k+h)| + \|f - f_k\|_\infty \\ &\quad + nh + \|f - f_k\|_\infty + |f(x_k) - f(x)|.\end{aligned}$$

Now, since f is continuous and $f_k \rightrightarrows f$, we just let $k \to \infty$ in our last estimate to arrive at $|f(x+h) - f(x)| \le nh$. That is, $f \in E_n$.

Notes and Remarks

Weierstrass's first theorem, on approximation by algebraic polynomials (Theorem 11.3), appeared in Weierstrass [1885, pp. 633–639]. His second theorem, on approximation by trigonometric polynomials (Theorem 11.8), appeared immediately after the first, in a paper under the same title, in Weierstrass [1885, pp. 789–805]. See Weierstrass [1886] for a French translation.

A great deal has been written about Weierstrass's approximation theorems and related questions. For a brief historical overview, see Shields [1987a] and Hedrick [1927]. More detailed discussions are given in Jackson [1920] and Fisher [1978]. For a short account of Weierstrass's life, see Polubarinova-Kochina [1966].

Three highly readable sources for detailed information on the approximation of functions are Natanson [1964], Cheney [1966], and Rivlin [1981].

The observation that the polygonal functions are dense in $C[a,b]$ (Theorem 11.2) is due to Lebesgue, as is the fact that this observation can be used to give an elementary proof of Weierstrass's first theorem (see Exercises 2 and 11). So is the elementary proof that Weierstrass's two theorems are, in fact, equivalent (the proof of Theorem 11.8 and the subsequent discussion). All this and more can be found in Lebesgue's first published paper, Lebesgue [1898]. The details, as given here, are based largely on the presentation in de la Vallée Poussin [1919].

Sergei Bernstein's proof of the Weierstrass theorem (Theorem 11.4) is from S. N. Bernstein [1912]. The curious fact that the proof of Bernstein's theorem rests on checking just three special cases, the polynomials $f_0(x) = 1$, $f_1(x) = x$, and $f_2(x) = x^2$, leads to a beautiful result of Korovkin on monotone (or positive) linear operators on $C[a, b]$. (A linear map $T : C[a, b] \to C[a, b]$ is *monotone* if $T(f) \le T(g)$ whenever $f \le g$.) Korovkin's theorem states that if any sequence (T_n) of monotone linear maps on $C[a, b]$ satisfies $T_n(f) \rightrightarrows f$ in each of the three cases $f = f_0$, $f = f_1$, and $f = f_2$, then $T_n(f) \rightrightarrows f$ for *every* $f \in C[a, b]$. Since the operators $B_n(f)$ are linear and positive (see Exercise 5), Bernstein's theorem is a special case of Korovkin's result. There is also a version of Korovkin's theorem for monotone linear maps on $C^{2\pi}$, in which case the "Korovkin set" $\{1, x, x^2\}$ now becomes $\{1, \cos x, \sin x\}$. For more details, see Cheney [1966], or Korovkin [1960]. For more recent developments along these lines, see Donner [1982].

Exercise 16 is taken from my classroom notes from W. B. Johnson's course in real analysis at The Ohio State University in 1974–75. The spaces $\text{Lip}\,\alpha$, for $0 < \alpha < 1$, in Exercises 20–24, 26 are sometimes referred to as the *Hölder continuous* functions.

The section on trigonometric polynomials, along with the proof of the equivalence of Weierstrass's first and second theorems, is based in part on the presentations found in de la Vallée Poussin [1919] and Natanson [1964] (and, to some extent, Jackson [1941] and Rogosinski [1950]) but, as already mentioned, is heavily influenced by Lebesgue's original presentation; see also Lebesgue [1906].

Several enlightening proofs of the Weierstrass theorems (especially, deductions of the first theorem from the second) can be found in Jackson [1941]. In one particularly direct approach, Jackson points out that if f is a polygonal function in $C^{2\pi}$, then the Fourier coefficients for f satisfy $|a_k|, |b_k| \le C/k^2$. (Compare this with the result in Exercise 40.) It follows (see Exercise 39) that each 2π-periodic polygonal function is the uniform limit of its Fourier series. Since the polygonal functions are clearly dense in $C^{2\pi}$, this observation gives a quick proof of Weierstrass's second theorem.

The constructions in Lemmas 11.10 and 11.11, along with Exercise 42, are based on the presentation in Beals [1973]. Lemma 11.12, Theorem 11.13, and Exercise 44 are based on the presentation in Pursell [1967].

The Italian mathematicians Ascoli and Arzelà were both interested in extending Cantor's set theory to sets whose elements were functions, sometimes referred to as "curves" or "lines," especially in regard to "functions of lines," or functions of functions, if you will. In particular, Arzelà examined the problems of finding necessary and sufficient conditions for the integrability of the pointwise limit of a sequence of integrable functions, of finding the correct mode of convergence that would preserve integrability, and of the validity of term-by-term integration of series.

Ascoli defined the notion of equicontinuity (at a point), and Arzelà used the concept at about the same time. It would seem that Ascoli proved the sufficiency of this new condition for compactness in Ascoli [1883] while Arzelà proved the necessity in Arzelà [1889] (for $C[0, 1]$ in either case). But Arzelà is generally credited for the first clear statement of Theorem 11.18 for $C[0, 1]$ in Arzelà [1895]. The metric space version is (once again) due to Fréchet; see Fréchet [1906]. For more details, see Dunford and Schwartz [1958] and Hawkins [1970]. Exercise 59 is based on a result in Dudley [1989].

A slightly different version of Theorem 11.20, concerning the set of points of uniform convergence of a pointwise convergent sequence of functions, was established in Osgood [1897]. For more on Osgood's approach, see Hobson [1927, Vol. II]. As stated here, Theorem 11.20 is part of Baire's thesis, Baire [1899]. The proof given here, along with Corollary 11.21 and Example 11.22, are taken from Oxtoby [1971]. For a discussion of related issues, see Hewitt [1960], Goffman [1960], and Myerson [1991].

Banach's clever application of the Baire category theorem to prove the existence of continuous nowhere differentiable functions is from Banach [1931]. The proof presented here is taken from Oxtoby [1971] (but see also Boas [1960]). Applications of the Baire category theorem to existence proofs are numerous; both Oxtoby and Boas provide several other curious examples. Two particular examples, though, are simply too curious to avoid mention. Compare "Most monotone functions are singular," Zamfirescu [1981] and "Most monotone functions are not singular," Cater [1982]. Katsuura [1991] offers an intriguing application of Banach's contraction mapping theorem to address the existence of nowhere differentiable functions.

CHAPTER TWELVE

The Stone–Weierstrass Theorem

Algebras and Lattices

We continue with our study of $B(X)$, the space of bounded real-valued functions on a set X. As we have seen, $B(X)$ is a Banach space when supplied with the norm $\|f\|_\infty = \sup_{x \in X} |f(x)|$. Moreover, convergence in $B(X)$ is the same as uniform convergence. Of course, if X is a metric space, we will also be interested in $C(X)$, the space of continuous real-valued functions on X, and its cousin $C_b(X) = C(X) \cap B(X)$, the closed subspace of bounded continuous functions in $B(X)$. Finally, if X is a compact metric space, recall that $C_b(X) = C(X)$.

But now we want to add a few more ingredients to the recipe: It's time we made use of the algebraic and lattice structures of $B(X)$. In this chapter we will make formal our earlier informal discussions of algebras and lattices. In particular, we will see how this additional structure leads to a generalization of the Weierstrass approximation theorem in $C(X)$, where X is a compact metric space.

To begin, an **algebra** is a vector space A on which there is defined a multiplication $(f, g) \mapsto fg$ (from $A \times A$ into A) satisfying

(i) $(fg)h = f(gh)$, for all $f, g, h \in A$;
(ii) $f(g + h) = fg + fh$, $(f + g)h = fh + gh$, for all $f, g, h \in A$;
(iii) $\alpha(fg) = (\alpha f)g = f(\alpha g)$, for all scalars α and all $f, g \in A$.

The algebra is called *commutative* if

(iv) $fg = gf$, for all $f, g \in A$.

And we say that A has an *identity* element if there is a vector $e \in A$ such that

(v) $fe = ef = f$, for all $f \in A$.

In the case where A is a normed vector space, we also require that the norm satisfy

(vi) $\|fg\| \le \|f\|\,\|g\|$

(this simplifies things a bit), and in this case we refer to A as a *normed algebra*. If a normed algebra is complete, we refer to it as a *Banach algebra*. Finally, a subset B of an algebra A is called a *subalgebra* (of A) if B is itself an algebra (under the same operations), that is, if B is a (vector) subspace of A that is closed under multiplication.

Examples 12.1

(a) $\mathbb{R}$, with the usual addition and multiplication, is a commutative Banach algebra with identity.

(b) If we define multiplication of vectors "coordinatewise," then $\mathbb{R}^n$ is a commutative Banach algebra with identity (the vector $(1, \ldots, 1)$) when equipped with the norm $\|x\|_\infty = \max_{1 \le i \le n} |x_i|$. We used this observation in Chapter Five.

(c) The collection $M_n(\mathbb{R})$ of all $n \times n$ real matrices, under the usual operations on matrices, is a noncommutative algebra with identity.

(d) Under the usual pointwise multiplication of functions, $B(X)$ is a commutative Banach algebra with identity (the constant 1 function). The constant functions in $B(X)$ form a subalgebra isomorphic (in every sense of the word) to $\mathbb{R}$.

(e) If X is a metric space, then $C(X)$ is a commutative algebra with identity (the constant 1 function) and $C_b(X)$ is a closed subalgebra of $B(X)$.

(f) The polynomials form a dense subalgebra of $C[a, b]$. The trig polynomials form a dense subalgebra of $C^{2\pi}$.

(g) $C^{(1)}[0, 1]$ and Lip 1 are dense subalgebras of $C[0, 1]$.

(h) $C^\infty(\mathbb{R})$ is a subalgebra of $C(\mathbb{R})$.

(i) A function $f : [a, b] \to \mathbb{R}$ is called a **step function** if there are finitely many points $a = t_0 < t_1 < \cdots < t_n = b$ such that f is constant on each of the open intervals (t_i, t_{i+1}). (And f is allowed to take on any arbitrary real values at the t_i.) We will write $S[a, b]$ for the collection of all step functions on $[a, b]$. Clearly, $S[a, b]$ is a subset of $B[a, b]$ but, in fact, $S[a, b]$ is also a subalgebra of $B[a, b]$. (Why?)

EXERCISES

▷ **1.** Let V be a normed vector space.

(a) Show that scalar multiplication, from $\mathbb{R} \times V$ into V, is continuous; that is, if $a_n \to a$ in $\mathbb{R}$, and if $x_n \to x$ in V, prove that $a_n x_n \to ax$ in V.

(b) Show that vector addition, from $V \times V$ into V, is continuous; that is, if $x_n \to x$ and $y_n \to y$ in V, prove that $x_n + y_n \to x + y$ in V.

(c) If W is a subspace of V, conclude that $\bar{W}$ is a subspace of V.

2. Let A be an algebra, and let B be a subset of A. Prove that B is a subalgebra of A if and only if B is a (vector) subspace of A that is also closed under multiplication.

▷ **3.** Let A be a normed algebra.

(a) Show that $\|fg - hk\| \le \|f\|\,\|g - k\| + \|k\|\,\|f - h\|$ for $f, g, h, k \in A$.

(b) Show that multiplication, from $A \times A$ into A, is continuous; that is, if $f_n \to f$ and $g_n \to g$ in A, prove that $f_n g_n \to fg$ in A.

(c) If B is a subalgebra of A, conclude that $\bar{B}$ is a subalgebra of A.

4. Show that the only subalgebras of $\mathbb{R}^2$, other than $\{(0, 0)\}$ and $\mathbb{R}^2$, are the sets $\{(x, 0) : x \in \mathbb{R}\}$, $\{(0, x) : x \in \mathbb{R}\}$ and $\{(x, x) : x \in \mathbb{R}\}$.

▷ **5.** Prove that $S[a, b]$ is a subalgebra of $B[a, b]$.

6. If X is infinite, show that $B(X)$ is not separable.

7. Prove that $C^{(1)}[a,b]$ is a Banach algebra when supplied with the norm $\|f\|_{C^{(1)}} = \|f\|_\infty + \|f'\|_\infty$. (See Exercise 10.18.)

8. Prove that $\text{Lip}\,\alpha$ is a Banach algebra when supplied with the norm $\|f\|_{\text{Lip}\,\alpha} = \|f\|_\infty + N_\alpha(f)$. (See Exercise 11.25.)

9. Let A be an algebra with identity e, and let $f \in A$. Given a polynomial $p(x) = \sum_{k=0}^{n} a_k x^k$ we (formally) define $p(f) \in A$ by $p(f) = \sum_{k=0}^{n} a_k f^k$, where $f^0 = e$, and we call $p(f)$ a *polynomial in* f. Show that the set of all polynomials in f forms a subalgebra of A. In fact, prove that the set of polynomials in f is the *smallest* subalgebra of A containing e and f. For this reason we refer to the set of polynomials in f as the subalgebra *generated by* e and f. Note that the set of (algebraic) polynomials in $C[a,b]$, for instance, is the subalgebra of $C[a,b]$ generated by the functions $e(x) = 1$ and $f(x) = x$.

The Weierstrass approximation theorem tells us that the subalgebra of polynomials in $C[a,b]$ is dense in $C[a,b]$. Using this language, it is now possible to reformulate the Weierstrass theorem in more general settings. In particular, our long-term goal in this chapter is to prove Stone's extension of the Weierstrass theorem, which characterizes the *dense subalgebras* of $C(X)$, where X is a compact metric space.

Our short-term goal will be to characterize $\overline{S[a,b]}$, the closure of the subalgebra of step functions $S[a,b]$ in the algebra of bounded functions $B[a,b]$. This will give us at least one nontrivial, and ultimately useful, example for later reference. Please note that it follows from Exercises 3 and 5 that $\overline{S[a,b]}$ is again a subalgebra of $B[a,b]$. To begin, let's check that $\overline{S[a,b]}$ contains the continuous functions.

Lemma 12.2. $C[a,b] \subset \overline{S[a,b]}$.

PROOF. Let $f \in C[a,b]$ and $\varepsilon > 0$. We need to find a step function $g \in S[a,b]$ such that $\|f - g\|_\infty < \varepsilon$.

Since f is uniformly continuous, there is a $\delta > 0$ such that $|f(x) - f(y)| < \varepsilon$ whenever $|x - y| < \delta$. Now take any partition $a = t_0 < t_1 < \cdots < t_n = b$ of $[a,b]$ for which $t_{i+1} - t_i < \delta$ for all i, and define g by $g(x) = f(t_i)$ for $t_i \le x < t_{i+1}$, and $g(b) = f(b)$ (see Figure 12.1). Then, $g \in S[a,b]$ and $|g(x) - f(x)| < \varepsilon$ for all x in $[a,b]$. □

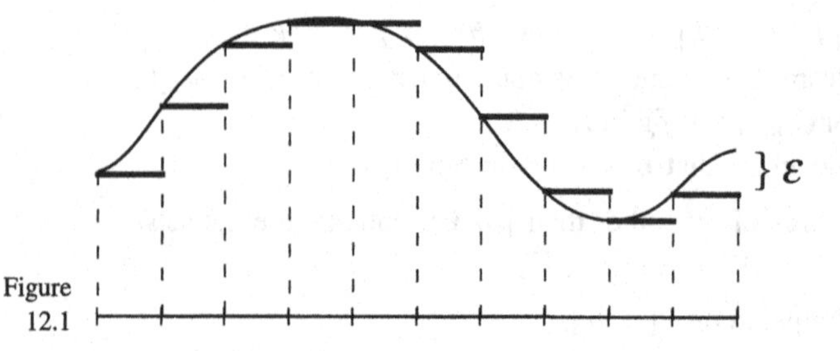

Figure 12.1

EXERCISES

10. Show that $\overline{S[a,b]}$ contains the monotone functions in $B[a,b]$. [Hint: "Slice up" the range of a monotone function to find an approximating step function.]

11. Let $f(x) = \sin(1/x)$, for $0 < x \le 1$, and $f(0) = 0$. Clearly, $f \in B[0,1]$. Show that $f \notin \overline{S[0,1]}$. [Hint: $f(0+)$ doesn't exist.]

12. Is $\chi_{\mathbb{Q}\cap[a,b]} \in \overline{S[a,b]}$? Explain.

What do Exercise 10 and Lemma 12.2 have in common? Well, recall that monotone functions have left- and right-hand limits at each point; that is, both $f(x+)$ and $f(x-)$ exist if f is monotone. This turns out to be precisely what is needed to be in the closure of the step functions.

Theorem 12.3. *Let $f \in B[a,b]$. Then, $f \in \overline{S[a,b]}$ if and only if $f(x+)$ and $f(x-)$ exist at each x in $[a,b]$ (but only $f(a+)$ and $f(b-)$, of course).*

PROOF. First suppose that $f \in \overline{S[a,b]}$, and let $a \le x < b$. We will show that $f(x+)$ exists (the other case is similar).

Let $\varepsilon > 0$, and choose $g \in S[a,b]$ such that $\|f - g\|_\infty < \varepsilon$. Now, since g is a step function, $g(x+)$ exists; in fact, there is a $\delta > 0$ such that g is *constant* on the interval $(x, x+\delta)$. (Why?) But then, for any $x < s, t < x + \delta$, we have $|f(s) - f(t)| \le |f(s) - g(s)| + |g(s) - g(t)| + |g(t) - f(t)| < 2\varepsilon$, and this is enough to imply that $f(x+)$ exists. Indeed, if (t_n) decreases to x, then this argument shows that $(f(t_n))$ is Cauchy (and hence converges).

Now suppose that $f \in B[a,b]$, that $f(x+)$ and $f(x-)$ exist for every x in $[a,b]$, and that $\varepsilon > 0$. For each x in $[a,b]$ there is a $\delta(x,\varepsilon) > 0$ such that

$$\left.\begin{array}{c} x - \delta(x,\varepsilon) < s, t < x \\ \text{or} \\ x < s, t < x + \delta(x,\varepsilon) \end{array}\right\} \Longrightarrow |f(s) - f(t)| < \varepsilon.$$

The intervals $\{(x - \delta(x,\varepsilon), x + \delta(x,\varepsilon)) : x \in [a,b]\}$ form an open cover for $[a,b]$. This means that we actually need only finitely many to do the job. After reducing to finitely many such intervals, we list the endpoints and midpoints of the intervals in their natural order; call them $a = t_0 < t_1 < \cdots < t_n = b$:

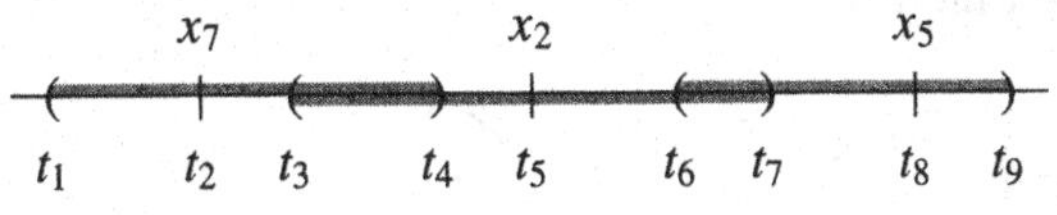

x_7 x_2 x_5

t_1 t_2 t_3 t_4 t_5 t_6 t_7 t_8 t_9

The important thing to notice here is that each interval (t_i, t_{i+1}) is a *subinterval* of some $(x - \delta(x,\varepsilon), x)$ or of some $(x, x + \delta(x,\varepsilon))$. In either case we have $|f(x) - f(t)| < \varepsilon$ whenever $s, t \in (t_i, t_{i+1})$.

Now we are ready to define our step function g. For each $i = 0, \dots, n-1$, choose $s_i \in (t_i, t_{i+1})$ and set $g(x) = f(s_i)$ for $x \in (t_i, t_{i+1})$. Finally, set $g(t_i) = f(t_i)$ for all $i = 0, \dots, n$. Clearly, $g \in S[a, b]$ and $\|f - g\|_\infty < \varepsilon$. □

We will say that a function possessing finite left- and right-hand limits at each point is **quasicontinuous**. Thus, $\overline{S[a,b]}$ is the algebra of quasicontinuous functions on $[a, b]$. A quasicontinuous function has only jump discontinuities. And, since a quasicontinuous function is the uniform limit of a sequence of step functions on each compact interval in $\mathbb{R}$, it follows from Exercise 10.14 (or Theorem 10.4) that a quasicontinuous function has at most countably many points of discontinuity.

EXERCISES

13. Fill in the missing details from the proof of Theorem 12.3.

14. If $f \in B[a, b]$ has only countably many points of discontinuity, does it follow that $f \in \overline{S[a,b]}$? Explain.

As it happens, the closed subalgebras of $B(X)$ inherit even more structure than one might guess. To explain this, it will help if we first formalize the order properties of $B(X)$.

A **lattice** is a set L, together with a partial order $\leq$, in which every pair of elements has both a least upper bound and a greatest lower bound (back in L). That is, given $f, g \in L$, there exist elements $f \vee g$ (the least upper bound of f and g) and $f \wedge g$ (the greatest lower bound of f and g) in L satisfying:

(i) If $f \leq h$ and $g \leq h$, for some $h \in L$, then $f \vee g \leq h$.
(ii) If $h \leq f$ and $h \leq g$, for some $h \in L$, then $h \leq f \wedge g$.

As you might expect, a *sublattice* is a subset of a lattice that is a lattice in its own right (under the same ordering).

A vector space that is also a lattice (under some given partial order) is called a *vector lattice*. In a vector lattice we may decompose each element into its *positive* and *negative* parts: $f = f^+ - f^-$, where

$$f^+ = f \vee 0 \qquad \text{and} \qquad f^- = -(f \wedge 0).$$

We may also define the *absolute value* of an element of a vector lattice using the formula $|f| = f^+ + f^-$. See Figure 12.2.

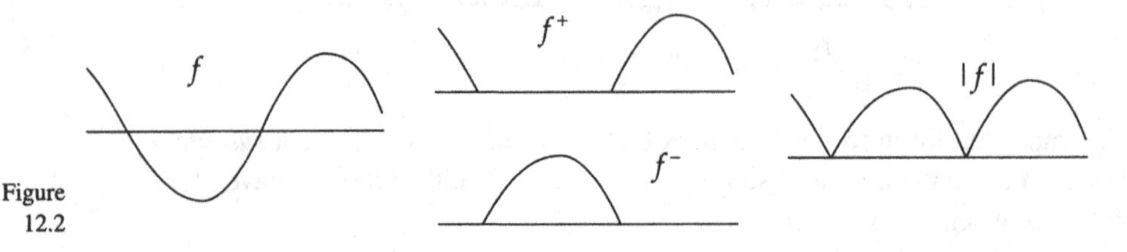

Figure 12.2

The notions of a *normed vector lattice* and a *Banach lattice* should be clear if you have read this far. In a normed lattice, we also require that the norm satisfy $\|f\| \le \|g\|$ whenever $|f| \le |g|$. (As in the case of normed algebras, this fact is used to show that the lattice operations are continuous.)

Examples 12.4

(a) Given any set X, ordinary set inclusion is a partial order on $\mathcal{P}(X)$, the power set of X; that is, we define $A \le B$ if and only if $A \subset B$. It is easy to see that $\mathcal{P}(X)$ is also a lattice under this ordering, and that $A \vee B = A \cup B$ and $A \wedge B = A \cap B$. For this reason, $A \vee B$ is sometimes read as "A *join* B," and $A \wedge B$ is sometimes read as "A *meet* B."

(b) $\mathbb{R}^n$, under "coordinatewise" ordering of vectors (i.e., $x \le y$ if and only if $x_i \le y_i$ for all i), is a Banach lattice when equipped with the norm $\|x\|_\infty = \max_{1 \le i \le n} |x_i|$.

(c) $B(X)$ is a Banach lattice under the usual pointwise ordering of functions: $f \le g$ if and only if $f(x) \le g(x)$ for all x. In this case, $(f \vee g)(x) = \max\{f(x), g(x)\}$ and $(f \wedge g)(x) = \min\{f(x), g(x)\}$. Notice, too, that $|f|(x) = |f(x)|$.

EXERCISES

15. Let L be a lattice, and let S be a subset of L. Show that S is a sublattice of L if and only if $f \vee g$ and $f \wedge g$ are in S whenever $f, g \in S$.

16. In a vector lattice L, show that $-(f \wedge g) = (-f) \vee (-g)$, and conclude that $f^- = (-f) \vee 0 = (-f)^+$.

▷ **17.** If $f, g \in B(X)$, prove that

(a) $f + g = f \vee g + f \wedge g$ and $|f - g| = f \vee g - f \wedge g$.

(b) $2(f \vee g) = f + g + |f - g|$ and $2(f \wedge g) = f + g - |f - g|$.

(c) $f^+ \wedge f^- = 0$ and $|f| = f \vee (-f) = f^+ \vee f^-$.

(d) $|f \vee g| \le |f| \vee |g| \le \max\{\|f\|_\infty, \|g\|_\infty\} \cdot \mathbf{1}$, where $\mathbf{1}$ stands for the constant 1 function.

[Hint: These are all just statements about real numbers.]

▷ **18.** Let A be a vector subspace of $B(X)$. Show that A is a sublattice of $B(X)$ if and only if $|f| \in A$ whenever $f \in A$. If X is a compact metric space, this gives an easy proof that $C(X)$ is a sublattice of $B(X)$.

19. If $f, g \in B(X)$, show that $\|f \vee g\|_\infty \le \max\{\|f\|_\infty, \|g\|_\infty\}$.

It follows from Exercise 18, for example, that $S[a, b]$ is a sublattice of $B[a, b]$. It would be nice to know whether the same holds for $\overline{S[a, b]}$. Our next result explains the claim, made earlier in this section, that the closed subalgebras of $B(X)$ inherit even more structure than one might guess.

Theorem 12.5. *Let A be a subalgebra of $B(X)$. Then, $\bar{A}$ is both a subalgebra and a sublattice of $B(X)$.*

PROOF. It follows from Exercise 3 that $\bar{A}$ is a subalgebra of $B(X)$. In particular, $\bar{A}$ is a subspace of $B(X)$. Thus, by Exercise 18, we need only show that $|f| \in \bar{A}$ whenever $f \in \bar{A}$.

Given $f \in \bar{A}$ and $\varepsilon > 0$, we will show that there is an element $g \in \bar{A}$ with $\| |f| - g \|_\infty < \varepsilon$ and, hence, that $|f| \in \bar{\bar{A}} = \bar{A}$.

Let $M = \|f\|_\infty$, and consider the function $|t|$ on the interval $[-M, M]$. By the Weierstrass approximation theorem (or by Exercise 11.11) there is a polynomial $p(t) = \sum_{k=0}^{n} a_k t^k$ such that $\big| |t| - p(t) \big| < \varepsilon$ for all t in $[-M, M]$. In particular, notice that $|a_0| = |p(0)| < \varepsilon$.

Now, since $|f(x)| \le M$ for all $x \in X$, it follows that $\big| |f(x)| - p(f(x)) \big| < \varepsilon$ for all $x \in X$. But $p(f(x)) = a_0 + a_1 f(x) + \cdots + a_n f^n(x) = a_0 + g(x)$, where the function $g = a_1 f + \cdots + a_n f^n \in \bar{A}$, because $\bar{A}$ is an algebra. Thus, $\big| |f(x)| - g(x) \big| \le \big| |f(x)| - p(f(x)) \big| + \big| p(f(x)) - g(x) \big| < \varepsilon + |a_0| < 2\varepsilon$ for all $x \in X$. In other words, for each $\varepsilon > 0$ we can supply an element $g \in A$ such that $\| \, |f| - g \|_\infty < 2\varepsilon$. Thus, $|f| \in \bar{A}$. □

Please note that the proof of Theorem 12.5 could be streamlined if we had also assumed, as some authors do, that A contains the constant functions. The import of this and other similar hypotheses will be made clear in the next section.

Corollary 12.6. *Let X be a compact metric space, and let A be a subalgebra of $C(X)$. Then, $\bar{A}$ is both a subalgebra and a sublattice of $C(X)$.*

Note that, from Exercise 11.11, the proof of Theorem 12.5 can be written without reference to the classical Weierstrass theorem. In particular, Corollary 12.6 can be proved without reference to Theorem 11.3.

EXERCISES

▷ **20.** Prove Corollary 12.6.

21. Show that the set of all even functions in $C[-1, 1]$ is a proper closed subalgebra of $C[-1, 1]$.

22. Let X be a compact metric space, and a let $x_0 \in X$. Show that the set $A = \{f \in C(X) : f(x_0) = 0\}$ is a proper closed subalgebra of $C(X)$.

The Stone–Weierstrass Theorem

Using our new terminology, we may restate the classical Weierstrass theorem to read: *If a subalgebra A of $C[a, b]$ contains the functions $e(x) = 1$ and $f(x) = x$, then A is dense in $C[a, b]$.* Any subalgebra of $C[a, b]$ containing 1 and x actually contains all of the polynomials; thus our restatement of Weierstrass's theorem amounts to the observation that any subalgebra containing a dense set is itself dense in $C[a, b]$.

Our goal in this section is to prove the analogue of this new version of the Weierstrass theorem for subalgebras of $C(X)$ where X is a compact metric space. In particular, we will want to extract the essence of the functions 1 and x from this statement. That is, we seek conditions on a subalgebra A of $C(X)$ that will force A to be dense in $C(X)$. The key role played by 1 and x, in the case of $C[a, b]$, is that a subalgebra containing these two functions must actually contain a much larger set of functions. But since we cannot be assured of anything remotely like polynomials living in the more general $C(X)$ spaces, we might want to change our point of view. What we really need is some requirement on a subalgebra A of $C(X)$ that will allow us to *construct* a wide variety of functions in A. And, if A contains a sufficiently rich variety of functions, it might just be possible to show that A is dense.

Since the two replacement conditions we have in mind have nothing to with the algebraic structure of $C(X)$, we state them in some generality.

Let A be a collection of real-valued functions on some set X. We say that A **separates points** in X if, given $x \neq y \in X$, there is some $f \in A$ such that $f(x) \neq f(y)$. We say that A **vanishes at no point** of X if, given $x \in X$, there is some $f \in A$ such that $f(x) \neq 0$.

Examples 12.7

(a) The single function $f(x) = x$ clearly separates points in $[a, b]$, and the function $e(x) = 1$ obviously vanishes at no point in $[a, b]$. Any subalgebra A of $C[a, b]$ containing these two functions will likewise separate points and vanish at no point in $[a, b]$.

(b) For any metric space X, the collection $C(X)$ separates points in X and vanishes at no point of X. Why?

(c) The set E of even functions in $C[-1, 1]$ fails to separate points in $[-1, 1]$; indeed, $f(x) = f(-x)$ for any even function. However, since the constant functions are even, E vanishes at no point of $[-1, 1]$. From Exercise 21, E is a proper closed subalgebra of $C[-1, 1]$. The set of odd functions will separate points (since $f(x) = x$ is odd), but the odd functions all vanish at 0. The set of odd functions is a proper closed subspace of $C[-1, 1]$, although not a subalgebra.

(d) The set of all functions $f \in C[-1, 1]$ for which $f(0) = 0$ is a proper closed subalgebra of $C[-1, 1]$. In fact, this set is a maximal (in the sense of containment) proper closed subalgebra of $C[-1, 1]$. We will see why shortly. Note, however, that this set of functions does separate points in $[-1, 1]$ (again, because it contains $f(x) = x$).

As these few examples illustrate, neither of our new conditions, taken separately, is sufficient to force a subalgebra of $C(X)$ to be dense. But, as we will see, both conditions together will do the job. To better appreciate the utility of these new conditions, let's isolate the key computational tool they permit within an algebra of functions.

Lemma 12.8. *Let A be an algebra of real-valued functions on some set X, and suppose that A separates points in X and vanishes at no point of X. Then, given $x \neq y \in X$ and $a, b \in \mathbb{R}$, we can find an $f \in A$ with $f(x) = a$ and $f(y) = b$.*

PROOF. Since A separates points in X and vanishes at no point of X, we can find $g, h, k \in A$ such that $g(x) \neq g(y)$, $h(x) \neq 0$, and $k(y) \neq 0$. Thus, both $u = gh - g(y)h$ and $v = gk - g(x)k$ are in A, since A is an algebra. Moreover, u and v satisfy $u(y) = 0 = v(x)$ and $u(x) \neq 0 \neq v(y)$. Finally, the function

$$f = \frac{a}{u(x)}\,u + \frac{b}{v(y)}\,v$$

is in A and satisfies $f(x) = a$, $f(y) = b$. □

Note that we were forced to be somewhat fussy in the proof of Lemma 12.8; it would not have been appropriate to write $u = [g - g(y)]h$, for example, since A need not contain the constant function $g(y) = g(y) \cdot \mathbf{1}$ and so need not contain the factor $g - g(y)$. To avoid just this sort of nuisance, some authors require that A contain the constant functions in place of the (weaker) condition that A vanish at no point of X.

A second, slick proof of Lemma 12.8 is based on the observation that, for any pair of distinct points $x \neq y \in X$, the set $\tilde{A} = \{(g(x), g(y)) : g \in A\}$ is a subalgebra of $\mathbb{R}^2$. (It is easy to list all of the subalgebras of $\mathbb{R}^2$; see Exercise 4.) If A separates points in X, then $\tilde{A}$ is apparently neither $\{(0,0)\}$ nor $\{(x,x) : x \in \mathbb{R}\}$. If A vanishes at no point, then both $\{(x,0) : x \in \mathbb{R}\}$ and $\{(0,x) : x \in \mathbb{R}\}$ are excluded. Thus $\tilde{A} = \mathbb{R}^2$, which is essentially the conclusion of Lemma 12.8.

Finally, we are ready for Stone's version of the Weierstrass theorem. It should be pointed out that the theorem, as stated, does not hold for algebras of complex-valued functions over $\mathbb{C}$. More on this later.

The Stone–Weierstrass Theorem, real scalars 12.9. *Let X be a compact metric space, and let A be a subalgebra of $C(X)$. If A separates points in X and vanishes at no point of X, then A is dense in $C(X)$.*

PROOF. First notice that we may assume that A is *closed* (and prove that $A = C(X)$). Indeed, if A satisfies the hypotheses of the theorem, then so does $\bar{A}$. (Why?) And if we are allowed to assume that A is closed, then, according to Corollary 12.6, we may also assume that A is a *sublattice* of $C(X)$. We would be foolish to do otherwise: Henceforth, A is a closed subalgebra and a sublattice of $C(X)$. We will break the remainder of the proof into two steps.

Step 1. Given $f \in C(X)$, $x \in X$, and $\varepsilon > 0$, there is an element $g_x \in A$ with $g_x(x) = f(x)$ and $g_x(y) > f(y) - \varepsilon$ for all $y \in X$.

From our "computational" lemma, Lemma 12.8, we know that for each $y \in X$, $y \neq x$, we can find an $h_y \in A$ so that $h_y(x) = f(x)$ and $h_y(y) = f(y)$, as in Figure 12.3.

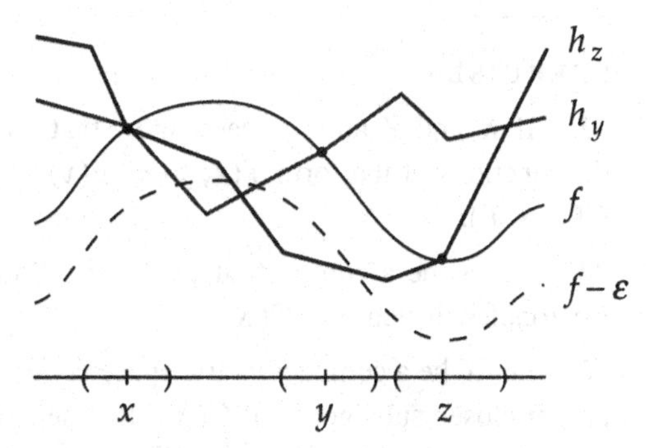

Figure 12.3

Next, since $h_y - f$ is continuous and vanishes at both x and y, the set $U_y = \{t \in X : h_y(t) > f(t) - \varepsilon\}$ is open and contains both x and y. Thus, the sets $(U_y)_{y \neq x}$ form an open cover for X. Since X is compact, finitely many U_y suffice, say $X = U_{y_1} \cup \cdots \cup U_{y_n}$. Now set $g_x = \max\{h_{y_1}, \ldots, h_{y_n}\}$. Because A is a lattice, we have $g_x \in A$. Note that $g_x(x) = f(x)$ since each h_{y_i} agrees with f at x. And $g_x > f - \varepsilon$ since, given $y \neq x$, we have $y \in U_{y_i}$ for some i, and hence $h_{y_i}(y) > f(y) - \varepsilon$.

Step 2. Given $f \in C(X)$ and $\varepsilon > 0$, there is an $h \in A$ with $\|f - h\|_\infty < \varepsilon$.

From Step 1, for each $x \in X$ we can find $g_x \in A$ such that $g_x(x) = f(x)$ and $g_x(y) > f(y) - \varepsilon$ for all $y \in X$, as in Figure 12.4. Now we reverse the process

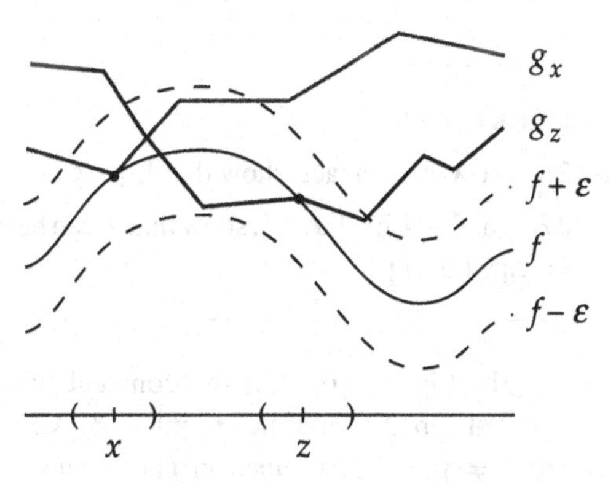

Figure 12.4

used in Step 1: For each x, the set $V_x = \{y \in X : g_x(y) < f(y) + \varepsilon\}$ is open and contains x. Again, since X is compact, $X = V_{x_1} \cup \cdots \cup V_{x_m}$. This time, set $h = \min\{g_{x_1}, \ldots, g_{x_m}\} \in A$. As before, $h(y) > f(y) - \varepsilon$ for all y since each $g_{x_i}(y)$ does so, and $h(y) < f(y) + \varepsilon$ for all y since at least one $g_{x_i}(y)$ does so. □

If we are careful to avoid reference to the classical Weierstrass theorem in the proof of the Stone–Weierstrass theorem (see the remarks following Corollary 12.6), then Theorem 11.3 may be considered a corollary to Theorem 12.9 (recall Example 12.7 (a)).

Corollary 12.10. *Given $f \in C[a, b]$ and $\varepsilon > 0$, there is a polynomial p such that $\|f - p\|_\infty < \varepsilon$.*

EXERCISES

23. If X and Y are compact, show that the subspace of $C(X \times Y)$ spanned by the functions of the form $f(x, y) = g(x)h(y)$, $g \in C(X)$, $h \in C(Y)$, is dense in $C(X \times Y)$.

24. Let K be a compact subset of $\mathbb{R}^n$. Show that the set of all polynomials (in n-variables) is dense in $C(K)$.

25. Let X be a compact metric space containing at least two points, and let A be a proper closed subalgebra of $C(X)$. If A separates points in X, show that there is an $x_0 \in X$ such that $A = \{f \in C(X) : f(x_0) = 0\}$.

We used the classical Weierstrass theorem to prove that $C[\,a, b\,]$ is separable. Likewise, the Stone–Weierstrass theorem can be used to show that $C(X)$ is separable where X is a compact metric space. While we do not have anything quite so convenient as polynomials at our disposal, we do, at least, have a familiar collection of functions to work with.

Given a metric space (X, d) and $0 \le K < \infty$, we will write $\mathrm{Lip}_K(X)$ to denote the collection of all real-valued Lipschitz functions on X, with constant at most K; that is, $f : X \to \mathbb{R}$ is in $\mathrm{Lip}_K(X)$ if $|f(x) - f(y)| \le K d(x, y)$ for all $x, y \in X$. And we will write $\mathrm{Lip}\,(X)$ to denote the set of functions that are in $\mathrm{Lip}_K(X)$ for some K; in other words, $\mathrm{Lip}\,(X) = \bigcup_{K=1}^{\infty} \mathrm{Lip}_K(X)$. It is easy to see that $\mathrm{Lip}\,(X)$ is a subspace of $C(X)$; in fact, if X is compact, then $\mathrm{Lip}\,(X)$ is even a subalgebra of $C(X)$.

EXERCISES

▷ **26.** If X is compact, show that $\mathrm{Lip}\,(X)$ is a subalgebra of $C(X)$.

27. If $f \in \mathrm{Lip}_K[\,a, b\,]$, show that f can be uniformly approximated by polynomials in $\mathrm{Lip}_K[\,a, b\,]$.

Clearly, $\mathrm{Lip}\,(X)$ contains the constant functions and so vanishes at no point of X. To see that $\mathrm{Lip}\,(X)$ separates point in X, we use the fact that the metric d is Lipschitz: Given $x_0 \ne y_0 \in X$, the function $f(x) = d(x, y_0)$ satisfies $f(x_0) > 0 = f(y_0)$. Moreover, $f \in \mathrm{Lip}\,(X)$ since

$$|f(x) - f(y)| = |d(x, y_0) - d(y, y_0)| \le d(x, y).$$

Thus, if X is compact, then $\mathrm{Lip}\,(X)$ is dense in $C(X)$.

Now, to see that $C(X)$ is separable for X compact, it suffices to show that $\mathrm{Lip}\,(X)$ is separable. To see this, first notice that $\mathrm{Lip}\,(X) = \bigcup_{K=1}^{\infty} E_K$, where

$$E_K = \{f \in C(X) : \|f\|_\infty \le K \text{ and } f \in \mathrm{Lip}_K(X)\}.$$

(Why?) The sets E_K are (uniformly) bounded and equicontinuous. Hence, by the Arzelà–Ascoli theorem, each E_K is compact in $C(X)$. Since compact sets are separable, as are countable unions of compact sets, it follows that $\mathrm{Lip}\,(X)$ is separable.

Corollary 12.11. *If X is a compact metric space, then $C(X)$ is separable.*

In many texts, the Stone–Weierstrass theorem is used to show that the trig polynomials are dense in $C^{2\pi}$. One approach here might be to identify $C^{2\pi}$ with the closed subalgebra of $C[0, 2\pi]$ consisting of those functions f that satisfy $f(0) = f(2\pi)$. Probably easier, though, is to identify $C^{2\pi}$ with the continuous functions on the unit circle $\mathbb{T}$ in the complex plane,

$$\mathbb{T} = \{e^{i\theta} : \theta \in \mathbb{R}\} = \{z \in \mathbb{C} : |z| = 1\},$$

by using the identification

$$f \in C^{2\pi} \quad \longleftrightarrow \quad g \in C(\mathbb{T}), \qquad \text{where } g(e^{it}) = f(t).$$

Under this correspondence, the trig polynomials in $C^{2\pi}$ match up with (certain) polynomials in $z = e^{it}$ and $\bar{z} = e^{-it}$. But, as we saw in Chapter Eleven, even if we start with real-valued trig polynomials, we will end up with polynomials in z and $\bar{z}$ having complex coefficients.

EXERCISE

28. The polynomials in z obviously separate points in $\mathbb{T}$ and vanish at no point of $\mathbb{T}$. Nevertheless, the polynomials in z (with complex coefficients) are not dense in the space of continuous complex-valued functions on $\mathbb{T}$. To see this, here is a proof that $f(z) = \bar{z}$ cannot be uniformly approximated by polynomials in z:

(a) If $p(z) = \sum_{k=0}^{n} c_k z^k$, show that $\int_0^{2\pi} \overline{f(e^{it})}\, p(e^{it})\, dt = 0$.

(b) Show that $2\pi = \int_0^{2\pi} \overline{f(e^{it})}\, f(e^{it})\, dt = \int_0^{2\pi} \overline{f(e^{it})}\, \big[f(e^{it}) - p(e^{it})\big]\, dt$.

(c) Conclude that $\| f - p \|_\infty \geq 1$ for any polynomial p. [Hint: Take absolute values in (b) and note that $|f| = 1$.]

Given the result in Exercise 28, it might make more sense to consider the complex-valued continuous functions on $\mathbb{T}$. We will write $C_{\mathbb{C}}(\mathbb{T})$ to denote the complex-valued continuous functions on $\mathbb{T}$ and $C_{\mathbb{R}}(\mathbb{T})$ to denote the real-valued continuous functions on $\mathbb{T}$. Similarly, $C_{\mathbb{C}}^{2\pi}$ is the space of complex-valued, 2π-periodic functions on $\mathbb{R}$ while $C_{\mathbb{R}}^{2\pi}$ stands for the real-valued, 2π-periodic functions on $\mathbb{R}$. Now, under the identification that we made earlier, we have $C_{\mathbb{C}}(\mathbb{T}) = C_{\mathbb{C}}^{2\pi}$ and $C_{\mathbb{R}}(\mathbb{T}) = C_{\mathbb{R}}^{2\pi}$. The complex-valued trig polynomials in $C_{\mathbb{C}}^{2\pi}$ now match up with the full set of polynomials, with complex coefficients, in $z = e^{it}$ and $\bar{z} = e^{-it}$. We will use the Stone–Weierstrass theorem to show that these polynomials are dense in $C_{\mathbb{C}}(\mathbb{T})$.

We might as well do this in some generality: Given a compact metric space X, we will write $C_{\mathbb{C}}(X)$ for the set of all continuous, complex-valued functions $f : X \to \mathbb{C}$, and we norm $C_{\mathbb{C}}(X)$ by $\|f\|_\infty = \sup_{x \in X} |f(x)|$ (where $|f(x)|$ is the modulus of the complex number $f(x)$, of course). $C_{\mathbb{C}}(X)$ is a Banach algebra over $\mathbb{C}$. To make it clear which field of scalars are involved, we will write $C_{\mathbb{R}}(X)$ for the real-valued members of $C_{\mathbb{C}}(X)$. Notice, though, that $C_{\mathbb{R}}(X)$ is nothing other than our old friend $C(X)$ with a new name.

More generally, we will write $A_{\mathbb{C}}$ to denote an algebra, over $\mathbb{C}$, of complex-valued functions and $A_{\mathbb{R}}$ to denote the real-valued members of $A_{\mathbb{C}}$. It is not hard to see that $A_{\mathbb{R}}$ is then an algebra, over $\mathbb{R}$, of real-valued functions.

Now, if f is in $C_{\mathbb{C}}(X)$, then so is the function $\bar{f}(x) = \overline{f(x)}$ (the complex conjugate of $f(x)$). This puts

$$\mathrm{Re}f = \frac{1}{2}(f + \bar{f}) \qquad \text{and} \qquad \mathrm{Im}f = \frac{1}{2i}(f - \bar{f}),$$

the real and imaginary parts of f, in $C_{\mathbb{R}}(X)$. Conversely, if $g, h \in C_{\mathbb{R}}(X)$, then $g + ih \in C_{\mathbb{C}}(X)$.

This simple observation gives us a hint as to how we might apply the Stone–Weierstrass theorem to subalgebras of $C_{\mathbb{C}}(X)$. Given a subalgebra $A_{\mathbb{C}}$ of $C_{\mathbb{C}}(X)$, suppose that we could prove that $A_{\mathbb{R}}$ is dense in $C_{\mathbb{R}}(X)$. Then, given any $f \in C_{\mathbb{C}}(X)$, we could approximate $\mathrm{Re}f$ and $\mathrm{Im}f$ by elements $g, h \in A_{\mathbb{R}}$. But since $A_{\mathbb{R}} \subset A_{\mathbb{C}}$, this means that $g + ih \in A_{\mathbb{C}}$ and $g + ih$ approximates f. That is, $A_{\mathbb{C}}$ is dense in $C_{\mathbb{C}}(X)$. Great! And what did we really use here? Well, we need $A_{\mathbb{R}}$ to contain the real and imaginary parts of "most" functions in $C_{\mathbb{C}}(X)$. If we insist that $A_{\mathbb{C}}$ separates points and vanishes at no point, then $A_{\mathbb{R}}$ will contain "most" of $C_{\mathbb{R}}(X)$. And to be sure that we get both the real and imaginary parts of each element of $A_{\mathbb{C}}$, we will insist that $A_{\mathbb{C}}$ contain the conjugates of each of its members: $\bar{f} \in A_{\mathbb{C}}$ whenever $f \in A_{\mathbb{C}}$. That is, we will require that $A_{\mathbb{C}}$ be **self-conjugate** (or, as some authors say, *self-adjoint*).

The Stone–Weierstrass Theorem, complex scalars 12.12. *Let X be a compact metric space, and let $A_{\mathbb{C}}$ be a subalgebra, over $\mathbb{C}$, of $C_{\mathbb{C}}(X)$. If $A_{\mathbb{C}}$ separates points in X, vanishes at no point of X, and is self-conjugate, then $A_{\mathbb{C}}$ is dense in $C_{\mathbb{C}}(X)$.*

PROOF. Again, write $A_{\mathbb{R}}$ for the set of real-valued members of $A_{\mathbb{C}}$. Since $A_{\mathbb{C}}$ is self-conjugate, $A_{\mathbb{R}}$ contains the real and imaginary parts of every $f \in A_{\mathbb{C}}$:

$$\mathrm{Re}f = \frac{1}{2}(f + \bar{f}) \in A_{\mathbb{R}} \qquad \text{and} \qquad \mathrm{Im}f = \frac{1}{2i}(f - \bar{f}) \in A_{\mathbb{R}}.$$

Moreover, $A_{\mathbb{R}}$ is a subalgebra, over $\mathbb{R}$, of $C_{\mathbb{R}}(X)$. In addition, $A_{\mathbb{R}}$ separates points in X and vanishes at no point of X. Indeed, given $x \neq y \in X$ and $f \in A_{\mathbb{C}}$ with $f(x) \neq f(y)$, we must have at least one of $\mathrm{Re}f(x) \neq \mathrm{Re}f(y)$ or $\mathrm{Im}f(x) \neq \mathrm{Im}f(y)$. Similarly, $f(x) \neq 0$ means that at least one of $\mathrm{Re}f(x) \neq 0$ or $\mathrm{Im}f(x) \neq 0$ holds. That is, $A_{\mathbb{R}}$ satisfies the hypotheses of the real-scalar version of the Stone–Weierstrass theorem, Theorem 12.9. Consequently, $A_{\mathbb{R}}$ is dense in $C_{\mathbb{R}}(X)$.

Now, given $f \in C_{\mathbb{C}}(X)$ and $\varepsilon > 0$, take $g, h \in A_{\mathbb{R}}$ with $\|g - \mathrm{Re}f\|_\infty < \varepsilon/2$ and $\|h - \mathrm{Im}f\|_\infty < \varepsilon/2$. Then, $g + ih \in A_{\mathbb{C}}$ and $\|f - (g + ih)\|_\infty < \varepsilon$. Thus, $A_{\mathbb{C}}$ is dense in $C_{\mathbb{C}}(X)$. □

Corollary 12.13. *The polynomials, with complex coefficients, in z and $\bar{z}$ are dense in $C_{\mathbb{C}}(\mathbb{T})$.*

Note that it follows from the proof of Theorem 12.11 that the real parts of the polynomials $\sum_{k=-n}^{n} c_k e^{ikx}$, that is, the real trig polynomials, are dense in $C_{\mathbb{R}}(\mathbb{T}) = C_{\mathbb{R}}^{2\pi}$.

Again, if we are careful to avoid using the classical Weierstrass theorem to prove the Stone–Weierstrass theorem (using Exercise 11.11 in place of Theorem 11.3 in the proof of Theorem 12.5), then we may consider Weierstrass's second theorem as a corollary to the complex-scalar version of the Stone–Weierstrass theorem.

Corollary 12.14. *Given $f \in C^{2\pi}$ and $\varepsilon > 0$, there is a trig polynomial T such that $\|f - T\|_\infty < \varepsilon$.*

Notes and Remarks

The foundations for the "algebraic" approach to the study of $C(X)$ are in Marshall Stone's landmark paper, Stone [1937]. It is here that Stone gives his version of the Weierstrass theorem, Theorem 12.9, but it is not easy to find among the dozens of important results in this mammoth, 106-page work! The premise that "$C(X)$ determines X" is taken to its logical conclusion. Specifically, Stone considered such questions as: If $C(X)$ and $C(Y)$ are isomorphic (as rings, or as Banach spaces, for example), does it follow that X and Y are homeomorphic? Which topological properties of X can be attributed to the structural properties of $C(X)$ (and conversely)? Paraphrasing a passage from his introduction: "We obtain a reasonably complete algebraic insight into the structure of $C_b(X)$ and its correlation with the structure of the underlying topological space." Stone later gave a less formal (but still formidable) summary in Stone [1962]. For an informal summary of related results, see Shields [1987a, 1989].

It would probably be fair to say that the study of lattices and their application to analysis (and much more) began with Riesz's address at the 1928 International Congress of Mathematics, Riesz [1930] (see also Riesz [1940]), and began in earnest with the appearance of G. Birkhoff's book *Lattice Theory* in 1940 (and later editions in 1948 and 1967; see Birkhoff [1940]). For a very brief introduction to the topic, see Birkhoff [1943] and Schaefer [1980].

For more details on algebras, lattices, and rings, as used in analysis and topology, see Simmons [1963], Goffman and Pedrick [1965], Jameson [1974], and the classic Gillman and Jerison [1960].

The proofs of Lemma 12.8 and Theorem 12.9 are largely based on the presentation in Rudin [1953], but see also Douglas [1965] , and Folland [1984] . The "slick" proof of Lemma 12.8 is taken from Folland [1984]. The material on Lip (X) and the Stone–Weierstrass theorem is based on the presentation in Dudley [1989].

CHAPTER THIRTEEN

Functions of Bounded Variation

Functions of Bounded Variation

Throughout this book we've encountered the theme that $C(X)$ determines X. Said another way, to fully understand X we want to understand $C(X)$ as well. Taking this one step further, though, raises a curious question: How are we to understand $C(X)$ without knowing something about $C\big(C(X)\big)$? If we want to be true to our principles, we will have to consider continuous real-valued functions on $C(X)$. If that sounds too esoteric to bother with, fear not. As it happens, we need only to consider the *continuous linear* real-valued functions on $C(X)$, and such functions have a simple and altogether user-friendly description: Definite integrals! But we're getting a little ahead of ourselves. We'll talk about integrals in the next chapter. For the present, we'll content ourselves with the study of a class of functions that turns out to be of paramount interest in this postponed discussion of integration.

To motivate the inevitable blur of definitions ahead of us, let's consider a simple example. Suppose that $\mathbf{f}(t) = \big(x(t), y(t)\big)$, for $a \le t \le b$, is a "nice" curve. What would we mean by the *length* of this curve?

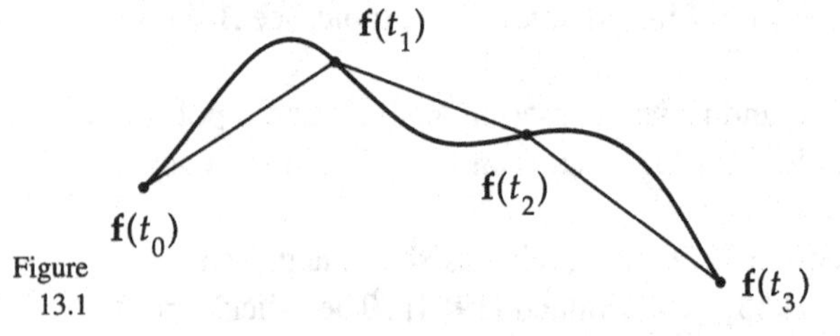

Figure 13.1

Well, we might consider a polygonal approximation to $\mathbf{f}$, with nodes at $a = t_0 < t_1 < \cdots < t_n = b$ (as in Figure 13.1), find the length of the approximating polygon: $\sum_{i=1}^{n} \|\mathbf{f}(t_i) - \mathbf{f}(t_{i-1})\|_2$, and then define the length of the curve as the limit, or supremum, of these approximate lengths as the partition $\{t_0, t_1, \ldots, t_n\}$ gets "bigger." And why not? If $x(t)$ and $y(t)$ are "reasonable" functions, this definition will work just fine. Keep this idea in mind as we proceed.

Given $f : [\,a, b\,] \to \mathbb{R}$ and a partition $P = \{a = t_0 < t_1 < \cdots < t_n = b\}$ of $[\,a, b\,]$, we define the **variation of f over P** by

$$V(f, P) = \sum_{i=1}^{n} |f(t_i) - f(t_{i-1})|.$$

Notice that this "one-dimensional" variation accounts only for the *vertical* changes in the graph of f between points in the partition P.

If Q is another partition of $[a, b]$ with $Q \supset P$, we say that Q **refines** P, or that Q is a **refinement** of P. In this case we have $V(f, Q) \geq V(f, P)$. To see why, first suppose that $Q = P \cup \{x\}$, where $t_k < x < t_{k+1}$. Then,

$$\begin{aligned} V(f, P) &= \sum_{i \neq k+1} |f(t_i) - f(t_{i-1})| + |f(t_{k+1}) - f(t_k)| \\ &\leq \sum_{i \neq k+1} |f(t_i) - f(t_{i-1})| + |f(x) - f(t_k)| + |f(t_{k+1}) - f(x)| \\ &= V(f, Q). \end{aligned}$$

The general case now follows by induction on the number of elements of $Q \setminus P$. In particular, since every partition contains the trivial partition $\{a, b\}$, we get $V(f, Q) \geq V(f, P) \geq |f(b) - f(a)|$, whenever $Q \supset P$.

We define the **total variation** of f over $[a, b]$ by

$$V_a^b f = \sup_P V(f, P).$$

If $V_a^b f < \infty$, we say that f is of **bounded variation** on $[a, b]$. In other words, f is of bounded variation on $[a, b]$ if the variations $V(f, P)$ are bounded above, independent of the partition P. This notation may remind you of the definition of the Riemann integral, and that is not entirely coincidental. As we will see, $V_a^b f$ behaves very much like an integral.

Examples 13.1

(a) If $f : [a, b] \to \mathbb{R}$ is monotone, then $V(f, P) = |f(b) - f(a)|$ for any partition P of $[a, b]$. (Why?) Thus, f is of bounded variation and $V_a^b f = |f(b) - f(a)|$.

(b) More generally, any piecewise monotone function $f : [a, b] \to \mathbb{R}$ is of bounded variation. This means that polygonal functions and polynomials, for example, are of bounded variation (over a bounded interval)..

(c) If $f : [a, b] \to \mathbb{R}$ satisfies $|f(x) - f(y)| \leq K|x - y|$ for all $x, y \in [a, b]$, then f is of bounded variation and $V_a^b f \leq K(b - a)$. (Why?)

(d) Every step function is of bounded variation. If f is a step function that is constant on each of the intervals (t_i, t_{i+1}), where $\{t_0, \ldots, t_n\}$ is a partition of $[a, b]$, then $V_a^b f$ is the sum of all of the left- and right-hand "jumps" in the graph of f, that is, the sum of $|f(t_i) - f(t_i+)|$ and $|f(t_i) - f(t_i-)|$ (where appropriate).

(e) We define the length of the curve $\mathbf{f}(t) = \big(x(t), y(t)\big)$, $a \leq t \leq b$, as the supremum of the (two-dimensional) variations $\sum_{i=1}^n \|\mathbf{f}(t_i) - \mathbf{f}(t_{i-1})\|_2$. Thus, the curve has finite length (or is *rectifiable*) if and only if both x and y are of bounded (one-dimensional) variation on $[a, b]$. This follows from the observation that $\max\{|x(t) - x(s)|, |y(t) - y(s)|\} \leq \|\mathbf{f}(t) - \mathbf{f}(s)\|_2 \leq |x(t) - x(s)| + |y(t) - y(s)|$.

We will write $BV[a, b]$ for the collection of all functions of bounded variation on $[a, b]$. You won't be surprised to learn that $BV[a, b]$ is both a Banach space and a Banach algebra, but you may find it curious that we have more than a little work ahead of us to establish these facts. In fact, it is probably not at all clear at this point that $BV[a,b] \subset B[a,b]$.

Lemma 13.2. *If $f : [a,b] \to \mathbb{R}$ is of bounded variation, then f is also bounded and satisfies $\|f\|_\infty \le |f(a)| + V_a^b f$.*

PROOF. Let $a \le x \le b$, and set $P = \{a, x, b\}$. Then, $|f(x) - f(a)| \le V(f, P) \le V_a^b f$. Consequently, $|f(x)| \le |f(a)| + V_a^b f$. □

But even bounded continuous functions need not be of bounded variation. Here's an example: Define $f(x) = x \sin(1/x)$ for $0 < x \le 1$ and $f(0) = 0$. Then, $f \in C[0,1] \subset B[0,1]$, but $f \notin BV[0,1]$. To see this, fix n, and let P be any partition of $[0,1]$ containing the points $t_k = 2/[(2k+1)\pi]$, for $k = 0, \ldots, n$. Notice that $f(t_k) = (-1)^k t_k$, and so

$$|f(t_{k+1}) - f(t_k)| = t_{k+1} + t_k \ge 2t_{k+1} \ge \frac{4}{3\pi} \cdot \frac{1}{k+1}.$$

Consequently,

$$V(f, P) \ge \frac{4}{3\pi} \sum_{k=0}^{n-1} \frac{1}{k+1} \to \infty \qquad \text{as } n \to \infty.$$

Now the point to these examples is that $BV[a,b]$ contains several subsets that we know to be dense in $C[a,b]$ (under the sup norm). Thus, $C[a,b]$ is contained in the closure of $BV[a,b]$ under uniform convergence but not in $BV[a,b]$ itself. That is, $BV[a,b]$ is evidently not closed under uniform convergence (and hence is not complete under uniform convergence). So, we might want to consider some norm other than the sup-norm on $BV[a,b]$. As it happens, the total variation $V_a^b f$ is "almost" a norm.

Lemma 13.3. *Let $f, g \in BV[a,b]$, and let $c \in \mathbb{R}$. Then:*

(i) $V_a^b f = 0$ *if and only if* f *is constant.*
(ii) $V_a^b(cf) = |c|\, V_a^b f$.
(iii) $V_a^b(f+g) \le V_a^b f + V_a^b g$.
(iv) $V_a^b(fg) \le \|f\|_\infty V_a^b g + \|g\|_\infty V_a^b f$.
(v) $V_a^b |f| \le V_a^b f$.
(vi) $V_a^b f = V_a^c f + V_c^b f$, *for* $a \le c \le b$.

PROOF. We will prove (iii) and (vi) and leave the rest as exercises. To begin, let P be a partition of $[a,b]$. By the triangle inequality, $V(f+g, P) \le V(f, P) + V(g, P)$. Hence, $V(f+g, P) \le V_a^b f + V_a^b g$, and (iii) follows.

Next, given any partition Q of $[a,c]$ and any partition R of $[c,b]$, then $P = Q \cup R$ is a partition of $[a,b]$. Moreover, $V(f, Q) + V(f, R) = V(f, P) \le V_a^b f$. Since Q and R were arbitrary, it follows that $V_a^c f + V_c^b f \le V_a^b f$. Conversely, if we are given a partition P of $[a,b]$, then $Q = (P \cup \{c\}) \cap [a,c]$ is a partition of $[a,c]$ and $R = (P \cup \{c\}) \cap [c,b]$ is a partition of $[c,b]$. Thus, $V(f, P) \le V(f, P \cup \{c\}) = V(f, Q) + V(f, R) \le V_a^c f + V_c^b f$. Hence, $V_a^b f \le V_a^c f + V_c^b f$, which proves (vi). □

EXERCISES

1. Show that $V_a^b(\chi_{\mathbb{Q}}) = +\infty$ on any interval $[a, b]$.

2. Show that $S[a, b] \subset BV[a, b]$, where $S[a, b]$ is the collection of step functions on $[a, b]$ (Example 12.1 (i)).

▷ **3.** If f has a bounded derivative on $[a, b]$, show that $V_a^b f \le \|f'\|_\infty (b - a)$.

4. If $f \in BV[a, b]$ and $[c, d] \subset [a, b]$, show that $f \in BV[c, d]$ and $V_c^d f \le V_a^b f$.

▷ **5.** Complete the proof of Lemma 13.3.

6. We can test several of the inclusions implicit in our discussion up to this point by means of a single family of functions. For $\alpha \in \mathbb{R}$ and $\beta > 0$, set $f(x) = x^\alpha \sin(x^{-\beta})$, for $0 < x \le 1$, and $f(0) = 0$. Show that:
(a) f is bounded if and only if $\alpha \ge 0$.
(b) f is continuous if and only if $\alpha > 0$.
(c) $f'(0)$ exists if and only if $\alpha > 1$.
(d) f' is bounded if and only if $\alpha \ge 1 + \beta$.
(e) If $\alpha > 0$, then $f \in BV[0, 1]$ for $0 < \beta < \alpha$ and $f \notin BV[0, 1]$ for $\beta \ge \alpha$.
[Hint: Try a few easy cases first, say $\alpha = \beta = 2$.]

7. Suppose that $f \in B[a, b]$. If $V_{a+\varepsilon}^b f \le M$ for all $\varepsilon > 0$, does it follow that f is of bounded variation on $[a, b]$? Is $V_a^b f \le M$? If not, what additional hypotheses on f would make this so?

8. If f is a polygonal function on $[a, b]$, or if f is a polynomial, show that $V_a^b f = \int_a^b |f'(t)|\, dt$. (This at least partly justifies our earlier claim that $V_a^b f$ behaves like an integral.) [Hint: In either case, f is piecewise monotone and piecewise differentiable. Thus we have $\int_c^d |f'(t)|\, dt = \pm(f(d) - f(c))$ over certain "pieces" $[c, d]$ of $[a, b]$.]

9. If f has a continuous derivative on $[a, b]$, and if P is any partition of $[a, b]$, show that $V(f, P) \le \int_a^b |f'(t)|\, dt$. Hence, $V_a^b f \le \int_a^b |f'(t)|\, dt$.

10. Suppose that $f_n \to f$ pointwise on $[a, b]$. If each f_n is increasing, show that f is increasing. If each f_n is of bounded variation, does it follow that f is of bounded variation? Explain.

▷ **11.** If $f_n \to f$ pointwise on $[a, b]$, show that $V(f_n, P) \to V(f, P)$ for any partition P of $[a, b]$. In particular, if we also have $V_a^b f_n \le K$ for all n, then $V_a^b f \le K$ too.

12. Here is a variation on Exercise 11: If (f_n) is a sequence in $BV[a, b]$, and if $f_n \to f$ pointwise on $[a, b]$, show that $V_a^b f \le \liminf_{n\to\infty} V_a^b f_n$.

Statements (ii) and (iii) of Lemma 13.3 tell us that $BV[a, b]$ is a vector space, while (iv) at least tells us that $BV[a, b]$ is closed under products (we will improve on this inequality later). Notice, too, that from (v) and Exercise 12.18 it follows that $BV[a, b]$ is a sublattice of $B[a, b]$. However, it is not true that $V_a^b f \le V_a^b g$ whenever $|f| \le |g|$.

For example, if $f(1/2) = 1$ and $f(x) = 0$ for $x \neq 1/2$, and if $g(x) = 1$ for all x, then $|f| \leq |g|$, but $V_0^1 f = 2$ while $V_0^1 g = 0$. In any case, it is clear that $V_a^b f$ defines a seminorm on $BV[a,b]$ (since $V_a^b(f - g) = 0$ only says that $f - g$ is constant). We won't need to make much of an adjustment to arrive at a norm. In fact, it is easy to check that

$$\|f\|_{BV} = |f(a)| + V_a^b f$$

defines a norm on $BV[a,b]$. From Lemma 13.2 we have $\|f\|_\infty \leq \|f\|_{BV}$, and hence convergence in $BV[a,b]$ implies uniform convergence.

Theorem 13.4. *$BV[a,b]$ is complete under $\|f\|_{BV} = |f(a)| + V_a^b f$.*

PROOF. Let (f_n) be a Cauchy sequence in $BV[a,b]$. Then, in particular, (f_n) is also Cauchy in $B[a,b]$. Thus, (f_n) converges uniformly (and pointwise) to some $f \in B[a,b]$. We need to show that $f \in BV[a,b]$ and that $\|f - f_n\|_{BV} \to 0$. We'll do both at once.

Let P be any partition of $[a,b]$, and let $\varepsilon > 0$. Now choose N such that $\|f_m - f_n\|_{BV} < \varepsilon$ whenever $m, n \geq N$. Then, from Exercise 11, for any $n \geq N$ we have

$$\begin{aligned} |f(a) - f_n(a)| + V(f - f_n, P) &= \lim_{m\to\infty} \big[\, |f_m(a) - f_n(a)| + V(f_m - f_n, P) \big] \\ &\leq \sup_{m \geq N} \|f_m - f_n\|_{BV} \leq \varepsilon. \end{aligned}$$

Since this estimate holds for all P, we have $\|f - f_n\|_{BV} \leq \varepsilon$ for any $n \geq N$. But if $f - f_n \in BV[a,b]$ and $f_n \in BV[a,b]$, then $f \in BV[a,b]$ too. Of course, our first estimate shows that $\|f - f_n\|_{BV} \to 0$. □

EXERCISES

13. Given a sequence of scalars (c_n) and a sequence of distinct points (x_n) in (a,b), define $f(x) = c_n$ if $x = x_n$ for some n, and $f(x) = 0$ otherwise. Under what condition(s) is f of bounded variation on $[a,b]$?

14. Let $I(x) = 0$ if $x < 0$ and $I(x) = 1$ if $x \geq 0$. Given a sequence of scalars (c_n) with $\sum_{n=1}^\infty |c_n| < \infty$ and a sequence of distinct points (x_n) in $(a,b]$, define $f(x) = \sum_{n=1}^\infty c_n I(x - x_n)$ for $x \in [a,b]$. Show that $f \in BV[a,b]$ and that $V_a^b f = \sum_{n=1}^\infty |c_n|$.

For the moment, let's put aside the "abstract" structure of $BV[a,b]$ and instead focus on a concrete, or intrinsic, characterization of the functions of bounded variation. This characterization will depend heavily on a knowledge of the function $V_a^x f$. Again, this should remind you of the Riemann integral (and the Fundamental Theorem of Calculus).

Theorem 13.5. *Fix $f \in BV[a,b]$ and set $v(x) = V_a^x f$, for $a < x \le b$, and $v(a) = 0$. Then, both v and $v - f$ are increasing. Consequently, $f = v - (v - f)$ is the difference of two increasing functions.*

PROOF. Although it is clear that v is increasing, the proof is still enlightening, especially if we are willing to go the extra mile.

Given $x < y$ in $[a,b]$, it follows from Lemma 13.3 (vi) that

$$v(y) - v(x) = V_a^y f - V_a^x f = V_x^y f \ge |f(y) - f(x)| \ge 0. \qquad (13.1)$$

Hence, v is increasing. But, in fact, $v(y) - v(x) \ge f(y) - f(x)$, too. That is, $v - f$ is also increasing. □

On the other hand, since monotone functions are of bounded variation, we get.

Corollary 13.6. (Jordan's Theorem) *A function $f : [a,b] \to \mathbb{R}$ is of bounded variation if and only if f can be written as the difference of two increasing functions.*

Corollary 13.7. *Each $f \in BV[a,b]$ is quasicontinuous. In particular, any $f \in BV[a,b]$ has at most countably many points of jump discontinuity.*

Corollary 13.8. $S[a,b] \subset BV[a,b] \subset \overline{S[a,b]}$, *where the closure is taken in* $B[a,b]$.

If we improve our first estimate (13.1), we will likewise improve our first corollary.

Theorem 13.9. *Fix $f \in BV[a,b]$, and let $v(x) = V_a^x f$. Then, f is right (left) continuous at x in $[a,b]$ if and only if v is right (left) continuous at x.*

PROOF. One direction is easy. If $x < y$, then $v(y) - v(x) \ge |f(y) - f(x)|$; hence, by taking limits as $y \to x$ or as $x \to y$, we get $v(x+) - v(x) \ge |f(x+) - f(x)|$ and $v(y) - v(y-) \ge |f(y) - f(y-)|$. Thus, if v is right (left) continuous at x, then so is f.

Next suppose that f is, say, right continuous at x, where $a \le x < b$. Then, given $\varepsilon > 0$, there is some $\delta > 0$ such that $|f(x) - f(t)| < \varepsilon/2$ whenever $x \le t < x + \delta$.

For this same ε, choose a partition P of $[x,b]$ such that $V_x^b f - \varepsilon/2 \le V(f, P)$. (How?) Now, since $V(f, P)$ would increase only by adding more points to P, we might as well assume that $P = \{x = t_0 < t_1 < \cdots < t_n = b\}$ satisfies $x < t_1 < x + \delta$. Then

$$\begin{aligned} V_x^b f - \varepsilon/2 &\le V(f, P) \\ &= |f(x) - f(t_1)| + V(f, \{t_1, \ldots, t_n\}) \\ &\le \frac{\varepsilon}{2} + V_{t_1}^b f. \end{aligned}$$

That is, $\varepsilon \geq V_x^b f - V_{t_1}^b = V_x^{t_1} f = v(t_1) - v(x) \geq 0$, for any $x < t_1 < x + \delta$. So, v is right-continuous at x, too. □

Corollary 13.10. *$f \in C[a,b] \cap BV[a,b]$ if and only if f can be written as the difference of two increasing continuous functions.*

EXERCISES

15. Show that $f \in C[a,b] \cap BV[a,b]$ if and only if f can be written as the difference of two *strictly* increasing continuous functions.

16. Given $f \in BV[a,b]$, define $g(x) = f(x+)$ for $a \leq x < b$ and $g(b) = f(b)$. Prove that g is right continuous and of bounded variation on $[a,b]$.

From our investigations into the structure of monotone functions in Chapter Two (see Exercise 2.36) it follows that each function of bounded variation can be written as the sum of a continuous function of bounded variation plus a **saltus**, or "pure jump," function. Specifically, let $f \in BV[a,b]$, and let (x_n) be an enumeration of the discontinuities of f. For each n, let $a_n = f(x_n) - f(x_n-)$ and $b_n = f(x_n+) - f(x_n)$ be the left and right "jumps" in the graph of f, where $a_n = 0$ if $x_n = a$ and $b_n = 0$ if $x_n = b$. Since f is of bounded variation, it follows that $\sum_{n=1}^{\infty} |a_n| < \infty$ and $\sum_{n=1}^{\infty} |b_n| < \infty$. (Why?) We obtain the "continuous part" of f by subtracting these jumps. To simplify our notation, we will define two auxiliary functions:

$$I(x) = \begin{cases} 0 & \text{if } x < 0 \\ 1 & \text{if } x \geq 0 \end{cases} \quad \text{and} \quad J(x) = \begin{cases} 0 & \text{if } x \leq 0 \\ 1 & \text{if } x > 0. \end{cases}$$

Now, let $h(x) = \sum_{n=1}^{\infty} a_n I(x - x_n) + \sum_{n=1}^{\infty} b_n J(x - x_n)$, and let $g = f - h$. From Exercise 14, h is of bounded variation, and hence so is g. Moreover, from Exercise 2.36, g is actually continuous. By design, $f = g + h$.

Returning to our discussion of Jordan's theorem, notice that the decomposition of a function of bounded variation into the difference of increasing functions is by no means unique: $f = g - h = (g+1) - (h+1)$. By making a clever choice, however, we can instill a certain amount of uniqueness into the decomposition.

Given $f \in BV[a,b]$ and $v(x) = V_a^x f$, we define the **positive variation** of f by

$$p(x) = \tfrac{1}{2}(v(x) + f(x) - f(a))$$

and the **negative variation** of f by

$$n(x) = \tfrac{1}{2}(v(x) - f(x) + f(a)).$$

Obviously, $v(x) = p(x) + n(x)$ and $f(x) = f(a) + p(x) - n(x)$. We will show that p and n are increasing, thus giving an alternate representation of f as the difference of increasing functions.

Proposition 13.11. *Let $f \in BV[a,b]$, and let v, p, and n be defined as above. Then:*

(i) $0 \le p \le v$ *and* $0 \le n \le v$.

(ii) p *and* n *are increasing functions on* $[a,b]$.

(iii) *If* g *and* h *are increasing functions on* $[a,b]$ *such that* $f = g - h$, *then* $V_x^y p \le V_x^y g$ *and* $V_x^y n \le V_x^y h$ *for all* $x < y$ *in* $[a,b]$.

PROOF. We will prove (i) and (ii) and leave (iii) as an exercise. The point to (iii) is that p and n give, in a sense, a minimal decomposition of f. To prove (i), recall that

$$v(x) = V_a^x f \ge |f(x) - f(a)| \ge \pm(f(x) - f(a)).$$

Thus, $p \ge 0$ and $n \ge 0$. Since $p + n = v$, we must also have $p \le v$ and $n \le v$.

To see that p is increasing, we essentially repeat this calculation. Take $x < y$ in $[a,b]$ and notice that

$$\begin{aligned} 2\big(p(y) - p(x)\big) &= v(y) - v(x) + f(y) - f(x) \\ &= V_x^y f + f(y) - f(x) \\ &\ge |f(y) - f(x)| + f(y) - f(x) \ge 0. \end{aligned}$$

And similarly for n. □

Since $f - f(a) = p - n$, it follows that $V_a^b f = V_a^b(f - f(a)) \le V_a^b p + V_a^b n$. We have taken "the" choice of p and n that give equality here:

$$V_a^b p + V_a^b n = p(b) + n(b) = v(b) = V_a^b f,$$

since p and n are increasing and vanish at a. Notice, too, that this gives $\|f\|_{BV} = |f(a)| + p(b) + n(b)$. We can use this fact to clean up an earlier, less than satisfactory estimate.

Proposition 13.12. $\|f_1 f_2\|_{BV} \le \|f_1\|_{BV}\,\|f_2\|_{BV}$.

PROOF. Write $f_1 = p_1 - n_1 + f_1(a)$ and $f_2 = p_2 - n_2 + f_2(a)$, as in Proposition 13.11. As pointed out above, this yields $\|f_1\|_{BV} = |f_1(a)| + p_1(b) + n_1(b)$ and $\|f_2\|_{BV} = |f_2(a)| + p_2(b) + n_2(b)$. Next, write

$$\begin{aligned} f_1 f_2 = p_1 p_2 &+ n_1 n_2 + f_1(a)\, p_2 + f_2(a)\, p_1 \\ &- n_1 p_2 - n_2 p_1 - f_1(a)\, n_2 - f_2(a)\, n_1 \\ &+ f_1(a)\, f_2(a). \end{aligned}$$

Each term save the last (a constant) is a monotone function vanishing at a. Finally, we apply the triangle inequality in $BV[a,b]$:

$$\begin{aligned} \|f_1 f_2\|_{BV} &= V_a^b(f_1 f_2) + |f_1(a)|\,|f_2(a)| \\ &\le V_a^b(p_1 p_2) + V_a^b(n_1 n_2) + \cdots + V_a^b(f_2(a)\, n_1) + |f_1(a)|\,|f_2(a)| \end{aligned}$$

$$\begin{aligned}
&= p_1(b)\, p_2(b) + n_1(b)\, n_2(b) + |f_1(a)|\, p_2(b) + |f_2(a)|\, p_1(b) \\
&\quad + n_1(b)\, p_2(b) + n_2(b)\, p_1(b) + |f_1(a)|\, n_2(b) + |f_2(a)|\, n_1(b) \\
&\quad + |f_1(a)|\, |f_2(a)| \\
&= \big(p_1(b) + n_1(b) + |f_1(a)| \big)\big(p_2(b) + n_2(b) + |f_2(a)| \big) \\
&= \|f_1\|_{BV}\, \|f_2\|_{BV}. \qquad \text{Phew!} \quad \square
\end{aligned}$$

EXERCISES

17. Prove part (iii) of Proposition 13.11. [Hint: If $f = g - h$, then $V_x^y f \le V_x^y g + V_x^y h = g(y) - g(x) + h(y) - h(x)$.]

18. In the notation of Proposition 13.11, show that each point of continuity for f is also a point of continuity for both p and n.

19. Suppose that f has a continuous derivative on $[\,a, b\,]$.

(a) Use the mean value theorem to show that $V(f, P)$ can be written as a Riemann sum for $|f\,'|$ over P.

(b) Show that $V_a^b f = \int_a^b |f\,'(t)|\, dt$.

(c) Conclude that $p(x) = \int_a^b \{f\,'\}^+(t)\, dt$ and $n(x) = \int_a^b \{f\,'\}^-(t)\, dt$, where $\{f\,'\}^+$ and $\{f\,'\}^-$ are the positive and negative parts of $f\,'$.

If $\mathbf{f}'(t) = \big(x\,'(t), y\,'(t)\big)$ is continuous on $[\,a, b\,]$, it follows from Exercise 19 that $\mathbf{f}$ is then a rectifiable curve and its length is given by a Riemann integral: $V_a^b \mathbf{f} = \int_a^b \|\mathbf{f}\,'\|_2\, dt$. In the parlance of calculus, $ds/dt = \|\mathbf{f}\,'\|_2$ defines the speed of a particle traveling along the path $\mathbf{f}$, and $V_a^b \mathbf{f}$ is the total distance traveled by the particle from time a to time b.

One of our goals is to make sense out of the formula in Exercise 19 in the case where $f\,'$ is not continuous, or fails to exist at several points, or, for that matter, fails to be Riemann integrable. But this raises two big questions: If f is of bounded variation, does $f\,'$ exist at enough points to at least be integrable? And what does it mean for a function to be integrable anyway? Our first attempt to salvage the formula will be to write it in the "differential" form $\int_a^b |df(t)|$, and, for this to make sense, we will need more detailed information about integrals.

Helly's First Theorem

Next we present a compactness result, of sorts, for $BV[\,a, b\,]$ that will prove useful in the next chapter (where we will also meet Helly's Second Theorem). We begin with two lemmas of independent interest. The first of these we have already encountered informally; the technique involved is sometimes called *diagonalization*.

Helly's Selection Principle 13.13. *Let (f_n) be a uniformly bounded sequence of real-valued functions defined on a set X, and let D be any countable subset of X. Then, there is a subsequence of (f_n) that converges pointwise on D.*

PROOF. Suppose that $|f_n(x)| \le K$ for all n and all $x \in X$, and let $D = \{x_k : k \ge 1\}$. Then, in particular, since the sequence $\big(f_n(x_1)\big)$ is bounded, we can pass to a subsequence $\big(f_n^{(1)}\big)$ of (f_n) such that $\big(f_n^{(1)}(x_1)\big)$ converges.

But now the sequence $\big(f_n^{(1)}(x_2)\big)$ is also bounded, so we can pass to a subsequence $\big(f_n^{(2)}\big)$ of $\big(f_n^{(1)}\big)$ such that $\big(f_n^{(2)}(x_2)\big)$ converges. Since we have taken care to choose a subsequence of $\big(f_n^{(1)}\big)$, we also have that $\big(f_n^{(2)}(x_1)\big)$ converges.

Next, since $\big(f_n^{(2)}(x_3)\big)$ is bounded, we can find a further subsequence $\big(f_n^{(3)}\big)$ of $\big(f_n^{(2)}\big)$ such that $\big(f_n^{(3)}(x_3)\big)$ converges. We necessarily also have that $\big(f_n^{(3)}(x_2)\big)$ and $\big(f_n^{(3)}(x_1)\big)$ converge. By induction, we can find a subsequence $\big(f_n^{(m+1)}\big)$ of $\big(f_n^{(m)}\big)$ such that $\big(f_n^{(m+1)}(x_k)\big)_{n=1}^{\infty}$ converges for each $k = 1, 2, \ldots, m+1$.

The claim is that the "diagonal" sequence $\big(f_n^{(n)}(x_k)\big)_{n=1}^{\infty}$ converges for every k. Why? Because, for any k, the tail sequence $\big(f_n^{(n)}(x_k)\big)_{n=k}^{\infty}$ is a subsequence of $\big(f_n^{(k)}(x_k)\big)_{n=1}^{\infty}$. □

The following lemma should remind you of our technique for extending the definition of the Cantor function.

Lemma 13.14. *Let D be a subset of $[\,a,b\,]$ with $a \in D$ and $b = \sup D$. If $f : D \to \mathbb{R}$ is increasing, then f extends to an increasing function on all of $[\,a,b\,]$.*

PROOF. For $x \in [\,a,b\,]$, define $g(x) = \sup\{f(t) : a \le t \le x,\ t \in D\}$. It is immediate that g is increasing and that $g(x) = f(x)$ whenever $x \in D$. □

We next apply these results to a sequence of increasing functions on an interval $[\,a,b\,]$.

Lemma 13.15. *If (f_n) is a uniformly bounded sequence of increasing functions on $[\,a,b\,]$, that is, if $|f_n(x)| \le K$ for all n and all x in $[\,a,b\,]$, then some subsequence of (f_n) converges pointwise to an increasing function f on $[\,a,b\,]$ (which also satisfies $|f(x)| \le K$).*

PROOF. Let D be the set of all rationals in $[\,a,b\,]$ together with the point a, if a is irrational. By applying Helly's Selection Principle to the sequence (f_n) and the countable set D, there is a subsequence (f_{n_k}) of (f_n) such that $\varphi(x) = \lim_{k\to\infty} f_{n_k}(x)$ exists for all $x \in D$. It is easy to see that this defines φ as an increasing function on D. By Lemma 13.14, we may suppose that φ has been extended to an increasing function on all of $[\,a,b\,]$.

We next show that $\varphi(x) = \lim_{k\to\infty} f_{n_k}(x)$ at any point x where φ is continuous. Given such an x and $\varepsilon > 0$, choose rationals p and q in $[\,a,b\,]$ such that $p < x < q$ and $\varphi(q) - \varphi(p) < \varepsilon/2$. Then, for all k sufficiently large, we have

$$f_k(x) \le f_k(q) \le \varphi(q) + \varepsilon/2 \le \varphi(x) + \varepsilon,$$

and, similarly, $f_k(x) \ge \varphi(x) - \varepsilon$. Thus, $\varphi(x) = \lim_{k\to\infty} f_{n_k}(x)$ for any $x \notin D(\varphi)$, the set of discontinuities of φ.

Since φ is increasing, $D(\varphi)$ is at most countable. Now here comes the clincher! Apply Helly's Selection Principle again, this time using the sequence (f_{n_k}) and the countable set $D(\varphi)$. We choose a further subsequence of (f_{n_k}), which we again label (f_{n_k}), such that $\lim_{k\to\infty} f_{n_k}(x)$ exists for all $x \in D(\varphi)$ and, hence, for all x in $[a, b]$. If we set $f(x) = \lim_{k\to\infty} f_{n_k}(x)$, then f is clearly increasing. □

Finally, we are ready to apply these techniques to $BV[a, b]$.

Helly's First Theorem 13.16. *Let (f_n) be a bounded sequence in $BV[a, b]$; that is, suppose that $\|f_n\|_{BV} \le K$ for all n. Then, some subsequence of (f_n) converges pointwise on $[a, b]$ to a function $f \in BV[a, b]$ (which also satisfies $\|f\|_{BV} \le K$).*

PROOF. First, note that since $\|f_n\|_\infty \le \|f_n\|_{BV} \le K$ for all n, the sequence (f_n) is uniformly bounded. Next, if we write $v_n(x) = V_a^x f_n$, then $|v_n(x)| \le V_a^b f_n \le K$ and $|v_n(x) - f_n(x)| \le 2K$ for all n. That is, (f_n) is the difference of two uniformly bounded sequences of increasing functions, (v_n) and $(v_n - f_n)$. By repeated application of Lemma 13.15, we can find a common subsequence (n_k) such that both $g(x) = \lim_{k\to\infty} v_{n_k}(x)$ and $h(x) = \lim_{k\to\infty} \big(v_{n_k}(x) - f_{n_k}(x)\big)$ exist at each point x in $[a, b]$. (How?) It is easy to see that g and h are increasing functions and, hence, that $f = g - h$ is of bounded variation. Of course, $f(x) = \lim_{k\to\infty} f_{n_k}(x)$ for all x in $[a, b]$. Finally, it follows from Exercise 11 that $\|f\|_{BV} \le K$. □

Helly's theorem is something of a compactness result in that it provides a convergent subsequence for any bounded sequence in $BV[a, b]$. Unfortunately, the convergence here is pointwise and not necessarily convergence in the metric of $BV[a, b]$ (recall that convergence in $BV[a, b]$ is even harder to come by than uniform convergence).

Notes and Remarks

According to Lakatos [1976], functions of bounded variation were discovered by Camille Jordan through a "critical re-examination" of Dirichlet's famous flawed proof that arbitrary functions can be represented by Fourier series; see Jordan [1881]. It was Jordan who gave the characterization of such functions as differences of increasing functions (Corollary 13.6), but, as pointed out by Hawkins [1970], the key observation that Dirichlet's proof was valid for differences of increasing functions had already been made by du Bois-Reymond [1880]. The connection between rectifiable curves and functions of bounded variation is also due to Jordan and can be found in Jordan [1893]. Curiously, the representation of arc length by means of a definite integral was considered inappropriate and overly restrictive. As Hawkins puts it: "Success in this direction required a more flexible definition of the integral and the genius of Lebesgue."

The results in Exercise 6 are (essentially) due to Lebesgue; see Hobson [1927, Vol. I] and Lebesgue [1928]. The proof of Proposition 13.12 is taken from Kuller [1969], but also see Bullen [1983] and Russell [1979]. Lemma 13.14 is taken from Łojasiewicz [1988].

Helly's theorems can be found in Helly [1912]. For more on saltus functions and Helly's theorem (Theorem 13.16), see Natanson [1955, Vol. I] or Łojasiewicz [1988]. For more on Eduard Helly, the Austrian mathematician whose work had a profound influence on Riesz and Banach, see Hochstadt [1980] and a follow-up letter from Monna [1980].

CHAPTER FOURTEEN

The Riemann–Stieltjes Integral

Weights and Measures

Several times throughout this book we've hinted at a physical basis for some of our notation. It's time that we made this more precise; a simple calculus problem will help explain.

Consider a thin rod, or wire, positioned along the interval $[a, b]$ on the x-axis and having a *nonuniform* distribution of mass. For example, the rod might vary slightly in thickness or in *density* (mass per unit length) as x varies. Our job is to compute the density (at a point) as a function $f(x)$, if at all possible.

What we can measure effectively is the *distribution* of mass along the rod. That is, we can easily measure the mass of any segment of the rod, and so we know the mass of the segment lying along the interval $[a, x]$ as a function $F(x)$. Said in slightly different terms, we are able to measure small, discrete "chunks" of mass as $dm = F(x+dx) - F(x) = dF$, and so we're led to define the density $f(x) = dm/dx = F'(x)$ as the derivative of the distribution $F(x)$, provided that F is differentiable, of course.

But F is an arbitrary increasing function – is every such function differentiable? And, if not, can we say anything meaningful about this problem? Could we, for example, still find the center of mass (the line $x = \mu$ through which the rod balances) when F is not differentiable?

As it happens, most of what we need to know about the rod, from a physical standpoint, depends not on differentiation but on integration. And integrals are easier to come by than derivatives. To see this, let's simply use the pure formalism of first calculus and continue to write dF as the mass of a small "chunk" of the rod. Given this, the total mass is then $m = \int_a^b dF(x) = F(b) - F(a)$. And, as you might recall, we can also compute various *moments* as integrals, too:

$$\mu = \frac{1}{m}\int_a^b x\, dF(x) \qquad \text{(center of mass)},$$

$$\sigma^2 = \frac{1}{m}\int_a^b (x-\mu)^2 dF(x) \qquad \text{(moment of inertia about } \mu\text{)},$$

and so on. We might even want to consider various *measurements* φ and compute expressions such as

$$\frac{1}{m}\int_a^b \varphi(x)\, dF(x) \qquad \text{(expected value of } \varphi\text{)}.$$

In other words, the claim here is that it is possible to make sense out of these "generalized" Riemann integrals without making any assumptions on the differentiability of F. If, however, F should have a density (i.e., if F' exists), then we would want our new integral to be consistent with the Riemann integral. In this case, we would expect to have

$$\int_a^b \varphi(x)\,dF(x) = \int_a^b \varphi(x)\,F'(x)\,dx.$$

In particular, we will see to it that the case $F(x) = x$ leads to the Riemann integral.

There are several issues at hand here. First, given an arbitrary increasing function F on $[a, b]$, we will attack the problem of interpreting integrals of the form $\int_a^b \varphi(x)\,dF(x)$. It won't surprise you to learn that we will define this new integral as the limit, in some appropriate sense, of Riemann-type sums of the form $\sum_{i=1}^n \varphi(t_i)\big[F(x_i)-F(x_{i-1})\big]$. What we will have, if we are careful, is a generalization of the Riemann integral. What may surprise you, though, is that there are a number of reasonable ways to accomplish this. Our first attempt at extending the integral will by no means be the most general, but it will suffice for now.

Next we will take up the more difficult question of when (or if) our new integral is actually a Riemann integral. For this we will want to know whether F is differentiable and, if so, whether F' is Riemann integrable. The answer, as we will see, lies in further refining the Riemann integral. In short, we will generalize our generalization. First things first, though.

The Riemann–Stieltjes Integral

We begin by fixing our notation. Throughout this section, we consider a nonconstant **increasing** function $\alpha : [a, b] \to \mathbb{R}$ and a *bounded* function $f : [a, b] \to \mathbb{R}$ (the function α is our "distribution" or "weight," F, and f is our "measurement," φ). We next set up the notation necessary to define the **Riemann–Stieltjes integral** $\int_a^b f\,d\alpha$.

Given a partition $P = \{a = x_0 < x_1 < \cdots < x_n = b\}$ of $[a, b]$, we write $\Delta\alpha_i = \alpha(x_i) - \alpha(x_{i-1})$, for $i = 1, \ldots, n$. Note that $\Delta\alpha_i \geq 0$ for all i, and that $\sum_{i=1}^n \Delta\alpha_i = \alpha(b) - \alpha(a)$. Next, for each $i = 1, \ldots, n$, we define

$$m_i = \inf\{f(x) : x_{i-1} \leq x \leq x_i\},$$
$$M_i = \sup\{f(x) : x_{i-1} \leq x \leq x_i\}.$$

We will also need

$$m = \inf\{f(x) : a \leq x \leq b\} = \min\{m_1, \ldots, m_n\},$$
$$M = \sup\{f(x) : a \leq x \leq b\} = \max\{M_1, \ldots, M_n\}.$$

Note that $m \leq m_i \leq M_i \leq M$ for any $i = 1, \ldots, n$.

We define the **lower Riemann–Stieltjes sum** of f over P, with respect to α, by $L(f, P) = \sum_{i=1}^n m_i\,\Delta\alpha_i$, and the **upper Riemann–Stieltjes sum** of f over P, with respect to α, by $U(f, P) = \sum_{i=1}^n M_i\,\Delta\alpha_i$. If we should need to refer to α, we will write

$L_\alpha(f, P)$ and $U_\alpha(f, P)$. For the time being at least, we will think of α as fixed and so ignore several of these additional quantifiers; we will refer to $L(f, P)$ and $U(f, P)$ as simply a lower sum and an upper sum. Clearly, $L(f, P) \le U(f, P)$ for any partition P. Notice, too, that $L(-f, P) = -U(f, P)$.

As you would imagine, we want to take "limits" of upper and lower sums to define our new integral. A few simple observations will clarify the process.

Proposition 14.1. *If $P \subset Q$ are partitions of $[a, b]$, then $L(f, P) \le L(f, Q)$ and $U(f, Q) \le U(f, P)$.*

PROOF. We first prove the inequality concerning lower sums. By induction (on the number of elements of $Q \setminus P$) it is enough to consider the case $Q = P \cup \{x_1\}$, and for this it is enough to establish $L(f, \{a, b\}) \le L(f, \{a, x_1, b\})$. (Why?) Now, if we set $m_1 = \inf\{f(x) : a \le x \le x_1\}$ and $m_2 = \inf\{f(x) : x_1 \le x \le b\}$, then

$$\begin{aligned} L(f, \{a, b\}) &= m\big[\alpha(b) - \alpha(a)\big] \\ &= m\big[\alpha(x_1) - \alpha(a)\big] + m\big[\alpha(b) - \alpha(x_1)\big] \\ &\le m_1\big[\alpha(x_1) - \alpha(a)\big] + m_2\big[\alpha(b) - \alpha(x_1)\big] \\ &= L(f, \{a, x_1, b\}). \end{aligned}$$

The proof for upper sums is similar but, since $U(f, P) = -L(-f, P)$, it actually follows from what we have already shown. □

Corollary 14.2. *$L(f, P) \le U(f, Q)$ for any partitions P, Q of $[a, b]$.*

PROOF. $L(f, P) \le L(f, P \cup Q) \le U(f, P \cup Q) \le U(f, Q)$. □

Here is where we stand: For any partitions P and Q, we have

$$m\big[\alpha(b) - \alpha(a)\big] \le L(f, P) \le U(f, Q) \le M\big[\alpha(b) - \alpha(a)\big].$$

As we increase the number of points in our partition, the lower sums increase while the upper sums decrease. Thus we are led to consider the **lower Riemann–Stieltjes integral** of f with respect to α over $[a, b]$ defined by

$$\underline{\int_a^b} f\, d\alpha = \sup_P L(f, P)$$

and the **upper Riemann–Stieltjes integral** of f with respect to α over $[a, b]$ defined by

$$\overline{\int_a^b} f\, d\alpha = \inf_P U(f, P).$$

Clearly,

$$m\big[\alpha(b) - \alpha(a)\big] \le \underline{\int_a^b} f\, d\alpha \le \overline{\int_a^b} f\, d\alpha \le M\big[\alpha(b) - \alpha(a)\big].$$

If the upper and lower integrals of f should agree, then we say that f is **Riemann–Stieltjes integrable** with respect to α over $[\,a, b\,]$, and we define the **Riemann–Stieltjes integral** of f with respect to α over $[\,a, b\,]$ to be their common value

$$\int_a^b f(x)\,d\alpha(x) = \int_a^b f\,d\alpha = \underline{\int_a^b} f\,d\alpha = \overline{\int_a^b} f\,d\alpha.$$

When $\alpha(x) = x$, this definition yields the **Riemann integral** of f over $[\,a, b\,]$. In this case, we will use the familiar notation $\int_a^b f(x)\,dx$ or, occasionally, just $\int_a^b f$.

Examples 14.3

(a) If $f(x) = c$ is a constant function, then f is Riemann–Stieltjes integrable with respect to every increasing α and $\int_a^b f\,d\alpha = c\big[\alpha(b) - \alpha(a)\big]$. (Why?) Likewise, if α is constant, then every bounded function is integrable with respect to α – but, of course, $\int_a^b f\,d\alpha = 0$ for any f. Not very interesting. Unless we need to specifically consider this trivial case, we will always assume that α is nonconstant.

(b) In general, not every bounded function is integrable. For example, $\chi_{\mathbb{Q}}$ is not Riemann integrable on any interval $[\,a, b\,]$. To see this, just check that $U(\chi_{\mathbb{Q}}, P) = b - a$ and $L(\chi_{\mathbb{Q}}, P) = 0$ for any partition P of $[\,a, b\,]$. That is, $\overline{\int_a^b} \chi_{\mathbb{Q}} = b - a$ while $\underline{\int_a^b} \chi_{\mathbb{Q}} = 0$. Essentially the same argument shows that $\chi_{\mathbb{Q}}$ is not integrable with respect to any (nonconstant) increasing α.

(c) A simple example of a Stieltjes integral, although not precisely of the type we have defined, is provided by a *contour integral*, or line integral. Such integrals are frequently used in complex analysis and might be written $\int_\Gamma f(z)\,dz$, where Γ is a curve in the complex plane. If $\gamma(t)$, $a \le t \le b$, is a parameterization of Γ, then we would write

$$\int_\Gamma f(z)\,dz = \int_a^b f\big(\gamma(t)\big)\,d\big(\gamma(t)\big).$$

In practice, of course, the contours that are actually used are often very simple. For instance, if Γ is the circle of radius r about 0, then $\gamma(t) = r\,e^{it}$, and our contour integral reduces to the Riemann integral $\int_0^{2\pi} f(re^{it})ri\,e^{it}\,dt$. In full generality, though, $\gamma(t)$ need not be everywhere differentiable, and so the generic contour integral is necessarily a Stieltjes integral.

We write $\mathcal{R}_\alpha[\,a, b\,]$ to denote the collection of all bounded functions on $[\,a, b\,]$ which are Riemann–Stieltjes integrable with respect to α. When $\alpha(x) = x$, we simply write $\mathcal{R}[\,a, b\,]$ for the space of Riemann integrable functions on $[\,a, b\,]$. In any case, notice that (by definition) $\mathcal{R}_\alpha[\,a, b\,] \subset B[\,a, b\,]$.

As you might imagine, we will eventually check that $\mathcal{R}_\alpha[\,a, b\,]$ is a vector space, an algebra, a lattice, a normed space, and so on. To begin, though, we need a simple criterion for Riemann–Stieltjes integrability.

Theorem 14.4. (Riemann's Condition) *Let $\alpha : [\,a, b\,] \to \mathbb{R}$ be increasing. A bounded function $f : [\,a, b\,] \to \mathbb{R}$ is in $\mathcal{R}_\alpha[\,a, b\,]$ if and only if, given $\varepsilon > 0$, there exists a partition P of $[\,a, b\,]$ such that $U(f, P) - L(f, P) < \varepsilon$.*

PROOF. First, suppose that $f \in \mathcal{R}_\alpha[a, b]$ and let $I = \int_a^b f\,d\alpha$. Given $\varepsilon > 0$, choose partitions P and Q of $[a, b]$ such that $I - \varepsilon/2 < L(f, P)$ and $U(f, Q) < I + \varepsilon/2$. The partition $P^* = P \cup Q$ will do the trick:

$$U(f, P^*) \le U(f, Q) < I + \frac{\varepsilon}{2} < L(f, P) + \varepsilon \le L(f, P^*) + \varepsilon.$$

Next, suppose that for every $\varepsilon > 0$ there is a partition P for which $U(f, P) - L(f, P) < \varepsilon$. Then, since

$$L(f, P) \le \underline{\int_a^b} f\,d\alpha \le \overline{\int_a^b} f\,d\alpha \le U(f, P)$$

for any partition P, we must have $\overline{\int_a^b} f\,d\alpha - \underline{\int_a^b} f\,d\alpha < \varepsilon$ for every $\varepsilon > 0$. That is, $\overline{\int_a^b} f\,d\alpha = \underline{\int_a^b} f\,d\alpha$. □

Riemann's condition makes short work of checking that continuous functions are integrable with respect to any increasing integrator.

Theorem 14.5. $C[a, b] \subset \mathcal{R}_\alpha[a, b]$ *for any increasing* α.

PROOF. Let $f : [a, b] \to \mathbb{R}$ be continuous and let $\varepsilon > 0$. Then, since f is actually uniformly continuous, we may choose a $\delta > 0$ so that $|f(x) - f(y)| < \varepsilon$ whenever $|x - y| < \delta$. Now if P is any partition of $[a, b]$ with $x_i - x_{i-1} < \delta$ for all i, then $M_i - m_i < \varepsilon$ for all i and hence

$$\begin{aligned} U(f, P) - L(f, P) &= \sum_{i=1}^n (M_i - m_i)\Delta\alpha_i \\ &< \varepsilon \sum_{i=1}^n \Delta\alpha_i = \varepsilon\,[\alpha(b) - \alpha(a)]. \quad □ \end{aligned}$$

EXERCISES

▷ **1.** If $f, g \in \mathcal{R}_\alpha[a, b]$ with $f \le g$, show that $\int_a^b f\,d\alpha \le \int_a^b g\,d\alpha$.

2. If $f, g \in \mathcal{R}_\alpha[a, b]$, show that $f + g \in \mathcal{R}_\alpha[a, b]$ and that $\int_a^b (f + g)\,d\alpha = \int_a^b f\,d\alpha + \int_a^b g\,d\alpha$.

3. If $f \in \mathcal{R}_\alpha[a, b]$, show that $|f| \in \mathcal{R}_\alpha[a, b]$ and that $\left|\int_a^b f\,d\alpha\right| \le \int_a^b |f|\,d\alpha$. [Hint: $U(|f|, P) - L(|f|, P) \le U(f, P) - L(f, P)$. Why?]

4. If $f, g \in \mathcal{R}_\alpha[a, b]$, is $fg \in \mathcal{R}_\alpha[a, b]$? How about f^2?

5. Give an example where $f^2 \in \mathcal{R}_\alpha[a, b]$ but $f \notin \mathcal{R}_\alpha[a, b]$.

▷ **6.** Define increasing functions α, β, and γ on $[-1, 1]$ by $\alpha = \chi_{(0,1]}$, $\beta = \chi_{[0,1]}$, and $\gamma = \frac{1}{2}(\alpha + \beta)$. Given $f \in B[-1, 1]$, show that:
(a) $f \in \mathcal{R}_\alpha[-1, 1]$ if and only if $f(0+) = f(0)$.
(b) $f \in \mathcal{R}_\beta[-1, 1]$ if and only if $f(0-) = f(0)$.
(c) $f \in \mathcal{R}_\gamma[-1, 1]$ if and only if f is continuous at 0.
(d) If $f \in \mathcal{R}_\gamma[-1, 1]$, then $\int_{-1}^1 f\,d\alpha = \int_{-1}^1 f\,d\beta = \int_{-1}^1 f\,d\gamma = f(0)$.

7. Let $P = \{x_0, \ldots, x_n\}$ be a (fixed) partition of $[a, b]$, and let α be an increasing step function on $[a, b]$ that is constant on each of the open intervals (x_{i-1}, x_i) and has jumps of size $\alpha_i = \alpha(x_i+) - \alpha(x_i-)$ at each of the x_i, where $\alpha_0 = \alpha(a+) - \alpha(a)$ and $\alpha_n = \alpha(b) - \alpha(b-)$. If $f \in B[a, b]$ is continuous at each of the x_i, show that $f \in \mathcal{R}_\alpha$ and $\int_a^b f\, d\alpha = \sum_{i=0}^n f(x_i)\alpha_i$.

8. If f is continuous on $[1, n]$, compute $\int_1^n f(x)\, d[x]$, where $[x]$ is the greatest integer in x. What is $\int_1^t f(x)\, d[x]$ if t is not an integer?

9. If f is monotone and α is continuous (and still increasing), show that $f \in \mathcal{R}_\alpha[a, b]$.

As a second application of Riemann's condition, we can now supply an integral formula for the total variation in at least one simple case.

Theorem 14.6. *Suppose that f' exists and is Riemann integrable on $[a, b]$. Then, $f \in BV[a, b]$ and $V_a^b f = \int_a^b |f'(t)|\, dt$.*

PROOF. First note that f is continuous on $[a, b]$. Thus, given a partition P of $[a, b]$, we can appeal to the mean value theorem and write

$$V(f, P) = \sum_{i=1}^n |f(x_i) - f(x_{i-1})| = \sum_{i=1}^n |f'(t_i)|\Delta x_i,$$

where $t_i \in (x_{i-1}, x_i)$ for each i. Consequently,

$$L(|f'|, P) \le V(f, P) \le U(|f'|, P).$$

Since $|f'|$ is Riemann integrable (see Exercise 3), it follows that f is of bounded variation and that $V_a^b f = \int_a^b |f'(t)|\, dt$. □

We can rephrase Riemann's condition to look more like the definition of a limit. Indeed, since

$$U(f, P) - L(f, P) \le U(f, P^*) - L(f, P^*) \qquad \text{for all } P \supset P^*,$$

we can say that $f \in \mathcal{R}_\alpha[a, b]$ if and only if, for each n, there is some partition P_n such that $U(f, P) - L(f, P) < (1/n)$ for all refinements $P \supset P_n$, that is, for all partitions "beyond" P_n. And we might as well assume that $P_{n+1} \supset P_n$ for all n. Thus, $f \in \mathcal{R}_\alpha[a, b]$ if and only if $U(f, P_n) - L(f, P_n) \to 0$ for some increasing sequence of partitions $P_1 \subset P_2 \subset \cdots$. In short, if $f \in \mathcal{R}_\alpha[a, b]$, then Riemann's condition supplies a particular selection of points from $[a, b]$ that refine our upper and lower estimates for the integral. In this case, $L(f, P_n)$ increases to $\int_a^b f\, d\alpha$ while $U(f, P_n)$ decreases to $\int_a^b f\, d\alpha$.

Riemann's condition not only supplies a simple criterion to test for integrability, it also tells us exactly which functions fail to be integrable. To see this, let f be a bounded function on $[a, b]$, let $P = \{x_0, \ldots, x_n\}$ be a partition of $[a, b]$, and write the difference

$$M_i - m_i = \sup_{[x_{i-1}, x_i]} f - \inf_{[x_{i-1}, x_i]} f = \omega(f; [x_{i-1}, x_i])$$

as the *oscillation* of f over $[x_{i-1}, x_i]$. Thus,

$$
\begin{aligned}
U(f,P) - L(f,P) &= \sum_{i=1}^{n} [M_i - m_i]\big[\alpha(x_i) - \alpha(x_{i-1})\big] \\
&= \sum_{i=1}^{n} \omega(f; [x_{i-1}, x_i])\, \omega(\alpha; [x_{i-1}, x_i]) \\
&\geq \sum_{i=1}^{n} \omega(f; (x_{i-1}, x_i))\, \omega(\alpha; (x_{i-1}, x_i)) \\
&\geq \omega_f(x)\, \omega_\alpha(x), \qquad \text{for } x \notin P.
\end{aligned}
$$

In order that $f \in \mathcal{R}_\alpha[a,b]$, then, we must have $\omega_f(x)\,\omega_\alpha(x) = 0$ for "most" values of x. In particular, if f and α share a common one-sided discontinuity, say both are discontinuous from the right at $x \in [a,b]$, then f will fail to be integrable with respect to α. (See Exercise 6 for several specific examples.)

EXERCISES

▷ **10.** If $f \in \mathcal{R}_\alpha[a,b]$, show that $f \in \mathcal{R}_\alpha[c,d]$ for every subinterval $[c,d]$ of $[a,b]$. Moreover, $\int_a^b f\,d\alpha = \int_a^c f\,d\alpha + \int_c^b f\,d\alpha$ for every $a < c < b$. In fact, if any two of the these integrals exist, then so does the third and the equation above still holds.

▷ **11.** If $f \in \mathcal{R}_\alpha[a,b]$ with $m \leq f \leq M$, show that $\int_a^b f\,d\alpha = c\big[\alpha(b) - \alpha(a)\big]$ for some c between m and M. If f is continuous, show that $c = f(x_0)$ for some x_0.

12. Given $f \in \mathcal{R}_\alpha[a,b]$, define $F(x) = \int_a^x f\,d\alpha$ for $a \leq x \leq b$. Show that $F \in BV[a,b]$. If α is continuous, show that $F \in C[a,b]$.

13. If $\int_a^b f\,d\alpha = 0$ for every $f \in C[a,b]$, show that α is constant.

▷ **14.** If $f \in \mathcal{R}_\alpha[a,b]$, and if $U(f,P) - L(f,P) < \varepsilon$ for some partition P, show that $\left|\sum_{i=1}^n f(t_i)\Delta\alpha_i - \int_a^b f\,d\alpha\right| < \varepsilon$, where t_i is any point in $[x_{i-1}, x_i]$.

15. Suppose there exists a number I with the property that, given any $\varepsilon > 0$, there is a partition P such that $\left|\sum_{i=1}^n f(t_i)\Delta\alpha_i - I\right| < \varepsilon$, where t_i is any point in $[x_{i-1}, x_i]$. Show that $f \in \mathcal{R}_\alpha[a,b]$ and $I = \int_a^b f\,d\alpha$.

16. If $U(f,P) - L(f,P) < \varepsilon$, show that $\sum_{i=1}^n |f(t_i) - f(s_i)|\Delta\alpha_i < \varepsilon$ for any choice of points $s_i, t_i \in [x_{i-1}, x_i]$.

▷ **17.** If f and α share a common-sided discontinuity in $[a,b]$, show that f is not in $\mathcal{R}_\alpha[a,b]$.

18. Show that $\bigcap \big\{\mathcal{R}_\alpha[a,b] : \alpha \text{ increasing}\big\} = C[a,b]$.

19. If $\mathcal{R}_\alpha[a,b] \supset S[a,b]$, show that α is continuous.

20. If α is continuous, show that $\int_a^b f\,d\alpha$ does not depend on the values of f at any *finite* number of points. Is this still true if we change "finite" to "countable"? Explain.

21. Given a sequence (x_n) of distinct points in (a, b) and a sequence (c_n) of positive numbers with $\sum_{n=1}^{\infty} c_n < \infty$, define an increasing function α on $[a, b]$ by setting $\alpha(x) = \sum_{n=1}^{\infty} c_n I(x - x_n)$, where $I(x) = 0$ for $x < 0$ and $I(x) = 1$ for $x \geq 0$. Show that $\int_a^b f\,d\alpha = \sum_{n=1}^{\infty} c_n f(x_n)$ for every continuous function f on $[a, b]$. [Hint: Given $\varepsilon > 0$, take N sufficiently large so that $\beta(x) = \sum_{n=N+1}^{\infty} c_n I(x - x_n)$ satisfies $\beta(b) - \beta(a) < \varepsilon$.]

22. If $f \in \mathcal{R}_\alpha[a, b]$ with $m \leq f \leq M$, and if φ is continuous on $[m, M]$, show that $\varphi \circ f \in \mathcal{R}_\alpha[a, b]$.

23. Suppose that φ is a strictly increasing continuous function from $[c, d]$ onto $[a, b]$. Given $f \in \mathcal{R}_\alpha[a, b]$, show that $g = f \circ \varphi \in \mathcal{R}_\beta[c, d]$, where $\beta = \alpha \circ \varphi$. Moreover, $\int_c^d g\,d\beta = \int_a^b f\,d\alpha$.

24. As we have seen, $\chi_{\mathbb{Q}}$ is not Riemann integrable on $[0, 1]$. The problem is that $\chi_{\mathbb{Q}}$ is "too discontinuous." But what might that mean? Here is another example with uncountably many points of discontinuity, but this time Riemann integrable: Show that the set of discontinuities of χ_Δ is precisely Δ (an uncountable set), but that χ_Δ is nevertheless Riemann integrable on $[0, 1]$. [Hint: Δ can be covered by finitely many intervals of arbitrarily small total length.]

The Space of Integrable Functions

In this section we will examine the algebraic structure of the space of integrable functions $\mathcal{R}_\alpha[a, b]$, where α is increasing. As you might imagine, this examination will reduce to a study of certain elementary properties of the integral. Most of these properties are both easy to guess and easy to check. For this reason, we will relegate many of the details to the exercises. On the other hand, whereas some accounts give these elementary properties as corollaries of a "metatheorem," we will give (or at least sketch) direct proofs wherever possible.

To begin, let's check that $\mathcal{R}_\alpha[a, b]$ is a vector space, a lattice, and an algebra!

Theorem 14.7. *Let $f, g \in \mathcal{R}_\alpha[a, b]$ and let $c \in \mathbb{R}$. Then:*

(i) *$cf \in \mathcal{R}_\alpha[a, b]$ and $\int_a^b cf\,d\alpha = c\int_a^b f\,d\alpha$.*

(ii) *$f + g \in \mathcal{R}_\alpha[a, b]$ and $\int_a^b (f + g)\,d\alpha = \int_a^b f\,d\alpha + \int_a^b g\,d\alpha$.*

(iii) *$\int_a^b f\,d\alpha \leq \int_a^b g\,d\alpha$ whenever $f \leq g$.*

(iv) *$|f| \in \mathcal{R}_\alpha[a, b]$ and $\left|\int_a^b f\,d\alpha\right| \leq \int_a^b |f|\,d\alpha \leq \|f\|_\infty\big[\alpha(b) - \alpha(a)\big]$.*

(v) *$fg \in \mathcal{R}_\alpha[a, b]$ and $\left|\int_a^b fg\,d\alpha\right| \leq \left(\int_a^b f^2\,d\alpha\right)^{1/2}\left(\int_a^b g^2\,d\alpha\right)^{1/2}$.*

PROOF. (i): If $c \geq 0$, then clearly $U(cf, P) = c\,U(f, P)$, and similarly for lower sums. If, however, $c < 0$, then

$$U(cf, P) = |c|\,U(-f, P) = -|c|\,L(f, P) = c\,L(f, P).$$

(Why?) Again, the lower sum version is similar. In either case we get

$$U(cf, P) - L(cf, P) = |c|\big[U(f, P) - L(f, P)\big],$$

and this should be enough to convince you that $cf \in \mathcal{R}_\alpha[a, b]$. Now, for the equality of integrals, notice that

$$\begin{aligned}\overline{\int_a^b} cf\, d\alpha &= c\,\overline{\int_a^b} f\, d\alpha \qquad \text{if } c \ge 0\\ &= c\,\underline{\int_a^b} f\, d\alpha \qquad \text{if } c < 0.\end{aligned}$$

(ii): Consider the following rather strange looking claim:

$$\begin{aligned}L(f, P) + L(g, Q) &\le L(f + g, P \cup Q)\\ &\le U(f + g, P \cup Q) \le U(f, P) + U(g, Q).\end{aligned}$$

(Why does this work?) Since we are allowed to make independent choices of P and Q, we can easily force $P \cup Q$ to "work" for $f + g$. Thus, $f + g \in \mathcal{R}_\alpha[a, b]$. And how about the integrals? Well, it follows from our claim that

$$\begin{aligned}\underline{\int_a^b} f\, d\alpha + \underline{\int_a^b} g\, d\alpha &\le \underline{\int_a^b} (f + g)\, d\alpha\\ &\le \overline{\int_a^b} (f + g)\, d\alpha \le \overline{\int_a^b} f\, d\alpha + \overline{\int_a^b} g\, d\alpha.\end{aligned}$$

The proof of (iii) is left as an exercise (see Exercise 1).

(iv): From the triangle inequality, $\big|\,|f(s)| - |f(t)|\,\big| \le |f(s) - f(t)|$, and so it follows that $\omega(|f|; I) \le \omega(f; I)$ for any interval I. In particular,

$$U(|f|, P) - L(|f|, P) \le U(f, P) - L(f, P).$$

Hence, $|f| \in \mathcal{R}_\alpha[a, b]$. Since $-f,\ f \le |f| \le \|f\|_\infty$, the integral inequality follows from (i) and (iii).

(v): We first show that $f^2 \in \mathcal{R}_\alpha[a, b]$. Indeed, since

$$f(x)^2 - f(y)^2 = \big(f(x) + f(y)\big)\big(f(x) - f(y)\big),$$

we have $\omega(f^2; I) \le 2\|f\|_\infty\, \omega(f; I)$ for any interval I. Consequently,

$$U(f^2, P) - L(f^2, P) \le 2\|f\|_\infty\big[U(f, P) - L(f, P)\big].$$

Thus, $f^2 \in \mathcal{R}_\alpha[a, b]$ whenever $f \in \mathcal{R}_\alpha[a, b]$. That $\mathcal{R}_\alpha[a, b]$ is closed under more general products now follows from a little sleight of hand: $4fg = (f+g)^2 - (f-g)^2$. Hence, by (i), (ii), and the first part of this proof, we have $fg \in \mathcal{R}_\alpha[a, b]$.

Finally, the integral inequality follows from the Cauchy–Schwarz inequality for sums and Exercise 14. Since all three integrals in the inequality exist, we can find a single partition $P = \{x_0, \ldots, x_n\}$ such that each integral is approximated, to within a given ε, by a finite sum of the form $\sum_{i=1}^{n} \varphi(t_i)\Delta\alpha_i$, where t_i is any point in $[x_{i-1}, x_i]$. Thus,

$$\begin{aligned}
\int_a^b fg\,d\alpha - \varepsilon &\le \sum_{i=1}^{n} f(t_i)g(t_i)\Delta\alpha_i \\
&= \sum_{i=1}^{n} f(t_i)\sqrt{\Delta\alpha_i}\cdot g(t_i)\sqrt{\Delta\alpha_i} \\
&\le \left(\sum_{i=1}^{n} f(t_i)^2\Delta\alpha_i\right)^{1/2}\left(\sum_{i=1}^{n} g(t_i)^2\Delta\alpha_i\right)^{1/2} \\
&\le \left(\int_a^b f^2\,d\alpha + \varepsilon\right)^{1/2}\left(\int_a^b g^2\,d\alpha + \varepsilon\right)^{1/2}. \quad \square
\end{aligned}$$

Please note that Theorem 14.7 (v) need not hold for unbounded functions (or "improper" integrals). Indeed, the improper Riemann integral $\int_0^1 (1/\sqrt{x})\,dx$ exists, while $\int_0^1 (1/x)\,dx$ does not.

Theorem 14.7 tells us that $\mathcal{R}_\alpha[a, b]$ is a vector space, an algebra, and a lattice; in fact, $\mathcal{R}_\alpha[a, b]$ is a subspace, a subalgebra, and a sublattice of $B[a, b]$. Moreover, there are at least two natural choices for a norm on $\mathcal{R}_\alpha[a, b]$. We might simply use the sup-norm, or we might want to consider $\|f\| = \int_a^b |f|\,d\alpha$. While the latter expression has most of the trademarks of a norm and will actually prove useful in certain settings, it falls just short of being a norm. It typically only defines a semi-norm (see Exercises 25 and 26).

For now, let's establish at least one good reason to consider the sup-norm: $\mathcal{R}_\alpha[a, b]$ is *closed* under uniform convergence. That is, $\mathcal{R}_\alpha[a, b]$ is a closed subspace of $B[a, b]$ and so is *complete* under the sup-norm.

Theorem 14.8. *Let (f_n) be a sequence in $\mathcal{R}_\alpha[a, b]$. If (f_n) converges uniformly to f on $[a, b]$, then $f \in \mathcal{R}_\alpha[a, b]$. Moreover, $\int_a^b f_n d\alpha \to \int_a^b f\,d\alpha$.*

PROOF. Given $\varepsilon > 0$, choose k such that $\|f - f_n\|_\infty < \varepsilon$ whenever $n \ge k$. Now, since f_k is integrable, we can find a partition P of $[a, b]$ such that $U(f_k, P) - L(f_k, P) < \varepsilon$. From this we want to estimate $U(f, P) - L(f, P)$.

Now for any pair of points $s, t \in [a, b]$, the triangle inequality gives $|f(s) - f(t)| \le |f_k(s) - f_k(t)| + 2\varepsilon$. It follows that $\omega(f; I) \le \omega(f_k; I) + 2\varepsilon$ for any interval $I \subset [a, b]$. Consequently,

$$\begin{aligned}
U(f, P) - L(f, P) &= \sum_{i=1}^{n} \omega(f; [x_{i-1}, x_i])\Delta\alpha_i \\
&\le \sum_{i=1}^{n} \omega(f_k; [x_{i-1}, x_i])\Delta\alpha_i + 2\varepsilon\sum_{i=1}^{n}\Delta\alpha_i \\
&= U(f_k, P) - L(f_k, P) + 2\varepsilon\big[\alpha(b) - \alpha(a)\big] \\
&< \varepsilon + 2\varepsilon\big[\alpha(b) - \alpha(a)\big].
\end{aligned}$$

Thus, since ε is arbitrary, $f \in \mathcal{R}_\alpha[\,a,b\,]$. To see that $\int_a^b f_n d\alpha \to \int_a^b f\,d\alpha$, we now just estimate

$$\left|\int_a^b (f_n - f)\,d\alpha\right| \le \int_a^b |f_n - f|\,d\alpha$$
$$\le \|f_n - f\|_\infty \big[\alpha(b) - \alpha(a)\big] \to 0 \qquad \text{as } n \to \infty. \quad \square$$

Notice that $C[\,a,b\,]$ is a subspace (as well as a subalgebra and a sublattice) of $\mathcal{R}_\alpha[\,a,b\,]$ for α increasing. It follows from Theorem 14.8 that $C[\,a,b\,]$ is *closed* in $\mathcal{R}_\alpha[\,a,b\,]$ when $\mathcal{R}_\alpha[\,a,b\,]$ is endowed with the sup-norm.

On the other hand, if α is continuous, and if we supply $\mathcal{R}_\alpha[\,a,b\,]$ with the semi-norm $\|f\| = \int_a^b |f|\,d\alpha$, then $C[\,a,b\,]$ is a *dense* subspace of $\mathcal{R}_\alpha[\,a,b\,]$.

Theorem 14.9. *Let α be continuous and increasing. Given $f \in \mathcal{R}_\alpha[\,a,b\,]$ and $\varepsilon > 0$, there exist*

(i) *a step function h on $[\,a,b\,]$ with $\|h\|_\infty \le \|f\|_\infty$ such that $\int_a^b |f - h|\,d\alpha < \varepsilon$, and*

(ii) *a continuous function g on $[\,a,b\,]$ with $\|g\|_\infty \le \|f\|_\infty$ such that $\int_a^b |f - g|\,d\alpha < \varepsilon$.*

PROOF. From Theorem 14.4, we can find a partition $P = \{x_0, \ldots, x_n\}$ such that

$$U(f,P) - L(f,P) = \sum_{i=1}^n \omega(f;[\,x_{i-1},x_i\,])\Delta\alpha_i < \varepsilon.$$

For each $i = 1, \ldots, n$, choose $t_i \in [\,x_{i-1}, x_i)$ and define a step function h by setting $h(x) = f(t_i)$ for $x_{i-1} \le x < x_i$, for $i = 1, \ldots, n$, and $h(x_n) = f(t_n)$. Clearly, $\|h\|_\infty \le \|f\|_\infty$. Since α is continuous, we have $h \in \mathcal{R}_\alpha[\,a,b\,]$. From Exercise 10 it follows that

$$\int_a^b |f - h|\,d\alpha = \sum_{i=1}^n \int_{x_{i-1}}^{x_i} |f(x) - f(t_i)|\,d\alpha(x)$$
$$\le \sum_{i=1}^n \omega(f;[\,x_{i-1},x_i\,])\,\Delta\alpha_i$$
$$= U(f,P) - L(f,P) < \varepsilon,$$

which proves (i).

To prove (ii), we use the fact that α is uniformly continuous. Since n is fixed, we may choose $0 < \delta < \min\{\Delta x_i/2 : i = 1, \ldots, n\}$ such that α has oscillation less than $\varepsilon/(n+1)$ on each of the intervals $[\,x_i - \delta, x_i + \delta\,] \cap [\,a,b\,]$. Now let g be the polygonal function that agrees with h at each of the nodes

$$x_0, \quad x_0 + \delta, \quad x_1 - \delta, \quad x_1 + \delta, \quad \ldots, \quad x_n - \delta, \quad x_n.$$

(Thus g is the piecewise linear continuous function that agrees with h on each of the intervals $[\,x_{i-1}+\delta, x_i - \delta\,]$ and is linear on each of the intervals $[\,x_i - \delta, x_i + \delta\,]$.)

Then, $\|g\|_\infty \le \|h\|_\infty \le \|f\|_\infty$ (why?), $g \in \mathcal{R}_\alpha[a,b]$, and

$$\int_a^b |h-g|\,d\alpha = \int_{x_0}^{x_0+\delta} |h-g|\,d\alpha + \sum_{i=1}^{n-1}\int_{x_i-\delta}^{x_i+\delta} |h-g|\,d\alpha + \int_{x_n-\delta}^{x_n} |h-g|\,d\alpha$$

$$< 2\|f\|_\infty \cdot \frac{\varepsilon}{n+1} + \cdots + 2\|f\|_\infty \cdot \frac{\varepsilon}{n+1} = 2\varepsilon\,\|f\|_\infty.$$

Finally, we use the triangle inequality to conclude that

$$\int_a^b |f-g|\,d\alpha \le \int_a^b |f-h|\,d\alpha + \int_a^b |h-g|\,d\alpha < \varepsilon + 2\varepsilon\,\|f\|_\infty. \quad \square$$

EXERCISES

25. Construct a nonconstant increasing function α and a nonzero continuous function $f \in \mathcal{R}_\alpha[a,b]$ such that $\int_a^b |f|\,d\alpha = 0$. Is it possible to choose α to also be continuous? Explain.

26. If f is continuous on $[a,b]$, and if $f(x_0) \ne 0$ for some x_0, show that $\int_a^b |f(x)|\,dx \ne 0$. Conclude that $\|f\| = \int_a^b |f(x)|\,dx$ defines a norm on $C[a,b]$. Does it define a norm on all of $\mathcal{R}[a,b]$? Explain.

27. Give an example of a sequence of Riemann integrable functions on $[0,1]$ that converges pointwise to a nonintegrable function.

Integrators of Bounded Variation

We next extend the definition of the Riemann–Stieltjes integral to accept integrators that are not necessarily increasing. In particular, we would like to use the *difference* of increasing weights, that is, functions of bounded variation. The only problem we face is that upper and lower sums will no longer be monotone. To generalize the integral, then, only requires that we take more general sums.

Throughout this section, unless otherwise specified, f and α will denote arbitrary, bounded, real-valued functions on $[a,b]$.

Given a partition $P = \{x_0, \ldots, x_n\}$ of $[a,b]$, let $T = \{t_1, \ldots, t_n\}$ denote an arbitrary selection of points from $[a,b]$ with $t_i \in [x_{i-1}, x_i]$. We call

$$S(f,P,T) = \sum_{i=1}^{n} f(t_i)\big[\alpha(x_i) - \alpha(x_{i-1})\big]$$

a **Riemann–Stieltjes sum** for f. If we need to display the dependence on α, we will write $S_\alpha(f,P,T)$.

In this general setting we say that f is **Riemann–Stieltjes integrable** with respect to α and write $f \in \mathcal{R}_\alpha[a,b]$ if and only if there exists a number $I \in \mathbb{R}$ such that, for every $\varepsilon > 0$, there is a partition P^* for which $|S(f,P,T) - I| < \varepsilon$ for all refinements

$P \supset P^*$ and all selections of points T. If such a number I exists, then it is easy to see that it must also be unique; in this case, we define $\int_a^b f\,d\alpha = I$. If we should need to distinguish this integral from an integral arising from another definition, we will write $(RS)\int_a^b f\,d\alpha$.

If α is increasing, then $L(f, P) \le S(f, P, T) \le U(f, P)$ for any P and any T. The fact that we have complete freedom in choosing the points T at which f is evaluated means that the sum $S(f, P, T)$ can be made arbitrarily close to either $L(f, P)$ or $U(f, P)$ by choosing T appropriately. Given this, it is not hard to see that the refinement definition of the integral coincides with our earlier definition when α is increasing (see Exercises 14 and 15).

For nonincreasing integrators, though, no such simple comparison of sums is available. If we permit $\Delta\alpha_i$ to take on negative values, then we sacrifice the monotonicity of upper and lower sums. The more general Riemann–Stieltjes sums $S(f, P, T)$ are needed in this case; the extra freedom in choosing T compensates for the lack of monotonicity of sums.

EXERCISES

▷ **28.** If α is increasing, show that the definition of the integral given above coincides with our previous definition (in terms of upper and lower sums).

▷ **29.** Show that $|S_\alpha(f, P, T)| \le \|f\|_\infty V(\alpha, P)$.

30. If α is a step function (and not necessarily increasing) and f is continuous, derive a formula for $(RS)\int_a^b f\,d\alpha$. [Hint: See Exercise 7.]

31. Let $a < c < b$, and suppose that $f \in \mathcal{R}_\alpha[\,a, c\,] \cap \mathcal{R}_\alpha[\,c, b\,]$. Show that $f \in \mathcal{R}_\alpha[\,a, b\,]$ and that $\int_a^b f\,d\alpha = \int_a^c f\,d\alpha + \int_c^b f\,d\alpha$. In fact, if any two of these integrals exist, then so does the third and the equation above still holds.

32. If $(RS)\int_a^b f\,d\alpha$ exists, and if $a < c < b$, does $(RS)\int_a^c f\,d\alpha$ exist? [Hint: The answer is "yes," but this is harder than the previous exercise.]

33. If $(RS)\int_a^b f\,d\alpha$ exists, show that it equals $\lim_{n\to\infty} S(f, P_n, T_n)$ for some increasing sequence of partitions (P_n) and any (T_n).

34. Just as with other limits, the refinement integral admits a "Cauchy criterion" for convergence: Show that $f \in \mathcal{R}_\alpha[\,a, b\,]$ if and only if, given $\varepsilon > 0$, there is a partition P^* such that $|S(f, P_1, T_1) - S(f, P_2, T_2)| < \varepsilon$ for any pair of refinements $P_1, P_2 \supset P^*$ and any T_1, T_2. [Hint: For the backward implication, choose a particular sequence of partitions for which $S(f, P_n, T_n)$ converges to, say, I. Now show that I "works" in the definition of $(RS)\int_a^b f\,d\alpha$.]

35. Let $P = \{x_0, \ldots, x_n\} \subset \{y_0, \ldots, y_m\} = P^*$ be partitions of $[\,a, b\,]$. Show that $S(f, P, T) - S(f, P^*, T^*) = \sum_{j=1}^m [f(s_j) - f(t_j^*)]\,[\alpha(y_j) - \alpha(y_{j-1})]$, where $s_j = t_i$ and t_j^* are in the same interval $[\,x_{i-1}, x_i\,]$. [Hint: Draw a picture!] Use this to give a direct proof, based on Exercise 34, that $C[\,a, b\,] \subset \mathcal{R}_\alpha[\,a, b\,]$ whenever α is of bounded variation.

Riemann–Stieltjes sums are easier to work with than you might suspect. For example, it is now quite easy to see that the integral is linear. Indeed, since the sums are linear, $S(cf + dg, P, T) = cS(f, P, T) + dS(g, P, T)$, we have

$$\left| S(cf + dg, P, T) - \left(c \int_a^b f\,d\alpha + d \int_a^b g\,d\alpha \right) \right|$$
$$\leq |c| \left| S(f, P, T) - \int_a^b f\,d\alpha \right| + |d| \left| S(f, P, T) - \int_a^b g\,d\alpha \right|.$$

That is, $I = c\int_a^b f\,d\alpha + d\int_a^b g\,d\alpha$ "works" and so becomes the only possible value for $\int_a^b (cf + dg)\,d\alpha$. Thus, $\mathcal{R}_\alpha[\,a, b\,]$ is at least a *vector space*.

Absolute values and products will not be so easy to come by, though. Again, we need the integral to be monotone (more or less), and it is not necessarily going to cooperate. In fact, one of our goals is to find an upper estimate for $\left|\int_a^b f\,d\alpha\right|$ in terms of $\|f\|_\infty$. This was simple for increasing weights α, but not so transparent in general. (Recall the proof of Theorem 14.7 (iv).)

On the other hand, certain other properties of the integral are still with us. For example, it is not at all hard to show that the integral is also "linear in α." That is, if $f \in \mathcal{R}_\alpha \cap \mathcal{R}_\beta$, then $f \in \mathcal{R}_{\alpha\pm\beta}$ and $\int_a^b f\,d(\alpha \pm \beta) = \int_a^b f\,d\alpha \pm \int_a^b f\,d\beta$. Rather than present several repetitious proofs, let's settle all such issues at once.

Theorem 14.10. (Integration by Parts) *$f \in \mathcal{R}_\alpha[\,a, b\,]$ if and only if $\alpha \in \mathcal{R}_f[\,a, b\,]$ and, in either case,*

$$\int_a^b f\,d\alpha + \int_a^b \alpha\,df = f(b)\alpha(b) - f(a)\alpha(a).$$

PROOF. The "if and only if" is a mirage! Since the statement is clearly symmetric in α and f, we need only establish the forward implication. So, suppose that $f \in \mathcal{R}_\alpha[\,a, b\,]$, and let $\varepsilon > 0$. Choose a partition P^* so that $\left|S_\alpha(f, P, T) - \int_a^b f\,d\alpha\right| < \varepsilon$ for all $P \supset P^*$ and all T.

Fix $P \supset P^*$ and a selection of points T. The idea is to write $S_f(\alpha, P, T)$ in terms of $S_\alpha(f, P', T')$, where $P' \supset P$ (and hence $P' \supset P^*$). First,

$$\begin{aligned} S_f(\alpha, P, T) &= \sum_{i=1}^n \alpha(t_i)\,[f(x_i) - f(x_{i-1})] \\ &= \sum_{i=1}^n f(x_i)\alpha(t_i) - \sum_{i=0}^{n-1} f(x_i)\alpha(t_{i+1}) \\ &= -\sum_{i=0}^n f(x_i)\,[\alpha(t_{i+1}) - \alpha(t_i)] - f(x_0)\alpha(t_0) + f(x_n)\alpha(t_{n+1}), \end{aligned}$$

where we have introduced $t_0 = a$ and $t_{n+1} = b$ (since a partition has to include a and b). That is, if we set $P' = \{t_0, t_1, \ldots, t_{n+1}\}$ and $T' = P$, then

$$S_f(\alpha, P, T) = f(b)\alpha(b) - f(a)\alpha(a) - S_\alpha(f, P', T'),$$

which is *almost* what we want. We wanted $P' \supset P$, and this is easy to fix:

$$\begin{aligned} S_\alpha(f, P', T') &= \sum_{i=0}^{n} f(x_i)\,[\alpha(t_{i+1}) - \alpha(t_i)] \\ &= \sum_{i=0}^{n} f(x_i)\,[\alpha(t_{i+1}) - \alpha(x_i)] + \sum_{i=0}^{n} f(x_i)\,[\alpha(x_i) - \alpha(t_i)] \\ &= S_\alpha(f, P'', T''), \end{aligned}$$

where $P'' = \{x_0, t_1, x_1, t_2, \ldots\} \supset P$ and $T'' = \{x_0, x_0, x_1, x_1, \ldots, x_n, x_n\}$. Hence,

$$\begin{aligned} &\left| S_f(\alpha, P, T) - \left[f(b)\alpha(b) - f(a)\alpha(a) - \int_a^b f\,d\alpha \right] \right| \\ &\qquad = \left| \int_a^b f\,d\alpha - S_\alpha(f, P'', T'') \right| < \varepsilon. \end{aligned}$$

That is, $\alpha \in \mathcal{R}_f[\,a, b\,]$ and $\int_a^b \alpha\,df = f(b)\alpha(b) - f(a)\alpha(a) - \int_a^b f\,d\alpha$. □

Now we just sit back and reap the benefits.

Corollary 14.11. *If $f \in \mathcal{R}_\alpha \cap \mathcal{R}_\beta$, then $f \in \mathcal{R}_{\alpha\pm\beta}$ and*

$$\int_a^b f\,d(\alpha \pm \beta) = \int_a^b f\,d\alpha \pm \int_a^b f\,d\beta.$$

Corollary 14.12. *If f is monotone and α is continuous on $[\,a, b\,]$, then $f \in \mathcal{R}_\alpha[\,a, b\,]$.*

Corollary 14.13. *If $\alpha \in BV[\,a, b\,]$, then $C[\,a, b\,] \subset \mathcal{R}_\alpha[\,a, b\,]$. Obversely, if $\alpha \in C[\,a, b\,]$, then $BV[\,a, b\,] \subset \mathcal{R}_\alpha[\,a, b\,]$. In particular, continuous functions and functions of bounded variation are Riemann integrable on $[\,a, b\,]$.*

Proof. If $\alpha = \beta - \gamma$, where β and γ are increasing, then

$$C[\,a, b\,] \subset \mathcal{R}_\beta[\,a, b\,] \cap \mathcal{R}_\gamma[\,a, b\,] \subset \mathcal{R}_{\beta-\gamma}[\,a, b\,] = \mathcal{R}_\alpha[\,a, b\,]. \quad \square$$

We would like to go one step further in the proof of Corollary 14.13 and ask whether $\mathcal{R}_\beta[\,a, b\,] \cap \mathcal{R}_\gamma[\,a, b\,] = \mathcal{R}_\alpha[\,a, b\,]$. This would truly reduce the study of bounded variation integrators to the case of increasing integrators. For example, since each of $\mathcal{R}_\beta$ and $\mathcal{R}_\gamma$ is closed under products, we would have that $\mathcal{R}_\alpha$ is closed under products, too. Unfortunately, the formula is not true for just any such splitting $\alpha = \beta - \gamma$ (take $\alpha = 0$ and $\beta = \gamma$, any nonconstant increasing function), but it is true for the canonical decomposition.

Theorem 14.14. *Let $\alpha \in BV[\,a, b\,]$, and let $\beta(x) = V_a^x \alpha$. (Recall that both β and $\beta - \alpha$ are increasing.) Then, $\mathcal{R}_\alpha[\,a, b\,] = \mathcal{R}_\beta[\,a, b\,] \cap \mathcal{R}_{\beta-\alpha}[\,a, b\,]$.*

Proof. From Corollary 14.11, it suffices to show that $\mathcal{R}_\alpha[\,a, b\,] \subset \mathcal{R}_\beta[\,a, b\,]$. So, let $\varepsilon > 0$, and let $f \in \mathcal{R}_\alpha[\,a, b\,]$.

We first make an observation about α and β. Since α is of bounded variation, we may choose a partition P^* so that $\beta(b) - \beta(a) = V_a^b\alpha \geq V(\alpha, P) \geq V_a^b\alpha - \varepsilon$ for all partitions $P \supset P^*$. That is, if $P = \{x_0, \ldots, x_n\} \supset P^*$, then

$$\begin{aligned} 0 &\leq \beta(b) - \beta(a) - V(\alpha, P) \\ &= \sum_{i=1}^{n}[\beta(x_i) - \beta(x_{i-1})] - \sum_{i=1}^{n}|\alpha(x_i) - \alpha(x_{i-1})| \\ &= \sum_{i=1}^{n}\{\Delta\beta_i - |\Delta\alpha_i|\} \leq \varepsilon. \end{aligned}$$

Since $f \in \mathcal{R}_\alpha[a, b]$, and since we are allowed to augment P^*, we may assume that P^* also satisfies $\left|S_\alpha(f, P, T) - \int_a^b f\,d\alpha\right| < \varepsilon/2$ for any $P \supset P^*$ and any T. In particular,

$$|S_\alpha(f, P, T) - S_\alpha(f, P, T^*)| < \varepsilon \qquad \text{for any } P \supset P^* \text{ and any } T,\ T^*.$$

Once P is fixed, we can force this difference to look like the difference of upper and lower sums for β by taking a suitable choice of T and T^*. Specifically, given P and $\varepsilon > 0$, choose T and T^* so that

$$\begin{aligned} S_\alpha(f, P, T) - S_\alpha(f, P, T^*) &= \sum_{i=1}^{n}[f(t_i) - f(t_i^*)]\,\Delta\alpha_i \\ &\geq \sum_{i=1}^{n}(M_i - m_i - \varepsilon)|\Delta\alpha_i| \\ &\geq \sum_{i=1}^{n}(M_i - m_i)|\Delta\alpha_i| - \varepsilon\, V_a^b\alpha. \end{aligned}$$

(Please note the absolute values! Why does this work?)

Combining these observations, we now compare $U_\beta(f, P) - L_\beta(f, P)$ and $S_\alpha(f, P, T) - S_\alpha(f, P, T^*)$:

$$\begin{aligned} U_\beta(f, P) - L_\beta(f, P) &= \sum_{i=1}^{n}(M_i - m_i)\Delta\beta_i \\ &= \sum_{i=1}^{n}(M_i - m_i)\{\Delta\beta_i - |\Delta\alpha_i|\} + \sum_{i=1}^{n}(M_i - m_i)|\Delta\alpha_i| \\ &\leq 2\|f\|_\infty\varepsilon + S_\alpha(f, P, T) - S_\alpha(f, P, T^*) + \varepsilon V_a^b\alpha \\ &\leq 2\|f\|_\infty\varepsilon + \varepsilon + \varepsilon\, V_a^b\alpha. \end{aligned}$$

Thus, $f \in \mathcal{R}_\beta[a, b]$. □

Corollary 14.15. *If $\alpha \in BV[a, b]$, then $\mathcal{R}_\alpha[a, b]$ is a vector space, an algebra, and a lattice.*

Although an upper estimate on $\left|\int_a^b f\,d\alpha\right|$ is hard to come by in general, an easy estimate is available when α is of bounded variation.

Theorem 14.16. *Let $\alpha \in BV[a,b]$ and let $\beta(x) = V_a^x \alpha$. Then, for any $f \in \mathcal{R}_\alpha[a,b]$,*

$$\left| \int_a^b f \, d\alpha \right| \leq \int_a^b |f| \, d\beta \leq \|f\|_\infty V_a^b \alpha.$$

PROOF. First notice that if $f \in \mathcal{R}_\alpha$, then $f \in \mathcal{R}_\beta$ and hence $|f| \in \mathcal{R}_\beta$ (since β is increasing). So, at least both integrals in the inequality exist. Next, recall that $|\alpha(y) - \alpha(x)| \leq \beta(y) - \beta(x)$ for any $x < y$. Thus,

$$|S_\alpha(f, P, T)| \leq \sum_{i=1}^n |f(t_i)| \, |\Delta \alpha_i| \leq \sum_{i=1}^n |f(t_i)| \, \Delta \beta_i = S_\beta(|f|, P, T).$$

It now follows that

$$\left| \int_a^b f \, d\alpha \right| \leq \int_a^b |f| \, d\beta \leq \|f\|_\infty [\beta(b) - \beta(a)] = \|f\|_\infty V_a^b \alpha. \quad \square$$

Corollary 14.17. *If $\alpha \in BV[a,b]$, then $f \mapsto \int_a^b f \, d\alpha$ is a continuous, linear map on $C[a,b]$. Dually, if $f \in C[a,b]$, then $\alpha \mapsto \int_a^b f \, d\alpha$ is a continuous, linear map on $BV[a,b]$. In short, $(f, \alpha) \mapsto \int_a^b f \, d\alpha$ is a continuous bilinear form on $C[a,b] \times BV[a,b]$.*

PROOF. The linearity, in either case, is obvious. To prove continuity, then, we only need to appeal to Theorem 8.20. That is, it suffices to note that each map is Lipschitz. But, $\left| \int_a^b f \, d\alpha \right| \leq \|f\|_\infty V_a^b \alpha \leq \|f\|_\infty \|\alpha\|_{BV}$. $\square$

Theorem 14.16 is an important result, so it couldn't hurt to sketch a second proof of the inequality. Recall that if p and n are the positive and negative variations of α, then $\alpha = p - n + \alpha(a)$ and $\beta = p + n$. Now see if you can fill in the details to the following short proof:

$$\begin{aligned} \left| \int_a^b f \, d\alpha \right| &= \left| \int_a^b f \, dp - \int_a^b f \, dn \right| \\ &\leq \int_a^b |f| \, dp + \int_a^b |f| \, dn = \int_a^b |f| \, d\beta. \end{aligned}$$

Since $d\alpha = dp - dn$ while $d\beta = dp + dn$, we might consider writing $d\beta = |d\alpha|$. With this suggestive notation, our integral inequality becomes

$$\left| \int_a^b f \, d\alpha \right| \leq \int_a^b |f| \, |d\alpha|.$$

If α' exists and is Riemann integrable, then Theorem 14.6 would further suggest that $|d\alpha(t)|$ should mean $|\alpha'(t)| \, dt$. Said another way, if α' exists and is Riemann integrable, then it seems reasonable to conjecture that β' also exists and equals $|\alpha'|$. We will have more to say about this conjecture later in the chapter.

EXERCISES

▷ **36.** If $\alpha \in BV[a,b]$ and $f \in \mathcal{R}_\alpha[a,b]$, show that $f \in \mathcal{R}_\alpha[c,d]$ for every subinterval $[c,d] \subset [a,b]$.

37. Assume that f' is continuous. Use integration by parts to prove:
(a) $\sum_{k=1}^{n} f(k) = [n]f(n) - \int_1^n f'(x)\,[x]\,dx$,
(b) $\sum_{k=1}^{2n}(-1)^k f(k) = \int_1^{2n} f'(x)([x] - 2[x/2])\,dx$,
where n is an integer and $[x]$ is the greatest integer in x.

38. Let $f \in BV[0, 2\pi]$ with $f(0) = f(2\pi)$. Show that both $\int_0^{2\pi} f(x)\sin nx\,dx$ and $\int_0^{2\pi} f(x)\cos nx\,dx$ exist and each integral is at most $(1/n)V_0^{2\pi} f$. (Conclusion: A periodic function of bounded variation has a Fourier series, and the terms of the series tend to 0.)

39. Given $\alpha \in BV[a,b]$, let p and n be the positive and negative variations of α. Show that $\mathcal{R}_\alpha = \mathcal{R}_p \cap \mathcal{R}_n$ and that $\int_a^b f\,d\alpha = \int_a^b f\,dp - \int_a^b f\,dn$ for any $f \in \mathcal{R}_\alpha$.

▷ **40.** If $\alpha \in BV[a,b]$, show that $\mathcal{R}_\alpha[a,b]$ is a closed subspace of $B[a,b]$. Specifically, if (f_n) is a sequence in $\mathcal{R}_\alpha[a,b]$ that converges uniformly to f on $[a,b]$, show that $f \in \mathcal{R}_\alpha[a,b]$ and that $\int_a^b f_n\,d\alpha \to \int_a^b f\,d\alpha$.

41. Suppose that (α_n) is a sequence in $BV[a,b]$ and that $V_a^b(\alpha_n - \alpha) \to 0$. Show that $\int_a^b f\,d\alpha_n \to \int_a^b f\,d\alpha$ for all $f \in C[a,b]$.

42. Suppose that φ is a strictly increasing continuous function from $[c,d]$ onto $[a,b]$. Given $\alpha \in BV[a,b]$ and $f \in \mathcal{R}_\alpha[a,b]$, show that $\beta = \alpha \circ \varphi \in BV[c,d]$ and that $g = f \circ \varphi \in \mathcal{R}_\beta[c,d]$. Moreover, $\int_c^d g\,d\beta = \int_a^b f\,d\alpha$.

43. Given a sequence (x_n) of distinct points in (a,b) and a sequence (c_n) of real numbers with $\sum_{n=1}^{\infty} |c_n| < \infty$, define $\alpha(x) = \sum_{n=1}^{\infty} c_n I(x - x_n)$. Show that $\int_a^b f\,d\alpha = \sum_{n=1}^{\infty} c_n f(x_n)$ for every $f \in C[a,b]$. [Hint: Write $\alpha_n(x) = \sum_{k=1}^{n} c_k I(x - x_k)$ and use Exercise 41.]

44. Given a sequence (x_n) of distinct points in (a,b) and a sequence (c_n) of real numbers with $\sum_{n=1}^{\infty} |c_n| < \infty$, define α by $\alpha(x) = c_n$ if $x = x_n$ and $\alpha(x) = 0$ otherwise. Show that $\alpha \in BV[a,b]$ and that $\int_a^b f\,d\alpha = 0$ for every $f \in C[a,b]$. Compare this result with Exercise 13.

45. Given $\alpha \in BV[a,b]$, show that there is a function $\beta \in BV[a,b]$ such that β is right-continuous on (a,b) and $\int_a^b f\,d\alpha = \int_a^b f\,d\beta$ for all $f \in C[a,b]$. [Hint: Define $\beta(a) = \alpha(a)$, $\beta(x) = \alpha(x+)$ for $a < x < b$, and $\beta(b) = \alpha(b)$. See Exercise 13.16.]

46. Suppose that α is differentiable, and that α' is a bounded, Riemann integrable function on $[a,b]$. Show that $f \in \mathcal{R}_\alpha[a,b]$ if and only if $f\alpha' \in \mathcal{R}[a,b]$. In this case, $\int_a^b f\,d\alpha = \int_a^b f(x)\alpha'(x)\,dx$. [Hint: α is of bounded variation. Why?]

47. Show that $\left|\int_a^b \alpha\,df\right| \le \|\alpha\|_{BV} \|f - f(b)\|_\infty$ for $\alpha \in BV[a,b]$ and $f \in C[a,b]$. [Hint: $df = d\big(f - f(b)\big)$, where $f(b)$ is a constant function.]

48. Suppose that (α_n) is a sequence in $BV[a,b]$ and that $f \in \mathcal{R}_{\alpha_n}$ for all n. If $V_a^b(\alpha_n - \alpha) \to 0$, show that $f \in \mathcal{R}_\alpha$ and that $\int_a^b f\,d\alpha_n \to \int_a^b f\,d\alpha$. [Hints: (i). First argue that $I = \lim_{n\to\infty} \int_a^b f\,d\alpha_n$ exists. (ii). Next show that $|S_{\alpha_n}(f,P,T) - S_\alpha(f,P,T)| \to 0$ for any P, T. (iii). Finally, an $\varepsilon/3$ argument shows that $|S_\alpha(f,P,T) - I| < \varepsilon$ for some suitable P.]

49. Let $f \in C[a,b]$. Given $\varepsilon > 0$, show that there exists a $\delta > 0$ such that $\left|\int_a^b f\,d\alpha - S(f,P,T)\right| < \varepsilon V_a^b \alpha$ for all partitions $P = \{x_0,\ldots,x_n\}$ with $\max_{1\le i\le n}(x_i - x_{i-1}) < \delta$, any T, and any $\alpha \in BV[a,b]$. [Hint: First show that $\int_a^b f\,d\alpha - S(f,P,T) = \sum_{i=1}^n \int_{x_{i-1}}^{x_i} (f(x) - f(t_i))\,d\alpha(x)$.]

The Riemann Integral

Let's put aside our discussion of esoteric topics for a moment and turn our attention to two concrete problems raised at the beginning of this chapter.

- Precisely which functions are Riemann integrable? If f is Riemann integrable, must f have a point of continuity?
- If α is increasing, does α' exist at all? Even at one point?

Now these are big questions. And, although it will take us a while, we will give complete answers to both. For now, let's see how we might take advantage of such information in connection with Stieltjes integrals. In this section we will give (incomplete) answers to the following questions.

- When does a Riemann–Stieltjes integral reduce to a Riemann integral? In particular, when is it true that $\int_a^b f\,d\alpha = \int_a^b f(x)\alpha'(x)\,dx$? (The first integral is a Stieltjes integral, while the second is a Riemann integral.)
- When does the formula $\int_a^b f'(x)\,dx = f(b) - f(a)$ hold?

The answer to both of these questions is contained in our next result.

Theorem 14.18. *Suppose that α' exists and is a (bounded) Riemann integrable function on $[a,b]$. Then, given a bounded function f on $[a,b]$, we have $f \in \mathcal{R}_\alpha[a,b]$ if and only if $f\alpha' \in \mathcal{R}[a,b]$. In either case,*

$$\int_a^b f\,d\alpha = \int_a^b f(x)\alpha'(x)\,dx.$$

PROOF. We want to compare $S_\alpha(f,P,T)$ and $S_x(f\alpha',P,T)$, where S_x denotes a Riemann sum (i.e., a Riemann–Stieltjes sum with respect to the weight $\beta(x) = x$).

Let $\varepsilon > 0$. Since α' is Riemann integrable, there is a partition P^* so that $U_x(\alpha',P) - L_x(\alpha',P) < \varepsilon$ for all $P \supset P^*$. (Again, U_x and L_x denote Riemann sums.) In particular, if $T = \{t_1, \dots, t_n\}$ and $T^* = \{s_1, \dots, s_n\}$ are any selections of points with $t_i, s_i \in [x_{i-1}, x_i]$, then

$$\sum_{i=1}^{n} |\alpha'(s_i) - \alpha'(t_i)|\Delta x_i < \varepsilon. \qquad \text{(Why?)}$$

Next, the mean value theorem allows us to write

$$\begin{aligned} S_\alpha(f, P, T) &= \sum_{i=1}^{n} f(t_i)[\alpha(x_i) - \alpha(x_{i-1})] \\ &= \sum_{i=1}^{n} f(t_i)\alpha'(s_i)\Delta x_i \end{aligned}$$

for some $s_i \in (x_{i-1}, x_i)$.

Finally,

$$\begin{aligned} \left|S_\alpha(f, P, T) - S_x(f\alpha', P, T)\right| &= \left|\sum_{i=1}^{n} f(t_i)\alpha'(s_i)\Delta x_i - \sum_{i=1}^{n} f(t_i)\alpha'(t_i)\Delta x_i\right| \\ &\le \|f\|_\infty \sum_{i=1}^{n} |\alpha'(s_i) - \alpha'(t_i)|\Delta x_i < \varepsilon\|f\|_\infty \end{aligned}$$

for any T and any $P \supset P^*$. Thus, if either integral exists, then so must the other – and they are necessarily equal. □

Theorem 14.18 gives us one-half of the Fundamental Theorem of Calculus. (Just take α and f in the formula above to be f and 1, respectively.)

Corollary 14.19. *If f is differentiable, and if f' is a (bounded) Riemann integrable function on $[a, b]$, then $\int_a^b f'(x)\,dx = f(b) - f(a)$.*

For the other half of the Fundamental Theorem, we want to show that the function $F(x) = \int_a^x f$ is differentiable, and that $F' = f$. Again, it wouldn't hurt to do this in some generality.

Theorem 14.20. *Let α be increasing, and let $f \in \mathcal{R}_\alpha[a, b]$. Define $F(x) = \int_a^x f\,d\alpha$ for $a \le x \le b$. Then:*

(i) *$F \in BV[a, b]$;*

(ii) *F is continuous at each point where α is continuous;*

(iii) *F is differentiable at each point where α is differentiable and f is continuous. At any such point, $F'(x) = f(x)\alpha'(x)$.*

PROOF. First note that, for $x < y$, we have

$$|F(y) - F(x)| = \left|\int_x^y f\,d\alpha\right| \le \|f\|_\infty[\alpha(y) - \alpha(x)].$$

And now (ii) plainly follows. The proof of (i) merely requires summing such differences; hence,

$$V(F, P) \le \|f\|_\infty V(\alpha, P) = \|f\|_\infty[\alpha(b) - \alpha(a)].$$

Finally, to prove (iii), we need to fine tune the first inequality to an equality. Specifically, the mean value theorem for integrals (Exercise 11) says that $\int_x^y f\,d\alpha = c[\alpha(y) - \alpha(x)]$, for some c (depending on x and y) between $\inf_{[x,y]} f$ and $\sup_{[x,y]} f$. If we divide by $y - x$, then

$$\frac{F(y) - F(x)}{y - x} = c\,\frac{\alpha(y) - \alpha(x)}{y - x} \to f(x)\alpha'(x), \qquad \text{as } y \to x,$$

provided that f is continuous at x and $\alpha'(x)$ exists. (Why?) □

Corollary 14.21. *Let $f \in \mathcal{R}[a,b]$, and let $F(x) = \int_a^x f(t)\,dt$. Then, $F \in C[a,b] \cap BV[a,b]$, and $F'(x) = f(x)$ at each point of continuity of f.*

Corollary 14.22. *Suppose that α' exists and is Riemann integrable on $[a,b]$. If $\beta(x) = V_a^x\alpha$ for $a \le x \le b$, then β is differentiable at each point where α' is continuous. At any such point, $\beta'(x) = |\alpha'(x)|$.*

At the risk of being repetitious, let's recall the two questions that we posed at the beginning of the section: If f is Riemann integrable, does f have any points of continuity at all? If α is increasing, does α' exist at all? Food for thought!

EXERCISES

50. If f is continuous on $[a,b]$, and if $\int_a^b |f(x)|\,dx = 0$, show that $f = 0$.

51. If f is continuous on $[a,b]$, and if $\int_a^x f(t)\,dt = 0$ for all x in $[a,b]$, show that $f = 0$.

The Riesz Representation Theorem

As pointed out in Corollary 14.17, if α is of bounded variation on $[a,b]$, then the map $f \mapsto \int_a^b f\,d\alpha$ is a continuous, linear, real-valued function on $C[a,b]$. As it happens, every continuous, linear, real-valued function on $C[a,b]$ is necessarily of this same form. In much the same way that a linear, real-valued map on $\mathbb{R}^n$ is represented by inner product against some fixed vector, a linear, real-valued map on $C[a,b]$ is represented by integration against some function in $BV[a,b]$. In part, *the Riesz representation theorem* states that if $L : C[a,b] \to \mathbb{R}$ is continuous and linear, then there exists an $\alpha \in BV[a,b]$ such that

$$L(f) = \int_a^b f\,d\alpha \qquad \text{for all } f \in C[a,b].$$

Now there are many proofs of Riesz's theorem. The proof that we will give is based largely on the observation made in Exercises 7 and 30 that finite sums of the form

$$L_n(f) = \sum_{i=1}^{n} c_i f(x_i)$$

can be represented as integration against a *step* function α_n. This gives us a plan of attack: We will approximate the linear map L by a finite sum of the form L_n, which is represented by a step function α_n, and argue that (α_n) converges to a function α of bounded variation that represents L.

This particular approach has the advantage that it is in keeping with the spirit of Riemann integration. After all, if $L(f)$ is supposed to be an integral, then it ought to be a limit of integrals of step functions. The only catch here is that we look for a "global" approximation to L itself rather than a "local" approximation to a particular f.

Before we can hope to give a proof of Riesz's theorem, then, we will want to review a few facts about linear maps, and we will also need to have a few more convergence results at our disposal.

Examples 14.23

(a) If V is a normed vector space, then a linear map $L : V \to \mathbb{R}$ is continuous precisely when it is Lipschitz (Theorem 8.20), that is, if and only if there is a constant K such that $|L(x)| \le K\|x\|$ for every $x \in V$. Said another way, L is continuous if and only if

$$\|L\| = \sup_{x \neq 0} \frac{|L(x)|}{\|x\|} < \infty.$$

The number $\|L\|$ is called the **norm** of L; it is the smallest constant K that works in the inequality above. In particular, $|L(x)| \le \|L\|\,\|x\|$ for every $x \in V$.

(b) Let's clarify our claim about linear maps on $\mathbb{R}^n$. Recall that every linear map $L : \mathbb{R}^n \to \mathbb{R}$ can be written as $L(x) = \langle x, y \rangle$ for some $y \in \mathbb{R}^n$. Moreover, the representing vector y is unique. Indeed, if $\langle x, y_1 \rangle = \langle x, y_2 \rangle$ for every x, then $\langle x, y_1 - y_2 \rangle = 0$ for every x and it is easy to see that this forces $y_1 - y_2 = 0$. What's more, the Cauchy–Schwarz inequality tells us that L is continuous, $|L(x)| = |\langle x, y \rangle| \le \|y\|_2 \|x\|_2$. That is, the constant $K = \|y\|_2$ "works," so we must have $\|L\| \le \|y\|_2$. But, in fact, $\|L\| = \|y\|_2$ since we also have $\|y\|_2^2 = \langle y, y \rangle = L(y) \le \|L\|\,\|y\|_2$.

(c) If $\alpha \in BV[a, b]$, then the map defined by $L(f) = \int_a^b f\,d\alpha$ for $f \in C[a, b]$ is continuous since $|L(f)| \le V_a^b \alpha\, \|f\|_\infty$ for every $f \in C[a, b]$. That is, $\|L\| \le V_a^b \alpha$. It is possible to show that we actually have $\|L\| = V_a^b \alpha$ (see Exercise 52). Note, however, that the map L has more than one representative in $BV[a, b]$. For any constant c we have $\int_a^b f\,d\alpha = \int_a^b f\,d(\alpha + c)$, and $V_a^b(\alpha + c) = V_a^b \alpha$, too. To instill a measure of uniqueness in Riesz's theorem, then, we will want to "nail down" our representative by insisting that $\alpha(a) = 0$, for instance. This alone will not quite do the trick, but it helps. (See Exercises 52 and 53.)

EXERCISES

52. Given $\alpha \in BV[a,b]$, define $\beta(a) = \alpha(a)$, $\beta(x) = \alpha(x+)$, for $a < x < b$ and $\beta(b) = \alpha(b)$. Show that β is right-continuous on (a,b), that $\beta \in BV[a,b]$, and that $\int_a^b f\,d\alpha = \int_a^b f\,d\beta$ for every $f \in C[a,b]$.

53. Given $\alpha \in BV[a,b]$, show that there is a unique $\beta \in BV[a,b]$ with $\beta(a) = 0$ such that β is right-continuous on (a,b) and $\int_a^b f\,d\alpha = \int_a^b f\,d\beta$ for every $f \in C[a,b]$.

54. Suppose that α is right-continuous and increasing. Given $\varepsilon > 0$ and $[c,d] \subset [a,b]$, construct a continuous function f with $0 \le f \le 1$ such that $\int_a^b f\,d\alpha \ge \alpha(d) - \alpha(c) - \varepsilon$. [Hint: f should "look like" $\chi_{[c,d]}$.]

55. Let $\alpha \in BV[a,b]$ be right-continuous. Given $\varepsilon > 0$ and a partition P of $[a,b]$, construct $f \in C[a,b]$ with $\|f\|_\infty \le 1$ such that $\int_a^b f\,d\alpha \ge V(\alpha, P) - \varepsilon$. Conclude that $V_a^b \alpha = \sup\left\{ \int_a^b f\,d\alpha : \|f\|_\infty \le 1 \right\}$.

Next we focus our attention on convergence. The particular result that we need is a companion to Helly's first theorem (Theorem 13.16).

Helly's Second Theorem 14.24. *Suppose that (α_n) is a sequence in $BV[a,b]$. If $\alpha_n \to \alpha$ pointwise on $[a,b]$, and if $V_a^b(\alpha_n) \le K$ for all n, then $\alpha \in BV[a,b]$ and $\int_a^b f\,d\alpha_n \to \int_a^b f\,d\alpha$ for all $f \in C[a,b]$.*

PROOF. The fact that $\alpha \in BV[a,b]$ follows from the observation that $V(\alpha, P) = \lim_{n\to\infty} V(\alpha_n, P)$ for any partition P. Thus, $V_a^b \alpha \le K$, too. Hence, if $f \in C[a,b]$, then $f \in \mathcal{R}_\alpha[a,b]$.

Now let $f \in C[a,b]$ and $\varepsilon > 0$. Since f is uniformly continuous, we can find a $\delta > 0$ such that $|f(x) - f(y)| < \varepsilon/(3K)$ whenever $|x - y| < \delta$. Thus, if we fix a partition $P = \{x_0, \ldots, x_n\}$ with $\max_{1\le i\le n}(x_i - x_{i-1}) < \delta$, then

$$\begin{aligned}
\left| \int_a^b f\,d\alpha - S_\alpha(f,P,T) \right| &= \left| \sum_{i=1}^n \left\{ \int_{x_{i-1}}^{x_i} f\,d\alpha - f(t_i) \int_{x_{i-1}}^{x_i} d\alpha \right\} \right| \\
&= \left| \sum_{i=1}^n \int_{x_{i-1}}^{x_i} \{f(x) - f(t_i)\}\,d\alpha(x) \right| \\
&\le \sum_{i=1}^n \left| \int_{x_{i-1}}^{x_i} \{f(x) - f(t_i)\}\,d\alpha(x) \right| \\
&\le \left(\frac{\varepsilon}{3K}\right) \sum_{i=1}^n V_{x_{i-1}}^{x_i} \alpha \\
&= \left(\frac{\varepsilon}{3K}\right) V_a^b \alpha < \frac{\varepsilon}{3}.
\end{aligned}$$

What's more, this same calculation applies equally well to any α_n, and hence we have $\left| \int_a^b f\,d\alpha_n - S_{\alpha_n}(f,P,T) \right| < \varepsilon/3$, too.

Next, notice that

$$|S_\alpha(f, P, T) - S_{\alpha_n}(f, P, T)| = |S_{\alpha-\alpha_n}(f, P, T)| \le \|f\|_\infty V(\alpha - \alpha_n, P).$$

Since P is fixed, we may choose n large enough so that $\|f\|_\infty V(\alpha - \alpha_n, P) < \varepsilon/3$. Thus, $\left| \int_a^b f\,d\alpha - \int_a^b f\,d\alpha_n \right| < \varepsilon$. □

EXERCISE

56. If $f \in C[0, 1]$, show that $(1/n)\sum_{k=2}^n f(\log k/\log n) \to f(1)$ as $n \to \infty$. [Hint: Consider $\alpha_n(x) = [n^x]/n$.]

Helly's second theorem allows us to further simplify our study of integration against functions of bounded variation. Recall from Chapter Thirteen that each function α of bounded variation may be written $\alpha = \alpha_c + \alpha_s$, where α_c is continuous and α_s is a saltus or "pure jump" function. Now it is easy to see that a saltus function α_s is the pointwise limit of a sequence of step functions, say (β_n), with $V_a^b \beta_n \le V_a^b \alpha_s$. (See, for example, Exercise 14.) Thus, for any $f \in C[a, b]$,

$$\int_a^b f\,d\alpha = \int_a^b f\,d\alpha_c + \lim_{n\to\infty} \int_a^b f\,d\beta_n.$$

Integration against step functions may be directly computed; the limit in the second term would yield an infinite series (see Exercise 43). Thus, we would only have to concern ourselves with integration against a continuous function of bounded variation. As we will see, this case has much in common with the Riemann integral.

For convenience, let's consolidate Helly's first and second theorems.

Corollary 14.25. *Suppose that (β_n) is a bounded sequence in $BV[a, b]$; that is, suppose that $\|\beta_n\|_{BV} \le K$ for all n. Then, some subsequence (α_n) of (β_n) converges pointwise to a function α on $[a, b]$ with $\|\alpha\|_{BV} \le K$. Moreover, $\int_a^b f\,d\alpha_n \to \int_a^b f\,d\alpha$ for every $f \in C[a, b]$.*

With all of this machinery at our disposal, we can make short work of the proof of Riesz's theorem.

The Riesz Representation Theorem 14.26. *Given a continuous, linear map $L : C[a, b] \to \mathbb{R}$, there exists an $\alpha \in BV[a, b]$ with $V_a^b \alpha = \|L\|$ such that*

$$L(f) = \int_a^b f\,d\alpha \qquad \text{for all } f \in C[a, b].$$

Moreover, we may take α to be right-continuous on (a, b) with $\alpha(a) = 0$. In this case, α is unique.

PROOF. We will prove only the existence of α; the uniqueness claim is left as an exercise (see Exercise 53).

First note that by Lemma 11.1 and Exercise 42 it is enough to prove the theorem for $[a, b] = [0, 1]$. Indeed, if $\varphi(t) = a+t(b-a), 0 \le t \le 1$, then $\tilde{L}(g) = L(g \circ \varphi)$, where $g \in C[0, 1]$, defines a continuous linear map. If we can find some $\beta \in BV[0, 1]$ such that $\tilde{L}(g) = \int_0^1 g\, d\beta$, then, since φ is strictly increasing, it follows that $\alpha = \beta \circ \varphi$ is in $BV[a, b]$ and that $L(g \circ \varphi) = \int_a^b g \circ \varphi\, d(\beta \circ \varphi)$ for all $g \in C[0, 1]$. That is, $L(f) = \int_a^b f\, d\alpha$ for all $f \in C[a, b]$.

Our motive for translating the problem to $[0, 1]$ is essentially cosmetic: We can now take advantage of the *Bernstein polynomials* (without introducing any additional translations). Recall that if we write

$$p_{n,k}(x) = \binom{n}{k} x^k (1-x)^{n-k}, \qquad \text{for } 0 \le k \le n,$$

then $B_n(f) = \sum_{k=0}^n f\left(\frac{k}{n}\right) p_{n,k} \rightrightarrows f$ on $[0, 1]$ for any $f \in C[0, 1]$. Thus, since L is continuous and linear, we have

$$L\left(B_n(f)\right) = \sum_{k=0}^n f\left(\frac{k}{n}\right) L(p_{n,k}) \to L(f)$$

for any $f \in C[0, 1]$. And here's the key: The numbers $L(p_{n,k})$ do not depend on f!

We next construct a sequence of step functions (α_n) such that

$$\int_0^1 f\, d\alpha_n = \sum_{k=0}^n f\left(\frac{k}{n}\right) L(p_{n,k})$$

for all $f \in C[0, 1]$. This is easy; just set

$$\begin{aligned}
&\alpha_n(0) = 0, \\
&\alpha_n(x) = L(p_{n,1}), \qquad \text{for } 0 < x < \frac{1}{n}, \\
&\alpha_n(x) = L(p_{n,k}), \qquad \text{for } \frac{k}{n} \le x < \frac{k+1}{n}, \quad k = 1, \ldots, n-1, \\
&\alpha_n(1) = L(p_{n,n}).
\end{aligned}$$

Then, α_n is a step function with a jump of size $L(p_{n,k})$ at k/n, $k = 0, 1, \ldots, n$. Thus, $\int_0^1 f\, d\alpha_n = L\big(B_n(f)\big) \to L(f)$ for all $f \in C[0, 1]$. Note that α_n is right continuous on $(0, 1)$ and $\alpha_n(0) = 0$.

All that remains, in light of Helly's theorems, is to show that $V_0^1 \alpha_n$ is bounded independent of n. To this end, recall that the binomial sequence $(p_{n,k})$ satisfies $\sum_{k=0}^n p_{n,k} = \sum_{k=0}^n |p_{n,k}| = 1$ on $[0, 1]$. Thus,

$$\begin{aligned}
V_0^1 \alpha_n = \sum_{k=0}^n |L(p_{n,k})| &= \left| \sum_{k=0}^n \pm L(p_{n,k}) \right| \\
&= \left| L\left(\sum_{k=0}^n \pm p_{n,k} \right) \right|
\end{aligned}$$

$$\leq \|L\| \left\| \sum_{k=0}^{n} \pm p_{n,k} \right\|_\infty$$

$$\leq \|L\| \left\| \sum_{k=0}^{n} |p_{n,k}| \right\|_\infty = \|L\|.$$

Here's where we stand: By Helly's theorems, we may suppose that (α_n) converges pointwise to a function α on $[0, 1]$ with

$$\int_0^1 f\,d\alpha = \lim_{n\to\infty} \int_0^1 f\,d\alpha_n = L(f)$$

for all $f \in C[0, 1]$ and with $V_0^1\alpha \leq \|L\|$. Finally, since L is integration against α, it follows that we actually have $\|L\| = V_0^1\alpha$. □

Other Definitions, Other Properties

In this section we briefly discuss a variation on our definition of the Riemann–Stieltjes integral. The emphasis here is on brevity, not on exhaustive generalization. For this reason, many of the details have been relegated to the exercises.

Throughout this section, f and α will denote arbitrary, bounded, real-valued functions on $[a, b]$.

We next compare our definition of the integral, which we will call the "refinement integral," to one given in terms of the *mesh* or **norm** of a partition P, defined by $\|P\| = \max_{1\leq i\leq n} |x_i - x_{i-1}|$. The **norm integral** is defined to be

$$(N)\int_a^b f\,d\alpha = \lim_{\|P\|\to 0} S(f, P, T),$$

provided that this limit exists. That is, the norm integral $(N)\int_a^b f\,d\alpha$ exists if and only if there is a number I with the property that, for every $\varepsilon > 0$, there exists a $\delta > 0$ so that $|S(f, P, T) - I| < \varepsilon$ for any partition P with $\|P\| < \delta$ and any choice of T. Again, if such a number I exists, then it is unique, and in this case we set $(N)\int_a^b f\,d\alpha = I$.

We will not require any notation for the space of norm-integrable functions; we will use $\mathcal{R}_\alpha[a, b]$ exclusively for the space of refinement-integrable functions.

It is easy to see that the existence of the norm integral $(N)\int_a^b f\,d\alpha$ implies the existence of the refinement integral $(RS)\int_a^b f\,d\alpha$. In fact, if you will recall the proof of Theorem 14.5, we showed that continuous functions were refinement-integrable by proving the existence of the norm integral. The converse is not typically true, however. Certain differences between the two integrals are described in the exercises. For our purposes, either integral will get us where we need to go, but more on this later.

EXERCISES

57. If $(N)\int_a^b f\,d\alpha$ exists, prove that the refinement integral $(RS)\int_a^b f\,d\alpha$ also exists, and the two are equal.

58. If $(N)\int_a^b f\,d\alpha$ exists, show that it equals $\lim_{n\to\infty} S(f, P_n, T_n)$, where (P_n) is any sequence of partitions with $\|P_n\| \to 0$, and where (T_n) is arbitrary. In particular, the sequence of "regular" partitions, consisting of n equally spaced points, will do nicely.

59. If f is continuous and α is increasing, show that $(N)\int_a^b f\,d\alpha$ exists. [Hint: Recall the proof of Theorem 14.5.]

60. If α is increasing, and if $I = (N)\int_a^b f\,d\alpha$ exists, show that

$$\lim_{\|P\|\to 0} L(f, P) = \lim_{\|P\|\to 0} U(f, P) = I.$$

61. In the notation of Exercise 6, show that

(a) $(RS)\int_{-1}^1 \beta\,d\alpha$ exists.

(b) Given $\delta > 0$, there are partitions Q and R, each having norm less than δ, such that $L_\alpha(\beta, Q) = 1$ and $L_\alpha(\beta, R) = 0$. In other words, $(N)\int_{-1}^1 \beta\,d\alpha$ does not exist.

(c) $(N)\int_0^1 \beta\,d\alpha$ and $(N)\int_{-1}^0 \beta\,d\alpha$ both exist (and both are 0).

62. If f and α share a common-sided discontinuity, show that the refinement integral $(RS)\int_a^b f\,d\alpha$ does not exist.

63. If f and α share a common point of discontinuity (of any kind), show that the norm integral $(N)\int_a^b f\,d\alpha$ does not exist.

64. Assuming that $(RS)\int_a^b f\,df$ exists, compute it! Under what conditions on f will this integral exist?

65. Show that $(N)\int_a^b f\,d\alpha$ exists if and only if, for every $\varepsilon > 0$, there exists a $\delta > 0$ such that $|S(f, P_1, T_1) - S(f, P_2, T_2)| < \varepsilon$ for any pair of partitions P_1, P_2 of norm less than δ and any T_1, T_2.

66. If f is continuous and α is of bounded variation, show that $(N)\int_a^b f\,d\alpha$ exists and equals $(RS)\int_a^b f\,d\alpha$.

Since our primary applications for the Riemann–Stieltjes integral require only continuous integrands f and bounded variation integrators α, the canonical (Jordan) decomposition of α into the difference of increasing functions (each having the same points of continuity as α itself) saves the day. By Exercise 66, the two definitions of $\int_a^b f\,d\alpha$ will agree in this case. We are free to use whichever definition suits our fancy without fear of ambiguity.

Exercises 62 and 63 highlight the difference between the refinement integral and the norm integral. The refinement integral admits a slightly larger class of integrable functions, in general. If, for example, α is both continuous and increasing, then both definitions coincide; that is, either both integrals exist (and are equal) or neither exists. In particular, both approaches are equally valid for defining the Riemann integral.

Theorem 14.27. *Suppose that α is continuous and increasing, and that f is bounded. Then,*

$$\lim_{\|P\|\to 0} U(f, P) = \inf_P U(f, P) \qquad \textit{and} \qquad \lim_{\|P\|\to 0} L(f, P) = \inf_P L(f, P).$$

In particular, if $(RS)\int_a^b f\,d\alpha$ exists, then so does $(N)\int_a^b f\,d\alpha$, and the two integrals are equal.

PROOF. Set $U = \inf_P U(f, P)$. We will show that $\lim_{\|P\|\to 0} U(f, P) = U$. That is, given $\varepsilon > 0$, we will show that there is a $\delta > 0$ such that $U \le U(f, P) < U + \varepsilon$ for any partition P with $\|P\| < \delta$.

To begin, let $\varepsilon > 0$, and choose $P^* = \{x_0^*, \ldots, x_k^*\}$ such that $U(f, P^*) < U + \varepsilon/2$. Now, since α is uniformly continuous, there is a $\delta^* > 0$ such that $|\alpha(x) - \alpha(y)| < \varepsilon/[4(k+1)\|f\|_\infty]$ whenever $|x - y| < \delta^*$. Finally, choose $0 < \delta \le \delta^*$ so that $\delta \le \min_{1\le i\le k}(x_i^* - x_{i-1}^*)$. The claim is that this δ works.

Let $P = \{x_0, \ldots, x_n\}$ be any partition with $\|P\| < \delta$. Since we already have that $U(f, P \cup P^*) \le U(f, P^*) < U + \varepsilon/2$, it is enough to show that $U(f, P) \le U(f, P \cup P^*) + \varepsilon/2$, or that $U(f, P) - U(f, P \cup P^*) \le \varepsilon/2$.

Suppose that we list the elements of $P \cup P^*$ in order, say,

$$x_0 = x_0^* < x_1 < x_2 < x_3 < x_2^* < x_4 < x_5 < x_3^* < \ldots < x_{n-1} < x_n = x_k^*.$$

Now, since $\max_{1\le i\le n}(x_i - x_{i-1}) < \delta < \min_{1\le j\le k}(x_j^* - x_{j-1}^*)$, it follows that a typical interval $[\,x_{i-1}, x_i\,]$ can contain at most one x_j^*. There are at most $k+1$ such intervals. We need not worry about those intervals $[\,x_{i-1}, x_i\,]$ that do not contain an x_j^*, because then P and $P \cup P^*$ will share $[\,x_{i-1}, x_i\,]$ as a "basic" subinterval, and so the common term in both $U(f, P)$ and $U(f, P \cup P^*)$ cancels upon subtraction. So, let's estimate a typical term in $U(f, P) - U(f, P \cup P^*)$ that is associated with an interval containing some x_j^*, say, $x_j^* \in [\,x_{i-1}, x_i\,]$. Let's write M_i for the supremum of f over $[\,x_{i-1}, x_i\,]$, as usual, M^* for the supremum of f over $[\,x_{i-1}, x_j^*\,]$, and M^{**} for the supremum of f over $[\,x_j^*, x_i\,]$. Then,

$$\begin{aligned} M_i[\alpha(x_i) - \alpha(x_{i-1})] &- M^*[\alpha(x_j^*) - \alpha(x_{i-1})] - M^{**}[\alpha(x_i) - \alpha(x_j^*)] \\ &= (M_i - M^*)[\alpha(x_j^*) - \alpha(x_{i-1})] + (M_i - M^{**})[\alpha(x_i) - \alpha(x_j^*)] \\ &\le 2\|f\|_\infty [\alpha(x_i) - \alpha(x_{i-1})] \\ &< 2\|f\|_\infty \frac{\varepsilon}{4(k+1)\|f\|_\infty} = \frac{\varepsilon}{2(k+1)}. \end{aligned}$$

Since there at most $k+1$ such terms, $U(f, P) - U(f, P \cup P^*) \le \varepsilon/2$. □

This is only the tip of the integral iceberg. There are several other variations on the Riemann–Stieltjes integral; the refinement integral and the norm integral are simply the two most common definitions. What's more, there is still room to move in other directions, too. For example, we might also consider unbounded intervals or unbounded integrands (i.e., "improper" integrals). The interested reader can find a wealth of information on such generalizations in the references given in the Notes and Remarks section at the end of this chapter.

Notes and Remarks

For more on the history of the development of the integral, see the books by Hawkins [1970], Hobson [1927], Kline [1972], and Lebesgue [1928], and the articles by Hildebrandt [1917, 1938].

An easy to read and informative synopsis of Stieltjes's own point of view is supplied by the selection "Stieltjes on the Stieltjes integral," in Birkhoff [1973]. In this short passage, translated from Stieltjes [1894], we find Stieltjes's description of the problem of moments, his proofs that increasing functions have left- and right-hand limits, and his definition of the integral that bears his name. Lebesgue had a great deal to say about the Stieltjes integral, too. He devoted 61 pages of his *Leçons* to the topic, including a discussion of Riesz's theorem (Theorem 14.26) and a tribute to the genius of Cauchy, who, according to Lebesgue, had already considered the notion of integration against weight functions. Lebesgue's insights on Cauchy's work and its relationship to the physical world are reason enough to read this particular passage (see Lebesgue [1928, Chap. XI]).

The notion of using upper and lower Riemann sums was independently introduced by several mathematicians in 1875, or thereabouts. These early approaches combined the features of the so-called "refinement" integral and the "norm" integral; rather than considering the supremum of lower sums, for example, one took the limit of $L(f, P)$ as $\|P\| \to 0$. The approach that we have taken is somewhat more modern and, according to Hildebrandt [1938], is due to Moore and Smith [1922] and Kolmogorov [1930]. For those who long for the "area under the graph" approach, see Bullock [1988].

Frigyes (Frédéric, Friedrich) Riesz first proved his representation theorem (Theorem 14.26) in Riesz [1909b]. It is fair to say that Riesz's result brought the Stieltjes integral to the attention of the general mathematical public. He was clearly fond of this particular result, as he later published three more proofs, along with several other related results. Important among these is Riesz [1911], in which he adds further detail to his initial result. Eduard Helly also gave a proof in Helly [1912]. Here you will find Helly's first and second theorems (Theorems 13.16 and 14.24) together with several clever observations used to prove Riesz's theorem. It is interesting to note here that Helly refers to Riesz in regard to the "principle of choice" (Helly's selection principle), and Riesz, in turn, refers to Fréchet's thesis, Fréchet [1906]. The proofs given here of Helly's second theorem (Theorem 14.23) and of the Riesz representation theorem (Theorem 14.25) are based largely on the presentation in Natanson [1955].

Both Helly and Riesz were interested in what has been variously called the Hausdorff or Stieltjes *moment problem*. In terms of Stieltjes integrals, the problem is to determine an increasing function α, all of whose moments have been specified in advance. That is, given a sequence of positive numbers (c_k), find an increasing function α for which $\int_a^b x^k \, d\alpha(x) = c_k$, where $k = 0, 1, \ldots$. The moment problem was of pivotal importance in the development of functional analysis and function spaces in general. If we interpret each of the integrals as a finite sum, then we are led to consider a system of infinitely many linear equations in infinitely many unknowns. This approach led to the study of abstract, infinite-dimensional vector spaces. If, on the other hand, we think of the integral as a linear operation on $C[a, b]$, then the problem asks whether a linear map whose value on each polynomial has been specified in advance may be represented as Stieltjes integration against some increasing function. This point of view led to the study of linear functions, or *operators*, between abstract vector spaces. For more information on the work of Helly and Riesz, especially with regard to its influence on the development of abstract spaces and functional analysis, see Bernkopf [1966, 1967], Monna [1973], and Dieudonné [1981]. For more details on the moment problem itself, see Shohat and Tamarkin [1943].

The Stieltjes integral is of value to probabilists and statisticians (you may have already surmised this from the similarity of nomenclature – a probability density function really is a density!). But do not take my word for it; just check out Volume 1 of the *Annals of Mathematical Statistics*. You will find two papers therein concerning the Stieltjes integral: Baten [1930] and Shohat [1930].

Work on the Stieltjes integral continues in modern times, too; witness Kenneth Ross [1980a]. Ross's approach seeks a middle ground between the norm integral and the refinement integral. A more complete discussion is available in his book, Ross [1980b].

Exercise 6 is taken, in part, from Rudin [1953]. Much of the flavor of Chapter Fourteen is borrowed from the tasty presentation in Apostol [1975]; Exercises 31, 37, 38, and 56 are based on Apostol exercises. Theorem 14.26 is taken from Wheeden and Zygmund [1977], a source of still more information about Stieltjes integrals. Also see Natanson [1955, Vol. I], Johnsonbaugh and Pfaffenberger [1981], and Lojasiewicz [1988]. Exercise 47 is taken from lecture notes on a course in real analysis given by W. B. Johnson at The Ohio State University in 1974–75.

CHAPTER FIFTEEN

Fourier Series

Preliminaries

In Chapter Ten we defined the **Fourier series** associated to a 2π-periodic function f, which is (bounded and Riemann) integrable on $[-\pi, \pi]$, by

$$\frac{a_0}{2} + \sum_{k=1}^{\infty} \big(a_k \cos kx + b_k \sin kx\big),$$

where the **Fourier coefficients** a_k and b_k are given by

$$a_k = \frac{1}{\pi}\int_{-\pi}^{\pi} f(t)\,\cos kt\,dt \qquad \text{and} \qquad b_k = \frac{1}{\pi}\int_{-\pi}^{\pi} f(t)\,\sin kt\,dt.$$

Note that each of these integrals is defined and finite; in fact, a_k and b_k satisfy

$$|a_k| \le \frac{1}{\pi}\int_{-\pi}^{\pi} |f(t)|\,dt \qquad \text{and} \qquad |b_k| \le \frac{1}{\pi}\int_{-\pi}^{\pi} |f(t)|\,dt.$$

Thus, since f is bounded, we even have $|a_k| \le 2\|f\|_\infty$ and $|b_k| \le 2\|f\|_\infty$. We denote the partial sums of this series by

$$s_n(f)(x) = \frac{a_0}{2} + \sum_{k=1}^{n} \big(a_k \cos kx + b_k \sin kx\big).$$

Please note that $s_n(f)$ is a trig polynomial of degree at most n; in symbols, $s_n(f) \in \mathcal{T}_n$.

While we will be interested in whether $s_n(f)$ converges to f, we will soon see that the Fourier series for f provides a useful *representation* for f even if the series should fail to converge pointwise to f. We mirror this in our notation by writing

$$f(x) \sim \frac{a_0}{2} + \sum_{k=1}^{\infty} \big(a_k \cos kx + b_k \sin kx\big).$$

Recall from our previous discussions that the key to the Fourier series representation is the fact that the functions $1, \cos x, \sin x, \cos 2x, \sin 2x, \ldots,$ are *orthogonal* on any interval of length 2π. Specifically, taking $[-\pi, \pi]$ as our interval of choice, it is not hard to check that

$$\begin{aligned}\int_{-\pi}^{\pi} \cos mx\,\cos nx\,dx &= \int_{-\pi}^{\pi} \sin mx\,\sin nx\,dx \\ &= \int_{-\pi}^{\pi} \cos mx\,\sin nx\,dx = 0\end{aligned}$$

for any $m \neq n$ (where the last equation holds even for $m = n$),

$$\int_{-\pi}^{\pi} \cos^2 mx\,dx = \int_{-\pi}^{\pi} \sin^2 mx\,dx = \pi$$

for any $m \neq 0$, and, of course, $\int_{-\pi}^{\pi} 1\,dx = 2\pi$. (The fact that this last integral equals 2π, rather than π, explains why we write the first Fourier coefficient as $a_0/2$.)

EXERCISES

▷ **1.** Let $f : \mathbb{R} \to \mathbb{R}$ be 2π-periodic and Riemann integrable on $[-\pi, \pi]$. If f is even (resp., odd), show that its Fourier series can be written using only cosine (resp., sine) terms.

2. Define $f(x) = \pi - x$ for $0 < x < 2\pi$, $f(0) = f(2\pi) = 0$, and extend f to a 2π-periodic function on $\mathbb{R}$ (in the obvious way). Show that the Fourier series for f is $2\sum_{n=1}^{\infty} \sin nx/n$.

3. Let $f \in BV[-\pi, \pi]$ with $f(-\pi) = f(\pi)$. Show that both $(1/\pi)\int_{-\pi}^{\pi} f(x) \sin nx\,dx$ and $(1/\pi)\int_{-\pi}^{\pi} f(x)\cos nx\,dx$ exist, and that each is at most $(1/n)V_{-\pi}^{\pi} f$.

The study of the pointwise convergence of Fourier series has a long and checkered history – to paraphrase Halmos, its history includes "almost 200 years of barking up the wrong tree." In all of its glory, pointwise convergence is a delicate and complex issue, arguably too complex to warrant thorough pursuit here. For this reason, we will be primarily concerned with the wealth of useful information that is already at hand. This "easy" approach will nevertheless provide some deep results of its own. Just watch!

Observations 15.1.

(a) If $T(x) = (\alpha_0/2) + \sum_{k=1}^{n} \big(\alpha_k \cos kx + \beta_k \sin kx\big)$ is a trig polynomial of degree n and if $m = 1, \ldots, n$, then

$$\int_{-\pi}^{\pi} T(x)\cos mx\,dx = \alpha_m \int_{-\pi}^{\pi} \cos^2 mx\,dx = \pi\alpha_m,$$

while if $m = 0$, then

$$\int_{-\pi}^{\pi} T(x)\,dx = \frac{\alpha_0}{2}\int_{-\pi}^{\pi} 1\,dx = \pi\alpha_0.$$

Similarly, for $m = 1, 2, \ldots, n$,

$$\int_{-\pi}^{\pi} T(x)\sin mx\,dx = \beta_m \int_{-\pi}^{\pi} \sin^2 mx\,dx = \pi\beta_m.$$

If $m > n$, then each of these integrals is 0. Thus, if $T \in \mathcal{T}_n$, then T is actually equal to its own Fourier series. Said another way: Given $T \in \mathcal{T}_n$, we have $s_m(T) = T$ whenever $m \geq n$.

(b) If f (and hence also f^2) is Riemann integrable on $[-\pi, \pi]$, then $s_n(f)$ minimizes the integral

$$\int_{-\pi}^{\pi} [f(x) - T(x)]^2\, dx$$

over all choices of trig polynomials T of degree at most n. To see this, let $T(x) = (\alpha_0/2) + \sum_{k=1}^{n} (\alpha_k \cos kx + \beta_k \sin kx)$ and first note that

$$\int_{-\pi}^{\pi} [f - T]^2 = \int_{-\pi}^{\pi} f^2 - 2\int_{-\pi}^{\pi} f\,T + \int_{-\pi}^{\pi} T^2.$$

By using the linearity of the integral and the orthogonality of the trig system, we can write each of the last two integrals in terms of the Fourier coefficients of f and T. Indeed, from (a),

$$\begin{aligned}\int_{-\pi}^{\pi} f(x)\,T(x)\,dx &= \frac{\alpha_0}{2}\int_{-\pi}^{\pi} f(x)\,dx + \sum_{k=1}^{n} \alpha_k \int_{-\pi}^{\pi} f(x)\cos kx\,dx \\ &\quad + \sum_{k=1}^{n} \beta_k \int_{-\pi}^{\pi} f(x)\sin kx\,dx \\ &= \pi\left[\frac{\alpha_0 a_0}{2} + \sum_{k=1}^{n} (\alpha_k a_k + \beta_k b_k)\right]\end{aligned}$$

and (after replacing f by T in the previous calculation)

$$\int_{-\pi}^{\pi} T(x)^2\,dx = \pi\left[\frac{\alpha_0^2}{2} + \sum_{k=1}^{n} (\alpha_k^2 + \beta_k^2)\right].$$

Now, since $\alpha_k^2 - 2\alpha_k a_k = (\alpha_k - a_k)^2 - a_k^2$, we get

$$\begin{aligned}\frac{1}{\pi}\int_{-\pi}^{\pi} [f(x) - T(x)]^2\,dx &= \frac{1}{\pi}\int_{-\pi}^{\pi} f(x)^2\,dx - \frac{a_0^2}{2} - \sum_{k=1}^{n} (a_k^2 + b_k^2) \\ &\quad + \frac{(\alpha_0 - a_0)^2}{2} + \sum_{k=1}^{n} \left((\alpha_k - a_k)^2 + (\beta_k - b_k)^2\right).\end{aligned}$$

The right-hand side is minimized precisely when $\alpha_k = a_k$ and $\beta_k = b_k$ for all k, in other words, precisely when $T = s_n(f)$. Please note that in this case we have

$$\begin{aligned}\frac{1}{\pi}\int_{-\pi}^{\pi} [f - s_n(f)]^2 &= \frac{1}{\pi}\int_{-\pi}^{\pi} f(x)^2\,dx - \frac{a_0^2}{2} - \sum_{k=1}^{n} (a_k^2 + b_k^2) \\ &= \frac{1}{\pi}\int_{-\pi}^{\pi} f(x)^2\,dx - \frac{1}{\pi}\int_{-\pi}^{\pi} s_n(f)(x)^2\,dx.\end{aligned}$$

(c) The calculation in (b) leads us to consider the L_2-norm, defined by

$$\|f\|_2 = \left(\frac{1}{\pi}\int_{-\pi}^{\pi} f(x)^2\,dx\right)^{1/2} \qquad (15.1)$$

where we assume here that f is Riemann integrable. The proof that this expression defines a (semi-)norm is essentially identical to the proof that we gave in Chapter Three for the ℓ_2-norm (Lemma 3.3 and Theorem 3.4); we will save the details for a later section (where we will prove an even more general result). Please note that if $f \in C^{2\pi}$, then $\|f\|_2 \leq \sqrt{2}\,\|f\|_\infty$.

Of greatest importance to us is the fact that we have a "continuous" analogue of the familiar "dot product" (or inner product; see the discussion preceding Lemma 3.3). In particular, if f and g are Riemann integrable, then the map

$$(f, g) \mapsto \langle f, g \rangle = \frac{1}{\pi}\int_{-\pi}^{\pi} f(x)\, g(x)\, dx$$

satisfies all of the familiar properties of the dot product in $\mathbb{R}^n$. Specifically, the map is linear in each of its arguments, satisfies the Cauchy–Schwarz inequality (see Theorem 14.7 (v)):

$$\left|\frac{1}{\pi}\int_{-\pi}^{\pi} f(x)\, g(x)\, dx\right| \leq \left(\frac{1}{\pi}\int_{-\pi}^{\pi} f(x)^2\, dx\right)^{1/2} \left(\frac{1}{\pi}\int_{-\pi}^{\pi} g(x)^2\, dx\right)^{1/2},$$

and is related to the L_2-norm by $\|f\|_2 = \sqrt{\langle f, f \rangle}$.

We can now clarify the claim made in (a): The functions $1, \cos x, \cos 2x, \ldots, \sin x, \sin 2x, \ldots,$ are *orthogonal* in the sense that any two distinct functions from the list have zero "dot product." Moreover, the functions $1/\sqrt{2}, \cos x, \cos 2x, \ldots, \sin x, \sin 2x, \ldots,$ are actually *orthonormal*; that is, they are mutually orthogonal and each has L_2-norm one (thanks to the extra factor $1/\pi$ in equation (15.1)).

(d) Observation (b) can now be rephrased: The partial sum $s_n(f)$ is the nearest point to f out of $\mathcal{T}_n$ relative to the L_2-norm. In other words,

$$\inf_{T \in \mathcal{T}_n} \|f - T\|_2 = \|f - s_n(f)\|_2.$$

Moreover,

$$\begin{aligned} \|f - s_n(f)\|_2^2 &= \frac{1}{\pi}\int_{-\pi}^{\pi} f(x)^2\, dx - \frac{a_0^2}{2} - \sum_{k=1}^{n} \left(a_k^2 + b_k^2\right) \\ &= \|f\|_2^2 - \|s_n(f)\|_2^2. \end{aligned} \tag{15.2}$$

Since $\|f - s_n(f)\|_2^2 \geq 0$, we have

$$\begin{aligned} \|s_n(f)\|_2^2 &= \frac{1}{\pi}\int_{-\pi}^{\pi} s_n(f)(x)^2\, dx \\ &= \frac{a_0^2}{2} + \sum_{k=1}^{n} \left(a_k^2 + b_k^2\right) \\ &\leq \frac{1}{\pi}\int_{-\pi}^{\pi} f(x)^2\, dx = \|f\|_2^2. \end{aligned}$$

In other words, $\|s_n(f)\|_2 \leq \|f\|_2$. This result is known as *Bessel's inequality*. Since n is arbitrary, it follows that the Fourier coefficients of a Riemann square-integrable function f are square-summable and satisfy

$$\frac{a_0^2}{2} + \sum_{k=1}^{\infty} \left(a_k^2 + b_k^2\right) \le \frac{1}{\pi} \int_{-\pi}^{\pi} f(x)^2 \, dx. \tag{15.3}$$

In particular, the Fourier coefficients of f must tend to zero:

$$\lim_{n\to\infty} \int_{-\pi}^{\pi} f(x) \cos nx \, dx = 0 = \lim_{n\to\infty} \int_{-\pi}^{\pi} f(x) \sin nx \, dx. \tag{15.4}$$

This fact is known as *Riemann's lemma* and will prove very useful in subsequent observations.

(e) For $f \in C^{2\pi}$ we have $\|f - s_n(f)\|_2 \to 0$; that is, f is the limit of its Fourier series in the L_2-norm. Indeed, given $\varepsilon > 0$, Weierstrass's second theorem (Theorem 11.8) supplies a trig polynomial T^*, of some finite degree m, with $\|f - T^*\|_\infty < \varepsilon$. Thus, for all $n \ge m$,

$$\|f - s_n(f)\|_2 = \inf_{T\in\mathcal{T}_n} \|f - T\|_2 \le \sqrt{2} \inf_{T\in\mathcal{T}_n} \|f - T\|_\infty < \varepsilon\sqrt{2},$$

since $T^* \in \mathcal{T}_m \subset \mathcal{T}_n$.

(f) If f and g are Riemann integrable on $[-\pi, \pi\,]$, then $s_n(f+g) = s_n(f) + s_n(g)$ for every n. In fact, each Fourier coefficient of the sum $f + g$ is the sum of the corresponding Fourier coefficients for f and g; for example,

$$\int_{-\pi}^{\pi} \big[f(x) + g(x)\big] \cos kx \, dx = \int_{-\pi}^{\pi} f(x) \cos kx \, dx + \int_{-\pi}^{\pi} g(x) \cos kx \, dx.$$

Essentially the same reasoning shows that $s_n(\alpha f + \beta g) = \alpha\, s_n(f) + \beta\, s_n(g)$ for any pair of real numbers α and β. In other words, the map $f \mapsto s_n(f)$ is linear.

(g) This linearity of s_n allows us to extend the result in observation (e): If f is Riemann integrable on $[-\pi, \pi\,]$, then $\|f - s_n(f)\|_2 \to 0$. It is in this sense that we justify the claim that f is *represented by* its Fourier series. To see this, let $\varepsilon > 0$ and choose a function $g \in C^{2\pi}$ satisfying

$$\|f - g\|_2 = \left(\frac{1}{\pi} \int_{-\pi}^{\pi} \big[f(x) - g(x)\big]^2 \, dx\right)^{1/2} < \varepsilon.$$

(How? See Exercise 5.) Next, since s_n is linear, we have

$$\|f - s_n(f)\|_2 \le \|f - g\|_2 + \|g - s_n(g)\|_2 + \|s_n(f - g)\|_2.$$

But, from Bessel's inequality, $\|s_n(f - g)\|_2 \le \|f - g\|_2 < \varepsilon$ and so

$$\|f - s_n(f)\|_2 \le 2\varepsilon + \|g - s_n(g)\|_2 < 3\varepsilon$$

for all n sufficiently large, from observation (e).

(h) Combining results (d)–(g) we arrive at *Parseval's equation*: If f is Riemann integrable on $[-\pi, \pi\,]$, then $\|f\|_2^2 = \lim_{n\to\infty} \|s_n(f)\|_2^2$; that is,

$$\frac{1}{\pi} \int_{-\pi}^{\pi} f(x)^2 \, dx = \frac{a_0^2}{2} + \sum_{k=1}^{\infty} \left(a_k^2 + b_k^2\right). \tag{15.5}$$

In other words, in light of equation (15.2), Parseval's equation is equivalent to the statement that $\|f - s_n(f)\|_2 \to 0$.

(i) It is immediate from Parseval's equation that distinct elements from $C^{2\pi}$ have different Fourier series. In other words, if $f, g \in C^{2\pi}$ satisfy $\int_{-\pi}^{\pi} [f(x) - g(x)] \cos nx\, dx = 0$ and $\int_{-\pi}^{\pi} [f(x) - g(x)] \sin nx\, dx = 0$ for all $n = 0, 1, 2, \ldots$, then we would also have $\int_{-\pi}^{\pi} [f(x) - g(x)]^2\, dx = 0$. But, since f and g are continuous, this easily implies that $f - g = 0$. (How?) Compare this approach to that used in Exercise 11.31.

(j) Here is an easy consequence of our discussion of uniform convergence in Chapter Ten: If the Fourier series of a function $f \in C^{2\pi}$ is uniformly convergent, then the series must actually converge to f. Of course, if a trigonometric series is uniformly convergent, then its sum defines a continuous function; let's call it $g \in C^{2\pi}$ in this case. All that remains is to notice that g has the same Fourier coefficients as f, and this is easy: If $s_n(f)$ converges uniformly to g, then $s_n(f)(x) \cos kx$ converges uniformly to $g(x) \cos kx$, for example, and so (interchanging limit and integral and using (a)) we have

$$\frac{1}{\pi}\int_{-\pi}^{\pi} g(x) \cos kx\, dx = \lim_{n\to\infty} \frac{1}{\pi}\int_{-\pi}^{\pi} s_n(f)(x) \cos kx\, dx = a_k.$$

Similarly, $(1/\pi)\int_{-\pi}^{\pi} g(x) \sin kx\, dx = b_k$. According to our last observation, this means that $f = g$.

(k) If the Fourier coefficients for f satisfy $\sum_n |a_n| < \infty$ and $\sum_n |b_n| < \infty$, then (as an easy consequence of the M-test, Lemma 10.9) the Fourier series for f is uniformly convergent on $\mathbb{R}$. Thus, if we are also given that $f \in C^{2\pi}$, it follows from (j) that the Fourier series for f converges uniformly to f.

The introduction of the L_2 norm is designed to make clear the sense in which a continuous function f is "represented by" its Fourier series: While f need not be the pointwise limit of its Fourier series (indeed the series may even diverge at certain points), f is nevertheless the limit of its Fourier series in some metric – and limits in metric spaces are unique. (See Exercise 4 for more on this.) Consequently, each $f \in C^{2\pi}$ is uniquely determined by its Fourier series.

EXERCISES

▷ **4.** If f is Riemann integrable on $[-\pi, \pi]$ and $\|f\|_2 = 0$, does it follow that $f = 0$? It is true if we assume, in addition, that f is continuous. Why? In other words, assuming the validity of the triangle inequality, verify that the L_2-norm is truly a norm on $C[-\pi, \pi]$.

5. Let f be Riemann integrable on $[-\pi, \pi]$, and let $\varepsilon > 0$.

(a) Show that there is a continuous function g on $[-\pi, \pi]$ satisfying $\|f - g\|_2 < \varepsilon$. [Hint: Mimic the proof of Theorem 14.9.]

(b) Show that there is a continuous, 2π-periodic function $h \in C^{2\pi}$ satisfying $\|f - h\|_2 < \varepsilon$.

(c) Show that there is a trig polynomial T with $\|f - T\|_2 < \varepsilon$.

6. Let $f : \mathbb{R} \to \mathbb{R}$ be 2π-periodic and Riemann integrable on $[-\pi, \pi]$. Prove that $\lim_{x\to 0} \int_{-\pi}^{\pi} |f(x+t) - f(t)|^2\, dt = 0$.

7. Define $f(x) = (\pi - x)^2$ for $0 \le x \le 2\pi$, and extend f to a 2π-periodic continuous function on $\mathbb{R}$ in the obvious way. Show that the Fourier series for f is $\pi^2/3 + 4\sum_{n=1}^{\infty} \cos nx/n^2$. Since the series is uniformly convergent, it actually converges to f. In particular, note that setting $x = 0$ yields the familiar formula $\sum_{n=1}^{\infty} 1/n^2 = \pi^2/6$.

Dirichlet's Formula

To better understand the pointwise convergence of Fourier series, it would be helpful to have a closed expression for $s_n(f)$ (that is, an expression not involving a sum). For this we will need a couple of trig identities; the first two need no explanation:

$$\cos kt \cos kx + \sin kt \sin kx = \cos k(t - x)$$

$$2\cos\alpha \sin\beta = \sin(\alpha + \beta) - \sin(\alpha - \beta)$$

$$\frac{1}{2} + \cos\theta + \cos 2\theta + \cdots + \cos n\theta = \frac{\sin\left(n + \frac{1}{2}\right)\theta}{2\sin\frac{1}{2}\theta}.$$

Here is a short proof for the third:

$$\begin{aligned}\sin\tfrac{1}{2}\theta + \sum_{k=1}^{n} 2\cos k\theta \,\sin\tfrac{1}{2}\theta &= \sin\tfrac{1}{2}\theta + \sum_{k=1}^{n}\left[\,\sin\left(k + \tfrac{1}{2}\right)\theta - \sin\left(k - \tfrac{1}{2}\right)\theta\,\right] \\ &= \sin\left(n + \tfrac{1}{2}\right)\theta.\end{aligned}$$

Now we are ready to rewrite our formula for $s_n(f)$:

$$\begin{aligned} s_n(f)(x) &= \frac{a_0}{2} + \sum_{k=1}^{n}(a_k \cos kx + b_k \sin kx) \\ &= \frac{1}{\pi}\int_{-\pi}^{\pi} f(t)\left[\frac{1}{2} + \sum_{k=1}^{n}(\cos kt\cos kx + \sin kt \sin kx)\right] dt \\ &= \frac{1}{\pi}\int_{-\pi}^{\pi} f(t)\left[\frac{1}{2} + \sum_{k=1}^{n}\cos k(t - x)\right] dt \\ &= \frac{1}{\pi}\int_{-\pi}^{\pi} f(t)\cdot\frac{\sin\left(n + \frac{1}{2}\right)(t - x)}{2\sin\frac{1}{2}(t - x)}\, dt. \end{aligned}$$

The function

$$D_n(t) = \frac{1}{2} + \sum_{k=1}^{n}\cos kt = \frac{\sin\left(n + \frac{1}{2}\right)t}{2\sin\frac{1}{2}t} \tag{15.6}$$

is called **Dirichlet's kernel**; note that $D_n \in \mathcal{T}_n$. In this notation, our formula for $s_n(f)$ reads

$$s_n(f)(x) = \frac{1}{\pi} \int_{-\pi}^{\pi} f(t)\, D_n(t - x)\, dt.$$

If f is 2π-periodic, then we may also write

$$s_n(f)(x) = \frac{1}{\pi} \int_{-\pi}^{\pi} f(x + t)\, D_n(t)\, dt.$$

While we know that $s_n(f)$ is a good approximation to f in the L_2-norm, a better understanding of its effectiveness as a uniform approximation will require a better understanding of the Dirichlet kernel D_n. Figure 15.1 displays the graph of D_n for $n = 30$, while the following are a few important observations about D_n and its integrals.

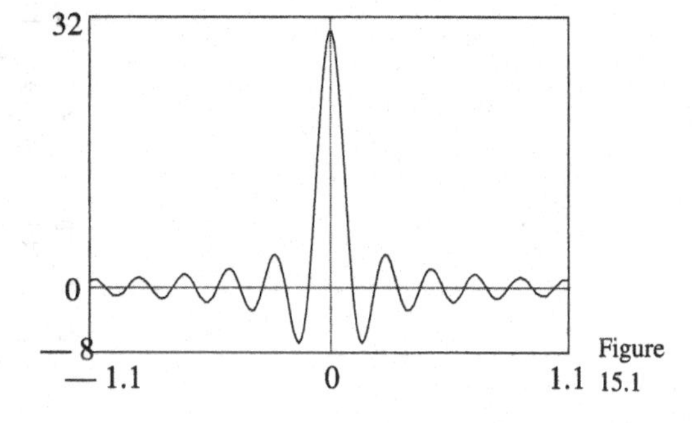

Figure 15.1

Lemma 15.2.

(a) D_n *is even,*

(b) $(1/\pi) \displaystyle\int_{-\pi}^{\pi} D_n(t)\, dt = (2/\pi) \int_0^{\pi} D_n(t)\, dt = 1,$

(c) $|D_n(t)| \le n + \frac{1}{2}$ *and* $D_n(0) = n + \frac{1}{2}$,

(d) $\left(\left|\sin(n + \frac{1}{2})\, t\,\right| / t\right) \le |D_n(t)| \le (\pi/2t)$ *for* $0 < t < \pi$,

(e) *If* $\lambda_n = (1/\pi) \displaystyle\int_{-\pi}^{\pi} |D_n(t)|\, dt$, *then* $(4/\pi^2) \log n \le \lambda_n \le 3 + \log n$.

PROOF. (a), (b), and (c) are relatively clear from the fact that

$$D_n(t) = \tfrac{1}{2} + \cos t + \cos 2t + \cdots + \cos nt.$$

(Notice, too, that (b) follows from the fact that $s_n(1) = 1$.) For (d) we use a more delicate estimate: Since $2\theta/\pi \le \sin\theta \le \theta$ for $0 < \theta < \pi/2$, it follows that $2t/\pi \le 2\sin(t/2) \le t$ for $0 < t < \pi$. Hence,

$$\frac{\pi}{2t} \ge \frac{\left|\sin(n + \frac{1}{2})\, t\,\right|}{2 \sin \frac{1}{2} t} \ge \frac{\left|\sin(n + \frac{1}{2})\, t\,\right|}{t}$$

for $0 < t < \pi$. Next, the upper estimate in (e) is easy:

$$\begin{aligned}\frac{2}{\pi}\int_0^{\pi} |D_n(t)|\,dt &= \frac{2}{\pi}\int_0^{\pi} \frac{\left|\sin\left(n+\frac{1}{2}\right)t\right|}{2\sin\frac{1}{2}t}\,dt \\ &\leq \frac{2}{\pi}\int_0^{1/n}\left(n+\frac{1}{2}\right)dt + \frac{2}{\pi}\int_{1/n}^{\pi}\frac{\pi}{2t}\,dt \\ &= \frac{2n+1}{\pi n} + \log\pi + \log n < 3 + \log n.\end{aligned}$$

The lower estimate takes more work:

$$\begin{aligned}\frac{2}{\pi}\int_0^{\pi} |D_n(t)|\,dt &= \frac{2}{\pi}\int_0^{\pi} \frac{\left|\sin\left(n+\frac{1}{2}\right)t\right|}{2\sin\frac{1}{2}t}\,dt \\ &\geq \frac{2}{\pi}\int_0^{\pi} \frac{\left|\sin\left(n+\frac{1}{2}\right)t\right|}{t}\,dt \\ &= \frac{2}{\pi}\int_0^{(n+(1/2))\pi} \frac{|\sin x|}{x}\,dx \\ &\geq \frac{2}{\pi}\int_0^{n\pi} \frac{|\sin x|}{x}\,dx \\ &= \frac{2}{\pi}\sum_{k=1}^{n}\int_{(k-1)\pi}^{k\pi} \frac{|\sin x|}{x}\,dx \\ &\geq \frac{2}{\pi}\sum_{k=1}^{n}\frac{1}{k\pi}\int_{(k-1)\pi}^{k\pi} |\sin x|\,dx \\ &= \frac{4}{\pi^2}\sum_{k=1}^{n}\frac{1}{k} \geq \frac{4}{\pi^2}\log n,\end{aligned}$$

because $\sum_{k=1}^{n}(1/k) \geq \log n$. □

The numbers $\lambda_n = (1/\pi)\int_{-\pi}^{\pi}|D_n(t)|\,dt$ are called the *Lebesgue numbers* and serve the following purpose:

Corollary 15.3. *If $f \in C^{2\pi}$, then*

$$|s_n(f)(x)| \leq \frac{1}{\pi}\int_{-\pi}^{\pi} |f(x+t)|\,|D_n(t)|\,dt \leq \lambda_n \|f\|_\infty. \tag{15.7}$$

In particular, $\|s_n(f)\|_\infty \leq \lambda_n\|f\|_\infty \leq (3+\log n)\|f\|_\infty$.

If we approximate the function $\operatorname{sgn} D_n$ by a continuous function f of norm one, then

$$s_n(f)(0) \approx \frac{1}{\pi}\int_{-\pi}^{\pi} |D_n(t)|\,dt = \lambda_n.$$

Thus, λ_n is the smallest constant that works in equation (15.7); see Exercise 8. The fact that $s_n(f)$ may have a very large sup-norm compared to f means, in particular, that

$s_n(f)$ is typically a poor approximation to f in the uniform norm. In sharp contrast, recall that in the L_2-norm we have $\|s_n(f)\|_2 \le \|f\|_2$. Of course, $s_n(f)$ is a very good approximation to f in the L_2-norm.

Now that we have Dirichlet's formula at our disposal, however, it is not difficult to find conditions under which $s_n(f)$ will converge uniformly to f.

Theorem 15.4. *Let f be a continuous function on $[-\pi, \pi]$ with $f(-\pi) = f(\pi)$ and suppose that f has a bounded, piecewise continuous derivative on $[-\pi, \pi]$. Then, the Fourier series for f converges uniformly to f on $[-\pi, \pi]$.*

PROOF. Since f' is piecewise continuous, we may use integration by parts to compare its Fourier coefficients, called a'_n and b'_n here, with those of f, which we will call a_n and b_n. Notice, for example, that

$$\begin{aligned} a'_n &= \frac{1}{\pi}\int_{-\pi}^{\pi} f'(x)\cos nx\,dx \\ &= -\frac{1}{\pi}\int_{-\pi}^{\pi} f(x)\,d(\cos nx) + [f(\pi)\cos n\pi - f(-\pi)\cos(-n\pi)] \\ &= \frac{n}{\pi}\int_{-\pi}^{\pi} f(x)\sin nx\,dx = nb_n \end{aligned}$$

(for $n \ge 1$). Similarly,

$$b'_n = \frac{1}{\pi}\int_{-\pi}^{\pi} f'(x)\sin nx\,dx = -\frac{n}{\pi}\int_{-\pi}^{\pi} f(x)\cos nx\,dx = -na_n.$$

Since the Fourier coefficients of f' are square-summable, we conclude that

$$\sum_{n=1}^{\infty} n^2 a_n^2 < \infty \quad\text{and}\quad \sum_{n=1}^{\infty} n^2 b_n^2 < \infty.$$

But now a simple application of the Cauchy–Schwarz inequality tells us that the Fourier coefficients of f must, in fact, be absolutely summable:

$$\sum_{n=1}^{\infty} |a_n| = \sum_{n=1}^{\infty} n|a_n| \cdot \frac{1}{n} \le \left(\sum_{n=1}^{\infty} n^2 a_n^2\right)^{1/2} \left(\sum_{n=1}^{\infty} \frac{1}{n^2}\right)^{1/2} < \infty.$$

Similarly, $\sum_{n=1}^{\infty} |b_n| < \infty$. An application of the Weierstrass M-test now shows that the Fourier series for f is uniformly convergent and hence must actually converge to f. □

Note, for example, that Theorem 15.4 holds for polygonal functions, or even for "piecewise polynomial" functions in $C^{2\pi}$, and these collections clearly form dense subsets of $C^{2\pi}$. But while Theorem 15.4 supplies a large class of functions for which $s_n(f)$ converges to f, there are examples available of continuous functions whose Fourier series fail to converge (in fact, we can even arrange for divergence on a dense set of points). In other words, $s_n(f)$ is typically not a good pointwise approximation to f, let alone a good uniform approximation. To approximate a continuous function f uniformly by trig polynomials, then, we will need to look for something better than the sequence $s_n(f)$. Said another way: We will need to replace D_n by a better kernel. And this is exactly what we will do.

EXERCISES

8. Fix $n \geq 1$ and $\varepsilon > 0$.

(a) Show that there is a continuous function $f \in C^{2\pi}$ satisfying $\|f\|_\infty = 1$ and $(1/\pi) \int_{-\pi}^{\pi} |f(t) - \operatorname{sgn} D_n(t)|\, dt < \varepsilon/(n+1)$.

(b) Show that $s_n(f)(0) \geq \lambda_n - \varepsilon$ and, hence, that $\|s_n(f)\|_\infty \geq \lambda_n - \varepsilon$.

Fejér's Theorem

To motivate our next result, we begin with a simple fact about numerical sequences. We suppose that we are given a sequence of real numbers (s_n) and we consider the sequence formed by their arithmetic means (or Cesàro sums)

$$\sigma_n = \frac{s_1 + s_2 + \cdots + s_n}{n}.$$

The claim here is that the sequence (σ_n) has better convergence properties than the original sequence (s_n).

Lemma 15.5. *If $s_n \to s$, then $\sigma_n \to s$.*

PROOF. If (s_n) is convergent, then it is also bounded; let's say that $|s_n| \leq B$ for all n. Next, given $\varepsilon > 0$, choose n such that $|s_k - s| < \varepsilon$ for all $k > n$. Fixing this n, now consider

$$\sigma_N = \frac{s_1 + s_2 + \cdots + s_N}{N} = \frac{s_1 + \cdots + s_n}{N} + \frac{s_{n+1} + \cdots + s_N}{N}.$$

Clearly, for $N > n$,

$$-\left(\frac{n}{N}\right) B + \left(\frac{N-n}{N}\right)(s - \varepsilon) \leq \sigma_n \leq \left(\frac{n}{N}\right) B + \left(\frac{N-n}{N}\right)(s + \varepsilon),$$

and hence

$$s - 2\varepsilon < \sigma_N < s + 2\varepsilon$$

for all N sufficiently large. □

The point to Lemma 15.5 is that averaging preserves convergence. In fact, it often enhances convergence: In the case of the nonconvergent sequence $s_n = (-1)^n$, it is not hard to check that the corresponding sequence (σ_n) converges to 0. In short, averaging cannot hurt and occasionally helps when considering nonconvergent sequences.

Now, since the sequence of partial sums $(s_n(f))$ of a Fourier series need not converge to f, we might try looking at the sequence of arithmetic means $(\sigma_n(f))$ defined by

$$\begin{aligned}\sigma_n(f)(x) &= \frac{1}{n}\big(s_0(f) + \cdots + s_{n-1}(f)\big)(x)\\ &= \frac{1}{\pi}\int_{-\pi}^{\pi} f(x+t)\left[\frac{1}{n}\sum_{k=0}^{n-1} D_k(t)\right]dt\\ &= \frac{1}{\pi}\int_{-\pi}^{\pi} f(x+t)\,K_n(t)\,dt,\end{aligned}$$

where $K_n = (1/n)(D_0 + D_1 + \cdots + D_{n-1})$ is called **Fejér's kernel**. The same techniques that we used earlier can be applied to find a closed form for $\sigma_n(f)$, which, of course, reduces to simplifying $(1/n)(D_0 + D_1 + \cdots + D_{n-1})$. As before, we begin with a trig identity:

$$\begin{aligned}2\sin\theta\sum_{k=0}^{n-1}\sin(2k+1)\theta &= \sum_{k=0}^{n-1}\big[\cos 2k\theta - \cos(2k+2)\theta\big]\\ &= 1 - \cos 2n\theta \;=\; 2\sin^2 n\theta.\end{aligned}$$

Thus,

$$K_n(t) = \frac{1}{n}\sum_{k=0}^{n-1}\frac{\sin(2k+1)t/2}{2\sin(t/2)} = \frac{\sin^2(nt/2)}{2n\sin^2(t/2)}. \tag{15.8}$$

Please note that K_n is an even, *nonnegative* trig polynomial of degree at most $n-1$ and satisfies $(1/\pi)\int_{-\pi}^{\pi} K_n(t)\,dt = 1$. (Why?) Figure 15.2 displays the graph of K_n for $n = 20$.

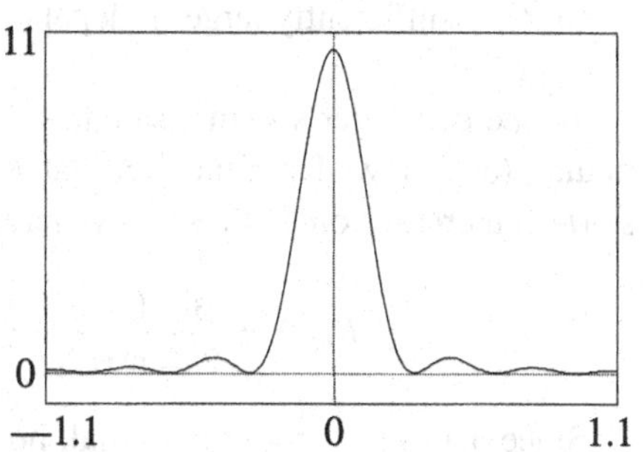

Figure 15.2

Now $\sigma_n(f)$ is still a good approximation to f in the L_2-norm. Indeed, from Lemma 15.5 we have

$$\|f - \sigma_n(f)\|_2 = \frac{1}{n}\left\|\sum_{k=0}^{n-1}(f - s_k(f))\right\|_2 \le \frac{1}{n}\sum_{k=0}^{n-1}\|f - s_k(f)\|_2 \to 0$$

as $n \to \infty$ (since $\|f - s_k(f)\|_2 \to 0$). But, more to the point, $\sigma_n(f)$ is actually a good *uniform* approximation to f, a fact that we will call *Fejér's theorem.*

Fejér's Theorem 15.6. *If $f \in C^{2\pi}$, then $\sigma_n(f)$ converges uniformly to f as $n \to \infty$.*

Now Fejér's theorem is but a single typical example of a larger class of convergence theorems. This point can be made most clear by isolating the key ingredients in its proof as a self-contained statement about certain "kernel operators."

Theorem 15.7. *Suppose that a sequence (k_n) in $C^{2\pi}$ satisfies*

(a) $k_n \geq 0$,

(b) $(1/\pi)\int_{-\pi}^{\pi} k_n(t)\,dt = 1$, *and*

(c) $\int_{\delta\leq|t|\leq\pi} k_n(t)\,dt \to 0$, *as* $n \to \infty$, *for every* $\delta > 0$.

Then, $(1/\pi)\int_{-\pi}^{\pi} f(x+t)\,k_n(t)\,dt \rightrightarrows f(x)$, *as* $n \to \infty$, *for each* $f \in C^{2\pi}$.

PROOF. Let $\varepsilon > 0$. Since f is uniformly continuous, we may choose $\delta > 0$ so that $|f(x) - f(x+t)| < \varepsilon$, for all x, whenever $|t| < \delta$. Next, we use the fact that k_n is nonnegative and integrates to 1 to write

$$\begin{aligned}\left| f(x) - \frac{1}{\pi}\int_{-\pi}^{\pi} f(x+t)\,k_n(t)\,dt \right| &= \frac{1}{\pi}\left|\int_{-\pi}^{\pi} \big[f(x) - f(x+t)\big]\,k_n(t)\,dt\right| \\ &\leq \frac{1}{\pi}\int_{-\pi}^{\pi} \big|f(x) - f(x+t)\big|\,k_n(t)\,dt \\ &\leq \frac{\varepsilon}{\pi}\int_{|t|<\delta} k_n + \frac{2\|f\|_\infty}{\pi}\int_{\delta\leq|t|\leq\pi} k_n \\ &< \varepsilon + \varepsilon = 2\varepsilon\end{aligned}$$

for all n sufficiently large (independent of x). □

To see that Fejér's kernel satisfies the conditions of Theorem 15.7 is easy. In particular, (c) follows from the fact that $K_n(t) \rightrightarrows 0$ on the set $\delta \leq |t| \leq \pi$. Indeed, since $\sin(t/2)$ increases on $\delta \leq t \leq \pi$, we have

$$0 \leq K_n(t) = \frac{\sin^2(nt/2)}{2n\sin^2(t/2)} \leq \frac{1}{2n\sin^2(\delta/2)} \to 0 \qquad (n \to \infty).$$

Since $\sigma_n(f)$ is a trig polynomial, notice that Fejér's theorem implies Weierstrass's second theorem. Here, then, is one of the independent proofs of Weierstrass's second theorem that we referred to in Chapter Eleven. (Notice too that although we used the Weierstrass theorem to facilitate our discussion of the L_2-theory, the proof of Fejér's theorem is self-contained.) As we pointed out in Chapter Eleven, the first Weierstrass theorem may then also be viewed as a consequence of Fejér's theorem.

Corollary 15.8. (Weierstrass's Second Theorem) *Given $f \in C^{2\pi}$ and $\varepsilon > 0$, there is a trig polynomial T such that $\|f - T\|_\infty < \varepsilon$.*

Corollary 15.9. (Weierstrass's First Theorem) *Given $f \in C[a,b]$ and $\varepsilon > 0$, there is a polynomial p such that $\|f - p\|_\infty < \varepsilon$.*

You might find it interesting to learn that Fejér was a fourth-year student, only 19 years old, when he proved his result (about 1900) while Weierstrass was 75 at the time he proved his approximation theorems (about 1885). It is especially interesting when you consider that, only a few years earlier, Fejér's teachers had decided he was a weak student and so should be charged extra tuition!

EXERCISES

9. Prove that $\|\sigma_n(f)\|_2 \le \|f\|_2$ and $\|\sigma_n(f)\|_\infty \le \|f\|_\infty$.

10.

(a) If $f, k \in C^{2\pi}$, prove that $g(x) = \int_{-\pi}^{\pi} f(x+t)\,k(t)\,dt$ is in $C^{2\pi}$.

(b) If we assume only that f is 2π-periodic and Riemann integrable (but still $k \in C^{2\pi}$), is $g(x) = \int_{-\pi}^{\pi} f(x+t)\,k(t)\,dt$ continuous?

(c) If we simply assume that f and k are 2π-periodic and Riemann integrable, is $g(x) = \int_{-\pi}^{\pi} f(x+t)\,k(t)\,dt$ still continuous? [Hint: See Exercise 6.]

11. Modify the proof of Theorem 15.7 to show that if f is Riemann integrable, then $(1/\pi)\int_{-\pi}^{\pi} f(x+t)\,k_n(t)\,dt \to f(x)$ pointwise, as $n \to \infty$, at each point of continuity of f. In particular, $\sigma_n(f)(x) \to f(x)$ at each point of continuity of f.

Complex Fourier Series

Lastly, a word or two about Fourier series involving complex coefficients. Most advanced textbooks consider a 2π-periodic function $f : \mathbb{R} \to \mathbb{C}$ and define the Fourier series for f by

$$\sum_{k=-\infty}^{\infty} c_k e^{ikt},$$

where now we have only one formula for the c_k:

$$c_k = \frac{1}{2\pi}\int_{-\pi}^{\pi} f(t)\,e^{-ikt}\,dt, \tag{15.9}$$

and where, of course, the c_k are now complex numbers. (The integral of a complex-valued function $g : \mathbb{R} \to \mathbb{C}$ is defined in terms of the real and imaginary parts of g, namely, $\int g = \int(\operatorname{Re} g) + i\int(\operatorname{Im} g)$. Thus, in our situation, if we require that both $\operatorname{Re} f$ and $\operatorname{Im} f$ are integrable on $[-\pi, \pi\,]$, then the integral in equation (15.9) will exist.)

This somewhat simpler approach has other advantages; for one, the exponentials e^{ikt} are now an *orthonormal* set relative to the normalizing constant $1/2\pi$. Specifically, we now define

$$\langle f, g \rangle = \frac{1}{2\pi}\int_{-\pi}^{\pi} f(t)\,\overline{g(t)}\,dt$$

and so have

$$\langle e^{int}, e^{imt} \rangle = \frac{1}{2\pi} \int_{-\pi}^{\pi} e^{int} e^{-imt} \, dt = \begin{cases} 0, & \text{for } m \neq n, \\ 1, & \text{for } m = n. \end{cases}$$

And, if we remain consistent with this choice and define the L_2-norm by

$$\|f\|_2 = \left(\frac{1}{2\pi} \int_{-\pi}^{\pi} |f(t)|^2 \, dt \right)^{1/2}, \tag{15.1'}$$

then we have the simpler estimate $\|f\|_2 \leq \|f\|_\infty$ for $f \in C^{2\pi}$.

The Dirichlet and Fejer kernels are essentially the same in this case, except that we now write $s_n(f)(x) = \sum_{k=-n}^{n} c_k e^{ikx}$. Given this, the Dirichlet and Fejér kernels can be written as

$$D_n(x) = \sum_{k=-n}^{n} e^{ikx} = 1 + \sum_{k=1}^{n} (e^{ikx} + e^{-ikx})$$
$$= 1 + 2 \sum_{k=1}^{n} \cos kx = \frac{\sin\left(n + \frac{1}{2}\right) x}{\sin \frac{1}{2} x} \tag{15.6'}$$

and

$$K_n(x) = \frac{1}{n} \sum_{m=0}^{n-1} D_m(x)$$
$$= \frac{1}{n} \sum_{m=0}^{n-1} \frac{\sin\left(m + \frac{1}{2}\right) x}{\sin \frac{1}{2} x} = \frac{\sin^2(nx/2)}{n \sin^2(x/2)}. \tag{15.8'}$$

In other words, each is twice its real coefficient counterpart. Since the choice of a normalizing constant ($1/\pi$ versus $1/2\pi$, and sometimes even $1/\sqrt{\pi}$ or $1/\sqrt{2\pi}$) has a (small) effect on these formulas, you may find some variation in other textbooks.

EXERCISE

12. Show that we may also write $K_n(x) = \sum_{k=-n}^{n} (1 - (|k|/n)) \, e^{ikx}$.

Notes and Remarks

The books by Carslaw [1930], Folland [1992], Jackson [1941], Körner [1988], Rogosinski [1950], Tolstov [1962], and Zygmund [1935] will supply you with a wealth of additional information about Fourier series and their applications. You will find discussions of Fourier series along with several related topics in Cheney [1966] and Natanson [1964]. For more on the history of Fourier series (and Fourier himself), see Birkhoff [1973], Carslaw [1930], Gibson [1893], González-Velasco [1992], Grattan-Guinness [1972], Hawkins [1970], Herivel [1975], Hobson [1927], Jeffery [1956], Kline [1972], Körner [1988], Langer [1947], Rogosinski [1950], and Van Vleck [1914].

The "barking up the wrong tree" quote is from Halmos [1978].

The early history of Fourier analysis was largely concerned with questions concerning existence, uniqueness, and pointwise convergence of the series. For example, Dirichlet [1829] proved that piecewise monotone functions are represented by their Fourier series. Jordan [1881] was able to generalize this result to functions of bounded variation (which were introduced for just this purpose). The *Dirichlet–Jordan theorem* states, in part, that if f is 2π-periodic and of bounded variation on $[-\pi, \pi]$, then $s_n(f)(\theta)$ converges to $[f(\theta-) + f(\theta+)]/2$, as $n \to \infty$, for each θ in $[-\pi, \pi]$. A similar result is given by the *Dini–Lipschitz theorem*, which, in part, states that if f is 2π-periodic and satisfies a Lipschitz condition of order $\alpha > 0$ on $[-\pi, \pi]$, then $s_n(f)$ converges pointwise to f. See, for example, Rogosinski [1950] for further details. Simple proofs of pointwise convergence (under various hypotheses) are given in Chernoff [1980], Franklin [1924], and Jackson [1926, 1934].

Real progress in these delicate matters would wait until the introduction of the Lebesgue integral in 1903. As one of the earliest applications of the new integral, F. Riesz [1906] introduced the L_2-theory. Once the L_2-theory was in place, the emphasis in Fourier analysis began to shift toward other issues. We will have much more to say about these issues later; the Lebesgue integral is the focus of the upcoming chapters. For a quick (and unusual) derivation of the Lebesgue integral based on what we already know about the L_2-theory, see Van Daele [1990].

The notation $f(x) \sim (a_0/2) + \sum_{k=1}^{\infty} \left(a_k \cos kx + b_k \sin kx\right)$, which is used to emphasize the fact that the Fourier series for f is a valid representation for f regardless of whether or not it actually converges pointwise to f, is apparently due to Hurwitz [1903]. The result in Exercise 2 is one of Fourier's original examples. Riemann's lemma is from his 1854 *Habilitationsshcrift*; see Riemann [1902, pp. 227–265]. Also see the excerpt, "Riemann on Fourier series and the Riemann integral," in Birkhoff [1973].

Corollary 15.3 (in a slightly different form) is due to Lebesgue [1906]. The proof of Theorem 15.4 is taken from Simon [1969]. There are several elementary convergence theorems of this type in Jackson [1941] and Rogosinski [1950]. Theorem 15.6 is, of course, due to the great Hungarian mathematician Lipót (Leopold) Fejér; his original proof in Fejér [1900] is amplified in Fejér [1904]. For more on Fejér himself, see Hersh and John-Steiner [1993] (and its references). Fejér's theorem fits the mold of Korovkin's theorem (see the notes at the end of Chapter Eleven); for a proof along these lines, see Cheney [1966].

PART THREE

LEBESGUE MEASURE AND INTEGRATION

CHAPTER SIXTEEN

Lebesgue Measure

The Problem of Measure

If you will recall Fourier's "proof" that every bounded function has a Fourier series, a central problem is to justify the term-by-term integration of a series of functions. Specifically, if we suppose that (f_n) is a sequence of integrable, or even continuous functions, is it true that

$$\int_a^b \left(\sum_{n=1}^{\infty} f_n(x) \right) dx = \sum_{n=1}^{\infty} \left(\int_a^b f_n(x)\, dx \right)?$$

For that matter, is $\sum f_n$ even integrable? And, as long as we're at it, what does it mean to say that a function is integrable?

These are a few of the questions that Bernhard Riemann set out to answer in "Über die Darstellbarkeit einer Function durch eine trigonometrische Reihe" (On the development of a function by a trigonometric series), a paper submitted in 1854 as part of his *Habilitationsschrift*, or "inauguration" examination. Riemann's work on the convergence of series, along with his concept of an integrable function, were in direct response to the problems posed by Fourier's proof. The paper was deemed incomplete in many ways, raising more questions than it answered, and it remained unpublished until 1867, one year after Riemann's death. Nevertheless, the publication of Riemann's paper is considered a landmark in the history of analysis. According to Grattan-Guinness:

> Soon Weierstrass's pupils were all working on problems in analysis inspired by Riemann; infinitely oscillatory and/or discontinuous functions; continuous non-differentiable functions; modes of uniform and nonuniform convergence; point discontinuities of Fourier series; and so on. This was the 1870s, the time of Hankel's contemporaries: The age of Bolzano's "pure analysis" had arrived with a vengeance.

In our time, the Riemann integral has surely become the workhorse of calculus. While this noble beast is a faithful and true servant, it is not without its shortcomings – not entirely flawed, mind you, just less than perfect. One such blemish, if you will, is that the Riemann integral is not defined for as many functions as we might hope. To better understand this, let's take another look at how the Riemann integral is computed.

Given a nonnegative, bounded function f on $[a, b]$ and a partition P of $[a, b]$, we effectively construct a step function g approximating f and estimate the area under the graph of f by the area under the graph of the step function (Figure 16.1).

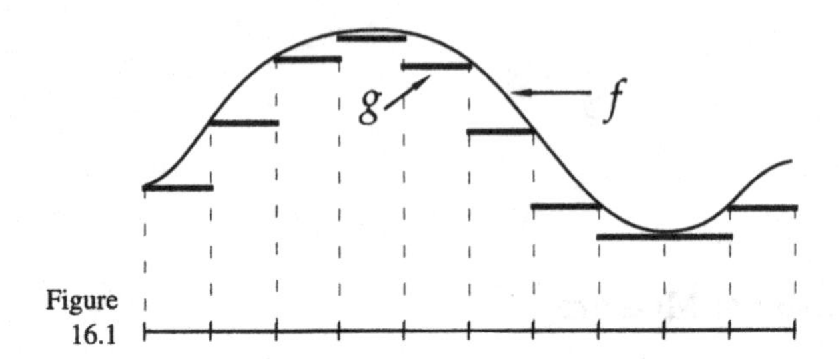

Figure 16.1

As we saw in the last chapter, the Riemann integral of f over $[a, b]$ exists if these approximate areas tend to a specific finite limit as $\max_{1 \le i \le n} \Delta x_i \to 0$. What's more, the existence of such a limit requires that the oscillation of f be relatively small at "most" points in the interval $[a, b]$. In short, the Riemann integral exists only for functions that are "almost continuous." We will make this notion precise later, but for now recall that the characteristic function of the rationals in $[a, b]$ fails to be Riemann integrable in spite of the fact that it differs from a continuous function, namely 0, at a mere handful of points. Evidently, "almost continuous" is a rather restrictive notion.

Said another way, if the difference of upper and lower sums for f is to tend to 0, then f will have to be the "almost uniform" limit of a sequence of step functions on $[a, b]$. Again, while a precise statement will have to wait, notice that the characteristic function of the rationals in $[a, b]$ is clearly the *pointwise* limit of a sequence of step functions – each having zero integral.

Either of these heuristic characterizations helps to explain a second shortcoming of the Riemann integral: While the Riemann integral easily commutes with *uniform* limits, it is very difficult to work with where *pointwise* limits are concerned. In this game of interchanging limits, it would be useful to have a more generous integral. Enter Lebesgue.

In 1902, Henri Lebesgue published his thesis, "Intégrale, longeur, aire," in which he presented an extension of the Riemann integral. The Lebesgue integral is defined for what are called "measurable" functions, a class that includes the Riemann integrable functions; the new integral reduces to the old in all of the familiar cases.

Lebesgue's ideas were influenced by the earlier works of Jordan and Borel, and were largely founded on preserving a geometric interpretation of length and area. He addressed a variety of issues that, at the time, were not associated with the integral, in particular, surface area and curve length. Moreover, Lebesgue's approach gave new insights on the differentiability of monotone functions and an extension of the fundamental theorem of calculus.

The Lebesgue integral overcomes at least one other shortcoming of the Riemann integral: It is very easy to establish the so-called "bounded convergence theorem" for the Lebesgue integral. Specifically, if (f_n) is a sequence of measurable functions such that $\sum_{k=1}^{n} f_k$ is a uniformly bounded sequence converging pointwise on $[a, b]$, then the limit $\sum f_n$ is necessarily also measurable and satisfies

$$\int_a^b \left(\sum_{n=1}^{\infty} f_n(x) \right) dx = \sum_{n=1}^{\infty} \left(\int_a^b f_n(x)\,dx \right).$$

While a similar result is known to hold for the Riemann integral, it is much harder to prove.

Lebesgue's theory of integration provided the ideal tool for research into the troublesome issues surrounding trigonometric series. Lebesgue himself would lead the way. By 1910, the Lebesgue integral was firmly established in the research community, and by 1930 it had found its way into several popular textbooks.

The problem of integration, as Lebesgue called it, is to assign to each bounded function f, defined on some interval, and each pair of real numbers a and b, a finite number $\int_a^b f(x)\,dx$ in such a way that the following six conditions are satisfied:

1. $\displaystyle\int_a^b f(x)\,dx = \int_{a+h}^{b+h} f(x-h)\,dx$, for any a, b, and h.
2. $\displaystyle\int_a^b f(x)\,dx + \int_b^c f(x)\,dx + \int_c^a f(x)\,dx = 0$, for any a, b, and c.
3. $\displaystyle\int_a^b [f(x)+g(x)]\,dx = \int_a^b f(x)\,dx + \int_a^b g(x)\,dx$, for any f and g.
4. $\displaystyle\int_a^b f(x)\,dx \geq 0$ whenever $a < b$ and $f \geq 0$.
5. $\displaystyle\int_0^1 1dx = 1$.
6. If $f_n(x)$ increases pointwise to $f(x)$, then $\displaystyle\int_a^b f_n(x)\,dx \to \int_a^b f(x)\,dx$.

These six conditions are what Lebesgue took to be the minimal set of requirements for a "reasonable" integral. And the six conditions are *independent*; that is, it is possible to define the number $\int_a^b f(x)\,dx$ in such a way that it will satisfy any five given conditions but fail to satisfy the remaining sixth condition.

Asking for a "reasonable" integral that is defined for all bounded functions may be optimistic, but is worth shooting for. The first five conditions are clearly desirable, and the Riemann integral already satisfies these. Thus, we are asking for an extension of the Riemann integral that is defined on as large a class of functions as we can manage, which preserves the "nice" properties of the Riemann integral and which will, in addition, commute with at least certain limits.

We can paraphrase Lebesgue's own description of his concept of integrability by making a slight revision to our simplistic description of Riemann's definition. To lift the burden of continuity from the integrand f, Lebesgue's approach is to again approximate f by a simpler function, but this time subdividing the y-axis rather than the x-axis! See Figure 16.2.

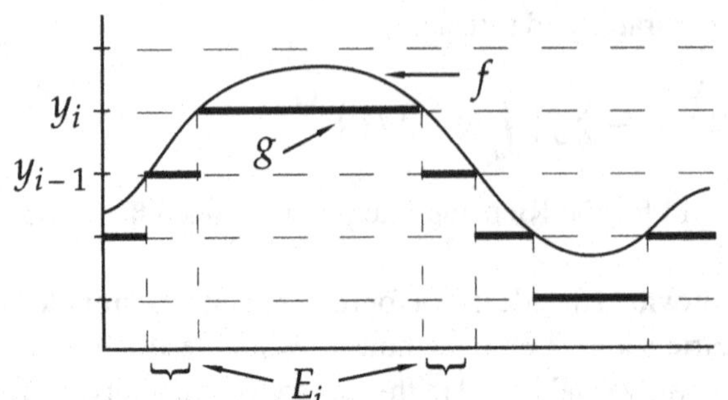

Figure 16.2

If $P = \{y_0, \ldots, y_n\}$ is a partition of an interval on the y-axis containing the range of f, then we might approximate f by the function

$$g = \sum_{i=1}^{n} c_i \, \chi_{\{y_{i-1} \le f < y_i\}},$$

where $c_i \in [\, y_{i-1}, y_i)$, and where $\{y_{i-1} \le f < y_i\}$ is shorthand for the set $E_i = \{x : y_{i-1} \le f(x) < y_i\}$. Now the premise here is that the integral of g is unambiguously defined; by rights it ought to be

$$\int_a^b g = \sum_{i=1}^{n} c_i \, m(E_i),$$

where $m(E)$ denotes the "length" or "measure" of a subset E of $[\,a, b\,]$. Assuming that we can do this, we would then define

$$\int_a^b f = \lim_{\|P\| \to 0} \sum_{i=1}^{n} c_i \, m(E_i).$$

And what do we gain by this new approach? Well, we are no longer speaking of changes in f relative to small changes in x (which suggests continuity); rather, we are speaking of changes in f arising from *measurable* changes in x.

What we will find is that Lebesgue's integral is defined for a larger class of functions than is Riemann's, indeed, for any bounded function for which sets of the form $\{c \le f < d\}$ are always measurable. The trade-off, of course, is that we will have to decide what is meant by the "measure" of a set, and which sets, if any, can be so measured.

Note that Lebesgue's approach reduces the problem of integration to that of defining the integral for two-valued functions of the form χ_E. That is, we need to find a suitable method of defining the number $m(E) = \int_a^b \chi_E$. In this way, the problem of integration becomes *the problem of measure*.

The problem is to assign to each subset E of $\mathbb{R}$ a nonnegative number $m(E)$, called the measure of E, in such a way that the following properties are satisfied:

1° $m([\,0, 1\,]) = 1$.

2° $m(E + h) = m(E) = m(-E)$, where $E + h = \{x + h : x \in E\}$ and $-E = \{-x : x \in E\}$; that is, geometrically congruent sets should have the same measure.

3° If (E_n) is any sequence, finite or infinite, of pairwise disjoint sets, then $m\left(\bigcup_{n\geq 1} E_n\right) = \sum_{n\geq 1} m(E_n)$.

Condition 1° obviously replaces condition 5 in the problem of integration, while condition 2° replaces condition 1. Condition 3° will ultimately replace condition 6. The three together will imply that the measure of an interval is simply its length, and that the measure of a bounded set is at least finite. It is the last condition, condition 3°, that marks Lebesgue's point of departure from what had gone on before. The geometric notions of length, area, and volume only call for those measures to be additive across *finitely many* disjoint objects. Based on Borel's work on the problem, though, Lebesgue knew that he must consider *countably additive* measures, for it is precisely this last condition that permits Lebesgue's integral to commute with certain pointwise limits.

Unfortunately, the three conditions are not only independent; they are also *inconsistent* with the Axiom of Choice. As we will see later, there is no solution to the problem of measure if we allow the Axiom of Choice (and we do!). Something has to give. For example, we might consider discarding condition 2°, or perhaps weakening condition 3° by only requiring finitely additive measures. But neither of these options is satisfactory. Assuming the Continuum Hypothesis, it can be shown that there is no countably additive measure defined on all subsets of $[0, 1]$ satisfying both $m([0, 1]) = 1$ and $m(\{x\}) = 0$ for every x in $[0, 1]$.

And the outlook is bleak even if we settle for only finitely additive measures, at least in $\mathbb{R}^3$. Consider *the Banach–Tarski paradox*:

Let U and V be nonempty, bounded, open sets in $\mathbb{R}^n$, where $n \geq 3$. Then, there exist a $k \in \mathbb{N}$ and partitions $E_1, \ldots, E_k$ and $F_1, \ldots, F_k$ of U and V, respectively, into an equal number of disjoint subsets such that E_i is congruent to F_i for each $i = 1, \ldots, k$.

Hence, an orange may be cut into finitely many pieces that could then be reassembled to form a citrus behemoth the size of the sun! Obviously, this result precludes the existence of a nonzero, finitely additive measure, defined on all subsets of $\mathbb{R}^3$, that assigns equal measure to congruent sets.

Well, OK. So we can't have everything. But rather than sacrifice any of the three geometrically aesthetic conditions that we have asked our measure to satisfy, we will instead restrict its domain. That is, we will not insist that m be defined on *all* subsets of $\mathbb{R}$. We'll ultimately settle for a measure defined only on certain "good" sets. What we will find is that there are plenty of "good" sets around to do analysis and that, after all, is what we came here for.

The problem of measure is important from a couple of points of view. For one, the concept of defining a measure in terms of a list of requirements, rather than by simply providing constructive examples and verifying their properties, was brand new in Lebesgue's time. Proclaiming in advance what properties are required of a solution lends a new dimension to the problem; by displaying the key issues, the problem becomes easier to generalize or to abstract. Although we are quite used to the axiomatic approach by now, it was still a novelty at the turn of the century. Equally important is the fact that a problem of calculus, of functions and integrals, has been transformed into a problem about abstract sets.

EXERCISE

1. Let f be a nonnegative bounded function on $[a, b]$ with $0 \le f \le M$. Let

$$E_{n,k} = \left\{ \frac{kM}{2^n} \le f < \frac{(k+1)M}{2^n} \right\},$$

for each $n = 1, 2, \ldots$, and $k = 0, 1, \ldots, 2^n$, and set $\varphi_n = \sum_{k=0}^{2^n} (kM/2^n) \chi_{E_{n,k}}$. Prove that $0 \le \varphi_n \le \varphi_{n+1} \le f$ and that $0 \le f - \varphi_n \le 2^{-n}M$ for each n. Thus, (φ_n) converges uniformly to f on $[a, b]$. [Hint: Notice that $E_{n,k} = E_{n+1,2k} \cup E_{n+1,2k+1}$.]

Lebesgue Outer Measure

In this section we take a first step toward extending the notion of length. To begin, let's agree that the word *interval* means a *bounded*, nonempty interval, that is, any one of the sets $[a, b]$, (a, b), $[a, b)$, or $(a, b]$, where a and b are finite real numbers with $a < b$. If I is any one of these four sets, we will use the shorthand $\ell(I) = b - a$ to denote the **length** of I. We will call sets of the form $(-\infty, b]$, (a, ∞), and so on, *unbounded* intervals and put $\ell(I) = \infty$ in any of these cases. In short, the word *interval*, with no additional quantifier, always means a *bounded interval.*

Now the notion of length obviously extends to finite unions of pairwise disjoint intervals. But, in fact, it extends unambiguously to all countable unions of pairwise disjoint intervals. Indeed, we simply take the sum of the lengths of the constituent intervals as the "total length" of the union. In general, though, given countably many intervals (I_n), not necessarily disjoint, the sum $\sum_{n=1}^{\infty} \ell(I_n)$ will be an overestimate for the total length of their union $\bigcup_{n=1}^{\infty} I_n$. The following lemma (which is obvious for finite collections of intervals) justifies this claim.

Proposition 16.1. *Let (I_n) and (J_k) be sequences of intervals such that $\bigcup_{n=1}^{\infty} I_n = \bigcup_{k=1}^{\infty} J_k$. If the I_n are pairwise disjoint, then $\sum_{n=1}^{\infty} \ell(I_n) \le \sum_{k=1}^{\infty} \ell(J_k)$. Thus, if the J_k are also pairwise disjoint, then the two sums are equal.*

PROOF. Suppose, to the contrary, that $\sum_{n=1}^{\infty} \ell(I_n) > \sum_{k=1}^{\infty} \ell(J_k)$. Then, for some N, we must have $\sum_{n=1}^{N} \ell(I_n) > \sum_{k=1}^{\infty} \ell(J_k)$. Of course, we also have $\bigcup_{n=1}^{N} I_n \subset \bigcup_{k=1}^{\infty} J_k$. But now, by expanding each J_k slightly and shrinking each I_n slightly, we may suppose that the J_k are *open* and the I_n are *closed*. (How?) Thus, the J_k form an open cover for the *compact* set $\bigcup_{n=1}^{N} I_n$. And here is the contradiction: Since we have $\sum_{n=1}^{N} \ell(I_n) > \sum_{k=1}^{M} \ell(J_k)$, for any M, the sets (J_k) form an open cover for $\bigcup_{n=1}^{N} I_n$ that admits no finite subcover. □

Now we are ready to extend ℓ to arbitrary subsets of $\mathbb{R}$. Given a subset E of $\mathbb{R}$, we define the (Lebesgue) **outer measure** of E by

$$m^*(E) = \inf\left\{\sum_{n=1}^{\infty} \ell(I_n) : E \subset \bigcup_{n=1}^{\infty} I_n\right\},$$

where the infimum is taken over all coverings of E by countable unions of intervals. Thus, the outer measure of E is the infimum of certain overestimates for the "length" of E. Before we say more, let's check a few simple properties of m^*.

Proposition 16.2.

(i) $0 \le m^*(E) \le \infty$ *for any* E.
(ii) *If* $E \subset F$, *then* $m^*(E) \le m^*(F)$.
(iii) $m^*(E + x) = m^*(E)$, *where* $E + x = \{e + x : e \in E\}$.
(iv) $m^*(E) = 0$ *for any countable set* E.
(v) $m^*(E) < \infty$ *for any bounded set* E.
(vi) $m^*(E) = \inf\left\{\sum_{n=1}^{\infty}(b_n - a_n) : E \subset \bigcup_{n=1}^{\infty}(a_n, b_n)\right\}$.

PROOF. The first three properties are nearly immediate from the definition of m^* and are left as exercises. For (iv), suppose that $E = \{e_1, e_2, \ldots\}$. Given $\varepsilon > 0$, notice that $E \subset \bigcup_{n=1}^{\infty}(e_n - 2^{-n}\varepsilon, e_n + 2^{-n}\varepsilon)$, and hence that $m^*(E) \le 2\varepsilon$. Next, for (v), note that if E is bounded, then $E \subset [a, b]$ for some (finite) $a < b$. Thus, $m^*(E) \le b - a < \infty$. Finally, given $E \subset \mathbb{R}$, notice that we always have

$$m^*(E) \le \inf\left\{\sum_{n=1}^{\infty}(b_n - a_n) : E \subset \bigcup_{n=1}^{\infty}(a_n, b_n)\right\}.$$

To establish the reverse inequality, then, it is enough to consider the case $m^*(E) < \infty$. (Why?) Now, given $\varepsilon > 0$, choose a sequence of intervals (I_n) covering E such that $\sum_{n=1}^{\infty} \ell(I_n) \le m^*(E) + \varepsilon$. For each n, let J_n be an open interval containing I_n with $\ell(J_n) \le \ell(I_n) + 2^{-n}\varepsilon$. Then, (J_n) covers E and $\sum_{n=1}^{\infty} \ell(J_n) \le m^*(E) + 2\varepsilon$. This proves (vi). □

Examples 16.3

(a) Please note that there are unbounded sets with finite outer measure. A rather spectacular example is $m^*(\mathbb{Q}) = 0$. There are also uncountable sets with outer measure zero; recall from Chapter Two that the Cantor set Δ has outer measure zero. Indeed, for each n, the Cantor set is contained in a finite union of intervals of total length $2^n/3^n$. Thus, $m^*(\Delta) \le 2^n/3^n \to 0$.

(b) Sets of outer measure zero, or *null sets*, play an important role in analysis; they provide another notion of "small" or "negligible" sets. Based on the two examples we have at hand, this makes for a curious comparison. From the point of view of cardinality, Δ is big (uncountable) while $\mathbb{Q}$ is small (countable); from a topological point of view, Δ is small (nowhere dense); while $\mathbb{Q}$ is big (dense); and from the point of view of measure, both Δ and $\mathbb{Q}$ are small (measure zero)! You will find further curiosities of this sort in the exercises.

(c) Quite often we encounter properties that hold everywhere *except* on a set of measure zero. We say that such a property holds **almost everywhere**, abbreviated "a.e." (Some authors use "almost all" or "almost always," abbreviated "a.a.," while probabilists use "almost surely," abbreviated "a.s." In some older books the abbreviation "p.p." is used, for the original French "presque partout.") By way of an example, notice that the Cantor function $f : [0, 1] \to [0, 1]$ satisfies $f' = 0$ almost everywhere, since f is constant on each subinterval of the complement of Δ.

(d) From Proposition 16.2 (iv), any countable set of exceptions would come under the almost everywhere banner. For instance, we might say that $\chi_{\mathbb{Q}} = 0$ almost everywhere, or that a monotone function f is continuous almost everywhere, that is, $m^*\big(D(f)\big) = 0$.

(e) The point to statement (vi) of Proposition 16.2 is that the definition of m^* has little to do with the particular type of intervals used; we might just as well have taken closed intervals. The advantage to using open intervals is that we now have a connection between the geometry of $\mathbb{R}$ (length) and the topology of $\mathbb{R}$ (open sets). We will have more to say about this observation later.

(f) Lebesgue originally defined outer measure for subsets E of a bounded interval $[a, b]$. In this case, he also defined the *inner measure* of E as $m_*(E) = b - a - m^*([a, b] \setminus E)$. It is not hard to see that $m_*(E) \le m^*(E)$; that is, inner measure is an underestimate of the "true" length of E while outer measure is an overestimate (see Exercise 7).

Next, let's check that outer measure truly is an extension of length.

Proposition 16.4. $m^*(I) = \ell(I)$ *for any interval* I, *bounded or not.*

PROOF. The heart of the matter is checking that the proposition holds for *compact* intervals, that is, $m^*\big([a, b]\big) = b - a$. Assuming that we have done this, let's see how this special case settles all other cases.

First, if I is unbounded, then I contains compact intervals of length n for any $n \ge 1$. By monotonicity (Proposition 16.2 (ii)), $m^*(I) \ge n$ for any n; hence, $m^*(I) = \infty = \ell(I)$.

Next, if I is a bounded, noncompact interval with endpoints $a < b$, then $[a + \varepsilon/2, b - \varepsilon/2] \subset I \subset [a, b]$ for any $\varepsilon > 0$. Again using monotonicity, it follows that $b - a - \varepsilon \le m^*(I) \le b - a$ for any $\varepsilon > 0$; that is, $m^*(I) = b - a = \ell(I)$.

So let's get to work! Let $I = [a, b]$. Since I is itself one of the candidate intervals used in computing $m^*(I)$, we certainly have $m^*(I) \le b - a$; we need to check that $m^*(I) \ge b - a$. Now, given $\varepsilon > 0$, Proposition 16.2 (vi) supplies a sequence of open intervals (a_n, b_n) such that $I \subset \bigcup_{n=1}^{\infty}(a_n, b_n)$ and $m^*(I) \ge \sum_{n=1}^{\infty}(b_n - a_n) - \varepsilon$. Since I is compact, we know that there are finitely many open intervals here that will cover I, say $I \subset \bigcup_{i=1}^{n}(a_i, b_i)$. By discarding any extraneous intervals and relabeling, if necessary, we may suppose that $a_1 < a_2 < \cdots < a_n$ and that $(a_i, b_i) \cap I \neq \emptyset$ for each $i = 1, \dots, n$. But I is connected! Thus, consecutive intervals from $(a_1, b_1), \dots, (a_n, b_n)$ must actually overlap; that is, $\bigcup_{i=1}^n (a_i, b_i)$ must be an *open interval* containing I. (Why?) Hence, $\sum_{i=1}^\infty (b_i - a_i) \geq \sum_{i=1}^n (b_i - a_i) \geq \ell(I) = b - a$ and so $m^*(I) \geq b - a - \varepsilon$. □

EXERCISES

2. Prove statements (i) and (ii) of Proposition 16.2.

3. Earlier attempts at defining the measure of a (bounded) set were similar to Lebesgue's, except that the infimum was typically taken over *finite* unions of intervals covering the set. Show that if $\mathbb{Q} \cap [0, 1]$ is contained in a finite union of open intervals $\bigcup_{i=1}^n (a_i, b_i)$, then $\sum_{i=1}^n (b_i - a_i) \geq 1$. Thus, $\mathbb{Q} \cap [0, 1]$ would have "measure" 1 by this definition.

▷ **4.** Given any subset E of $\mathbb{R}$ and any $h \in \mathbb{R}$, show that $m^*(E + h) = m^*(E)$, where $E + h = \{x + h : x \in E\}$.

5. If we define $rE = \{rx : x \in E\}$, what is $m^*(rE)$ in terms of $m^*(E)$?

6. If E has nonempty interior, show that $m^*(E) > 0$.

7. Referring to Example 16.3 (f), show that $m_*(E) \leq m^*(E)$ for any $E \subset [a, b]$.

▷ **8.** Given $\delta > 0$, show that $m^*(E) = \inf \sum_{n=1}^\infty \ell(I_n)$ where the infimum is taken over all coverings of E by sequences of intervals (I_n), where each I_n has diameter less than δ.

▷ **9.** If $E = \bigcup_{n=1}^\infty I_n$ is a countable union of pairwise disjoint intervals, prove that $m^*(E) = \sum_{n=1}^\infty \ell(I_n)$.

10. Prove that $m^*\left(\bigcup_{n=1}^\infty U_n\right) = \sum_{n=1}^\infty m^*(U_n)$ for any sequence (U_n) of pairwise disjoint open sets.

11. Prove that $m^*(E) = \inf \sum_{n=1}^\infty \ell(I_n)$ where the infimum is taken over all coverings of E by sequences of pairwise disjoint open intervals (I_n).

12. Prove that $m^*(E) = \inf\{m^*(U) : U \text{ is open and } E \subset U\}$.

13. Show that $m^*(E \cup F) \leq m^*(E) + m^*(F)$ for any sets E, F.

14. If E and F are countable unions of pairwise disjoint intervals, prove that $m^*(E \cup F) + m^*(E \cap F) = m^*(E) + m^*(F)$. [Hint: First verify the formula when E and F are finite unions of pairwise disjoint intervals. How does this help?]

15. Prove that a subset of a set of outer measure zero is again a set of outer measure zero. Prove that a finite union of sets of outer measure zero has outer measure zero.

▷ **16.** If $m^*(E) = 0$, show that $m^*(E \cup A) = m^*(A) = m^*(A \setminus E)$ for any A.

17. If $E \subset [a, b]$ and $m^*(E) = 0$, show that E^c is dense in $[a, b]$.

18. If E is a compact set with $m^*(E) = 0$, and if $\varepsilon > 0$, prove that E can be covered by finitely many open intervals, $I_1, \dots, I_n$, satisfying $\sum_{j=1}^n m^*(I_j) < \varepsilon$.

19. For $E \subset [a, b]$, show that $m^*(E) = 0$ if and only if E can be covered by a sequence of intervals (I_n) such that $\sum_{n=1}^{\infty} m^*(I_n) < \infty$, and such that each $x \in E$ is in *infinitely many* I_n.

20. If $m^*(E) = 0$, prove that $m^*(E^2) = 0$, where $E^2 = \{x^2 : x \in E\}$. [Hint: First consider the case where E is bounded.]

21. If $f : \mathbb{R} \to \mathbb{R}$ satisfies $|f(x) - f(y)| \leq K|x - y|$ for all x and y, show that $m^*(f(E)) \leq Km^*(E)$ for any $E \subset \mathbb{R}$.

We have come a long way toward solving the problem of measure. We now have an extension of the notion of length that is defined for any subset of $\mathbb{R}$ and that, according to Proposition 16.2 (iii), is *translation-invariant*. All that is missing is the countable additivity and here, as we'll see, is where outer measure falls short. We can come close, though: m^* is at least **countably subadditive**.

Proposition 16.5. $m^*\left(\bigcup_{n=1}^{\infty} E_n\right) \leq \sum_{n=1}^{\infty} m^*(E_n)$ *for any sequence* (E_n) *of subsets of* $\mathbb{R}$.

PROOF. We may clearly suppose that $m^*(E_n) < \infty$ for each n, for otherwise there is nothing to show. Now, let $\varepsilon > 0$. For each n, choose intervals $(I_{n,i})$ with

$$E_n \subset \bigcup_{i=1}^{\infty} I_{n,i} \qquad \text{and} \qquad \sum_{i=1}^{\infty} m^*(I_{n,i}) \leq m^*(E_n) + \frac{\varepsilon}{2^n}.$$

Then $\bigcup_{n=1}^{\infty} E_n \subset \bigcup_{n=1}^{\infty} \bigcup_{i=1}^{\infty} I_{n,i}$, and so

$$m^*\left(\bigcup_{n=1}^{\infty} E_n\right) \leq \sum_{n=1}^{\infty} \sum_{i=1}^{\infty} m^*(I_{n,i}) \leq \sum_{n=1}^{\infty} m^*(E_n) + \varepsilon,$$

which proves the Proposition. □

Corollary 16.6. *If* $m^*(E_n) = 0$ *for each* n, *then* $m^*\left(\bigcup_{n=1}^{\infty} E_n\right) = 0$.

Corollary 16.7. *Given a subset* E *of* $\mathbb{R}$ *and* $\varepsilon > 0$, *there is an open set* G *containing* E *such that* $m^*(G) \leq m^*(E) + \varepsilon$. *Consequently,*

$$m^*(E) = \inf\{m^*(U) : U \text{ is open and } E \subset U\}.$$

PROOF. According to Proposition 16.2 (vi), we may choose a sequence of open intervals (I_n) covering E such that $\sum_{n=1}^{\infty} m^*(I_n) \leq m^*(E) + \varepsilon$. But then, $G = \bigcup_{n=1}^{\infty} I_n$ is an open set containing E and $m^*(G) \leq \sum_{n=1}^{\infty} m^*(I_n) \leq m^*(E) + \varepsilon$. Since $m^*(E) \leq m^*(G)$ whenever $E \subset G$, the second assertion now follows. □

Although we cannot hope to show that m^* is countably additive, in general, we can at least spell out one easy case where m^* is finitely additive.

Proposition 16.8. *If* E *and* F *are disjoint compact sets, then* $m^*(E \cup F) = m^*(E) + m^*(F)$.

PROOF. If E and F are disjoint compact sets, then

$$d(E,F) = \inf\{|x-y| : x \in E, y \in F\} > 0.$$

Thus, no interval of diameter less than $\delta = d(E, F)$ will hit both E and F.

Now, given $\varepsilon > 0$, we can choose a sequence of open intervals (I_n) covering $E \cup F$ such that each I_n has diameter less than δ, and such that $\sum_{n=1}^{\infty} m^*(I_n) \le m^*(E \cup F) + \varepsilon$. (How?) Note that a given I_n can hit at most one of E or F. Thus, if (I'_n) and (I''_n) denote those I_n that hit E and those that hit F, respectively, then $E \subset \bigcup_{n=1}^{\infty} I'_n$ and $F \subset \bigcup_{n=1}^{\infty} I''_n$. Hence,

$$\begin{aligned} m^*(E) + m^*(F) &\le \sum_{n=1}^{\infty} m^*(I'_n) + \sum_{n=1}^{\infty} m^*(I''_n) \\ &\le \sum_{n=1}^{\infty} m^*(I_n) \le m^*(E \cup F) + \varepsilon. \end{aligned}$$

That is, $m^*(E) + m^*(F) \le m^*(E \cup F)$. Since $m^*(E \cup F) \le m^*(E) + m^*(F)$ follows from Proposition 16.5, we are done. □

Corollary 16.9. *If $E_1, \ldots, E_n$ are pairwise disjoint compact sets, then $m^*\left(\bigcup_{i=1}^{n} E_i\right) = \sum_{i=1}^{n} m^*(E_i)$.*

EXERCISES

▷ **22.** Let $E = \bigcup_{n=1}^{\infty} E_n$. Show that $m^*(E) = 0$ if and only if $m^*(E_n) = 0$ for every n.

23. Given a bounded open set G and $\varepsilon > 0$, show that there is a compact set $F \subset G$ such that $m^*(F) > m^*(G) - \varepsilon$.

▷ **24.** Given a subset E of $\mathbb{R}$, prove that there is a G_δ-set G containing E such that $m^*(G) = m^*(E)$.

25. Suppose that $m^*(E) > 0$. Given $0 < \alpha < 1$, show that there exists an open interval I such that $m^*(E \cap I) > \alpha\, m^*(I)$. [Hint: It is enough to consider the case $m^*(E) < \infty$. Now suppose that the conclusion fails.]

26. Given $E \subset \mathbb{R}$, show that the set of points x for which $m^*(E \cap I) > 0$, for all open intervals I containing x, is a *closed* set.

27. For each n, let G_n be an open subset of $[0, 1]$ containing the rationals in $[0, 1]$ with $m^*(G_n) < 1/n$, and let $H = \bigcap_{n=1}^{\infty} G_n$. Prove that $m^*(H) = 0$ and that $[0, 1] \setminus H$ is a first category set in $[0, 1]$. Thus, $[0, 1]$ is the disjoint union of two "small" sets!

▷ **28.** Fix α with $0 < \alpha < 1$ and repeat our "middle thirds" construction for the Cantor set except that now, at the nth stage, each of the 2^{n-1} open intervals we discard from $[0, 1]$ is to have length $(1 - \alpha)3^{-n}$. (We still want to remove each open interval from the "middle" of a closed interval in the current level – it is important that the closed intervals that remain turn out to be nested.) The limit of this process, a set that we will name Δ_α, is called a **generalized Cantor set** and is very much like the ordinary Cantor set. Note that Δ_α is uncountable, compact, nowhere dense, and so on, but has nonzero outer measure. Indeed, check that $m^*(\Delta_\alpha) = \alpha$. (See Chapter Two for an example.) [Hint: You only need upper estimates for $m^*(\Delta_\alpha)$ and $m^*(\Delta_\alpha^c)$.]

29. In the notation of Exercise 28, check that $\bigcup_{n=1}^\infty \Delta_{1-(1/n)}$ has outer measure 1. Use this to give another proof that $[0, 1]$ can be written as the disjoint union of a set of first category and a set of measure zero.

30. Here is a related construction: Let (I_n) be an enumeration of all of the closed subintervals of $[0, 1]$ having rational endpoints (this is a countable collection). In each I_n, build a generalized Cantor set K_n having measure $m^*(K_n) = m^*(I_n)/2^n$. Now let $K = \bigcup_{n=1}^\infty K_n$. Prove that both K and its complement are dense in $[0, 1]$ and that both have positive outer measure.

Riemann Integrability

Rather than generate more properties of m^*, let's take a break for an important application: We next present Lebesgue's criterion for Riemann integrability (which is a restatement of Riemann's own criterion).

Theorem 16.10. *Let $f : [a, b] \to \mathbb{R}$ be bounded. Then, f is Riemann integrable on $[a, b]$ if and only if $m^*\big(D(f)\big) = 0$, that is, if and only if f is continuous at almost every point in $[a, b]$.*

Before we dive into the proof, please note that the condition "continuous at almost every point" or, briefly, "continuous a.e.," means something very different from the condition "almost everywhere equal to a continuous function." Indeed, the characteristic function of the rationals is almost everywhere equal to 0 (a continuous function) but is not continuous at *any* point. Moreover, note that the characteristic function of $[0, 1/2]$ is continuous a.e. in $[0, 1]$ but is clearly not equal a.e. to any continuous function. (Why?) Thus, the two conditions are incomparable in spite of their apparent similarity.

Next, let's recall our notation. First,

$$D(f) = \{x \in [a, b] : \omega_f(x) > 0\} = \bigcup_{n=1}^\infty \left\{x \in [a, b] : \omega_f(x) \geq \frac{1}{n}\right\},$$

where

$$\omega_f(x) \;=\; \inf_{I \ni x} \omega(f; I) \;=\; \inf_{I \ni x} \sup_{s,t \in I} |f(s) - f(t)|,$$

and where I denotes an *open* interval containing x. Recall, too, that the set $\{x : \omega_f(x) \geq (1/n)\}$ is closed for each n. We will refer to this set using the abbreviated notation $\{\omega_f \geq (1/n)\}$. Now, since $D(f)$ is written as a countable union, we may rephrase the conclusion of Lebesgue's theorem:

$$\begin{aligned} f \in \mathcal{R}[a,b] &\iff m^*\big(D(f)\big) = 0 \\ &\iff m^*\left(\left\{\omega_f \geq \tfrac{1}{n}\right\}\right) = 0 \qquad \text{for all } n. \end{aligned}$$

(Why?) Finally, recall that the difference between an upper and a lower sum can be written in terms of the oscillation of f:

$$U(f, P) - L(f, P) = \sum_{i=1}^{n} \omega(f; [x_{i-1}, x_i])\, \Delta x_i,$$

where, in our new terminology, $\Delta x_i = m^*([x_{i-1}, x_i])$. The fact that $\omega_f(x)$ is defined in terms of *open* intervals while $U(f, P) - L(f, P)$ is written in terms of *closed* intervals is a minor nuisance, but nothing we can't handle.

Since this is essentially all that is needed to prove the forward direction of Lebesgue's theorem, let's get that out of the way.

PROOF (of Theorem 16.10, forward implication). Let $f \in \mathcal{R}[a,b]$, and fix $k \geq 1$. We will show that $m^*\big(\{\omega_f \geq (1/k)\}\big) = 0$ and, hence, that $m^*\big(D(f)\big) = 0$.

Given $\varepsilon > 0$, choose a partition $P = \{x_0, \ldots, x_n\}$ such that

$$U(f, P) - L(f, P) = \sum_{i=1}^{n} \omega(f; [x_{i-1}, x_i])\, \Delta x_i < \frac{\varepsilon}{k}.$$

Notice that if $x \in \{\omega_f \geq (1/k)\} \cap (x_{i-1}, x_i)$, then $\omega(f; [x_{i-1}, x_i]) \geq \omega_f(x) \geq (1/k)$. Now, since the open intervals (x_{i-1}, x_i), $i = 1, \ldots, n$, cover all but finitely many points of $[a, b]$, it follows that those that hit $\{\omega_f \geq (1/k)\}$ will cover all but finitely many points of $\{\omega_f \geq (1/k)\}$. But finite sets have outer measure 0; hence

$$\frac{\varepsilon}{k} > \sum_{i=1}^{n} \omega(f; [x_{i-1}, x_i])\, \Delta x_i \geq \frac{1}{k} \sum{}' \Delta x_i \geq \frac{1}{k}\, m^*\left(\left\{\omega_f \geq \frac{1}{k}\right\}\right),$$

where $\sum{}'$ denotes the sum over those i for which $\{\omega_f \geq (1/k)\} \cap (x_{i-1}, x_i) \neq \emptyset$. Thus, $m^*\big(\{\omega_f \geq (1/k)\}\big) < \varepsilon$. □

The backward direction of Lebesgue's theorem is somewhat harder. We begin, though, with an easy observation.

Lemma 16.11. *If $\omega_f(x) < \delta$ for all x in some compact interval J, then there is a partition $Q = \{t_0, \ldots, t_n\}$ of J such that $\omega(f; [t_{i-1}, t_i]) < \delta$ for all $i = 1, \ldots, n$. Hence, $U(f, Q) - L(f, Q) < \delta\, m^*(J)$.*

PROOF. For each $x \in J$, choose an open interval I_x containing x such that $\omega(f; I_x) < \delta$ and a second open interval J_x with $x \in J_x \subset \bar{J}_x \subset I_x$. Note that $\omega(f; \bar{J}_x) < \delta$, too. The intervals J_x form an open cover for the compact set J, and so finitely many will do the job, say, $J \subset \bigcup_{i=1}^{k} J_i$, where $\omega(f; \bar{J}_i) < \delta$ for each $i = 1, \ldots, k$.

Now let $Q = \{t_0, \ldots, t_n\}$ be any partition of J containing the endpoints of each of the intervals $J \cap \bar{J}_i$. Then, since each interval $[t_{i-1}, t_i]$ is contained in some

$\bar{J}_m$, we have $\omega(f;[t_{i-1},t_i]) < \delta$. Hence,

$$U(f,Q) - L(f,Q) = \sum_{i=1}^{n} \omega(f;[t_{i-1},t_i])\,\Delta t_i$$
$$< \delta \sum_{i=1}^{n} \Delta t_i = \delta\, m^*(J). \quad \square$$

Finally, we are ready to finish the proof of Theorem 16.10.

PROOF (of Theorem 16.10, reverse implication). Suppose that $m^*\big(D(f)\big) = 0$; that is, suppose that $m^*\big(\{\omega_f \geq (1/k)\}\big) = 0$ for all k. We must show that $f \in \mathcal{R}[a,b]$.

Given $\varepsilon > 0$, we first choose a positive integer k with $(1/k) < \varepsilon$. Next, since $\{\omega_f \geq (1/k)\}$ is compact, we can find finitely many open intervals $I_1,\ldots,I_n$ such that $\{\omega_f \geq (1/k)\} \subset \bigcup_{j=1}^n I_j$ and $\sum_{j=1}^n m^*(I_j) < \varepsilon$. (How?)

Now $[a,b] \setminus \bigcup_{j=1}^n I_j$ is a finite union of closed intervals, say $J_1,\ldots,J_r$, and $\omega_f(x) < (1/k) < \varepsilon$ at each point $x \in \bigcup_{i=1}^r J_i$. In this way, $[a,b]$ has been decomposed into two sets of intervals: the I_j, which have small total length, and the J_i, on which f has small oscillation. We may apply Lemma 16.11 to find partitions $Q_1,\ldots,Q_r$ of $J_1,\ldots,J_r$ such that $U(f,Q_i) - L(f,Q_i) < \varepsilon\, m^*(J_i)$ for each $i = 1,\ldots,r$.

If we define a partition of $[a,b]$ by setting $P = \{a,b\} \cup \big(\bigcup_{i=1}^r Q_i\big)$, then

$$U(f,P) - L(f,P) = \sum_{i=1}^{r} \big[U(f,Q_i) - L(f,Q_i)\big] + \sum_{j=1}^{n} \omega(f;\bar{I}_j)\, m^*(I_j)$$
$$< \varepsilon \sum_{j=1}^{r} m^*(J_i) + 2\,\|f\|_\infty \sum_{j=1}^{n} m^*(I_j)$$
$$\leq \varepsilon(b-a) + 2\varepsilon\,\|f\|_\infty .$$

Hence, $f \in \mathcal{R}[a,b]$. $\square$

Combining Lebesgue's criterion with Theorem 14.19 yields two useful corollaries (see also Exercise 14.50).

Corollary 16.12. *If $f \in \mathcal{R}[a,b]$ and $F(x) = \int_a^x f$, then $F' = f$ a.e. (In particular, F' exists a.e.)*

Corollary 16.13. *If $f \in \mathcal{R}[a,b]$ and $\int_a^b |f| = 0$, then $f = 0$ a.e.*

EXERCISES

▷ **31.** For which subsets $A \subset [a,b]$ is χ_A Riemann integrable?

32. Prove Corollary 16.12.

33. Give a direct proof of Corollary 16.13. [Hint: If f is continuous at x_0, and if $f(x_0) \neq 0$, show that $\int_a^b |f| > 0$.]

34. If $f \in \mathcal{R}[a, b]$ and $\int_a^x f = 0$ for all x, prove that $f = 0$ a.e.

▷ **35.** If $f \in \mathcal{R}[a, b]$ and $f = g$ a.e., does it follow that $g \in \mathcal{R}[a, b]$? What if "a.e." is weakened to "except at countably many points"? Or to "except at finitely many points"?

36. If $f, g \in \mathcal{R}[a, b]$ and $f = g$ a.e., does it follow that $\int_a^b f = \int_a^b g$?

37. Let G be an open set containing the rationals in $[0, 1]$ with $m^*(G) < 1/2$. Prove that $f = \chi_G$ is *not* Riemann integrable on $[0, 1]$. Moreover, prove that f cannot be equal a.e. to any Riemann integrable function on $[0, 1]$; in other words, f is "substantially different" from any Riemann integrable function.

Measurable Sets

Let's briefly summarize our progress thus far. We have successfully defined a nonnegative function m^*, defined on all subsets of $\mathbb{R}$, that satisfies:

- m^* extends the notion of length; if I is an interval, then $m^*(I)$ is the length of I.
- m^* is translation invariant; $m^*(E + x) = m^*(E)$ for all E and all $x \in \mathbb{R}$.
- m^* is countably subadditive; $m^*\left(\bigcup_{n=1}^\infty E_n\right) \leq \sum_{n=1}^\infty m^*(E_n)$ for any sequence of sets (E_n).
- m^* is countably additive in certain cases; if (G_n) is a sequence of pairwise disjoint open sets, then $m^*\left(\bigcup_{n=1}^\infty G_n\right) = \sum_{n=1}^\infty m^*(G_n)$. (Why?)
- m^* is completely determined by its values on open sets; indeed, $m^*(E) = \inf\{m^*(U) : U \text{ is open and } E \subset U\}$.

The rumored failure of m^* to be countably additive, in general, will have to be taken on faith for just a bit longer – we will see an example later in this chapter. For now, let's concentrate on the good news: By taking a closer look at our last two observations, it is possible to isolate a large class of sets on which m^* is countably additive. The secret is to consider sets that are, in a sense, "approximately open."

Specifically, we say that a set E is (Lebesgue) **measurable** if, for each $\varepsilon > 0$, we can find a closed set F and an open set G with $F \subset E \subset G$ such that $m^*(G \setminus F) < \varepsilon$.

Please note that if E is measurable, then so is E^c, since $G^c \subset E^c \subset F^c$ and $F^c \setminus G^c = G \setminus F$. In fact, we might paraphrase the measurability condition by saying that *both* E and E^c are required to be "approximately open." In any case, notice that E is measurable if and only if E^c is measurable.

It is very easy to see that any interval, bounded or otherwise, is measurable. Equally simple is that any null set is measurable. Indeed, if $m^*(E) = 0$, then, for any $\varepsilon > 0$, we may choose an open set G containing E such that $m^*(G) < \varepsilon$. Since $F = \emptyset$ is a perfectly legitimate closed subset of E, it follows that E is measurable.

It is less clear that every open (closed) set is measurable. To help us with this task, let's first legitimize the usual operations with measurable sets.

Lemma 16.14. *If E_1 and E_2 are measurable sets, then so are $E_1 \cup E_2$, $E_1 \cap E_2$, and $E_1 \setminus E_2$.*

PROOF. Since $E_1 \cap E_2 = (E_1^c \cup E_2^c)^c$ and $E_1 \setminus E_2 = E_1 \cap E_2^c$, it is enough to check that $E_1 \cup E_2$ is measurable whenever E_1 and E_2 are measurable. (Why?)

Let $\varepsilon > 0$. Choose closed sets F_1, F_2 and open sets G_1, G_2, with $F_1 \subset E_1 \subset G_1$ and $F_2 \subset E_2 \subset G_2$, and such that $m^*(G_1 \setminus F_1) < \varepsilon/2$ and $m^*(G_2 \setminus F_2) < \varepsilon/2$. Then $F = F_1 \cup F_2$ is closed, $G = G_1 \cup G_2$ is open, $F \subset E_1 \cup E_2 \subset G$, and $G \setminus F \subset (G_1 \setminus F_1) \cup (G_2 \setminus F_2)$. Thus,

$$m^*(G \setminus F) \leq m^*(G_1 \setminus F_1) + m^*(G_2 \setminus F_2) < \varepsilon. \quad \square$$

We will write $\mathcal{M}$ for the collection of all measurable subsets of $\mathbb{R}$. Our goal in this section is to show that $\mathcal{M}$ contains a wealth of sets. From what we have just shown, we know that $\mathcal{M}$ is an **algebra of sets** (sometimes called a *Boolean algebra* or *Boolean lattice*). Specifically, this means that $E^c \in \mathcal{M}$ whenever $E \in \mathcal{M}$ and $E \cup F \in \mathcal{M}$ whenever $E, F \in \mathcal{M}$. By induction (and De Morgan's laws), it is easy to see that $\mathcal{M}$ is actually closed under any *finite* string of set operations.

The hard work comes in showing that $\mathcal{M}$ is closed under *countable* unions and intersections, too. From this it will follow that $\mathcal{M}$ contains the open sets, the closed sets, the G_δ-sets, the F_σ-sets, and so on. That may sound like a lot of sets, but all of these constitute a mere drop in the bucket! (All of the sets that we have listed so far, for example, form a collection having cardinality only $\mathfrak{c}$, whereas there are $2^{\mathfrak{c}}$ subsets of $\mathbb{R}$ altogether.)

In fact, the simple observation that $\Delta \in \mathcal{M}$ already implies that $\mathcal{M}$ is a huge collection of sets. How? Well, since Δ is a null set, so is every subset of Δ. Consequently, Δ and all of its subsets are measurable; thus, $\mathcal{P}(\Delta) \subset \mathcal{M} \subset \mathcal{P}(\mathbb{R})$. But Δ has the same cardinality as $\mathbb{R}$, and hence $\mathcal{M}$ has the same cardinality as $\mathcal{P}(\mathbb{R})$. Given this, it may surprise you to learn that there are, in fact, nonmeasurable subsets of $\mathbb{R}$. On the other hand, it will now come as no surprise that finding an example of a nonmeasurable set is by no means easy. This strange example awaits us later in this chapter, where we will solve the mystery of the lost countable additivity of m^*.

But for now, back to work! We still need to establish that open sets are measurable. We will begin by showing that bounded open sets and bounded closed sets (i.e., compact sets) are measurable.

Lemma 16.15.

(i) *If G is a bounded open set, then, for every $\varepsilon > 0$, there exists a closed set $F \subset G$ such that $m^*(F) > m^*(G) - \varepsilon$.*

(ii) *If F is a bounded closed set, then, for every $\varepsilon > 0$, there exists a bounded open set $G \supset F$ such that $m^*(G) < m^*(F) + \varepsilon$.*

(iii) *If F is a closed subset of a bounded open set G, then $m^*(G \setminus F) = m^*(G) - m^*(F)$.*

PROOF. Let G be a bounded open set and write $G = \bigcup_{n=1}^{\infty} I_n$, where (I_n) is a sequence of pairwise disjoint open intervals. Then (from Exercise 9), $\sum_{n=1}^{\infty} m^*(I_n) = m^*(G) < \infty$. Now, given $\varepsilon > 0$, choose N such that $\sum_{n=N+1}^{\infty} m^*(I_n) < \varepsilon/2$. For each $n = 1, \ldots, N$, choose a closed subinterval $J_n \subset I_n$ with $m^*(J_n) > m^*(I_n) - \varepsilon/(2N)$. Then, $F = \bigcup_{n=1}^{N} J_n$ is a closed subset of G and, from Corollary 16.9,

$$m^*(F) = \sum_{n=1}^{N} m^*(J_n) > \sum_{n=1}^{N} m^*(I_n) - \varepsilon/2 \geq m^*(G) - \varepsilon.$$

This proves (i).

Next, suppose that F is a bounded closed set, and let $\varepsilon > 0$. Since F is a compact set of finite outer measure, we may choose finitely many open intervals $I_1, \ldots, I_n$ such that $G = \bigcup_{j=1}^{n} I_j$ is an open set containing F, and such that $m^*(G) \leq \sum_{j=1}^{n} m^*(I_j) \leq m^*(F) + \varepsilon$. This proves (ii).

Finally, suppose that F is a closed subset of a bounded open set G. Then $G \setminus F$ is also a bounded open set. Hence, by (i), for any $\varepsilon > 0$, there is a closed set $E \subset G \setminus F$ such that $m^*(E) > m^*(G \setminus F) - \varepsilon$. But then, E and F are disjoint compact sets and so

$$\begin{aligned} m^*(G) &\leq m^*(G \setminus F) + m^*(F) \\ &< m^*(E) + \varepsilon + m^*(F) \\ &= m^*(E \cup F) + \varepsilon \leq m^*(G) + \varepsilon. \end{aligned}$$

Since this holds for any ε, we must have $m^*(G) = m^*(G \setminus F) + m^*(F)$. This completes the proof. □

Our next lemma shows that it is enough to consider bounded sets when testing measurability.

Lemma 16.16. *E is measurable if and only if $E \cap (a, b)$ is measurable for every bounded open interval (a, b).*

PROOF. The forward implication is clear from Lemma 16.14. So, suppose that $E \cap (a, b)$ is measurable for any (a, b), and let $\varepsilon > 0$. Then, in particular, for each integer $n \in \mathbb{Z}$ we can find a closed set F_n and an open set G_n with $F_n \subset E \cap (n, n+1) \subset G_n$ and such that $m^*(G_n \setminus F_n) < 2^{-|n|}\varepsilon$. By enlarging G_n slightly, if necessary, we may also suppose that both $n, n+1 \in G_n$. In this way, $G = \bigcup_{n\in\mathbb{Z}} G_n$ is an open set containing E.

Now, $F = \bigcup_{n\in\mathbb{Z}} F_n$ is certainly a subset of E, but is it closed? Well, sure! A convergent sequence from F must eventually lie in some open interval of the form $(n - 1, n + 1)$. Thus, all but finitely many terms are in $F_{n-1} \cup F_n$ for some n. Since $F_{n-1} \cup F_n$ is closed, the limit must be in one of the two; in particular, the limit must be back in F. Thus, F is closed.

Finally, $G \setminus F \subset \bigcup_{n\in\mathbb{Z}}(G_n \setminus F_n)$, and hence $m^*(G \setminus F) \leq \sum_{n\in\mathbb{Z}} m^*(G_n \setminus F_n) < \sum_{n\in\mathbb{Z}} 2^{-|n|}\varepsilon = 3\varepsilon$. □

Corollary 16.17. *Open sets, and hence also closed sets, are measurable.*

Finally we are ready to show that $\mathcal{M}$ is closed under countable disjoint unions. At the same time, we will show that m^* is countably additive when applied to pairwise disjoint measurable sets.

Theorem 16.18. *If (E_n) is a sequence of pairwise disjoint measurable sets, then $E = \bigcup_{n=1}^{\infty} E_n$ is measurable and $m^*(E) = \sum_{n=1}^{\infty} m^*(E_n)$.*

PROOF. We first suppose that E is contained in some bounded open interval I and, in particular, that $m^*(E) < \infty$. Of course, this means that $E_n \subset I$ for all n, too. Now, given $\varepsilon > 0$, choose closed sets (F_n) and open sets (G_n) such that $F_n \subset E_n \subset G_n \subset I$ and such that $m^*(G_n \setminus F_n) < 2^{-n}\varepsilon$ for all n. Next, since the E_n are pairwise disjoint and bounded, so are the F_n. Hence, for any K, we have

$$\begin{aligned}\sum_{n=1}^{K} m^*(E_n) &\le \sum_{n=1}^{K} m^*(F_n) + \varepsilon \qquad \text{(Why?)}\\ &= m^*\left(\bigcup_{n=1}^{K} F_n\right) + \varepsilon\\ &\le m^*(E) + \varepsilon.\end{aligned}$$

Since K and ε are arbitrary, it follows that $\sum_{n=1}^{\infty} m^*(E_n) \le m^*(E)$. Thus, $m^*(E) = \sum_{n=1}^{\infty} m^*(E_n)$, since the other inequality is supplied by countable subadditivity.

Next, notice that we also have

$$\sum_{n=1}^{\infty} m^*(G_n) \le \sum_{n=1}^{\infty} m^*(E_n) + \varepsilon = m^*(E) + \varepsilon < \infty.$$

In particular, we may choose N such that $\sum_{n=N+1}^{\infty} m^*(G_n) < \varepsilon$. Finally, $G = \bigcup_{n=1}^{\infty} G_n$ is an open set containing E and $F = \bigcup_{n=1}^{N} F_n$ is a closed set contained in E with

$$m^*(G \setminus F) \le \sum_{n=1}^{N} m^*(G_n \setminus F_n) + \sum_{n=N+1}^{\infty} m^*(G_n) < 2\varepsilon.$$

Hence, E is measurable.

Lastly, suppose that E is unbounded. We know that E is measurable from the first part of the proof (and Lemma 16.16), but we still have to check countable additivity in this case. To this end, consider

$$E_{n,k} = E_n \cap (k, k+1] \qquad \text{and} \qquad A_k = E \cap (k, k+1], \qquad \text{for } k \in \mathbb{Z}.$$

The sets $(E_{n,k})$ and (A_k) are measurable and pairwise disjoint and, of course,

$$E_n = \bigcup_{k=-\infty}^{\infty} E_{n,k} \qquad \text{and} \qquad E = \bigcup_{k=-\infty}^{\infty} A_k = \bigcup_{k=-\infty}^{\infty} \bigcup_{n=1}^{\infty} E_{n,k}.$$

By countable subadditivity and the first part of the proof we have

$$\sum_{n=1}^{\infty} m^*(E_n) \le \sum_{n=1}^{\infty} \sum_{k=-\infty}^{\infty} m^*(E_{n,k}) = \sum_{k=-\infty}^{\infty} \sum_{n=1}^{\infty} m^*(E_{n,k}) = \sum_{k=-\infty}^{\infty} m^*(A_k),$$

since each A_k is a bounded measurable set. But, for any N, the first part of the proof also tells us that

$$\sum_{k=-N}^{N} m^*(A_k) = m^*\left(\bigcup_{k=-N}^{N} A_k \right) \le m^*(E).$$

Putting the pieces together, we get $\sum_{n=1}^{\infty} m^*(E_n) \le m^*(E)$, which is all we need. □

Theorem 16.18 tells us that $\mathcal{M}$ is closed under countable *disjoint* unions, but what about arbitrary countable unions? Well, as luck would have it, since $\mathcal{M}$ is an algebra of sets, disjoint unions are the rule and not the exception.

Lemma 16.19. *Given any sequence (A_i) of measurable sets, we can find a sequence (B_i) of disjoint measurable sets such that $B_i \subset A_i$ for all i and $\bigcup_{i=1}^{\infty} A_i = \bigcup_{i=1}^{\infty} B_i$.*

PROOF. Let $B_1 = A_1$, and for each $n > 1$ let $B_n = A_n \setminus \bigcup_{i=1}^{n-1} A_i$. Then $B_n \in \mathcal{M}$, since $\mathcal{M}$ is an algebra. Clearly, $B_i \subset A_i$ for all i, $B_i \cap B_j = \emptyset$ for $i \ne j$, and $\bigcup_{i=1}^{n} A_i = \bigcup_{i=1}^{n} B_i$ for all n. □

Corollary 16.20. *If (E_n) is an arbitrary sequence of measurable sets, then $\bigcup_{n=1}^{\infty} E_n$ and $\bigcap_{n=1}^{\infty} E_n$ are measurable.*

An algebra of sets that is closed under countable unions (or intersections) is called a **σ-algebra**. Thus, we have shown that the collection $\mathcal{M}$ of measurable sets is a σ-algebra and that the restriction of m^* to $\mathcal{M}$ is countably additive (and so is a solution, of sorts, to the problem of measure).

Lebesgue measure m is defined to be the restriction of m^* to $\mathcal{M}$. If E is measurable, we write $m(E)$ in place of $m^*(E)$, and we refer to $m(E)$ as the (Lebesgue) measure of E.

EXERCISES

38. Prove that E is measurable if and only if $E \cap K$ is measurable for every compact set K.

39. If $A \supset B$ are measurable, show that $m(A \setminus B) = m(A) - m(B)$ whenever $m(B) < \infty$.

40. If A and B are measurable sets, show that $m(A \cup B) + m(A \cap B) = m(A) + m(B)$.

41. Let E denote the set of all real numbers in $[\,0, 1\,]$ whose decimal expansions contain no 5's or 7's. Prove that E is measurable and compute $m(E)$. [Hint: There are only a few "ambiguous" numbers; it does not matter whether they are included. Why?]

▷ **42.** Suppose that E is measurable with $m(E) = 1$. Show that:

(a) There is a measurable set $F \subset E$ such that $m(F) = 1/2$. [Hint: Consider the function $f(x) = m(E \cap (-\infty, x\,]).$]

(b) There is a closed set F, consisting entirely of irrationals, such that $F \subset E$ and $m(F) = 1/2$.

(c) There is a compact set F with empty interior such that $F \subset E$ and $m(F) = 1/2$.

43. Let $E \subset [\,a, b\,]$. According to Lebesgue's original definition, E is measurable if and only if $m_*(E) = m^*(E)$. (See Example 16.3 (f).) Check that Lebesgue's definition is the same as ours in this case. [Hint: It is easy to see that our notion of measurability implies Lebesgue's. If, on the other hand, E is measurable according to Lebesgue's definition, note that an open superset of $[\,a, b\,] \setminus E$ supplies a closed subset of E.]

44. Let E be a measurable set with $m(E) > 0$. Prove that $E - E = \{x - y : x, y \in E\}$ contains an interval centered at 0. [This is a famous result due to Steinhaus. There are several proofs available; here is a particularly simple one: Take I as in Exercise 25 for $\alpha = 3/4$. If $|x| < m(I)/2$, note that $I \cup (I + x)$ has measure at most $3m(I)/2$. Thus, $E \cap I$ and $(E \cap I) + x$ cannot be disjoint. (Why?) Finally, $(E + x) \cap E \neq \emptyset$ means that $x \in E - E$; that is, $E - E \supset (-m(I)/2, m(I)/2)$.]

45. Let $f : X \to Y$ be any function.

(a) If $\mathcal{B}$ is a σ-algebra of subsets of Y, show that $\mathcal{A} = \{f^{-1}(B) : B \in \mathcal{B}\}$ is a σ-algebra of subsets of X.

(b) If $\mathcal{A}$ is a σ-algebra of subsets of X, show that $\mathcal{B} = \{B : f^{-1}(B) \in \mathcal{A}\}$ is a σ-algebra of subsets of Y.

46. Let $\mathcal{A}$ be an algebra of sets. Show that the following are equivalent:

(i) $\mathcal{A}$ is closed under arbitrary countable unions; that is, if $E_n \in \mathcal{A}$ for all n, then $\bigcup_{n=1}^{\infty} E_n \in \mathcal{A}$.

(ii) $\mathcal{A}$ is closed under countable disjoint unions; that is, if (E_n) is a sequence of pairwise disjoint sets from $\mathcal{A}$, then $\bigcup_{n=1}^{\infty} E_n \in \mathcal{A}$.

(iii) $\mathcal{A}$ is closed under increasing countable unions; that is, if $E_n \in \mathcal{A}$ for all n, and if $E_n \subset E_{n+1}$ for all n, then $\bigcup_{n=1}^{\infty} E_n \in \mathcal{A}$.

47. $\{\emptyset, \mathbb{R}\}$ and $\mathcal{P}(\mathbb{R})$ are both σ-algebras, and $\{\emptyset, \mathbb{R}\} \subset \mathcal{A} \subset \mathcal{P}(\mathbb{R})$ holds for any other σ-algebra of subsets of $\mathbb{R}$.

▷ **48.** Let $\mathcal{E}$ be any collection of subsets of $\mathbb{R}$. Show that there is always a smallest σ-algebra $\mathcal{A}$ containing $\mathcal{E}$. [Hint: Show that the intersection of σ-algebras is again a σ-algebra.]

▷ **49.** The smallest σ-algebra containing $\mathcal{E}$ is called **the σ-algebra generated by $\mathcal{E}$** and is denoted by $\sigma(\mathcal{E})$. If $\mathcal{E} \subset \mathcal{F}$, prove that $\sigma(\mathcal{E}) \subset \sigma(\mathcal{F})$.

50. Prove that $\mathcal{A} = \{E \subset \mathbb{R} : \text{either } E \text{ or } E^c \text{ is countable}\}$ is a σ-algebra; in fact, $\mathcal{A}$ is the σ-algebra generated by the singletons.

51. Let $\mathcal{A} = \{E \subset \mathbb{R} : \text{either } E \text{ or } E^c \text{ is finite}\}$. Is $\mathcal{A}$ an algebra? Is $\mathcal{A}$ a σ-algebra? Explain.

52. Show that $\mathcal{A} = \{E \subset \mathbb{R} : \text{either } m(E) = 0 \text{ or } m(E^c) = 0\}$ is a σ-algebra; in fact, $\mathcal{A}$ is the σ-algebra generated by the null sets.

The **Borel σ-algebra** $\mathcal{B}$ is defined to be the smallest σ-algebra of subsets of $\mathbb{R}$ containing the open sets; equivalently, $\mathcal{B}$ is the σ-algebra generated by the (open) intervals (see Exercise 53). The elements of $\mathcal{B}$ are called the **Borel sets**. Notice that closed sets, G_δ-sets, F_σ-sets, $G_{\delta\sigma}$-sets, and so on, are all Borel sets. From Corollaries 16.17 and 16.20, every Borel set is measurable; that is, $\mathcal{B} \subset \mathcal{M}$.

▷ **53.** Show that $\mathcal{B}$ is generated by each of the following:

(i) The open intervals $\mathcal{E}_1 = \{(a, b) : a < b\}$.

(ii) The closed intervals $\mathcal{E}_2 = \{[a, b] : a < b\}$.

(iii) The half-open intervals $\mathcal{E}_3 = \{(a, b], [a, b) : a < b\}$.

(iv) The open rays $\mathcal{E}_4 = \{(a, \infty), (-\infty, b) : a, b \in \mathbb{R}\}$.

(v) The closed rays $\mathcal{E}_5 = \{[a, \infty), (-\infty, b] : a, b \in \mathbb{R}\}$.

[Hint: It is easy to see that $\mathcal{B} = \sigma(\mathcal{E}_1)$. In each of the remaining cases, you just need to show that $\mathcal{E}_1 \subset \sigma(\mathcal{E}_i)$ for $i = 2, 3, 4, 5$. Why?]

54. Prove that the collection of all open subsets of $\mathbb{R}$ has cardinality $\mathfrak{c}$. What is the cardinality of the collection of all G_δ subsets of $\mathbb{R}$?

The Structure of Measurable Sets

At this point we know that the collection $\mathcal{M}$ of measurable sets is a σ-algebra containing the open sets, and hence all of the Borel sets $\mathcal{B}$, and we know that Lebesgue measure m, the restriction of Lebesgue outer measure m^* to $\mathcal{M}$, is countably additive on $\mathcal{M}$. Moreover, we know that m^*, and hence also m, is completely determined by its values on open sets. In this section, we will pursue this last observation still further and, in so doing, arrive at a connection between the Borel sets $\mathcal{B}$ and the Lebesgue measurable sets $\mathcal{M}$.

To begin, we note that a Lebesgue measurable set differs from a Borel set by a set of measure zero.

Theorem 16.21. *For a subset E of $\mathbb{R}$, the following are equivalent:*

(i) *E is measurable.*

(ii) *For every $\varepsilon > 0$, there exists an open set $G \supset E$ such that $m^*(G \setminus E) < \varepsilon$.*

(iii) *For every $\varepsilon > 0$, there exists a closed set $F \subset E$ such that $m^*(E \setminus F) < \varepsilon$.*

(iv) *$E = G \setminus N$, where G is a G_δ-set and N is a null set.*

(v) *$E = F \cup N$, where F is an F_σ-set and N is a null set.*

PROOF. If E is measurable, then certainly both (ii) and (iii) hold. Also, since null sets and Borel sets are measurable, either (iv) or (v) implies that E is measurable. Thus, it is enough to show that (ii) implies (iv) and that (iii) implies (v). (Why?)

So, suppose that (ii) holds. Then, for each n, there is an open set G_n such that $E \subset G_n$ and $m^*(G_n \setminus E) < 1/n$. Let $G = \bigcap_{n=1}^{\infty} G_n$. Clearly, G is a G_δ-set; moreover, $G \setminus E$ is a null set because it is contained in $G_n \setminus E$ and so has measure at most $1/n$ for any n. That is, (iv) holds. The proof that (iii) implies (v) is very similar. □

Corollary 16.22. *If $m(E) = 0$, then E is contained in a Borel set G with $m(G) = 0$.*

The conclusion to be drawn here is this: A Lebesgue measurable set is a Borel set plus (or minus) a *subset* of a Borel set of measure zero. While a subset of a Borel set need not be a Borel set (as we will see later), a subset of a null set is always a null set. Thus there are more measurable sets than Borel sets. In fact, it can be shown, by using transfinite induction, that the Borel σ-algebra $\mathcal{B}$ has cardinality $\mathfrak{c}$ while, as we have seen, the Lebesgue σ-algebra $\mathcal{M}$ has cardinality $2^{\mathfrak{c}}$.

The Lebesgue measurable sets are said to be **complete** because every subset of a null set is again measurable. In fact, the Lebesgue measurable sets are the *completion* of the Borel sets (see Exercises 56 and 57).

EXERCISES

55. Complete the proof of Theorem 16.21.

56. Given a σ-algebra $\mathcal{A}$ of subsets of $\mathbb{R}$, let

$$\bar{\mathcal{A}} = \{E \cup N : E \in \mathcal{A} \text{ and } N \subset F \in \mathcal{A} \text{ with } m(F) = 0\}.$$

$\bar{\mathcal{A}}$ is called the **completion** of $\mathcal{A}$ (with respect to m). Show that $\bar{\mathcal{A}}$ is a σ-algebra. [Hint: First show that $\bar{\mathcal{A}}$ is an algebra.]

57. Prove Corollary 16.22, thus showing that $\mathcal{M} = \bar{\mathcal{B}}$, the completion of the Borel σ-algebra.

▷ **58.** Suppose that $m^*(E) < \infty$. Prove that E is measurable if and only if, for every $\varepsilon > 0$, there is a finite union of bounded intervals A such that $m^*(E \triangle A) < \varepsilon$ (where $E \triangle A$ is the symmetric difference of E and A).

▷ **59.** If E is a Borel set, show that $E + x$ and rE are Borel sets for any $x, r \in \mathbb{R}$. [Hint: Show, for example, that $\mathcal{A} = \{E : E + x \in \mathcal{B}\}$ is a σ-algebra containing the intervals.]

▷ **60.** If E is a measurable set, show that $E + x$ and rE are measurable for any x, $r \in \mathbb{R}$. [Hint: Use Theorem 16.21.]

Our next result should be viewed as a continuity property of Lebesgue measure.

Theorem 16.23. *Let (E_n) be a sequence of measurable sets.*

(i) *If $E_n \subset E_{n+1}$ for each n, then $m\left(\bigcup_{n=1}^{\infty} E_n\right) = \lim_{n\to\infty} m(E_n)$.*

(ii) *If $E_n \supset E_{n+1}$ for each n, and if some E_k has $m(E_k) < \infty$, then $m\left(\bigcap_{n=1}^{\infty} E_n\right) = \lim_{n\to\infty} m(E_n)$.*

PROOF. Please note that, in either case, $\bigcup_{n=1}^{\infty} E_n$ and $\bigcap_{n=1}^{\infty} E_n$ are measurable. The "trick" in each case is to manufacture a disjoint union of sets and appeal to the countable additivity of m.

First, suppose that $E_n \subset E_{n+1}$ for each n. Then, $m(E_n) \le m(E_{n+1})$ for all n, and hence $\lim_{n\to\infty} m(E_n) = \sup_n m(E_n)$ exists and is at most $m\big(\bigcup_{n=1}^{\infty} E_n\big)$. Of course, if some E_n has infinite measure, then so does $\bigcup_{n=1}^{\infty} E_n$; thus, we may assume that each E_n has finite measure. Next, notice that

$$\bigcup_{n=1}^{\infty} E_n = E_1 \cup \bigcup_{n=1}^{\infty} (E_{n+1} \setminus E_n),$$

and hence, since $m(E_n) < \infty$ for all n, we get

$$\begin{aligned} m\left(\bigcup_{n=1}^{\infty} E_n\right) &= m(E_1) + \sum_{n=1}^{\infty} m\big(E_{n+1} \setminus E_n\big) \\ &= m(E_1) + \sum_{n=1}^{\infty} \big[m(E_{n+1}) - m(E_n)\big] \\ &= \lim_{n\to\infty} m(E_{n+1}). \end{aligned}$$

Next, suppose that $E_n \supset E_{n+1}$ for each n. Then, $m(E_n) \ge m(E_{n+1})$ for all n and, again, $\lim_{n\to\infty} m(E_n) = \inf_n m(E_n)$ exists and is at least $m\big(\bigcap_{n=1}^{\infty} E_n\big)$. Now, if some E_k has finite measure, then, by relabeling, we may simply suppose that E_1 has finite measure. (Why does this work?) Then, since

$$E_1 \setminus \bigcap_{n=1}^{\infty} E_n = \bigcup_{n=1}^{\infty} (E_n \setminus E_{n+1}),$$

we have

$$\begin{aligned} m(E_1) - m\left(\bigcap_{n=1}^{\infty} E_n\right) &= m\left(E_1 \setminus \bigcap_{n=1}^{\infty} E_n\right) \\ &= \sum_{n=1}^{\infty} m\big(E_n \setminus E_{n+1}\big) \\ &= \sum_{n=1}^{\infty} \big[m(E_n) - m(E_{n+1})\big] \\ &= m(E_1) - \lim_{n\to\infty} m(E_n). \end{aligned}$$

Hence, $m\big(\bigcap_{n=1}^{\infty} E_n\big) = \lim_{n\to\infty} m(E_n)$. □

If we think of $\mathcal{M}$ as a lattice, where $A \le B$ means that $A \subset B$, then $\bigcup_{n=1}^{\infty} E_n$ is the same as $\sup_n E_n$ for an increasing sequence of sets (E_n). Likewise, $\bigcap_{n=1}^{\infty} E_n$ is the same as $\inf_n E_n$ for a decreasing sequence of sets (E_n). Thus, the conclusion of the theorem is that $m\big(\sup_n E_n\big) = \sup_n m(E_n)$ for an increasing sequence of measurable sets (E_n) and $m\big(\inf_n E_n\big) = \inf_n m(E_n)$ for a decreasing sequence of measurable sets (E_n), provided that $\inf_n m(E_n)$ is finite. From this point of view, Theorem 16.23 is a continuity result.

In particular, notice that if (E_n) decreases to the empty set $\emptyset$, and if some E_k has finite measure, then $m(E_n)$ decreases to 0. This says that m is "continuous at $\emptyset$" as a function on $\mathcal{M}$ (for more details, see Exercise 66). Also, note that if E is any measurable set, then $m(E) = \lim_{n\to\infty} m\big(E \cap [-n, n\,]\big)$. If, in addition, $m(E) < \infty$, then we could also write $\lim_{n\to\infty} m\big(E \setminus [-n, n\,]\big) = 0$.

As a corollary to Theorem 16.23, we have *the Borel–Cantelli lemma.*

Corollary 16.24. *If each E_n is measurable, and if $\sum_{n=1}^{\infty} m(E_n) < \infty$, then*

$$m\left(\bigcap_{n=1}^{\infty}\bigcup_{k=n}^{\infty} E_k\right) = m\left(\limsup_{n\to\infty} E_n\right) = 0.$$

Corollary 16.25. *For any set $E \subset \mathbb{R}$, we have*

$$m^*(E) = \inf\{m(G) : E \subset G \text{ and } G \text{ is open}\}.$$

If E is measurable, then we also have

$$m(E) = \sup\{m(K) : K \subset E \text{ and } K \text{ is compact}\}.$$

PROOF. The first formula follows from Corollary 16.6. For the second, suppose that E is measurable. For each n, choose a compact set $K_n \subset E \cap [-n, n\,]$ such that $m(K_n) > m(E \cap [-n, n\,]) - 1/n$. Since $m(E \cap [-n, n\,])$ increases to $m(E)$, it follows that

$$\begin{aligned} m(E) &\ge \sup\{m(K) : K \subset E \text{ and } K \text{ is compact}\} \\ &\ge \sup_n m(K_n) \\ &\ge \limsup_{n\to\infty} m(K_n) = m(E). \quad \square \end{aligned}$$

Our continuity result also allows us to "fine tune" the characterization of measurable sets given by Theorem 16.22 in the case of sets with finite outer measure (or bounded sets).

Corollary 16.26. *Suppose that $m^*(E) < \infty$. Then, E is measurable if and only if, for every $\varepsilon > 0$, there exists a compact set $F \subset E$ such that $m(F) > m^*(E) - \varepsilon$.*

EXERCISES

61. Find a sequence of measurable sets (E_n) that decrease to $\emptyset$, but with $m(E_n) = \infty$ for all n.

62. If E_n is measurable for each n, show that $m(\liminf_{n\to\infty} E_n) \le \liminf_{n\to\infty} m(E_n)$ and also that $m(\limsup_{n\to\infty} E_n) \ge \limsup_{n\to\infty} m(E_n)$, provided that $m(\bigcup_{n=k}^{\infty} E_n) < \infty$ for some $k \ge 1$.

▷ **63.** Prove Corollary 16.24.

▷ **64.** Prove Corollary 16.26.

65. Let $\mathcal{M}_1$ denote the measurable subsets of $[\,0, 1\,]$. Given $E, F \in \mathcal{M}_1$, define $E \sim F$ if $m(E \Delta F) = 0$. Prove that $\sim$ is an equivalence relation.

66. In the notation of Exercise 65, define $d(E, F) = m(E \Delta F)$ for $E, F \in \mathcal{M}_1$. Prove that d defines a pseudometric on $\mathcal{M}_1$. (That is, d induces a metric on $\mathcal{M}_1/\sim$, the set of equivalence classes under equality a.e.)

67. In the notation of Exercise 65, show that m is continuous as a function on $(\mathcal{M}_1, d)$. [Hint: Since m is additive, you only need to check continuity at one point; $\emptyset$ is a convenient choice.]

68. Prove that $(\mathcal{M}_1, d)$ is complete. [Hint: If (E_n) is d-Cauchy, then, by passing to a subsequence, you may assume that $d(E_n, E_{n+1}) < 2^{-n}$. Now argue that (E_n) converges to, say, $\limsup_{n\to\infty} E_n$.]

For our final topic in this section, we further demonstrate the interplay between Lebesgue measure and the topology of $\mathbb{R}$ by presenting an important result concerning coverings by families of intervals.

We say that a collection $\mathcal{C}$ of closed, nontrivial intervals in $\mathbb{R}$ forms a **Vitali cover** for a subset E of $\mathbb{R}$ if, for any $x \in E$ and any $\varepsilon > 0$, there is an interval $I \in \mathcal{C}$ with $x \in I$ and $m(I) < \varepsilon$. In other words, $\mathcal{C}$ is a Vitali cover for E if, for every $\varepsilon > 0$,

$$E \subset \bigcup \{I : I \in \mathcal{C} \text{ and } m(I) < \varepsilon\}.$$

In particular, notice that if $\mathcal{C}$ is a Vitali cover for E, then so is the collection

$$\{I \in \mathcal{C} : m(I) < \varepsilon\}$$

for any (fixed) $\varepsilon > 0$. Loosely speaking, the intervals in $\mathcal{C}$ form a neighborhood base for the points in E; that is, given a point $x \in E$ and any open set U containing x, we can always find an interval I from $\mathcal{C}$ with $x \in I \subset U$. (How?)

Vitali's Covering Theorem 16.27. *Let E be a set of finite outer measure, and let $\mathcal{C}$ be a Vitali cover for E. Then, there exist countably many pairwise disjoint intervals (I_n) in $\mathcal{C}$ such that*

$$m\left(E \setminus \bigcup_{n=1}^{\infty} I_n\right) = 0.$$

PROOF. We can simplify things a bit by making two observations: First, since $m^*(E) < \infty$, there is an open set U containing E with $m(U) < \infty$. Next, given $x \in E \subset U$ and $\varepsilon > 0$, there is an interval $I \in \mathcal{C}$ such that $x \in I \subset U$ and $m(I) < \varepsilon$. Thus, the collection $\{I \in \mathcal{C} : I \subset U\}$ is still a Vitali cover for E. Since it is enough to prove the theorem for this collection, we may simply suppose that each element of $\mathcal{C}$ is already contained in U.

To begin, choose any interval I_1 in $\mathcal{C}$. If $m(E \setminus I_1) = 0$, we are done; otherwise, we continue to choose intervals from $\mathcal{C}$ according to the following scheme: Suppose that pairwise disjoint, closed intervals $I_1, \ldots, I_n$ have been constructed with $m^*\left(E \setminus \bigcup_{k=1}^{n} I_k\right) > 0$. We want to choose I_{n+1} so that it is the "next biggest" interval in $\mathcal{C}$ that is disjoint from $I_1, \ldots, I_n$. To accomplish this, consider the intervals in $\mathcal{C}$ that are completely contained in the *open set*

$$G_n = U \setminus \bigcup_{k=1}^{n} I_k.$$

Since $E \setminus \bigcup_{k=1}^{n} I_k \neq \emptyset$, and since $\mathcal{C}$ is a Vitali cover for E, such intervals exist; notice that any such interval J will also satisfy $0 < m(J) \leq m(U)$ (since the intervals in $\mathcal{C}$ are nontrivial). Setting

$$k_n = \sup\{m(J) : J \in \mathcal{C} \text{ and } J \subset G_n\},$$

it is clear that $0 < k_n < \infty$. We now choose $I_{n+1} \in \mathcal{C}$ with $m(I_{n+1}) > k_n/2$ and $I_{n+1} \subset G_n = U \setminus \bigcup_{k=1}^{n} I_k$. Obviously, I_{n+1} is disjoint from $I_1, \ldots, I_n$. If $m\left(E \setminus \bigcup_{k=1}^{n+1} I_k\right) = 0$, the construction terminates and the theorem is proved; otherwise we continue, choosing I_{n+2}, and so on.

If our construction does not terminate in finitely many steps, then it yields a sequence (I_k) of pairwise disjoint intervals in $\mathcal{C}$ with $\bigcup_{k=1}^{\infty} I_k \subset U$ and, of course, $\sum_{k=1}^{\infty} m(I_k) \leq m(U) < \infty$. It only remains to show that $m\left(E \setminus \bigcup_{k=1}^{\infty} I_k\right) = 0$. To this end, first notice that each $J \in \mathcal{C}$ must hit some I_n. Indeed, if $J \cap \left(\bigcup_{k=1}^{n} I_k\right) = \emptyset$ for all n, then we would have $m(J) \leq k_n < 2m(I_{n+1}) \to 0$ (as $n \to \infty$), which contradicts the fact that $m(J) > 0$.

Finally, let $\varepsilon > 0$ and choose N so that $\sum_{k=N+1}^{\infty} m(I_k) < \varepsilon$. Given $x \in E \setminus \bigcup_{k=1}^{N} I_k \subset G_N$, choose an interval $J \in \mathcal{C}$ with $x \in J$ and $J \cap \left(\bigcup_{k=1}^{N} I_k\right) = \emptyset$. By our observation above, we know that there is a smallest n such that $J \cap I_n \neq \emptyset$. Necessarily, $n > N$ and $m(J) < 2m(I_n)$. (Why?) Thus, if we let J_n be the closed interval having the same midpoint as I_n but with radius five times that of I_n, that is, with $m(J_n) = 5m(I_n)$, then it is easy to see that $J \subset J_n$. (Why?) In other words, what we have shown is that

$$E \setminus \bigcup_{k=1}^{\infty} I_k \subset E \setminus \bigcup_{k=1}^{N} I_k \subset \bigcup_{k=N+1}^{\infty} J_k,$$

and so

$$\begin{aligned} m^*\left(E \setminus \bigcup_{k=1}^{\infty} I_k\right) \leq m^*\left(E \setminus \bigcup_{k=1}^{N} I_k\right) &\leq \sum_{k=N+1}^{\infty} m(J_k) \\ &= 5 \sum_{k=N+1}^{\infty} m(I_k) < 5\varepsilon. \end{aligned}$$

Since ε is arbitrary, we get $m\left(E \setminus \bigcup_{k=1}^{\infty} I_k\right) = 0$. □

Corollary 16.28. *Let E be a set of finite outer measure, and let $\mathcal{C}$ be a Vitali cover for E. Given $\varepsilon > 0$, there are finitely many pairwise disjoint intervals $I_1, \ldots, I_n$ in $\mathcal{C}$ such that*

$$m^*\left(E \setminus \bigcup_{k=1}^{n} I_k\right) < \varepsilon.$$

Corollary 16.29. *An arbitrary union of intervals is measurable. That is, if $(I_\alpha)_{\alpha \in A}$ is any collection of intervals in $\mathbb{R}$, then the set $E = \bigcup_{\alpha \in A} I_\alpha$ is measurable.*

EXERCISES

69. Let E be a set of finite outer measure, and suppose that for some sequence of intervals (I_n) we have $m\left(E \setminus \bigcup_{n=1}^\infty I_n\right) = 0$. Show that $m^*(E) \le \sum_{n=1}^\infty m(I_n)$.

70. Prove Corollary 16.29. [Hint: Let $\mathcal{C}$ be the collection of all closed intervals J such that $J \subset I_\alpha$ for some α.]

A Nonmeasurable Set

Well, now for the bad news: There exist nonmeasurable sets. In this section we will present an example due to Vitali, dating back to 1905. You may find it easier to follow the example if you first know where it comes from. We identify the interval $[0, 1)$ with the unit circle in $\mathbb{C}$ (or in $\mathbb{R}^2$) under the map: $x \mapsto 2\pi x \mapsto e^{2\pi i x}$ (or $(\cos 2\pi x, \sin 2\pi x)$). That is, $[0, 1)$ is identified with $[0, 2\pi)$, and then $[0, 2\pi)$ is wrapped around the circle, in the usual way, by identifying each angle in $[0, 2\pi)$ with the point it determines on the circle (see Figure 16.3).

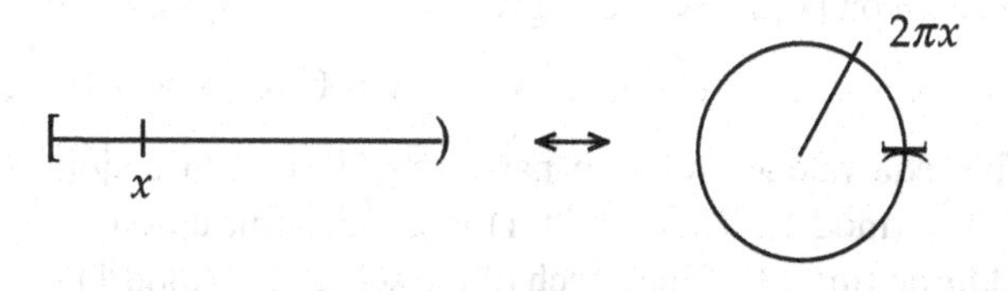

Figure 16.3

Under this identification, the addition of angles corresponds to **addition (mod 1)**. Specifically, given $x, y \in [0, 1)$, we define

$$x + y \ (\mathrm{mod}\ 1) = \begin{cases} x + y, & \text{if } x + y < 1 \\ x + y - 1, & \text{if } x + y \ge 1. \end{cases}$$

Given a subset E of $[0, 1)$, we also define the translate of E under addition (mod 1) by

$$E + x \ (\mathrm{mod}\ 1) = \{a + x \ (\mathrm{mod}\ 1) : a \in E\}.$$

In this way, translation by x (mod 1) in the interval $[0, 1)$ corresponds to rotation through an angle $2\pi x$ on the circle (see Figure 16.4).

It is easy to see that addition (mod 1) is reasonably well behaved; for example, $x + y$ (mod 1) $= y + x$ (mod 1). Better still, Lebesgue measure is invariant under translation (mod 1).

Lemma 16.30. *Let $E \subset [0, 1)$ and $x \in [0, 1)$. If E is measurable, then so is $E + x$ (mod 1). Moreover, in this case, $m\big(E + x \ (\mathrm{mod}\ 1)\big) = m(E)$.*

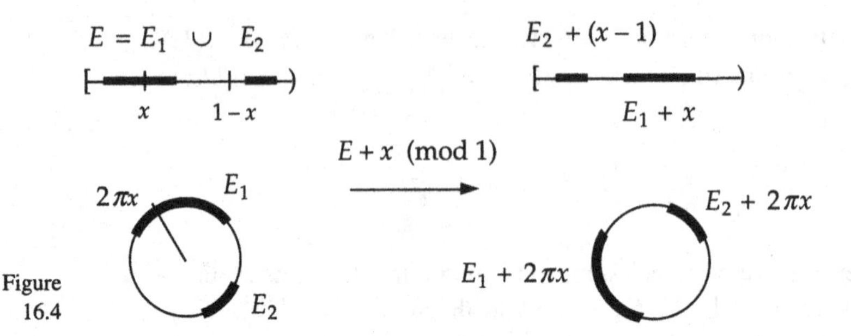

Figure 16.4

PROOF. Put $E_1 = E \cap [0, 1 - x)$ and $E_2 = E \setminus E_1 = E \cap [1 - x, 1)$. Clearly, E_1 and E_2 are measurable and disjoint, and so $m(E) = m(E_1) + m(E_2)$. Now it is easy to check that

$$\begin{aligned} E + x \text{ (mod 1)} &= \big[E_1 + x \text{ (mod 1)}\big] \cup \big[E_2 + x \text{ (mod 1)}\big] \\ &= \big[E_1 + x\big] \cup \big[E_2 + (x - 1)\big], \end{aligned}$$

where the last two sets are ordinary translates. What's more, these last two sets are measurable (see Exercise 60) and disjoint, so $E + x$ (mod 1) is measurable. Also, by translation invariance,

$$\begin{aligned} m\big(E + x \text{ (mod 1)}\big) &= m(E_1 + x) + m(E_2 + (x - 1)) \\ &= m(E_1) + m(E_2) = m(E). \quad \square \end{aligned}$$

We have introduced arithmetic (mod 1) so that we may consider a curious *equivalence relation* on $[0, 1)$. Namely, given $x, y \in [0, 1)$, we define

$$x \sim y \iff x - y \in \mathbb{Q} \iff y \in \mathbb{Q} + x \text{ (mod 1)}.$$

This equivalence relation partitions $[0, 1)$ into disjoint *equivalence classes* $[x]_\sim = \mathbb{Q} + x$ (mod 1). That is, $[0, 1)$ is the disjoint union of the distinct *cosets* of $\mathbb{Q}$ under addition (mod 1). Since each of the sets $\mathbb{Q} + x$ (mod 1) is countable, there are evidently *uncountably* many distinct equivalence classes.

We next call on *the Axiom of Choice* to choose a full set N of distinct coset representatives for our equivalence relation. That is, N contains precisely one element from each equivalence class and no more. Thus, given any $x \in [0, 1)$, there is a unique $y \in N$ such that $x \sim y$. Moreover, for $x, y \in N$, we have $x \sim y \iff x = y$. Please note that N is necessarily an uncountable set.

The idea here is that we now reverse the process described above and write $[0, 1)$ as a union of cosets, or translates (mod 1) of N. Indeed, if, for each rational $r \in \mathbb{Q} \cap [0, 1)$, we set $N_r = N + r$ (mod 1), then

$$[0, 1) = \bigcup_{r \in \mathbb{Q} \cap [0,1)} N_r \qquad \text{and} \qquad N_r \cap N_s = \emptyset \qquad \text{for } r \neq s.$$

The first claim is easy: Given $x \in [0, 1)$, we know that $x \sim y$ for some $y \in N$, and hence $x = y + r$ (mod 1) for some $r \in \mathbb{Q} \cap [0, 1)$; that is, $x \in N_r$ for some $r \in \mathbb{Q} \cap [0, 1)$. The other containment is obvious since $N_r \subset [0, 1)$ for any $r \in \mathbb{Q} \cap [0, 1)$. Next, to see that the N_r are pairwise disjoint, note that if $x \in N_r \cap N_s$, then we would have

$$y + r \text{ (mod 1)} = x = z + s \text{ (mod 1)},$$

for some $y, z \in N$ and some $r, s \in \mathbb{Q} \cap [0, 1)$. But then, $y - z \in \mathbb{Q}$; that is,

$$y \sim z \implies y = z \implies r = s,$$

since $0 \le r, s < 1$. Thus, either $N_r = N_s$ (for $r = s$) or $N_r \cap N_s = \emptyset$ (for $r \ne s$).

Finally, putting all of these observations to work, we have

Theorem 16.31. *N is nonmeasurable.*

PROOF. If N were measurable, then all of the N_r would be measurable too, by Lemma 16.30. Moreover, we would have $m(N_r) = m(N)$ for all r. Consequently,

$$1 = m([0, 1)) = m\left(\bigcup_{r \in \mathbb{Q} \cap [0,1)} N_r\right) = \sum_{r \in \mathbb{Q} \cap [0,1)} m(N_r) = \sum_{r \in \mathbb{Q} \cap [0,1)} m(N).$$

Oops! We cannot assign any value at all to $m(N)$ without arriving at a contradiction! Thus, N is nonmeasurable. □

Notice that by repeating the argument above, using m^* and countable subadditivity in place of m and countable additivity, we must have $0 < m^*(N) \le 1$. (Why?) That is, we now have our example showing that m^* is not countably additive on all of $\mathcal{P}(\mathbb{R})$.

Corollary 16.32. *There exists a sequence of pairwise disjoint subsets (E_n) of $[0, 1)$ with $m^*\left(\bigcup_{n=1}^{\infty} E_n\right) < \sum_{n=1}^{\infty} m^*(E_n)$.*

This construction of a nonmeasurable set used only the countable additivity and the translation invariance of Lebesgue measure, and so we have actually proved something more.

Theorem 16.33. *Suppose that $\mathcal{A}$ is a σ-algebra of subsets of $[0, 1)$, and that $\mu : \mathcal{A} \to [0, \infty]$ is countably additive and translation-invariant. If $N \in \mathcal{A}$, then we must have either $\mu([0, 1)) = 0$ or $\mu([0, 1)) = \infty$. In other words, if $\mu([0, 1)) = 1$, then $N \notin \mathcal{A}$ and hence $\mathcal{A} \ne \mathcal{P}([0, 1))$.*

EXERCISES

71. Prove Corollary 16.32.

72. Find a decreasing sequence of sets $E_1 \supset E_2 \supset \cdots$, such that $m^*(E_1) < \infty$ and $m\left(\bigcap_{n=1}^{\infty} E_n\right) < \lim_{n\to\infty} m^*(E_n)$.

▷ **73.** If E is a measurable subset of the nonmeasurable set N (constructed in this section), prove that $m(E) = 0$. [Hint: Consider $E_r = E + r \pmod 1$, for $r \in \mathbb{Q} \cap [0, 1)$.]

▷ **74.** If $m^*(A) > 0$, show that A contains a nonmeasurable set. [Hint: We must have $m^*(A \cap [n, n+1)) > 0$ for some $n \in \mathbb{Z}$, and so we may suppose that $A \subset [0, 1)$. (How?) It follows from Exercise 73 that one of the sets $E_r = A \cap N_r$ is nonmeasurable. (Why?)]

75. Measurable sets aren't necessarily preserved by continuous maps, not even sets of measure zero. Here's an old example: Recall that the Cantor function $f : [0, 1] \to [0, 1]$ maps the Cantor set Δ onto $[0, 1]$. That is, the Cantor function takes a set of measure zero and "spreads it out" to a set of measure one. Conclude that f maps some measurable set onto a nonmeasurable set.

Other Definitions

There are several popular approaches to defining Lebesgue measurable sets. The approach that we have adopted takes full advantage of the topology of the real line, along with certain intrinsic properties of outer measure $m*$, to arrive at the notion of a measurable set. The disadvantage to this approach is that it is hard to generalize to the case of an "abstract" measure. For this reason, many authors prefer a different approach, one that was first suggested by Carathéodory. In this section we will give a brief overview of Carathéodory's definition.

To begin, let's recall Lebesgue's original definition: Given a subset E of $[a, b]$, Lebesgue would say that E measurable if

$$b - a = m^*(E) + m^*\big([a, b] \setminus E\big).$$

Lebesgue's definition extends to unbounded sets E using the same observation that we used earlier: It is enough to know that $E \cap [a, b]$ is measurable for any bounded interval $[a, b]$. Thus, we could rephrase the requirement as

$$m^*\big([a, b]\big) = m^*\big([a, b] \cap E\big) + m^*\big([a, b] \cap E^c\big)$$

for every interval $[a, b]$. Written this way, the requirement for measurability is that E and E^c should split every interval into two pieces whose outer measures add up to be the full measure of the interval. Carathéodory's idea is to replace intervals by arbitrary subsets of $\mathbb{R}$. That is, Carathéodory calls a set E measurable if

$$m^*(A) = m^*(A \cap E) + m^*(A \cap E^c) \tag{16.1}$$

for every subset A of $\mathbb{R}$. In other words, a measurable set is required to split *every* set "nicely."

Now Carathéodory's requirement is stronger than Lebesgue's, and hence a set that is measurable by Carathéodory's standard is measurable by Lebesgue's (and, hence, by ours too). It may seem surprising that the two definitions are actually equivalent – at least until you recall that outer measure is completely determined by its values on intervals.

The hard work in using Carathéodory's definition is cut in half by two simple observations: For one, it is only necessary to test

$$m^*(A) \geq m^*(A \cap E) + m^*(A \cap E^c),$$

since countable subadditivity always gives the other inequality. For another, it is now clear that we only have to consider sets A with $m^*(A) < \infty$. (Why?) From here, we would start down the same road that we traveled earlier: We would check that this definition yields an algebra of measurable sets (this is the easy part) and, in fact, a σ-algebra of sets (and this is where the real fighting takes place). Ultimately, we would arrive at the same conclusion: Measurable sets are Borel sets plus or minus null sets. In any case, using the machinery of Theorem 16.20, it is a simple matter to check that Carathéodory's notion of measurability coincides with our own.

Theorem 16.34. *Let $E \subset \mathbb{R}$. Then, E is measurable if and only if $m^*(A) = m^*(A \cap E) + m^*(A \cap E^c)$ for every subset A of $\mathbb{R}$.*

PROOF. First suppose that E is measurable. Given A, choose a G_δ-set G containing A such that $m^*(A) = m(G)$. (How?) Then, since both E and G are measurable,

$$\begin{aligned} m^*(A) = m(G) &= m(G \cap E) + m(G \cap E^c) \\ &\geq m^*(A \cap E) + m^*(A \cap E^c). \end{aligned}$$

Hence, equation (16.1) holds.

Next, suppose that $m^*(A) = m^*(A \cap E) + m^*(A \cap E^c)$ for every subset A of $\mathbb{R}$. If $m^*(E) < \infty$, choose a G_δ-set G containing E such that $m^*(E) = m(G)$. Then (putting $A = G$ in equation (16.1)),

$$m(G) = m^*(G \cap E) + m^*(G \cap E^c) = m^*(E) + m^*(G \setminus E).$$

Hence, $m^*(G \setminus E) = 0$ and, in particular, $G \setminus E$ is measurable. It follows that $E = G \setminus (G \setminus E)$ is measurable, too. If $m^*(E) = \infty$, we apply the first part of this argument to each of the sets $E_n = E \cap [-n, n]$, where $n \in \mathbb{N}$. For each n, we choose a G_δ-set G_n containing E_n with $m^*(G_n \setminus E_n) = 0$. Then, E is contained in the measurable set $G = \bigcup_{n=1}^{\infty} G_n$ and $m^*(G \setminus E) \leq \sum_{n=1}^{\infty} m^*(G_n \setminus E) = 0$. As before, it follows that E is measurable. □

EXERCISES

76. If $m^*(E) = 0$, check that E satisfies Carathéodory's condition (16.1).

77. If both E and F satisfy Carathéodory's condition, prove that $E \cup F$, $E \cap F$, and $E \setminus F$ do too. [Hint: It is only necessary to check $E \cup F$. (Why?) For this, use the fact that $A \cap (E \cup F) = (A \cap E) \cup (A \cap E^c \cap F)$.]

78. If E is a measurable subset of A, show that $m^*(A) = m(E) + m^*(A \setminus E)$. Thus, $m^*(A \setminus E) = m^*(A) - m(E)$ provided that $m(E) < \infty$.

◊

Notes and Remarks

The passage quoted at the beginning of the chapter is taken from Grattan-Guinness [1970].

The interchange of limits and integrals, as in the formula $\int_a^b \sum_{n=1}^\infty f_n = \sum_{n=1}^\infty \int_a^b f_n$, can be handled successfully by using the Riemann integral in several important cases. However, the proofs of such convergence theorems are typically rather difficult. For more details, see Eberlein [1957], Kestelman [1970], Lewin [1986], Luxemburg [1971], and Riesz [1917].

Lebesgue's thesis [1902] was based on a series of five short papers, or research announcements, published between the years 1899 and 1901. During the academic year 1902–3, Lebesgue gave the Course Peccot at the Collège de France; these lectures were published in 1904 in Borel's monograph series as *Leçons sur l'Intégration et la Recerche des Fonctions Primitives*. The second edition of the *Leçons* appeared in 1928 and included several important new results; see Lebesgue [1928]. Lebesgue's *Leçons* continues to be an important work. A substantial portion of the notes are devoted to the history of the development of the integral before Lebesgue – over 100 pages in the second edition.

Lebesgue was a prolific expository writer, too. He published several essays on the teaching of mathematics and several expository articles describing his own work. Two of the latter, "Sur le développement de la notion d'intégrale" and "Sur la mesure des grandeurs," have been translated into English by Kenneth May and appear, along with a short biographical essay, in Lebesgue [1966]. Other expository articles of interest here include Ulam [1943] and Riesz [1920, 1949].

For more on the history of the development of the Lebesgue integral see Hawkins [1970], Hobson [1927, Vol. I], Bliss [1917], and Hildebrandt [1917].

During the 1920s, the newly formed Polish school of mathematicians, headed by Wacław Sierpiński, went a long way toward resolving the various questions associated with the problem of measure. Indeed, the early volumes of *Fundamenta Mathematicae* contain dozens of important papers on measure theory, analysis, and the foundations of topology and descriptive set theory. Of particular interest here are Banach [1923], Banach and Kuratowski [1930], and Banach and Tarski [1924]. For more on the history of this important journal, see Kuzawa [1970].

For a down-to-earth discussion of the Banach–Tarski paradox, see French [1988]. A detailed proof of the Banach–Tarski theorem in $\mathbb{R}^3$ is given in Stromberg [1977]. As Stromberg points out, an excellent paper related to extensions of Lebesgue measure is Bruckner and Ceder [1974].

Many of the results in this chapter have been adapted from, or at least influenced by, de La Vallée Poussin [1934] and Oxtoby [1971]. It would seem that de La Vallée Poussin was the first to define a measurable set as one that could be well approximated, in terms of outer measure, by open sets and closed sets (although Theorem 16.21 (iii) and (iv) were known to Lebesgue). This approach has the distinct pedagogical advantage of being "hands on"; that is, most of what we need to know can be deduced from first principles without appealing to unintuitive definitions or to the "sleight of hand" of σ-algebra arguments. As a matter of curiosity, and some small nuisance, de La Vallée Poussin is a difficult name to track down in most library catalogs; you may find selections under any of the four initial letters "D," "L," "V," or "P." According to Burkill [1964], the most appropriate choice here is "L."

The "measure" described in Exercise 3, called the *outer content* of a set, was introduced by Peano [1887] and later by Jordan [1892].

Theorem 16.10 is due to Lebesgue [1902], but Guiseppe Vitali [1904] and W. H. Young [1904] independently discovered the theorem at about the same time. This discovery led Vitali and Young to develop their own theories of measure, which closely mirrored Lebesgue's; see Hawkins [1970]. The proof of Theorem 16.10 given here is based on the presentation in Oxtoby [1971]. For a proof requiring only advanced calculus, see Botsko [1988].

Exercise 19 is based on the discussion in Riesz and Sz.-Nagy [1955]. Exercises 27 and 37 are based on the discussion in Wilansky [1953a], but see also Wilansky [1953b] and Rudin [1983]. Exercise 42 is cribbed from W. B. Johnson's lectures on real analysis given at The Ohio State University in 1974–75.

A *ring* of sets $\mathcal{R}$ is a collection of subsets of a fixed set X that is closed under differences and finite unions. It is easy to see that if the ring $\mathcal{R}$ contains X itself, then $\mathcal{R}$ is an algebra of sets. For a short proof that a ring of sets actually is a ring (in the algebraic sense), see Wilker [1982].

The so-called Steinhaus lemma, Exercise 44, is from Steinhaus [1920] and appears in the first volume of *Fundamenta*. The elegant proof outlined here is from the *Annexe* of the same volume. Please compare this result with the observation made back in Chapter Two, also due to Steinhaus, that $\Delta - \Delta = [-1, 1]$. For variations on this theme, along with a few applications, see Chae [1980]. Still more variations and extensions are given in Oxtoby [1971] and Kominek [1983].

Theorem 16.27 is due to Vitali [1905a]. The proof presented here is due to Banach [1924] by way of Natanson [1955].

According to most sources, the first, and simplest, construction of a nonmeasurable set, presented here as Theorem 16.31, was given by Vitali [1905a]. See Van Vleck [1908] for a similar construction. Thomas [1985] provides an unusual graph-theoretic construction of a nonmeasurable set. Other, less elementary constructions are given in Oxtoby [1971]. Theorem 16.33 is from Folland [1984]. See also Mauldin [1979] and Briggs and Schaffter [1979].

The definition of measurability given in the last section, along with Theorem 16.34, is due to Constantin Carathéodory, who was among the first to develop a general theory of "abstract" measures; see Carathéodory [1918].

CHAPTER SEVENTEEN

Measurable Functions

Measurable Functions

Recall from our discussion in Chapter Sixteen that Lebesgue's approach to the integral applies to functions f for which the sets $\{x : a \le f(x) < b\}$ are measurable for every $a < b$. In this chapter we will pursue this notion (and then some). What we will find is that such functions are "almost" continuous, but in a somewhat weaker sense than was the case for Riemann integrable functions. This is as it should be, since we expect the class of Lebesgue integrable functions to be larger than the class of Riemann integrable functions.

Given a function $f : D \to \mathbb{R}$, defined on some domain D, we say that f is (Lebesgue) **measurable** if D is measurable and if, for each real α, the set

$$\{f > \alpha\} = \{x \in D : f(x) > \alpha\} = f^{-1}\big((\alpha, \infty)\big)$$

is measurable. In particular, notice that if D is a null set, then *every* function $f : D \to \mathbb{R}$ is measurable.

The requirement that D be measurable is actually redundant, since

$$D = f^{-1}(\mathbb{R}) = f^{-1}\left(\bigcup_{n=1}^{\infty}(-n, \infty)\right) = \bigcup_{n=1}^{\infty} f^{-1}\big((-n, \infty)\big) = \bigcup_{n=1}^{\infty}\{f > -n\},$$

but there are nevertheless good reasons for repeating this requirement.

As you might expect, we want the collection of measurable functions to be a vector space, an algebra, and so on. Most of these properties will follow easily from what we know about measurable sets (the fact that $\mathcal{M}$ is a σ-algebra, for example). Before we start on this project, though, let's first note that we could use any one of several similar definitions for the measurability of functions.

Proposition 17.1. *Let $f : D \to \mathbb{R}$, where D is measurable. Then, f is measurable if and only if any one of the following holds:*

(i) *$\{f \ge \alpha\}$ is measurable for all real α;*
(ii) *$\{f < \alpha\}$ is measurable for all real α;*
(iii) *$\{f \le \alpha\}$ is measurable for all real α.*

PROOF. First suppose that f is measurable. Then,

$$\{f \geq \alpha\} = f^{-1}([\alpha, \infty)) = f^{-1}\left(\bigcap_{k=1}^{\infty}(\alpha - \tfrac{1}{k}, \infty)\right)$$
$$= \bigcap_{k=1}^{\infty} f^{-1}((\alpha - \tfrac{1}{k}, \infty))$$
$$= \bigcap_{k=1}^{\infty} \{f > \alpha - \tfrac{1}{k}\},$$

which is measurable. Thus, (i) holds.

Now, that (i) implies (ii) is obvious, since $\{f < \alpha\} = D \setminus \{f \geq \alpha\} \in \mathcal{M}$. That (ii) implies (iii) follows the same lines as our first observation; in this case, $\{f \leq \alpha\} = \bigcap_{k=1}^{\infty} \{f < \alpha + (1/k)\}$. Finally, that (iii) implies f is measurable is obvious, since $\{f > \alpha\} = D \setminus \{f \leq \alpha\}$. □

Now if f is measurable, it is easy to see that the set $\{f = \alpha\}$ is measurable for every real α; but this condition alone is not sufficient to ensure measurability (see Exercise 5). Instead, notice that if f is measurable, then the set $\{a < f < b\}$ is measurable for any $a < b$. In fact, we can use this to manufacture another equivalent formulation to include in Proposition 17.1: f is measurable if and only if the set $\{a < f < b\}$ is measurable for any pair of real numbers $a < b$. But why stop there?

Corollary 17.2. *Let $f : D \to \mathbb{R}$, where D is measurable. Then, f is measurable if and only if $f^{-1}(U)$ is measurable for every open set $U \subset \mathbb{R}$.*

The class of functions that give relatively "nice" sets as inverse images of open sets is quite large, as we will see. In fact, there are several familiar classes of functions that are easily seen to be measurable.

Corollary 17.3. *Continuous functions, monotone functions, step functions, and semicontinuous functions (all defined on some interval in $\mathbb{R}$) are measurable.*

EXERCISES

▷ **1.** Prove Corollary 17.2.

2. Prove Corollary 17.3. In which cases, if any, is it necessary to assume that the domain D is an interval?

▷ **3.** Let $f : D \to \mathbb{R}$, where D is measurable. Show that f is measurable if and only if the function $g : \mathbb{R} \to \mathbb{R}$ is measurable, where $g(x) = f(x)$ for $x \in D$ and $g(x) = 0$ for $x \notin D$.

▷ **4.** Prove that χ_E is measurable if and only if E is measurable.

5. Let N be a nonmeasurable subset of $(0, 1)$, and let $f(x) = x \cdot \chi_N(x)$. Show that f is nonmeasurable, but that each of the sets $\{f = \alpha\}$ is measurable.

6. Suppose that $f : D \to \mathbb{R}$, where D is measurable. Show that f is measurable if and only if $\{f > \alpha\}$ is measurable for each *rational* α.

7. If $f : D \to \mathbb{R}$ is measurable and $g : \mathbb{R} \to \mathbb{R}$ is continuous, show that $g \circ f$ is measurable.

8. Suppose that $D = A \cup B$, where A and B are measurable. Show that $f : D \to \mathbb{R}$ is measurable if and only if $f|_A$ and $f|_B$ are measurable (relative to their respective domains A and B, of course).

With just a bit more work, we can improve on Corollary 17.3 and, at the same time, confirm a conjecture that is implicit in our discussions of Lebesgue integration.

Theorem 17.4. *If $f : [a, b] \to \mathbb{R}$ is a Riemann integrable function, then f is Lebesgue measurable.*

PROOF. Recall that $D(f)$, the set of points of discontinuity of f, is a Borel set, and so is measurable. The same is true of $C(f) = [a, b] \setminus D(f)$, the set of points where f is continuous. What's more, if f is Riemann integrable, then $m\big(D(f)\big) = 0$, which means that every subset of $D(f)$ is measurable.

Now, let's compute the inverse image $f^{-1}(U)$ of an open set U:

$$f^{-1}(U) = \big(f^{-1}(U) \cap C(f)\big) \cup \big(f^{-1}(U) \cap D(f)\big).$$

The first of these is an open set, relative to $C(f)$; that is, $f^{-1}(U) \cap C(f) = V \cap C(f)$, where V is open in $\mathbb{R}$. Thus, $f^{-1}(U) \cap C(f)$ is even a Borel set. The second set, $f^{-1}(U) \cap D(f)$, is a subset of a set of measure zero, and so is necessarily measurable. Consequently, $f^{-1}(U)$ is measurable. $\square$

Corollary 17.5. *Every function $f : [a, b] \to \mathbb{R}$ of bounded variation is measurable.*

Please note that the collection of measurable functions is evidently strictly larger than the collection of Riemann integrable functions. Indeed, $\chi_{\mathbb{Q}}$ is measurable (why?), but not Riemann integrable.

We can continue with our "fine tuning" of Corollary 17.2 by introducing another level of classification of functions. What this amounts to is simply naming a class of functions that is intermediate to continuous functions and measurable functions.

We say that $f : D \to \mathbb{R}$ is **Borel measurable** if D is a Borel set and if, for each real α, the set $\{f > \alpha\}$ is a Borel set. Equivalently, f is Borel measurable if the set $f^{-1}(U)$ is always a Borel set for any open set U.

Continuous	$\iff$	f^{-1}(open) is open,
Borel measurable	$\iff$	f^{-1}(open) is a Borel set,
Lebesgue measurable	$\iff$	f^{-1}(open) is measurable.

Clearly, a continuous function is Borel measurable, and a Borel measurable function is Lebesgue measurable. It is not hard to see that neither of these statements can be reversed: There are Borel measurable functions that are not continuous, and there are Lebesgue measurable functions that are not Borel measurable. For example, note that monotone functions, step functions, and semicontinuous functions (defined on some interval in $\mathbb{R}$) are actually Borel measurable. And, since we know that there are Lebesgue measurable sets that are not Borel sets, there are necessarily Lebesgue measurable functions that are not Borel measurable. (Why?)

Henceforth, if there is no danger of ambiguity, the word "measurable" (with no additional quantifiers) will be understood to mean "Lebesgue measurable." In other words, if we are interested in the more restrictive notion of Borel measurability, we will specify the extra quantifier "Borel."

EXERCISES

9. Prove that monotone functions are Borel measurable when we take the domain D to be an interval.

10. If $f : [a, b] \to \mathbb{R}$ is quasicontinuous, show that f is measurable. Is f Borel measurable?

11. Let G be an open subset of $[0, 1]$ containing the rationals in $[0, 1]$ and having $m(G) < 1/2$. Prove that $f = \chi_G$ is Borel measurable but is *not* Riemann integrable on $[0, 1]$. Moreover, prove that f cannot be equal a.e. to any Riemann integrable function on $[0, 1]$; in other words, f is substantially different from any Riemann integrable function.

12. If $f : [a, b] \to \mathbb{R}$ is Lipschitz with constant K, and if $E \subset [a, b]$, show that $m^*(f(E)) \le K\, m^*(E)$. In particular, f maps null sets to null sets.

13. If $f : [a, b] \to \mathbb{R}$ is continuous, prove that the following are equivalent, where $E \subset [a, b]$:

(a) $m(f(E)) = 0$ whenever $m(E) = 0$.

(b) $f(E)$ is measurable whenever E is measurable. [Hint: Show that f maps F_σ-sets to F_σ-sets.]

▷ **14.** If f is measurable and B is a Borel set, show that $f^{-1}(B)$ is measurable. [Hint: $\{A : f^{-1}(A) \in \mathcal{M}\}$ is a σ-algebra containing the open sets.]

▷ **15.** If f is Borel measurable and B is a Borel set, show that $f^{-1}(B)$ is a Borel set. In particular, this holds for continuous f.

16.

(a) If E is a Borel set, show that $E + x$ and rE are Borel sets.

(b) If E is measurable, show that $E + x$ and rE are measurable.

17. If $f, g : \mathbb{R} \to \mathbb{R}$ are Borel measurable, show that $f \circ g$ is Borel measurable. If f is Borel measurable and g is Lebesgue measurable, show that $f \circ g$ is Lebesgue measurable.

18. Let $f : [0, 1] \to [0, 1]$ be the Cantor function, and set $g(x) = f(x) + x$. Prove that:

(a) g is a homeomorphism of $[0, 1]$ onto $[0, 2]$. In particular, $h = g^{-1}$ is continuous.

(b) $g(\Delta)$ is measurable and $m(g(\Delta)) = 1$. In particular, $g(\Delta)$ contains a nonmeasurable set A.

(c) g maps some measurable set onto a nonmeasurable set.

(d) $B = g^{-1}(A)$ is Lebesgue measurable but *not* a Borel set.

(e) There is a Lebesgue measurable function F and a continuous function G such that $F \circ G$ is not Lebesgue measurable.

The proof of Theorem 17.4 suggests the following observation:

Lemma 17.6. *If f is measurable, and if $g = f$ a.e., then g is measurable, too. Moreover, $m(\{g > \alpha\}) = m(\{f > \alpha\})$ for all $\alpha \in \mathbb{R}$.*

PROOF. Suppose that $f : D \to \mathbb{R}$ and that $g : E \to \mathbb{R}$. Then $f = g$ a.e. means that

$$\{f \neq g\} = (D \Delta E) \cup \{x \in D \cap E : f(x) \neq g(x)\}$$

is a null set and hence is measurable. Thus,

$$\{f = g\} = \{x \in D \cap E : f(x) = g(x)\} = D \setminus \{f \neq g\}$$

is measurable. And, because $\{f \neq g\}$ is a null set, we also have that $E = \{f = g\} \cup \big(E \cap \{f \neq g\}\big)$ is measurable. Finally,

$$\{g > \alpha\} = \big(\{f > \alpha\} \setminus \{f \neq g\}\big) \cup \big(\{g > \alpha\} \cap \{f \neq g\}\big)$$

is measurable since $\{f > \alpha\}$ is measurable and $\{f \neq g\}$ is a null set. For these same reasons, we get $m(\{g > \alpha\}) = m(\{f > \alpha\})$. □

One of our goals is to characterize the Lebesgue measurable functions in much the same way that we did the Lebesgue measurable sets. For example, we will show that a Lebesgue measurable function f is almost everywhere equal to a Borel measurable function g. Along the way, we will actually show that f is "almost" equal to a continuous function. But notice, please, how very different measurable functions are from continuous functions: A measurable function may be altered on any set of measure zero without sacrificing its measurability, while altering a continuous function at even a single point can easily destroy its continuity. At any rate, the premise here is the same as before: Lebesgue measurable functions should be well approximated by some simpler type of function. This project will take some time, but it will be all the easier to complete if we take advantage of the arithmetic of measurable functions. It is about time we checked whether the measurable functions form an algebra.

Theorem 17.7. *Let $c \in \mathbb{R}$, and let $f, g : D \to \mathbb{R}$ be measurable. Then, each of cf, $f + g$ and fg are measurable.*

PROOF. The first claim is nearly obvious:

$$\begin{aligned}
\{cf > \alpha\} &= \{f > \alpha/c\}, && \text{for } c > 0,\\
&= \{f < \alpha/c\}, && \text{for } c < 0,\\
&= D \text{ or } \emptyset, && \text{for } c = 0.
\end{aligned}$$

In any case, the set $\{cf > \alpha\}$ is measurable.

For $f + g$ we use a simple trick: Two real numbers a, b satisfy $a > b$ if and only if there is some rational r with $a > r > b$. Consequently,

$$\begin{aligned}\{f+g>\alpha\} &= \{f > \alpha - g\} \\ &= \bigcup_{r\in\mathbb{Q}} (\{f > r\} \cap \{r > \alpha - g\}) \\ &= \bigcup_{r\in\mathbb{Q}} (\{f > r\} \cap \{g > \alpha - r\}).\end{aligned}$$

Since we have written $\{f + g > \alpha\}$ as a countable union of measurable sets, it is measurable too.

To prove that fg is measurable, we will use a gimmick that we have seen before: We will first check that f^2 is measurable:

$$\begin{aligned}\{f^2 > \alpha\} &= \{f > \sqrt{\alpha}\} \cup \{f < -\sqrt{\alpha}\}, && \text{if } \alpha \geq 0, \\ &= D, && \text{if } \alpha < 0.\end{aligned}$$

Thus, f^2 is measurable. It now follows that $fg = \frac{1}{4}\left[(f+g)^2 - (f-g)^2\right]$ is measurable. □

Theorem 17.7 allows us to clarify a few more of the details that are implicit in our discussion of Lebesgue integration. In particular, it is now clear that the natural building blocks for the Lebesgue integral are measurable functions.

A **simple function** is a finite linear combination of characteristic functions of measurable sets. That is, φ is simple if

$$\varphi = \sum_{i=1}^{n} a_i \chi_{E_i}, \qquad a_i \text{ real}, \qquad E_i \text{ measurable}.$$

Clearly, every simple function is measurable. (In truth, what we have actually defined here is a *measurable* simple function – some authors allow for nonmeasurable sets E_i – but we are only interested in measurable functions, so we will insist on measurable sets.) Notice, too, that any step function is a simple function, but not conversely.

Now there are lots of representations for a given simple function. Indeed, we could introduce bogus terms such as $0 \cdot \chi_E$, or we might split up a given set and so introduce extra terms: $\chi_E = \chi_{E\cap A} + \chi_{E\setminus A}$. If we want some measure of uniqueness in the representation, we should rephrase our definition slightly. The key here is that a function φ is simple if (and only if) it takes on only finitely many, distinct, real values $a_1, \ldots, a_n \in \mathbb{R}$ and if, for each i, the set where each value occurs $A_i = \{\varphi = a_i\}$ is measurable. If the a_i are distinct, then the A_i are pairwise disjoint; thus,

$$\varphi = \sum_{i=1}^{n} a_i \chi_{A_i}, \qquad a_i \text{ distinct}, \qquad A_i \text{ disjoint, measurable}.$$

We will call this representation the **standard representation** for φ. Notice that in this case the sets $A_1, \ldots, A_n$ partition $\mathbb{R}$ into finitely many, pairwise disjoint, measurable sets.

EXERCISES

19. If $f, g : D \to \mathbb{R}$ are measurable, show that $\{f > g\}$ is measurable.

20. Let $f_n : D \to \mathbb{R}$ be measurable for each n, and suppose that $f_n(x) \le f_{n+1}(x)$ for each n and each $x \in D$. If $f(x) = \lim_{n\to\infty} f_n(x)$ exists (in $\mathbb{R}$) for each $x \in D$, prove that f is measurable. Thus, the measurable functions are closed under monotone limits.

21. Let f be a nonnegative, bounded, measurable function on $[\,a, b\,]$ with $0 \le f \le M$. Let

$$E_{n,k} = \left\{ \frac{kM}{2^n} \le f < \frac{(k+1)M}{2^n} \right\},$$

for each $n = 1, 2, \ldots$, and $k = 0, 1, \ldots, 2^n$, and set

$$\varphi_n = \sum_{k=0}^{2^n} \frac{kM}{2^n} \chi_{E_{n,k}}.$$

Prove that $0 \le \varphi_n \le \varphi_{n+1} \le f$ and that $0 \le f - \varphi_n \le 2^{-n}M$ for each n. Thus, (φ_n) is a sequence of simple functions that converges uniformly to f on $[\,a, b\,]$. [Hint: Notice that $E_{n,k} = E_{n+1,2k} \cup E_{n+1,2k+1}$.]

22. Check that the conclusion of Theorem 17.7 still holds (with the same proof) if we everywhere replace the word "measurable" by the words "Borel-measurable" (and "measurable set" by "Borel set," of course).

23. If $f \in BV[\,a, b\,]$, show that f is Borel measurable.

24. Does Lemma 17.6 hold for Borel measurable functions? How about if we take "a.e." to mean "except for a Borel set of measure zero"?

Extended Real-Valued Functions

We must occasionally consider functions that take on the values $\pm\infty$, that is, functions with values in the **extended real numbers** $\bar{\mathbb{R}} = [-\infty, \infty\,]$. A good example of a situation where infinite values are virtually unavoidable is when considering derivatives; even relatively tame functions, say monotone functions, can easily have infinite derivatives.

But at least we do not have to alter our definition of measurability. Given $f : D \to [-\infty, \infty\,]$, where D is a measurable subset of $\mathbb{R}$, we still say that f is measurable if, for each real α, the set

$$\{f > \alpha\} = \{x \in D : f(x) > \alpha\} = f^{-1}\big((\alpha, \infty\,]\big)$$

is measurable. Note that if f is measurable, then so are $\{f = +\infty\} = \bigcap_{n=1}^{\infty}\{f > n\}$ and $\{f = -\infty\} = D \setminus \{f > -\infty\} = D \setminus \big(\bigcup_{n=1}^{\infty}\{f > -n\}\big)$. In particular, the set where f is finite is measurable: $\{-\infty < f < +\infty\} = D \setminus \big(\{f = +\infty\} \cup \{f = -\infty\}\big)$.

Since we have taken the same formal definition for measurability as in the real-valued case, the various equivalent definitions given by Lemma 17.1 are still valid for extended real-valued functions. In fact, even Corollary 17.2 is still good, provided that we take sets of the form $(\alpha, +\infty]$ and $[-\infty, \alpha)$ as "neighborhoods of $\pm\infty$" (respectively), and this is just what we will do. Thus, the open sets in $\bar{\mathbb{R}}$ are open sets in $\mathbb{R}$, together with neighborhoods of $-\infty$ and $+\infty$ and unions of such sets. It follows that the Borel subsets of $\bar{\mathbb{R}}$ are Borel sets in $\mathbb{R}$, together with $\{-\infty\}$, $\{+\infty\}$, and unions of such sets.

Defining an appropriate arithmetic for extended real-valued functions is problematic: We need to define expressions such as $\infty \pm \infty$ and $0 \cdot (\pm\infty)$. Convention dictates that

$$0 \cdot (\pm\infty) = 0,$$
$$\infty \cdot (\pm\infty) = \pm\infty, \qquad -\infty \cdot (\pm\infty) = \mp\infty,$$
$$\infty + \infty = \infty, \qquad -\infty - \infty = -\infty,$$

while expressions such as $\infty - \infty$ and $-\infty + \infty$ are ambiguous (and should be avoided). With some care, however, we can still patch together an amended version of Theorem 17.7 for extended real-valued functions. We will relegate the details to the exercises.

In actual practice, the extended real-valued functions that we will encounter will be allowed to take infinite values only on sets of measure zero. We say that a measurable function $f : D \to [-\infty, \infty]$ is **finite almost everywhere** if it happens that $m(\{|f| = \infty\}) = 0$. If f and g are finite a.e., then any ambiguities arising from expressions such as $f + g$ occur only on sets of measure zero. This means that we are free to define $f + g$ in any way we please in the uncertain cases (see Lemma 17.6). Again, we will leave the details to the exercises.

EXERCISES

25. Suppose that $D = A \cup B$, where A and B are measurable. Show that $f : D \to [-\infty, \infty]$ is measurable if and only if both $f|_A$ and $f|_B$ are measurable. In particular, if D is measurable, then $f : D \to [-\infty, \infty]$ is measurable if and only if both of the sets $\{f = +\infty\}$ and $\{f = -\infty\}$ are measurable and $f|_{\{|f|<\infty\}}$ is measurable.

26. Suppose that $f, g : D \to [-\infty, \infty]$ are measurable. Show that fg is always measurable, where we take $0 \cdot (\pm\infty) = 0$.

27. Suppose that $f,\ g : D \to [-\infty, \infty]$ are measurable. Show that $f + g$ is measurable, provided that we define $f(x)+g(x)$ to be the same value, say 5, whenever it is of the form $\infty - \infty$ or $-\infty + \infty$.

28. Suppose that $f, g : D \to [-\infty, \infty]$ are measurable and finite a.e.; that is, $m(\{f = \pm\infty\}) = 0 = m(\{g = \pm\infty\})$. Show that $f + g$ is measurable no matter how it is defined when it has the form $\infty - \infty$ or $-\infty + \infty$.

29. Let $f : [a, b] \to [-\infty, \infty]$ be measurable and finite a.e. Given $\varepsilon > 0$, show that there is some finite M such that $m(\{|f| > M\}) < \varepsilon$.

Sequences of Measurable Functions

We now know that the collection of measurable functions sharing a common domain form a vector space and an algebra of functions. But of course we can't stop there! We want max's and min's and absolute values, too. With just a little extra effort, we can handle all of these cases, and more, at one and the same time. The key here is that the collection of measurable functions is closed under monotone limits, and, as we'll see, this means that the collection is closed under all pointwise limits.

Throughout this section, unless otherwise specified, we will assume that *all functions are defined on a common measurable domain* D, and that all functions take values in the extended real numbers $\bar{\mathbb{R}} = [-\infty, +\infty]$.

Theorem 17.8. *Let* (f_n) *be a sequence (finite or infinite) of measurable functions. Then, both* $\sup_n f_n$ *and* $\inf_n f_n$ *are measurable.*

PROOF. If $\alpha \in \mathbb{R}$, then $\sup_n f_n(x) > \alpha$ means that $f_n(x) > \alpha$ for some n, and conversely. That is,

$$\left\{\sup_n f_n > \alpha\right\} = \bigcup_{n=1}^{\infty}\{f_n > \alpha\},$$

which is measurable, provided that every f_n is measurable. The argument for $\inf_n f_n$ is easy, too; for example,

$$\left\{\inf_n f_n \geq \alpha\right\} = \bigcap_{n=1}^{\infty}\{f_n \geq \alpha\}.$$

Alternatively, note that $\inf_n f_n = -\sup_n(-f_n)$, and so inf's are measurable because sup's are.

The arguments for $\max\{f_1, \ldots, f_n\}$ and $\min\{f_1, \ldots, f_n\}$ are essentially the same (just take finite unions and intersections). □

Corollary 17.9. *If* f *and* g *are measurable, then* $\max\{f, g\}$, $\min\{f, g\}$, $f^+ = \max\{f, 0\}$, $f^- = -\min\{f, 0\}$, *and* $|f| = \max\{f, -f\} = f^+ + f^-$ *are all measurable.*

Since $f = f^+ - f^-$, we actually have something more:

Corollary 17.10. f *is measurable if and only if both* f^+ *and* f^- *are measurable.*

It also follows from Theorem 17.8 that the collection of measurable functions is closed under pointwise limits, and this is the best evidence we have that the class of measurable functions is quite large, surely larger than any we have seen thus far.

Corollary 17.11. *Let* (f_n) *be a sequence of measurable functions. Then, both* $\limsup_{n\to\infty} f_n$ *and* $\liminf_{n\to\infty} f_n$ *are measurable.*

PROOF. All we have to do is write each in terms of inf's and sup's:

$$\limsup_{n\to\infty} f_n = \inf_n \left(\sup_{k\geq n} f_k \right) \qquad \text{and} \qquad \liminf_{n\to\infty} f_n = \sup_n \left(\inf_{k\geq n} f_k \right). \quad \square$$

Corollary 17.12. *If (f_n) is a sequence of measurable functions, and if $f(x) = \lim_{n\to\infty} f_n(x)$ exists (in $\mathbb{R}$) for all $x \in D$, then f is measurable. In fact, f is measurable even if we only have $f(x) = \lim_{n\to\infty} f_n(x)$ a.e. on D (regardless of how f might be defined otherwise).*

EXERCISES

▷ **30.** Prove Corollary 17.12.

31. Let (f_n) be a sequence of measurable functions, all defined on some measurable set D. Show that the set $C = \{x \in D : \lim_{n\to\infty} f_n(x) \text{ exists}\}$ is measurable. [Hint: C is the set where $(f_n(x))$ is Cauchy.]

32. Check that the conclusion of Theorem 17.8 holds (with the same proof) if "measurable" is everywhere interpreted as "Borel measurable" (and "measurable set" as "Borel set," of course). Do the same for the four corollaries. What modifications, if any, are needed in Corollary 17.12?

33. If $f : (a, b) \to \mathbb{R}$ is differentiable, show that f' is Borel measurable. If f is only differentiable a.e., show that f' is still Lebesgue measurable. [Hint: Write f' as the limit of a sequence of continuous functions.]

We say that (f_n) converges **pointwise a.e.** to f if $f(x) = \lim_{n\to\infty} f_n(x)$ for almost every x in D, that is, if (f_n) converges pointwise to f on $D \setminus E$, where $m(E) = 0$. Thus, Corollary 17.12 says that the collection of measurable functions is closed even under pointwise a.e. limits.

Remarkably, pointwise a.e. convergence on a set of *finite* measure is actually equivalent to a slightly stronger form of convergence.

Egorov's Theorem. 17.13. *Let (f_n) be a sequence of measurable functions converging pointwise a.e. to a* real-valued *function f on a measurable set D of* finite *measure. Then, given $\varepsilon > 0$, there is a measurable set $E \subset D$ such that $m(E) < \varepsilon$ and such that (f_n) converges* uniformly *to f on $D \setminus E$.*

PROOF. We may obviously assume that $f_n \to f$ everywhere on D. Now, for each n and k, consider

$$E(n,k) = \bigcup_{m=n}^{\infty} \left\{ x \in D : |f_m(x) - f(x)| \geq \frac{1}{k} \right\}.$$

If k is fixed, then the sets $E(n, k)$ clearly decrease as n increases; moreover, $\bigcap_{n=1}^{\infty} E(n, k) = \emptyset$, since $f_n \to f$ everywhere on D. (Why?)

Since $m(D) < \infty$, we have $m\big(E(n,k)\big) \to 0$ as $n \to \infty$. Consequently, we may choose a subsequence (n_k) for which $m\big(E(n_k,k)\big) < \varepsilon/2^k$. (How?) Now, if we set $E = \bigcup_{k=1}^{\infty} E(n_k,k)$, then $m(E) < \varepsilon$. What's more, for $x \notin E$, we have $x \notin E(n_k,k)$ for any k and, in particular, $|f_m(x) - f(x)| < 1/k$ for all $m \geq n_k$. Thus, $f_n \rightrightarrows f$ on $D \setminus E$. □

We say that (f_n) converges **almost uniformly** to f on D if, for each $\varepsilon > 0$, there is a measurable subset E of D, with $m(E) < \varepsilon$, such that (f_n) converges uniformly to f on $D \setminus E$. Now it is easy to see that almost uniform convergence implies convergence pointwise almost everywhere; thus, on a set of finite measure, Egorov's theorem tells us that the two notions are equivalent. The requirement that f be real-valued (or, at worst, finite a.e.) cannot be dropped, nor can the requirement that $m(D) < \infty$, in general. We will leave the proofs of these various claims to the exercises.

EXERCISES

34. Give an example showing that the requirement that f be finite, at least a.e., cannot be dropped from the statement of Egorov's theorem.

▷ **35.** Give an example showing that the requirement that $m(D) < \infty$ cannot be dropped from Egorov's theorem.

▷ **36.** If (f_n) converges almost uniformly to f, prove that (f_n) converges almost everywhere to f. [Hint: For each k, choose a set E_k such that $m(E_k) < 1/k$ and $f_n \rightrightarrows f$ off E_k. Then $m\left(\bigcap_{k=1}^{\infty} E_k\right) = 0$.]

37. Clearly, if (f_n) converges uniformly to f except, possibly, on a set of measure zero, then (f_n) converges almost uniformly to f. On the other hand, give an example showing that almost uniform convergence does *not* imply uniform convergence except on a set of measure zero.

38. Let (f_n) be a sequence of measurable functions converging pointwise a.e. to a real-valued function f on a measurable set D of arbitrary measure. Show that there exist measurable sets $E_1 \subset E_2 \subset \cdots \subset D$ such that (f_n) converges uniformly to f on each E_k and $m\left(\left(\bigcup_{k=1}^{\infty} E_k\right)^c\right) = 0$.

Approximation of Measurable Functions

Our long-term goal is to improve on the result in Corollary 17.12 and to actually characterize measurable functions as the almost everywhere limits of certain "nice" functions. The first step in this process is extremely important. Watch closely. Better still, draw a few pictures!

Basic Construction 17.14. *If $f : D \to [0,\infty]$ is a nonnegative measurable function, then we can find an increasing sequence of nonnegative simple functions (φ_n) with $0 \leq \varphi_1 \leq \varphi_2 \leq \cdots \leq f$, such that (φ_n) converges pointwise to f* everywhere *on D, and such that (φ_n) converges* uniformly *to f on any set where f is bounded.*

PROOF. For each $n = 1, 2, \ldots$, define $F_n = \{x \in D : f(x) \geq 2^n\}$ and

$$E_{n,k} = \{x \in D : k2^{-n} \leq f(x) < (k+1)2^{-n}\} \qquad \text{for} \quad k = 0, 1, \ldots, 2^{2n} - 1.$$

Since f is measurable, so are F_n and $E_{n,k}$. Now, for each $n = 1, 2, \ldots$, define a (measurable) simple function by

$$\varphi_n = 2^n \chi_{F_n} + \sum_{k=0}^{2^{2n}-1} k\,2^{-n} \chi_{E_{n,k}}.$$

Please note that φ_n vanishes outside of D, that $0 \leq \varphi_n \leq f$, and that $0 \leq f - \varphi_n \leq 2^{-n}$ on the set $\{f < 2^n\}$. Since $D = \bigcup_{n=1}^{\infty}\{f < 2^n\} \cup \{f = \infty\}$, and since $\{f < 2^n\} \subset \{f < 2^{n+1}\}$ for any n, we get that $\varphi_n \to f$ pointwise on D (notice that $\varphi_n = 2^n$ on the set $\{f = \infty\}$). What's more, it is obvious that $\varphi_n \rightrightarrows f$ on any set of the form $\{f \leq M\}$. (Why?)

All that remains is to check that the φ_n increase. But

$$E_{n,k} = \{2k/2^{n+1} \leq f < (2k+2)/2^{n+1}\} = E_{n+1,2k} \cup E_{n+1,2k+1}.$$

On $E_{n+1,2k}$ we have $\varphi_n = k/2^n = 2k/2^{n+1} = \varphi_{n+1}$, while on $E_{n+1,2k+1}$ we have $\varphi_n = k/2^n < (2k+1)/2^{n+1} = \varphi_{n+1}$. Finally, on the set

$$F_n = \{f \geq 2^n\} = \{f \geq 2^{2n+1}\, 2^{-(n+1)}\},$$

it is clear that $\varphi_n = 2^n = 2^{2n+1}/2^{n+1} \leq \varphi_{n+1}$. Thus, $\varphi_n \leq \varphi_{n+1}$ everywhere on D. □

Given a measurable function $f : D \to [-\infty, \infty]$, we apply the basic construction to each of f^+ and f^- to conclude:

Corollary 17.15. *If $f : D \to [-\infty, \infty]$ is measurable, then there exists a sequence of simple functions (φ_n) such that $0 \leq |\varphi_1| \leq |\varphi_2| \leq \cdots \leq |f|$ and $\varphi_n \to f$ everywhere on D. Moreover, $\varphi_n \rightrightarrows f$ on any set where $|f|$ is bounded.*

It is interesting to note that this construction works for *any* function $f : D \to [-\infty, \infty]$, provided that we no longer require a simple function to be based on measurable sets. In other words, the measurability of f was only needed to ensure the measurability of the φ_n.

Corollary 17.16. *Let $f : D \to [-\infty, \infty]$, where D is measurable. Then, f is measurable if and only if f is the pointwise (everywhere) limit of a sequence of (measurable) simple functions.*

EXERCISES

39. Modify the Basic Construction in the following way: For each n and k, choose a Borel subset of $E_{n,k}$ of equal measure, call it $A_{n,k}$, and choose a Borel subset of F_n of equal measure, and call it B_n. Now define $\psi_n = 2^n \chi_{B_n} + \sum_{k=0}^{2^{2n}-1} k\,2^{-n} \chi_{A_{n,k}}$. Note that ψ_n is Borel measurable. Argue that (ψ_n) converges pointwise to f on D except, possibly, on a set of measure zero.

40. If f is Lebesgue measurable, prove that there is a Borel measurable function g such that $f = g$ except, possibly, on a Borel set of measure zero. [Hint: Every null set is contained in a Borel set of measure zero.]

The point to Corollary 17.16 is that the collection of measurable functions is the closure of the (measurable) simple functions under pointwise limits. We could have easily taken this as our definition of measurability.

If we consider measurable functions defined on an interval, it is possible to modify our construction to involve step functions, or even continuous functions, in place of simple functions (at the price of an extra "a.e." here and there). This is the next item on the agenda.

For the remainder of this section, then, we will suppose that we are given a measurable, finite almost everywhere function $f : [a, b] \to [-\infty, \infty]$ and an $\varepsilon > 0$.

Lemma 17.17. *There is a finite constant K (depending on ε) such that $|f| \le K$ except, possibly, on a set of measure less than $\varepsilon/2$.*

PROOF. The sets $\{|f| > n\}$ decrease as n increases, each has finite measure, and $\bigcap_{n=1}^{\infty}\{|f| > n\} = \{f = \pm\infty\}$ is a set of measure zero. Thus, $m\big(\{|f| > n\}\big) \to 0$ as $n \to \infty$. In particular, $m\big(\{|f| > n\}\big) < \varepsilon/2$ for some n. □

The next step follows immediately from our Basic Construction.

Lemma 17.18. *There is a* simple *function φ, vanishing outside of $[a, b]$, such that $|\varphi| \le |f|$, and such that $|f - \varphi| < \varepsilon$ except, possibly, on the set where $|f| > K$ (a set of measure less than $\varepsilon/2$).*

At this point, f has been well approximated by a simple function φ based on measurable sets. We next replace each of these underlying measurable sets by "nice" sets, and so build a new approximation for f. As with the Basic Construction itself, you may find it helpful to sketch a few pictures to go along with the refinements presented below.

Lemma 17.19. *There is a* continuous *function g on $\mathbb{R}$, vanishing outside of $[a, b]$, such that $g = \varphi$ except, possibly, on a set of measure less than $\varepsilon/2$.*

PROOF. Write $\varphi = \sum_{i=1}^{n} a_i \chi_{A_i}$, where each $a_i \in \mathbb{R}$, and where $A_1, \ldots, A_n$ are pairwise disjoint measurable subsets of $[a, b]$ with $\bigcup_{i=1}^{n} A_i = [a, b]$. For each i, choose a closed set $F_i \subset A_i \cap (a, b)$ such that $m(A_i \setminus F_i) < \varepsilon/(2n)$, and consider the function $\psi = \sum_{i=1}^{n} a_i \chi_{F_i}$. We clearly have $\psi = \varphi$ on the set $F = \bigcup_{i=1}^{n} F_i$, where $[a, b] \setminus F = \bigcup_{i=1}^{n}(A_i \setminus F_i)$ is a set of measure less than $\varepsilon/2$.

To finish the proof, then, it suffices to show that the function g defined by $g = a_i$ on the set F_i, for $i = 1, \ldots, n$, that is, $g = \psi|_F$, can be extended to a continuous function on $\mathbb{R}$ that vanishes outside $[a, b]$. The fact that $F \cup \{a, b\}$ is closed makes this easy: Since the open set $G = \mathbb{R} \setminus (F \cup \{a, b\})$ can be written as the countable union of pairwise disjoint intervals (with endpoints in $F \cup \{a, b\}$), we may extend g linearly on each of the constituent intervals in G, taking $g = 0$

on $(-\infty, a\,]$ and $[\,b, \infty)$. (How?) It is easy to see that this defines g as a continuous function on $\mathbb{R}$ (see Exercise 41). □

Combining these results gives us *Borel's theorem* (see also Exercise 43).

Theorem 17.20. *Let $f : [\,a, b\,] \to [-\infty, \infty\,]$ be measurable and finite a.e. Then, for each $\varepsilon > 0$, there is a continuous function g on $[\,a, b\,]$ such that $|f - g| < \varepsilon$ except, possibly, on a set of measure less than ε. If $k \le f \le K$, for some constants k and K, then we can arrange for $k \le g \le K$, too.*

PROOF. The first assertion follows easily from the previous three lemmas. To prove the second assertion, note that if $k \le f \le K$, then the function

$$\tilde{g} = K \wedge (k \vee g) = \min\{K, \max\{k, g\}\}$$

is continuous, satisfies $k \le \tilde{g} \le K$, and, in addition, has $|f - \tilde{g}| \le |f - g|$. (Why?) □

It is convenient to use the shorthand $m\{|f - g| \ge \varepsilon\} < \varepsilon$ in place of the more cumbersome phrase "$|f - g| < \varepsilon$ except, possibly, on a set of measure less than ε." Similar abbreviations could be used to shorten other statements; for example, $m\{g \ne \varphi\} < \varepsilon$ is an obvious replacement for "$g = \varphi$ except, possibly, on a set of measure less than ε."

EXERCISES

▷ **41.** Let E be a closed subset of $\mathbb{R}$, and let $f : E \to \mathbb{R}$ be continuous. Prove that f extends to a continuous function on all of $\mathbb{R}$. That is, prove that there is a continuous function $g : \mathbb{R} \to \mathbb{R}$ such that $g(x) = f(x)$ for $x \in E$. Moreover, g can be chosen to satisfy $\sup_{x\in\mathbb{R}} |g(x)| \le \sup_{x\in E} |f(x)|$.

42.

(a) Given a compact set K and a bounded open set $U \supset K$, show that there is a continuous function $f : \mathbb{R} \to \mathbb{R}$ such that $f = 1$ on K, $f = 0$ on U^c, and $0 \le f \le 1$ everywhere.

(b) Given a measurable set E with $m(E) < \infty$, and $\varepsilon > 0$, show that there is a continuous function $f : \mathbb{R} \to \mathbb{R}$, vanishing outside some compact set, such that $0 \le f \le 1$ everywhere, and $m\{f \ne \chi_E\} < \varepsilon$.

43. Let $f : [\,a, b\,] \to [-\infty, \infty\,]$ be measurable and finite a.e., and let $\varepsilon > 0$. Modify the proof of Borel's theorem to show that there is a *polynomial* p such that $m\{|f - p| \ge \varepsilon\} < \varepsilon$.

44. Let $f : [\,a, b\,] \to [-\infty, \infty\,]$ be measurable and finite a.e. Prove that there is a sequence of continuous functions (g_n) on $[\,a, b\,]$ such that $g_n \to f$ a.e. on $[\,a, b\,]$. In fact, the g_n can be taken to be polynomials. [Hint: For each n, choose g_n so that $E_n = \{|f - g_n| \ge 2^{-n}\}$ has $m(E_n) < 2^{-n}$. Now argue that $g_n \to f$ off the set $E = \limsup_{n\to\infty} E_n$.]

45. Let $f : [a, b] \to \bar{\mathbb{R}}$ be measurable and finite a.e., and let $\varepsilon > 0$. Show that there is a continuous function g on $[a, b]$ with $m\{f \neq g\} < \varepsilon$. [Hint: Combine Exercises 41 and 44 and Egorov's theorem to find continuous functions (g_n) and a *closed* set F with $m([a, b] \setminus F) < \varepsilon$ and $g_n \rightrightarrows f$ on F. Now argue that $f|_F$ extends to a continuous function g.]

46. (Luzin's Theorem) Show that $f : \mathbb{R} \to \mathbb{R}$ is measurable if and only if, for each $\varepsilon > 0$, there is a measurable set E with $m(E) < \varepsilon$ such that the restriction of f to $\mathbb{R} \setminus E$ is continuous (relative to $\mathbb{R} \setminus E$).

47. Show that $f : \mathbb{R} \to \mathbb{R}$ is measurable if and only if, for each $\varepsilon > 0$, there is a continuous function $g : \mathbb{R} \to \mathbb{R}$ such that $m\{f \neq g\} < \varepsilon$.

48. Luzin's theorem does *not* say that a measurable function is continuous on the complement of a null set. Indeed, show that there is a measurable set $K \subset [0, 1]$ such that χ_K is everywhere discontinuous in $[0, 1] \setminus N$ for any null set N.

49.

(a) Given a simple function $\varphi : [a, b] \to \mathbb{R}$ and $\varepsilon > 0$, show that there is a step function g on $[a, b]$ such that $m\{g \neq \varphi\} < \varepsilon$. [Hint: Write $\varphi = \sum_{i=1}^n a_i \chi_{A_i}$. For each i, choose a finite union of intervals B_i with $m(A_i \triangle B_i) < \varepsilon/n$. Now let $g = \sum_{i=1}^n a_i \chi_{B_i}$.]

(b) Let $f : [a, b] \to [-\infty, \infty]$ be measurable and finite a.e., and let $\varepsilon > 0$. Show that there is a step function g on $[a, b]$ such that $m\{|f - g| \geq \varepsilon\} < \varepsilon$. If, in addition, $k \leq f \leq K$, show that g can be chosen to satisfy $k \leq g \leq K$, too.

50. Let (f_n) be a sequence of real-valued measurable functions on $[0, 1]$. Show that there exists a sequence of positive real numbers (a_n) such that $a_n f_n \to 0$ a.e.

The various approximation results in this section, along with certain of the exercises, allow us to summarize our findings:

f is measurable and finite a.e.

$\iff$ f is the limit of a sequence of (measurable) simple functions;

$\iff$ f is the a.e. limit of a sequence of step functions;

$\iff$ f is the a.e. limit of a sequence of continuous functions;

$\iff$ given $\varepsilon > 0$, there is a continuous function g such that $m\{f \neq g\} < \varepsilon$.

Notes and Remarks

Lebesgue's approach to integration is intimately tied to the notion of measurable functions. Indeed, according to Hawkins [1970], "it was the properties of measurable functions and the structure of the sets $[\{x : a \leq f(x) < b\}]$ that guided Lebesgue's reasoning and led to his major results." However, it is also fair to say that Lebesgue had little interest in the formalities of measure and of measurable functions; his primary interest was integration. The formal discussions of measurable sets and measurable functions occupy but a few pages in the *Leçons* (Lebesgue [1928]).

Exercise 11 is based on the discussion in Wilansky [1953a]. Exercise 18 can be traced to Hille and Tamarkin [1929].

Theorem 17.13 is due to D. F. Egorov [1911]. The clever proof presented here is due to F. Riesz [1928b]. Necessary and sufficient conditions for almost uniform convergence are given in R. G. Bartle [1980a]. Other variations, generalizations, and examples can be found in Luther [1967], Rozycki [1965], Suckau [1935], and Weston [1959, 1960].

Much of the last section is adapted from, or at least influenced by, Sierpiński [1922] (and its references). Herein Sierpiński proves the theorems of Borel (Theorem 17.20; see Exercise 43 for a result that is closer in spirit to Borel's original theorem), Fréchet (Exercise 44), and Luzin (Exercises 46 and 47).

N. N. Luzin (sometimes spelled "Lusin") was a student of D. F. Egorov; not surprisingly, Luzin's proof of his result is based on Egorov's theorem. For an elementary proof of Luzin's theorem, independent of Egorov's theorem, see Oxtoby [1971]. For more on this student–adviser pair, see Allen Shields [1987b]. Shields's article is highly recommended to any student with an adviser, and, likewise, to any adviser with a student: See Egorov's letter to Luzin, quoted on p. 24 of the article, for a taste of a time gone by.

Exercise 41 is a simple version of Tietze's extension theorem, whereas Exercise 42 (a) is an easy version of Urysohn's lemma. See, for example, Folland [1984] for more general versions of these two theorems.

CHAPTER EIGHTEEN

The Lebesgue Integral

We've set the stage for the Lebesgue integral in the previous two chapters; now it's time for the star to make her entrance. By way of a reminder, recall that we want our new integral to satisfy at least the following few, loosely stated properties:

- $\int \chi_E = m(E)$, whenever E is measurable.
- The integral should be linear: $\int(\alpha f + \beta g) = \alpha \int f + \beta \int g$.
- The integral should be positive (or monotone): $f \geq 0 \Longrightarrow \int f \geq 0$ (or $f \geq g \Longrightarrow \int f \geq \int g$). In the presence of linearity, these are the same.
- The integral should be defined for a large class of functions, including at least the bounded Riemann integrable functions, and it should coincide with the Riemann integral whenever appropriate.

The first two properties tell us how to define the integral for simple functions. Once we know how to integrate simple functions, the third property suggests how to define the integral for nonnegative measurable functions: If $f \geq 0$ is measurable, then we can find a sequence (φ_n) of simple functions that *increase* to f. Now set $\int f = \lim_{n\to\infty} \int \varphi_n$. Finally, linearity supplies the appropriate definition for the general case: If f is measurable, then f^+ and f^- are nonnegative, measurable, and $f = f^+ - f^-$. So, set $\int f = \int f^+ - \int f^-$, provided that this expression makes sense (we wouldn't want $\infty - \infty$, for example).

These few steps outline our plan of attack. If all goes well, we'll find that the new integral is defined (and finite) for any bounded measurable function defined on a bounded interval – more than enough functions to recover the Riemann integral.

Meanwhile, we will take some care to distinguish between this new integral and the Riemann integral; in particular, the abbreviated notation $\int f$ in place of $\int_a^b f(x)\,dx$ is not simply an example of laziness, but rather is intended to further highlight this distinction.

There are, of course, a few details to check along the way. We begin with the "obvious" case of defining the Lebesgue integral for simple functions.

Simple Functions

We say that a simple function φ is (Lebesgue) **integrable** if the set $\{\varphi \neq 0\}$ has finite measure (in short, if φ has *finite support*). In this case, we may write the standard representation for φ as $\varphi = \sum_{i=0}^n a_i \chi_{A_i}$, where $a_0 = 0, a_1, \ldots, a_n$ are distinct real numbers, where $A_0 = \{\varphi = 0\}$, $A_1, \ldots, A_n$ are pairwise disjoint and measurable, and where only A_0 has infinite measure. Once φ is so written, there is an obvious definition for $\int \varphi$, namely,

$$\int \varphi = \int_{\mathbb{R}} \varphi = \int_{-\infty}^{\infty} \varphi(x)\,dx = \sum_{i=1}^n a_i\, m(A_i).$$

In other words, by adopting the convention that $0 \cdot \infty = 0$, we define the **Lebesgue integral** of φ by

$$\int \left(\sum_{i=0}^n a_i \chi_{A_i} \right) = \sum_{i=0}^n a_i\, m(A_i).$$

Please note that $a_i\, m(A_i)$ is a product of real numbers for $i \neq 0$, and it is $0 \cdot \infty = 0$ for $i = 0$; that is, $\int \varphi$ is a finite real number.

In brief, if φ is an integrable simple function, then

$$\int \varphi = \sum_{a \in \mathbb{R}} a\, m\{\varphi = a\},$$

where the sum on the right actually involves only finitely many nonzero terms, each of which is finite, provided that we take $0 \cdot \infty = 0$.

By way of an easy example, note that $\chi_{\mathbb{Q}}$ is Lebesgue integrable and that $\int \chi_{\mathbb{Q}} = 0$.

Our first chore is to check that the definition of $\int \varphi$ does not actually depend on any particular representation of φ. This requires a couple of easy calculations.

Lemma 18.1. *Let φ be an integrable simple function, and let $\varphi = \sum_{i=1}^n b_i \chi_{E_i}$ be any representation with $E_1, \ldots, E_n$ disjoint and measurable. Then, $\int \varphi = \sum_{i=1}^n b_i\, m(E_i)$.*

PROOF. First note that for any $a \in \mathbb{R}$ we have $\{\varphi = a\} = \bigcup_{b_i = a} E_i$, where the union is over the set $\{i : b_i = a \text{ for some } 1 \le i \le n\}$. In particular, notice that $a\, m\{\varphi = a\} = \sum_{b_i = a} b_i\, m(E_i)$, and that this is good even for $a = 0$. Consequently,

$$\int \varphi = \sum_{a \in \mathbb{R}} a\, m\{\varphi = a\} = \sum_{a \in \mathbb{R}} \sum_{b_i = a} b_i\, m(E_i) = \sum_{i=1}^n b_i\, m(E_i). \quad \square$$

Using Lemma 18.1, we can easily check that the integral is both linear and positive on integrable simple functions.

Proposition 18.2. *If φ and ψ are integrable simple functions, then for $\alpha, \beta \in \mathbb{R}$ we have $\int(\alpha\varphi + \beta\psi) = \alpha \int \varphi + \beta \int \psi$. If $\varphi \ge \psi$ a.e., then $\int \varphi \ge \int \psi$.*

PROOF. The heart of the matter here is to find representations for φ and ψ based on a *common* partition of $\mathbb{R}$ so that we can readily combine and compare integrals, and this is easy.

Write $\varphi = \sum_{i=0}^n a_i \chi_{A_i}$ and $\psi = \sum_{j=0}^k b_j \chi_{B_j}$, where $a_0 = 0, a_1, \ldots, a_n$ are distinct, $b_0 = 0, b_1, \ldots, b_k$ are distinct, $A_0, \ldots, A_n$ are disjoint and measurable, and $B_0, \ldots, B_k$ are disjoint and measurable. Then $\bigcup_{i=0}^n A_i = \mathbb{R} = \bigcup_{j=0}^k B_j$, both being disjoint unions, and all but A_0 and B_0 have finite measure. Now we can write $\mathbb{R} = \bigcup_{i=0}^n \bigcup_{j=0}^k (A_i \cap B_j)$. This is again a disjoint union, and all but $A_0 \cap B_0$ have finite measure.

Using this new partition of $\mathbb{R}$ we may write

$$\varphi = \sum_{i=0}^n \sum_{j=0}^k a_i \chi_{A_i \cap B_j}, \qquad \psi = \sum_{i=0}^n \sum_{j=0}^k b_j \chi_{A_i \cap B_j},$$

and so

$$\alpha\varphi + \beta\psi = \sum_{i=0}^n \sum_{j=0}^k (\alpha a_i + \beta b_j) \chi_{A_i \cap B_j}.$$

The linearity of the integral is now an immediate consequence of Lemma 18.1:

$$\begin{aligned}
\int (\alpha\varphi + \beta\psi) &= \sum_{i=0}^n \sum_{j=0}^k (\alpha a_i + \beta b_j)\, m(A_i \cap B_j) \\
&= \alpha \sum_{i=0}^n \sum_{j=0}^k a_i\, m(A_i \cap B_j) + \beta \sum_{i=0}^n \sum_{j=0}^k b_j\, m(A_i \cap B_j) \\
&= \alpha \int \varphi + \beta \int \psi.
\end{aligned}$$

Finally, if $\varphi - \psi \ge 0$ a.e., then $\int \varphi - \int \psi = \int (\varphi - \psi) \ge 0$, since any negative values of $\varphi - \psi$ occur only on null sets. □

Corollary 18.3. *Given $a_1, \ldots, a_n \in \mathbb{R}$ and measurable sets $E_1 \ldots, E_n$, each with* **finite** *measure, we have*

$$\int \left(\sum_{i=1}^n a_i \chi_{E_i} \right) = \sum_{i=1}^n a_i\, m(E_i).$$

If φ is an integrable simple function, and if E is a measurable set, we also define

$$\int_E \varphi = \int \varphi \cdot \chi_E.$$

This makes sense since $\varphi \cdot \chi_E$ is again an integrable simple function. When $E = [\,a, b\,]$, though, we usually just write $\int_a^b \varphi$.

Nonnegative Functions

We next define the integral for nonnegative measurable functions. There is a bit of "upper and lower integral" going on here (which we will pursue later) but, in essence, the definition is based only on the monotonicity of the integral and what we already know about simple functions.

If $f : \mathbb{R} \to [0, \infty]$ is measurable, we define the **Lebesgue integral** of f over $\mathbb{R}$ by

$$\int f = \sup\left\{\int \varphi : 0 \le \varphi \le f,\ \varphi \text{ simple and integrable}\right\}.$$

We are not excluding the possibility that $\int f = \infty$ here. If $\int f < \infty$, then we will say that f is (Lebesgue) **integrable** on $\mathbb{R}$. Please note that in any case we obviously have $\int f \ge 0$.

This definition is consistent with our first one. That is, if ψ is a nonnegative, integrable, simple function, then

$$\int \psi = \sup\left\{\int \varphi : 0 \le \varphi \le \psi,\ \varphi \text{ simple and integrable}\right\}.$$

(Why?) But the new definition says more: It defines $\int \psi$ for any *nonnegative* simple function. In particular, if E is any measurable set, then $\int \chi_E = m(E)$. This is clear if $m(E) < \infty$, and when $m(E) = \infty$, we have

$$\int \chi_E \ge \sup_n \int \chi_{E\cap[-n,n]} = \sup_n m(E \cap [-n, n]) = m(E) = \infty.$$

It is easy to see that if f and g are nonnegative measurable functions with $f \le g$, then $\int f \le \int g$. And it is virtually effortless to check that $\int (cf) = c \int f$ for $c \ge 0$. Additivity is harder to check; in fact, we will stall the proof until we have gathered more equipment for the task.

If E is a measurable set, and if f is nonnegative and measurable, we define

$$\int_E f = \int f \cdot \chi_E.$$

When f is defined only on E, we simply take $f = 0$ outside of E. From our earlier remarks, this, too, is consistent with the case for simple functions. Again, if $E = [a, b]$, we will stick to the familiar notation $\int_a^b f$.

In our search for new machinery, an extremely important observation is that the expression $\int_E f$ is a well-behaved function of the set E. For example, notice that if $m(E) = 0$, then $\int_E f = 0$. Indeed, if φ is an integrable simple function with $0 \le \varphi \le f\chi_E$, then we must have $\varphi = 0$ a.e., and hence $\int \varphi = 0$. (Why?) Also note that if $f \ge 0$ and if $E \subset F$ are measurable, then $\int_E f \le \int_F f$, since $f\chi_E \le f\chi_F$.

Along similar lines, if f is bounded above on E, say $0 \le f \le K$ on E, then $\int_E f \le K\, m(E)$, since $f\chi_E \le K\chi_E$ (see Figure 18.1). A somewhat more interesting observation is that $f \ge \alpha\, \chi_{\{f\ge\alpha\}}$ for any $\alpha \ge 0$, and hence $\int f \ge \alpha\, m\{f \ge \alpha\}$ (see Figure 18.2). This timid little inequality ranks right up there with the triangle inequality for utility per pound. It certainly merits stating again.

Chebyshev's Inequality 18.4. *If f is nonnegative and measurable, then $\int f \ge \alpha\, m\{f \ge \alpha\}$ for any $\alpha \ge 0$.*

Here is an immediate application:

Corollary 18.5. *If f is nonnegative and integrable, then f is finite a.e.*

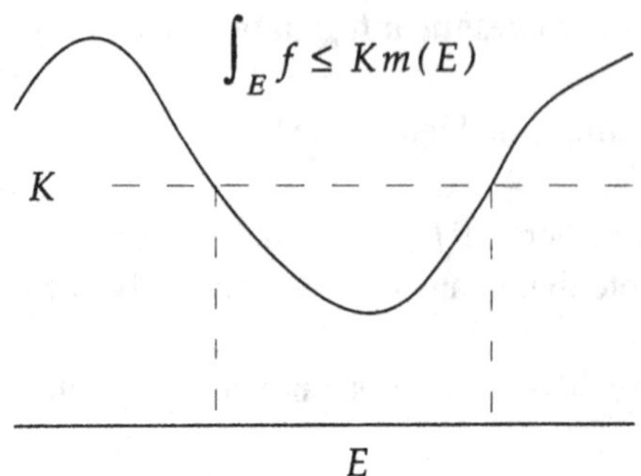

Figure 18.1

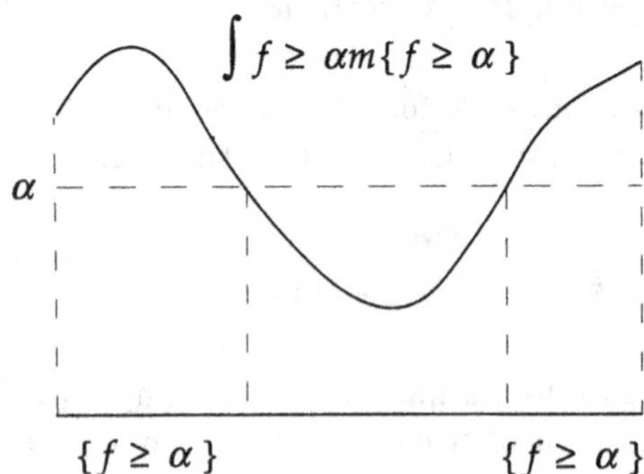

Figure 18.2

PROOF. Recall that $\{f = \infty\} = \bigcap_{n=1}^{\infty}\{f \ge n\}$. The sets $\{f \ge n\}$ decrease as n increases and, from Chebyshev's inequality, $m\{f \ge n\} \le (1/n)\int f \to 0$, as $n \to \infty$, since f is integrable. Thus, $m\{f = \infty\} = \lim_{n\to\infty} m\{f \ge n\} = 0$. □

EXERCISES

▷ **1.** If ψ is a nonnegative simple function, check that

$$\int \psi = \sup\left\{\int \varphi : 0 \le \varphi \le \psi,\ \varphi \text{ simple and integrable}\right\}.$$

2. Let $f : \mathbb{R} \to [0, \infty]$ be integrable and define $F : [0, \infty) \to [0, \infty]$ by $F(\alpha) = m\{f > \alpha\}$. Show that F is decreasing and right-continuous, and that $F(\alpha) \to 0$ as $\alpha \to \infty$. [Hint: f is finite a.e.]

▷ **3.** Prove that $\int_1^{\infty}(1/x)\,dx = \infty$ (as a Lebesgue integral).

We next roll up our sleeves and tackle the question of additivity of the integral. As was suggested earlier, we will consider $\int_E f$ as a function of the set E. What we will find is that the function $\mu(E) = \int_E f$, $E \in \mathcal{M}$, is a **measure** on $\mathcal{M}$. This means that μ is nonnegative, monotone, $\mu(\emptyset) = 0$, and, most importantly, that μ is countably additive. We have already checked a few of these properties; the hard work comes in establishing countable additivity. We begin with a special case:

Lemma 18.6. *Let φ be an integrable simple function. If $E_1 \subset E_2 \subset \cdots$ is an increasing sequence of measurable sets, and if $E = \bigcup_{n=1}^{\infty} E_n$, then $\int_E \varphi = \lim_{n\to\infty} \int_{E_n} \varphi_n$.*

PROOF. Write $\varphi = \sum_{i=1}^{k} a_i \chi_{A_i}$, where each $a_i \neq 0$ and where the A_i are pairwise disjoint measurable sets, each having finite measure. Now, let (E_n) be an increasing sequence of measurable sets, and let $E = \bigcup_{n=1}^{\infty} E_n$. Then, $\int_E \varphi = \int \varphi \cdot \chi_E = \sum_{i=1}^{k} a_i\, m(A_i \cap E)$. And now we appeal to the fact that Lebesgue measure is countably additive, à la Lemma 16.23 (i), to write

$$\int_E \varphi = \sum_{i=1}^{k} a_i\, m(A_i \cap E) = \sum_{i=1}^{k} \lim_{n\to\infty} a_i\, m(A_i \cap E_n)$$

$$= \lim_{n\to\infty} \sum_{i=1}^{k} a_i\, m(A_i \cap E_n) = \lim_{n\to\infty} \int_{E_n} \varphi. \quad \square$$

We used the fact that Lebesgue measure is countably additive to establish the "continuity" results of Lemma 16.23. It is not hard to see, though, that the conclusion of Lemma 16.23 (i) is actually equivalent to the countable additivity of m. In the same way, Lemma 18.6 actually shows that the map $\mu(E) = \int_E \varphi$ is a measure on $\mathcal{M}$. See Exercise 8 for more details.

We will use Lemma 18.6 to prove a result of fundamental importance:

Monotone Convergence Theorem 18.7. *If $0 \le f_1 \le f_2 \le \cdots$ is an increasing sequence of nonnegative measurable functions, then*

$$\int \left(\lim_{n\to\infty} f_n\right) = \lim_{n\to\infty} \int f_n.$$

PROOF. Since the f_n increase, note that $f = \lim_{n\to\infty} f_n = \sup_n f_n$ exists and is also nonnegative and measurable. And since we also have $\int f_n \le \int f_{n+1} \le \int f$ for all n, we have that $\lim_{n\to\infty} \int f_n$ exists and satisfies $\lim_{n\to\infty} \int f_n \le \int f$.

We need to show that $\lim_{n\to\infty} \int f_n \ge \int f$. Of course, given $\varepsilon > 0$, it would be enough to show that $\lim_{n\to\infty} \int f_n \ge (1-\varepsilon) \int f$. To do this, it is enough to show that $\lim_{n\to\infty} \int f_n \ge (1-\varepsilon)\int \varphi$ for any integrable simple function φ with $0 \le \varphi \le f$. (Why?)

Let φ be an integrable simple function with $0 \le \varphi \le f$, and consider the sets $E_n = \{f_n \ge (1-\varepsilon)\varphi\}$. Note that E_n is measurable and that, since $f_n \le f_{n+1}$, we have $E_n \subset E_{n+1}$. Also, since $f_n \to f > (1-\varepsilon)\varphi$, we have that $\bigcup_{n=1}^{\infty} E_n = \mathbb{R}$. (Why?)

Now we apply Lemma 18.6. Since

$$\int f_n \ge \int_{E_n} f_n \ge \int_{E_n} (1-\varepsilon)\varphi = (1-\varepsilon) \int_{E_n} \varphi$$

for all n, we have

$$\lim_{n\to\infty} \int f_n \ge (1-\varepsilon) \lim_{n\to\infty} \int_{E_n} \varphi = (1-\varepsilon) \int \varphi. \quad \square$$

The fact that the integral commutes with increasing limits allows us to put an interesting twist on our Basic Construction.

Corollary 18.8. *If f is a nonnegative measurable function, then there is an increasing sequence of integrable simple functions $0 \le \varphi_1 \le \varphi_2 \le \cdots \le f$ such that $f = \lim_{n\to\infty} \varphi_n$ and $\int f = \lim_{n\to\infty} \int \varphi_n$.*

PROOF. Let (ψ_n) be any sequence of nonnegative simple functions that increase pointwise to f. For example, take (ψ_n) to be the sequence of simple functions given by the Basic Construction. We need to show that the ψ_n can be replaced by a sequence of *integrable* simple functions. But this is easy: Just take $\varphi_n = \psi_n \cdot \chi_{[-n,n]}$. Each φ_n is now supported on a set of finite measure, and hence is integrable, and (φ_n) increases pointwise to f since $\chi_{[-n,n]}$ increases pointwise to $\chi_{\mathbb{R}}$, the constant 1 function. It follows from the Monotone Convergence Theorem that $\int \varphi_n$ increases to $\int f$. □

The point to Corollary 18.8 is that both f and $\int f$ are completely determined by the sequence (φ_n). We might have even used this fact to define $\int f$. In any case, the additivity of the integral is now a piece of cake: We already know that the integral is additive over simple functions, and we know that limits are additive. The rest is easy.

Corollary 18.9. *If f and g are nonnegative measurable functions, then $\int(f + g) = \int f + \int g$. In particular, $\int_{E\cup F} f = \int_E f + \int_F f$ for any disjoint measurable sets E, F.*

PROOF. Choose two sequences of nonnegative, integrable simple functions: (φ_n) increasing to f and (ψ_n) increasing to g. Then, $(\varphi_n + \psi_n)$ increases to $f + g$ and so, by applying the Monotone Convergence Theorem (no less than three times!), we have

$$\begin{aligned}\int (f+g) &= \lim_{n\to\infty} \int (\varphi_n + \psi_n) \\ &= \lim_{n\to\infty} \int \varphi_n + \lim_{n\to\infty} \int \psi_n = \int f + \int g. \quad \square\end{aligned}$$

EXERCISES

▷ **4.** Find a sequence (f_n) of nonnegative measurable functions such that $\lim_{n\to\infty} f_n = 0$, but $\lim_{n\to\infty} \int f_n = 1$. In fact, show that (f_n) can be chosen to converge *uniformly* to 0.

5. Suppose that f and g are measurable with $0 \le f \le g$. If g is integrable, show that f and $g - f$ are integrable and that $\int(g - f) = \int g - \int f$. In fact, the formula is still true even if we assume only that f is integrable.

6. Suppose that f and (f_n) are nonnegative measurable functions, that (f_n) *decreases* pointwise to f, and that $\int f_k < \infty$ for some k. Prove that $\int f =$

$\lim_{n\to\infty} \int f_n$. [Hint: Consider $(f_k - f_n)$ for $n > k$.] Give an example showing that this fails without the assumption that $\int f_k < \infty$ for some k.

We are halfway home: The set function $\mu(E) = \int_E f$ is nonnegative, monotone, and finitely additive. We next consider the null sets for μ. Here, finally, we will see a connection with the underlying function f. In brief, our next result tells us that the integral ignores the letters "a.e."

Lemma 18.10. *Let f be nonnegative and measurable. Then, $\int f = 0$ if and only if $f = 0$ a.e.*

PROOF. First suppose that $f = 0$ a.e. Then, $m\{f > 0\} = 0$ and, hence,

$$\int f = \int_{\{f=0\}} f + \int_{\{f>0\}} f = 0 + 0. \quad \text{(Why?)}$$

Next suppose that $\int f = 0$. To compute $m\{f > 0\}$, we first use Chebyshev's inequality to note that

$$m\left\{f \geq \frac{1}{n}\right\} \leq n \int f = 0$$

for all n. Since $\{f > 0\} = \bigcup_{n=1}^{\infty} \{f \geq (1/n)\}$, we get $m\{f > 0\} = 0$. □

Our two applications of Chebyshev's inequality provide some insight into how integrable functions are "built." If f is nonnegative and integrable, then $m\{f = \infty\} = 0$ since $m\{f \geq n\} \leq (1/n) \int f \to 0$. What's more, the *support* of f, that is, the set $\{f \neq 0\}$, can be written as an increasing union of sets of finite measure: $\{f > 0\} = \bigcup_{n=1}^{\infty} \{f \geq (1/n)\}$ and $m\{f \geq (1/n)\} \leq n \int f < \infty$. (This still allows $m\{f > 0\} = \infty$, of course.) Once we bring the Monotone Convergence Theorem into the picture, we can say even more. Consider the following string of equations:

$$\int_{-\infty}^{\infty} f = \lim_{n\to\infty} \int_{-n}^{n} f = \lim_{n\to\infty} \int_{\{f \geq (1/n)\}} f = \lim_{n\to\infty} \int_{\{f \leq n\}} f.$$

The first two limits are good for any nonnegative measurable function f. In order that the third limit equal $\int f$, it is necessary that f be finite a.e. (Why?)

The Monotone Convergence Theorem easily allows us to consider series of nonnegative functions. The following corollary is actually *equivalent* to the Monotone Convergence Theorem, but it's well worth the effort of a separate statement. In this form it's often called *the Beppo Levi theorem*, after its creator.

Corollary 18.11. *If (f_n) is a sequence of nonnegative measurable functions, then*

$$\int \left(\sum_{n=1}^{\infty} f_n \right) = \sum_{n=1}^{\infty} \int f_n.$$

PROOF. Note that since the f_n are nonnegative, both infinite sums exist: The partial sums $\sum_{n=1}^{N} f_n$ increase to $\sum_{n=1}^{\infty} f_n$, while from monotonicity and additivity of the integral we have that $\int \left(\sum_{n=1}^{N} f_n \right) = \sum_{n=1}^{N} \int f_n$ increases to $\sum_{n=1}^{\infty} \int f_n$. The Monotone Convergence Theorem finishes the job:

$$\int \left(\sum_{n=1}^{\infty} f_n \right) = \int \left(\lim_{N\to\infty} \sum_{n=1}^{N} f_n \right) = \lim_{N\to\infty} \int \left(\sum_{n=1}^{N} f_n \right)$$
$$= \lim_{N\to\infty} \sum_{n=1}^{N} \int f_n = \sum_{n=1}^{\infty} \int f_n. \quad \square$$

Here, finally, is the result we were looking for:

Corollary 18.12. *If f is nonnegative and measurable, then the map $E \mapsto \int_E f$ is a measure on $\mathcal{M}$. In particular, if (E_n) is a sequence of pairwise disjoint measurable sets, then*

$$\int_{\bigcup_{n=1}^{\infty} E_n} f = \sum_{n=1}^{\infty} \int_{E_n} f.$$

Again, the upshot of this observation is that the map $E \mapsto \int_E f$ has certain "continuity" properties. See Exercise 17 for a particularly striking result along these lines.

EXERCISES

7. Let $\mu : \mathcal{A} \to [\,0, \infty\,]$ be a nonnegative, finitely additive, set function defined on a σ-algebra $\mathcal{A}$. Prove that:

(i) $\mu(E) \le \mu(F)$ whenever $E, F \in \mathcal{A}$ satisfy $E \subset F$.

(ii) if $\mu(\emptyset) \ne 0$, then $\mu(E) = \infty$ for all $E \in \mathcal{A}$.

8. Let $\mu : \mathcal{A} \to [\,0, \infty\,]$ be a nonnegative, finitely additive, set function defined on a σ-algebra $\mathcal{A}$. Prove that the following are equivalent:

(i) $\mu\left(\bigcup_{n=1}^{\infty} E_n\right) = \sum_{n=1}^{\infty} \mu(E_n)$ for every sequence of pairwise disjoint sets (E_n) in $\mathcal{A}$.

(ii) $\mu\left(\bigcup_{n=1}^{\infty} E_n\right) = \lim_{n\to\infty} \mu(E_n)$ for every increasing sequence of sets (E_n) in $\mathcal{A}$.

▷ **9.** Let f be measurable with $f > 0$ a.e. If $\int_E f = 0$ for some measurable set E, show that $m(E) = 0$.

▷ **10.** If f is nonnegative and measurable, show that $\int_{-\infty}^{\infty} f = \lim_{n\to\infty} \int_{-n}^{n} f = \lim_{n\to\infty} \int_{\{f \ge (1/n)\}} f$.

▷ **11.** If f is nonnegative and integrable, show that $\int_{-\infty}^{\infty} f = \lim_{n\to\infty} \int_{\{f \le n\}} f$.

▷ **12.** True or False? If f is nonnegative and integrable, then $\lim_{x\to\pm\infty} f(x) = 0$. Explain.

13. Let $f : [\,0, 1\,] \to [\,0, 1\,]$ be the Cantor function. Show that $\int_0^1 f = 1/2$. [Hint: f is constant on each interval in the complement of Δ.]

14. Define $f : [\,0, 1\,] \to [\,0, \infty)$ by $f(x) = 0$ if x is rational and $f(x) = 2^n$ if x is irrational with exactly $n = 0, 1, 2, \ldots$ leading zeros in its decimal expansion. Show that f is measurable, and find $\int_0^1 f$.

15. Let f be nonnegative and measurable. Prove that $\int f < \infty$ if and only if $\sum_{k=-\infty}^{\infty} 2^k m\{f > 2^k\} < \infty$.

16. Let f be nonnegative and integrable. Given $\varepsilon > 0$, show that there is a measurable set E with $m(E) < \infty$ such that $\int_E f > \int f - \varepsilon$. Moreover, show that E can be chosen so that f is bounded (above) on E.

17. If f is nonnegative and integrable, prove that the function $F(x) = \int_{-\infty}^{x} f$ is continuous. In fact, even more is true: Given $\varepsilon > 0$, show that there is a $\delta > 0$ such that $\int_E f < \varepsilon$ whenever $m(E) < \delta$. [Hint: This is easy if f is bounded; see Exercise 16.]

By now you've noticed how effortlessly we've been able to exchange limits and integrals, at least in certain cases. If you'll take it on faith, temporarily, that the Lebesgue integral includes the Riemann integral as a special case, then you'll certainly agree that we've improved on our old integral. Of course, as the exercises point out, even the Lebesgue integral won't commute with *all* limits. Nevertheless, we can always at least *compare* $\int \lim_{n\to\infty} f_n$ and $\lim_{n\to\infty} \int f_n$. Our next result tells us how; it's a useful little gem!

Fatou's Lemma 18.13. *If (f_n) is a sequence of nonnegative measurable functions, then*

$$\int \left(\liminf_{n\to\infty} f_n\right) \le \liminf_{n\to\infty} \int f_n.$$

PROOF. Let $g_n = \inf\{f_n, f_{n+1}, \ldots\}$. Then g_n is nonnegative, measurable, and (g_n) increases to $\liminf_{k\to\infty} f_k$. From the Monotone Convergence Theorem, $\int (\liminf_{n\to\infty} f_n) = \lim_{n\to\infty} \int g_n$. It remains only to estimate $\lim_{n\to\infty} \int g_n$. But,

$$g_n = \inf_{k\ge n} f_k \implies \int g_n \le \int f_k \qquad \text{for } k \ge n$$

$$\implies \int g_n \le \inf_{k\ge n} \int f_k.$$

Thus,

$$\lim_{n\to\infty} \int g_n \le \lim_{n\to\infty} \inf_{k\ge n} \int f_k = \liminf_{n\to\infty} \int f_n. \quad \square$$

Just for good measure, here's the proof of Fatou's Lemma in one line:

$$\int \liminf_{n\to\infty} f_n = \lim_{n\to\infty} \int \inf_{k\ge n} f_k \le \lim_{n\to\infty} \inf_{k\ge n} \int f_k = \liminf_{n\to\infty} \int f_n.$$

Of course, should both $\lim_{n\to\infty} f_n$ and $\lim_{n\to\infty} \int f_n$ exist, then Fatou's Lemma assures us that $\int \lim_{n\to\infty} f_n \le \lim_{n\to\infty} \int f_n$.

EXERCISES

18. Show that strict inequality is possible in Fatou's Lemma. [Hint: Consider $f_n = \chi_{[n,n+1)}$.]

19. If (f_n) is a sequence of nonnegative measurable functions, is it true that $\limsup_{n\to\infty} \int f_n \le \int (\limsup_{n\to\infty} f_n)$? What if (f_n) is uniformly bounded?

20. If f and (f_n) are nonnegative measurable functions, and if $f_n \to f$ a.e., prove that $\int f \le \liminf_{n\to\infty} \int f_n$.

21. Suppose that f and (f_n) are nonnegative measurable functions, that $f = \lim_{n\to\infty} f_n$, and that $f_n \le f$ for all n. Show that $\int f = \lim_{n\to\infty} \int f_n$.

▷ **22.** Suppose that f and (f_n) are nonnegative measurable functions, that $f = \lim_{n\to\infty} f_n$, and that $\int f = \lim_{n\to\infty} \int f_n < \infty$. Prove that $\int_E f = \lim_{n\to\infty} \int_E f_n$ for any measurable set E. [Hint: Consider both $\int_E f$ and $\int_{E^c} f$.] Give an example showing that this need not be true if $\int f = \lim_{n\to\infty} \int f_n = \infty$.

The General Case

We are now ready to define the Lebesgue integral for the general measurable function $f : \mathbb{R} \to [-\infty, \infty]$. As you will recall, if f is measurable, then so are the positive and negative parts of f:

$$f^+ = f \vee 0 \qquad \text{and} \qquad f^- = -f \wedge 0.$$

Recall, too, that f^+ and f^- satisfy

$$f = f^+ - f^- \qquad \text{and} \qquad |f| = f^+ + f^-$$

and also $(f^+)(f^-) = 0 = f^+ \wedge f^-$ (that is, f^+ and f^- are *disjointly supported*).

We now define the Lebesgue integral of f in the only way we can! *If at least one of f^+ or f^- is integrable*, we define

$$\int f = \int f^+ - \int f^-;$$

otherwise, $\int f$ is not defined (after all, we cannot allow $\infty - \infty$). *If both f^+ and f^- are integrable*, then we say that f is (Lebesgue) **integrable**. This is precisely the condition that is needed to force $\int f$ to be a real number. But please note that this differs substantially from Riemann integrability; in fact, it is worth repeating:

$$\begin{aligned} & f \text{ is Lebesgue integrable} \\ &\iff \text{both } f^+ \text{ and } f^- \text{ are integrable} \\ &\iff \textstyle\int f^+ < \infty \text{ and } \int f^- < \infty \\ &\iff \textstyle\int f^+ + \int f^- < \infty \text{ (since each is } \ge 0) \\ &\iff \textstyle\int |f| < \infty \\ &\iff |f| \text{ is Lebesgue integrable.} \end{aligned}$$

By way of a quick example, note that $f = 2\chi_{\mathbb{Q}\cap[0,1]} - 1$ is *not* Riemann integrable on $[0, 1]$ while $|f| = 1$ is.

If E is a measurable set, then, as before, we define

$$\int_E f = \int f \cdot \chi_E = \int_E f^+ - \int_E f^-,$$

provided that this makes sense, of course. As usual, if f is defined only on E, we simply extend f to all of $\mathbb{R}$ by setting $f = 0$ outside E. In this case, $\int_E f = \int f$. In either case, $\int_E f$ depends only on the restriction of f to E. If $\int_E |f| < \infty$, then we will say that f is *integrable on* E.

High on our list of projects is to show that the collection of integrable functions is a vector space and that the integral is a linear real-valued function on this space. Before we attack these issues, though, let's make a few simple observations.

Observations 18.14

(a) One more time: f is integrable if and only if $|f|$ is integrable and, in either case, $\left|\int f\right| \le \int |f|$. (Why?)

(b) If f is integrable, then f is finite a.e.; that is, $m\{|f| = \infty\} = 0$.

(c) If f is integrable and $m(E) = 0$, then $\int_E f = 0$. Together with our second observation, this says that we might as well consider an integrable function to be real-valued. Notice, too, that our new definition is in accord with our previous definition; in fact, if $f \ge 0$ a.e., then $f = f^+$ a.e., and so $\int f = \int f^+ \ge 0$.

(d) If f and g are measurable with $|f| \le |g|$ a.e., and if g is integrable, then f is integrable, too, and $\int |f| \le \int |g|$. In particular, if we also have $f = g$ a.e., then f is integrable and $\int f = \int g$.

(e) If $f : [a, b] \to \mathbb{R}$ is bounded and measurable, then f is integrable on $[a, b]$. In particular, if f is a bounded Riemann integrable function on $[a, b]$, then f is also Lebesgue integrable on $[a, b]$. What's more, as we will see shortly, the two integrals agree; that is, $(R)\int_a^b f(x)\,dx = (L)\int_a^b f$. For the time being, or at least until we prove that the Lebesgue integral subsumes the Riemann integral, we will distinguish between the two integrals, when necessary, by using the prefixes (R) and (L).

(f) Given a Lebesgue measurable function f, recall that there is a Borel measurable function g with $f = g$ a.e. If we were only interested in computing integrals, this means that we would only need Borel measurable functions.

We denote the collection of Lebesgue integrable functions f defined on $\mathbb{R}$ by $L_1(\mathbb{R})$. Given a measurable set E, we will also consider the collection of functions integrable on E, which we denote by $L_1(E)$. More precisely, $L_1(E)$ is the collection of measurable functions f defined on E for which $\int_E |f| < \infty$. Equivalently, $L_1(E)$ consists of all functions of the form $f\chi_E$, where $f : \mathbb{R} \to [-\infty, \infty]$ is measurable and $f\chi_E$ is integrable. The point here is that when considering the collection $L_1(E)$, we do not care much about what goes on outside the set E; the elements of $L_1(E)$ need not be defined outside of E.

As you might imagine, we are most interested in the case where E is an interval; that is, we are interested in the spaces $L_1[a, b]$, $L_1[0, \infty)$, $L_1(\mathbb{R})$, and so on. But, as it happens, the vector and metric space properties that we are concerned with will not actually depend on the underlying set E. For this reason, we may occasionally just write L_1 to denote a typical space $L_1(E)$. In fact, there is no real harm in thinking of $L_1 = L_1(\mathbb{R})$ as the typical case (and this is precisely what we will do).

Before we compare our new integral to the Riemann integral, let's at least establish a few of the familiar properties of the integral in the case of the Lebesgue integral. As you have no doubt grown accustomed to by now, we will interpret the elementary properties of the integral in terms of the vector space and lattice structure of the entire collection of integrable functions.

Proposition 18.15 *L_1 is a vector space and a lattice, under the usual pointwise operations on functions, and the Lebesgue integral is a positive, linear, real-valued function on the space.*

PROOF. Given $f, g \in L_1$ and $a, b \in \mathbb{R}$, we have $|af + bg| \le |a|\,|f| + |b|\,|g|$ a.e. (at least where f and g are real-valued). Thus, $af + bg \in L_1$.

That L_1 is a lattice now follows from the fact that it is a linear space containing the absolute values of its elements. Specifically, if f and g are integrable, then

$$f \vee g = \tfrac{1}{2}(|f - g| + f + g) \quad \text{a.e.}$$

and so $|f \vee g| \le |f| + |g|$ a.e. Thus, $|f \vee g| \in L_1$. Similarly for $f \wedge g$.

Now, to show that integration is linear and positive is easy. First notice that $(af)^{\pm} = af^{\pm}$ for $a \ge 0$, and that $(af)^{\pm} = -af^{\mp}$ for $a < 0$. From this it is easy to check that $\int(af) = a\int f$ for any $f \in L_1$ and any $a \in \mathbb{R}$. Next, if $f, g \in L_1$, then

$$(f + g)^+ - (f + g)^- = f + g = f^+ - f^- + g^+ - g^- \quad \text{a.e.,}$$

or at least where $f^{\pm}$ and $g^{\pm}$ are all real-valued. Thus,

$$(f + g)^+ + f^- + g^- = (f + g)^- + f^+ + g^+ \quad \text{a.e.,}$$

and both sides represent nonnegative measurable functions. Since the integral is additive for nonnegative functions (and since it ignores that "a.e."),

$$\int (f + g)^+ + \int f^- + \int g^- = \int (f + g)^- + \int f^+ + \int g^+.$$

Now, since each of these integrals is finite, we can rearrange them to read

$$\begin{aligned}\int (f + g) &= \int (f + g)^+ - \int (f + g)^- \\ &= \int f^+ - \int f^- + \int g^+ - \int g^- = \int f + \int g.\end{aligned}$$

Finally, as we have already observed, if $f \in L_1$ and if $f \geq 0$ a.e., then $\int f \geq 0$. Combining this with the linearity of the integral, we have $\int f \geq \int g$ whenever f, $g \in L_1$ satisfy $f \geq g$ a.e. □

Please take note of the fact that in each of the various calculations in Proposition 18.15 we were able to draw conclusions based only on the almost everywhere validity of equations and inequalities. We will have more to say about this fact later.

For now, let's recover all of elementary calculus in one stroke:

Theorem 18.16 *If $f : [a,b] \to \mathbb{R}$ is a bounded Riemann integrable function, then f is also Lebesgue integrable on $[a,b]$ and the two integrals agree:*

$$(R)\int_a^b f(x)\,dx = (L)\int_a^b f.$$

PROOF. From Theorem 17.4 and Observation 18.14 (e), all we need to establish here is the equality of the integrals. What makes this possible is the fact that the two integrals clearly coincide for *step functions*. (Why?)

Since f is Riemann integrable, we can find two sequences of step functions (ℓ_n) and (u_n) with $\ell_n \leq f \leq u_n$ such that

$$\sup_n \int_a^b \ell_n = (R)\int_a^b f(x)\,dx = \inf_n \int_a^b u_n.$$

(Notice that we do not need to distinguish between the Riemann and the Lebesgue integrals for either the ℓ_n or the u_n.) But, from the monotonicity of the Lebesgue integral, we have

$$\sup_n \int_a^b \ell_n \leq (L)\int_a^b f \leq \inf_n \int_a^b u_n.$$

Thus, $(R)\int_a^b f(x)\,dx = (L)\int_a^b f$.

At the price of a few more lines, we can actually show something more. If we define $\ell = \sup_n \ell_n$ and $u = \inf_n u_n$, then ℓ and u are bounded, measurable functions on $[a,b]$ satisfying $\ell \leq f \leq u$. Moreover,

$$\sup_n \int_a^b \ell_n \leq (L)\int_a^b \ell \leq (L)\int_a^b u \leq \inf_n \int_a^b u_n.$$

Thus, $(L)\int_a^b \ell = (L)\int_a^b u$. It follows that $\ell = f = u$ a.e. (How?) This gives another proof that f is Lebesgue measurable. □

While the Lebesgue integral subsumes the *proper* Riemann integral, we will see some differences in the case of the *improper* Riemann integral. For now we will settle for a single example:

Example 18.17

The improper Riemann integral $(IR)\int_0^\infty (\sin x/x)\,dx$ exists, while the Lebesgue integral $(L)\int_0^\infty (\sin x/x)\,dx$ does *not*.

PROOF. The improper Riemann integral can be written as an alternating series:

$$
\begin{aligned}
(IR)\int_0^\infty \frac{\sin x}{x}\,dx &= \sum_{n=1}^\infty \int_{(n-1)\pi}^{n\pi} \frac{\sin x}{x}\,dx \\
&= \sum_{n=1}^\infty (-1)^{n-1} \int_{(n-1)\pi}^{n\pi} \frac{|\sin x|}{x}\,dx \\
&= \sum_{n=1}^\infty (-1)^{n-1} \int_0^{\pi} \frac{|\sin x|}{x+(n-1)\pi}\,dx.
\end{aligned}
$$

To show that the series converges, we only have to show that the terms tend monotonically to zero. But $|\sin x|/(x+(n-1)\pi)$ clearly decreases as n increases (for x fixed), and

$$
\int_0^{\pi} \frac{|\sin x|}{x+(n-1)\pi}\,dx \le \frac{1}{n-1} \to 0.
$$

In order that the Lebesgue integral exist, on the other hand, it is necessary to have $(L)\int_0^\infty (|\sin x|/x)\,dx < \infty$. But, from the Monotone Convergence Theorem,

$$
\begin{aligned}
(L)\int_0^\infty \frac{|\sin x|}{x}\,dx &= \sum_{n=1}^\infty \int_{(n-1)\pi}^{n\pi} \frac{|\sin x|}{x}\,dx \\
&\ge \sum_{n=1}^\infty \frac{1}{n\pi} \int_0^{\pi} |\sin x|\,dx = \infty. \quad \square
\end{aligned}
$$

As the last example demonstrates, the difference between the improper Riemann integral and the Lebesgue integral is roughly the same as the difference between conditionally convergent series and absolutely convergent series. The improper Riemann integral may exist due to the effect of delicate cancellations, while the Lebesgue integral does not permit such issues to arise. In any event, please note that there is no such thing as an "improper" Lebesgue integral: We made no special assumptions about the boundedness of our integrand, or about the boundedness of the set over which we integrate. We will say more about the comparisons between the Riemann (and even the Riemann – Stieltjes) integral and the Lebesgue integral later.

While we have shown that L_1 is both a vector space and a lattice, it is easy to see that L_1 is not an algebra under the usual pointwise multiplication of functions. For example, if we set $f(x) = x^{-1/2}$ for $0 < x < 1$ and $f(x) = 0$ otherwise, then f is integrable while f^2 is not. See Exercise 26 for a variation on this example.

For the remainder of this book, we will assume that all integrals are Lebesgue integrals, unless otherwise specified. With very few exceptions, this should cause no problems. If f is a nonnegative, continuous function, for example, then the Riemann and Lebesgue integrals of f over any interval either both exist (and are equal) or both fail to exist.

EXERCISES

▷ **23.** If (f_n) is a sequence of Lebesgue integrable functions on $[\,a, b\,]$, and if $f_n \rightrightarrows f$ on $[\,a, b\,]$, prove that f is integrable and that $\int_a^b |f_n - f| \to 0$.

24. Prove that $\int_0^\infty e^{-x}\,dx = \lim_{n\to\infty} \int_0^n (1-(x/n))^n\,dx = 1$. [Hint: For x fixed, $(1-(x/n))^n$ increases to e^{-x} as $n \to \infty$.]

25. Compute $\lim_{n\to\infty} \int_0^n (1-(x/n))^n\, e^{x/2}\,dx$, justifying your calculations.

26. Let $f(x) = x^{-1/2}$ for $0 < x < 1$ and $f(x) = 0$ otherwise. Let (r_n) be an enumeration of $\mathbb{Q}$, and let $g(x) = \sum_{n=1}^\infty 2^{-n} f(x - r_n)$. Show that:

(a) $g \in L_1$ and, in particular, g is finite a.e.

(b) g is discontinuous at every point and is unbounded on every interval; it remains so even after modification on an arbitrary set of measure zero.

(c) g^2 is finite a.e., but g^2 is not integrable on any interval.

27. Suppose that $E \subset [\,0, 2\pi\,]$ is measurable and that $\int_E x^n \cos x\,dx = 0$ for all $n = 0, 1, 2, \ldots$. Show that $m(E) = 0$.

28. Suppose that f, g, and h are measurable and that $f \le g \le h$ a.e. If f and h are Lebesgue integrable, does it follow that g is Lebesgue integrable? Explain.

29. If $f, (f_n)$ are Lebesgue integrable, and if (f_n) increases pointwise to f, does it follow that $\int f_n \to \int f$? Explain.

30. Construct a sequence of integrable functions (f_n) such that $f_n \to 0$ a.e., but such that $\int |f_n| \not\to 0$. Construct a sequence of integrable functions (g_n) such that $\int |g_n| \to 0$, but such that $g_n \not\to 0$ a.e.

31. Let (f_n), f be integrable. If $\int |f_n - f| \to 0$, show that $\int f_n \to \int f$ and $\int |f_n| \to \int |f|$.

32. Let (f_n), f be integrable, and suppose that $\int |f_n - f| \to 0$. Show that $\int_E f_n \to \int_E f$ for all measurable sets E, and that $\int f_n^+ \to \int f^+$.

33. Let f be measurable. Prove that f is Lebesgue integrable if and only if $\sum_{k=-\infty}^{\infty} 2^k m\{|f| > 2^k\} < \infty$.

34. Let f be Lebesgue integrable. Given $\varepsilon > 0$, show that there is a measurable set E with $m(E) < \infty$ such that $\int_E |f| > \int |f| - \varepsilon$. Moreover, show that E can be chosen so that f is bounded on E.

35. If f is Lebesgue integrable, prove that the function $F(x) = \int_{-\infty}^x f$ is continuous. In fact, even more is true: Given $\varepsilon > 0$, show that there is a $\delta > 0$ such that $\int_E |f| < \varepsilon$ whenever $m(E) < \delta$. [Hint: This is easy if f is bounded; see Exercise 34.]

36. Suppose that f, (f_n) are measurable and uniformly bounded on $[\,a, b\,]$. If $f_n \to f$ on $[\,a, b\,]$, prove that $\int_a^b |f_n - f| \to 0$. [Hint: Egorov's theorem.]

We are almost ready to define a norm on L_1; one final observation will come in handy.

Lemma 18.18 *For $f, g \in L_1$, the following are equivalent:*

(i) $f = g$ *a.e.*

(ii) $\int |f - g| = 0$.

(iii) $\int_E f = \int_E g$ *for every measurable set E.*

PROOF. That (i) and (ii) are equivalent is easy: From Lemma 18.10, we have

$$\int |f - g| = 0 \iff |f - g| = 0 \text{ a.e.} \iff f = g \text{ a.e.}$$

Now, for (ii) $\Longrightarrow$ (iii), note that

$$\left| \int_E f - \int_E g \right| \le \int_E |f - g| \le \int |f - g| = 0.$$

Finally, for (iii) $\Longrightarrow$ (ii), let $E = \{f - g \ge 0\}$. Then,

$$\int |f - g| = \int_E (f - g) + \int_{E^c} (g - f) = 0. \quad \square$$

We have a natural choice for a norm on L_1, namely, $\|f\|_1 = \int |f|$. In other words, $d(f, g) = \int |f - g|$ would appear to be a good guess for a metric on L_1. Unfortunately, this will not quite work since $d(f, g) = 0$ only means that $f = g$ a.e. To remedy this, we will simply *identify* functions that are almost everywhere equal.

Formally, we define an equivalence relation on L_1 by taking $f \sim g$ to mean that $f = g$ a.e., and we denote the equivalence classes under $\sim$ by $[f]$. It is easy to check that the collection of all equivalence classes is again a vector space and a lattice under the operations $a[f] = [af]$ for $a \in \mathbb{R}$, $[f] + [g] = [f + g]$, and $[f] \le [g]$ whenever $f \le g$ a.e. What's more, $\|\,[f]\,\|_1 = \int |f|$ now defines a norm on the collection of all equivalence classes.

For all practical purposes, we need not bother with the formalities outlined above; after all, we are typically interested in specific, concrete functions. But, if we want to consider L_1 as a normed linear space under its natural "norm" (and we do!), then we will want to modify our definition of L_1. Henceforth, we will consider L_1 to be the collection of all *equivalence classes* of integrable functions under equality almost everywhere. In symbols, we identify f with $[f]$, and we define $\|f\|_1 = \|\,[f]\,\|_1 = \int |f|$. It is not hard to see that this "new L_1" is, indeed, a normed vector space and a normed lattice under $\|\cdot\|_1$. In particular, notice that $\|f\|_1 \le \|g\|_1$ whenever $|f| \le |g|$ a.e.

EXERCISE

37. Check that the operations $a[f] = [af]$ for $a \in \mathbb{R}$, $[f] + [g] = [f + g]$, and $[f] \le [g]$ whenever $f \le g$ a.e. are well defined, and that the collection of equivalence classes is a vector lattice when supplied with this arithmetic. What is $|\,[f]\,|$ in this lattice? Is it $[\,|f|\,]$?

Lebesgue's Dominated Convergence Theorem

Now that we have a norm on L_1, the next question is whether L_1 is complete. The first step that we will take in this direction is to prove a truly wonderful theorem, one that de La Vallée Poussin called the "crowning achievement" of Lebesgue's work.

Dominated Convergence Theorem 18.19 *Let (f_n) be a sequence in L_1 and suppose that $f_n \to f$ pointwise. If the f_n are all dominated by a single L_1 function, that is, if $|f_n| \le g$ for all n, where $g \in L_1$, then we have $f \in L_1$ and $\int f_n \to \int f$ as $n \to \infty$.*

PROOF. Since $f_n \to f$, then we must also have $|f| \le g$. Since $g \in L_1$, this means that $f \in L_1$ and $\int |f| \le \int g$.

The proof that $\int f_n \to \int f$ consists of a clever application of Fatou's lemma. Notice that each of the sequences $(g + f_n)$ and $(g - f_n)$ is comprised of *nonnegative* functions, and that $g + f_n \to g + f$ and $g - f_n \to g - f$. Now we unleash Fatou: First,

$$\int g + \int f = \int (g + f) \le \liminf_{n\to\infty} \int (g + f_n)$$

$$= \int g + \liminf_{n\to\infty} \int f_n;$$

thus, $\int f \le \liminf_{n\to\infty} \int f_n$. Next,

$$\int g - \int f = \int (g - f) \le \liminf_{n\to\infty} \int (g - f_n)$$

$$= \int g - \limsup_{n\to\infty} \int f_n.$$

Thus, $\limsup_{n\to\infty} \int f_n \le \int f$. Aha! $\limsup_{n\to\infty} \int f_n \le \int f \le \liminf_{n\to\infty} \int f_n$, and so $\lim_{n\to\infty} \int f_n$ exists and equals $\int f$. □

Note that the "domination condition," $|f_n| \le g$ for all n, where $g \in L_1$, is equivalent to the requirement that $\sup_n |f_n|$ be integrable.

By discarding countably many sets of measure zero, we may weaken the hypotheses of the Dominated Convergence Theorem by requiring only that $|f_n| \le g$ a.e. and that $f_n \to f$ a.e. What's more, by applying the theorem instead to the sequence $(|f_n - f|)$, noting that $|f_n - f| \to 0$ a.e. and $|f_n - f| \le 2g$ a.e., we actually get a stronger conclusion:

Corollary 18.20 *Let (f_n) be a sequence in L_1. Suppose that $f_n \to f$ a.e. and that $|f_n| \le g$ a.e. for all n, where $g \in L_1$. Then, $f \in L_1$ and $\int |f_n - f| \to 0$ as $n \to \infty$. That is, $\|f_n - f\|_1 \to 0$ as $n \to \infty$.*

We will have many opportunities to use the Dominated Convergence Theorem. Here are three quick applications that demonstrate its utility (compare the first of these with Exercise 35).

Corollary 18.21 *If $f \in L_1$, then $F(x) = \int_{-\infty}^{x} f$ is continuous.*

PROOF. If $x_n \to x$, then $f\chi_{(-\infty,x_n]} \to f\chi_{(-\infty,x]}$ a.e. (Why?) Also, $\left|f\chi_{(-\infty,x_n]}\right| \le |f| \in L_1$. Thus, by the Dominated Convergence Theorem, $\int_{-\infty}^{x_n} f \to \int_{-\infty}^{x} f$; that is, $F(x_n) \to F(x)$. □

Corollary 18.22 *If $f \in L_1$, then*

$$\int_{-\infty}^{\infty} f = \lim_{a\to-\infty}\lim_{b\to\infty}\int_a^b f = \lim_{b\to\infty}\lim_{a\to-\infty}\int_a^b f,$$

and for $a, b \in \mathbb{R}$, $a < b$,

$$\int_a^b f = \lim_{\varepsilon\to 0^+}\int_{a+\varepsilon}^b f = \lim_{\varepsilon\to 0^+}\int_a^{b-\varepsilon} f.$$

This "continuous parameter" application of the Dominated Convergence Theorem is proved by applying the theorem to all possible *sequential* limits. Along similar lines, we can easily derive another comparison with the improper Riemann integral.

Corollary 18.23 *Suppose that $f : [0,\infty) \to \mathbb{R}$ is Riemann integrable on $[a,b]$ for every $0 < a < b < \infty$, and that*

$$(IR)\int_0^{\infty} |f(t)|\,dt = \lim_{a\to 0^+}\lim_{b\to\infty}\int_a^b |f|$$

exists. Then, $(IR)\int_0^{\infty} f(t)\,dt$ and $(L)\int_0^{\infty} f$ both exist and are equal.

PROOF. Since $f \in \mathcal{R}[a,b]$, we know that $f \in L_1[a,b]$, that $(R)\int_a^b f = (L)\int_a^b f$, and that $(R)\int_a^b |f| = (L)\int_a^b |f|$ for any $0 < a < b < \infty$. Moreover, since the restriction of f to $[a,b]$ is measurable for all $0 < a < b < \infty$, then f is clearly measurable on $[0,\infty)$. An appeal to the Monotone Convergence Theorem now shows that $f \in L_1[0,\infty)$:

$$(L)\int_0^{\infty} |f| = \lim_{n\to\infty}\int_{1/n}^{n} |f| \le (IR)\int_0^{\infty} |f| < \infty.$$

It now follows from Corollary 18.22 that

$$(IR)\int_0^{\infty} f = \lim_{a\to 0^+}\lim_{b\to\infty}\int_a^b f = (L)\int_0^{\infty} f.$$

In fact, Corollary 18.22 also shows that $(L)\int_0^{\infty}|f| = (IR)\int_0^{\infty}|f|$. □

Just as with the Monotone Convergence Theorem, it is useful to have a version of the Dominated Convergence Theorem written in terms of series of functions.

Theorem 18.24 *Let (f_n) be a sequence of integrable functions such that $\sum_{n=1}^{\infty}\int |f_n| < \infty$. Then, $\sum_{n=1}^{\infty} f_n$ converges a.e. to an integrable function. Moreover,*

$$\left|\int\left(\sum_{n=1}^{\infty} f_n\right)\right| \le \int\left|\sum_{n=1}^{\infty} f_n\right| \le \sum_{n=1}^{\infty}\int |f_n|.$$

and

$$\int\left(\sum_{n=1}^{\infty} f_n\right) = \sum_{n=1}^{\infty}\int f_n.$$

PROOF. Consider $g = \sum_{n=1}^{\infty} |f_n|$. From the Monotone Convergence Theorem we know that $\int g = \sum_{n=1}^{\infty} \int |f_n| < \infty$. Thus, g is integrable and, most importantly, g is finite a.e. What this means is that

$$\left| \sum_{n=1}^{\infty} f_n \right| \le \sum_{n=1}^{\infty} |f_n| < \infty \quad \text{a.e.}$$

That is, $\sum_{n=1}^{\infty} f_n$ converges absolutely a.e. to a finite limit f that satisfies $|f| \le g$ a.e. And so f is integrable. Of course, $\left|\int f\right| \le \int |f| \le \int g$, which proves the first assertion of the theorem.

Notice, too, that the series $\sum_{n=1}^{\infty} \int f_n$ converges; in fact, it is even absolutely summable:

$$\sum_{n=1}^{\infty} \left| \int f_n \right| \le \sum_{n=1}^{\infty} \int |f_n| = \int g < \infty.$$

To prove the second claim, we apply the Dominated Convergence Theorem to the sequence of partial sums. Notice that $\left|\sum_{n=1}^{N} f_n\right| \le g$ a.e. and $\sum_{n=1}^{N} f_n \to \sum_{n=1}^{\infty} f_n$ a.e. Hence,

$$\begin{aligned} \int \left(\sum_{n=1}^{\infty} f_n \right) &= \int \left(\lim_{N\to\infty} \sum_{n=1}^{N} f_n \right) \\ &= \lim_{N\to\infty} \sum_{n=1}^{N} \int f_n = \sum_{n=1}^{\infty} \int f_n. \quad \square \end{aligned}$$

By applying Theorem 18.24 a second time to $\sum_{n=N+1}^{\infty} f_n$, the tail of the series, we get a much stronger result. If $f = \sum_{n=1}^{\infty} f_n$, then, as $N \to \infty$,

$$\int \left| f - \sum_{n=1}^{N} f_n \right| = \int \left| \sum_{n=N+1}^{\infty} f_n \right| \le \sum_{n=N+1}^{\infty} \int |f_n| \to 0.$$

That is, the series $\sum_{n=1}^{\infty} f_n$ converges in the norm of L_1. In brief, Theorem 18.24 shows that

$$\sum_{n=1}^{\infty} \|f_n\|_1 < \infty \Longrightarrow \sum_{n=1}^{\infty} f_n \quad \text{converges in } L_1.$$

By Theorem 7.12, this proves:

Corollary 18.25 *L_1 is complete.*

It also follows from Theorem 18.24 that if $f \in L_1$, then the set function $\mu(E) = \int_E f$, $E \in \mathcal{M}$, is countably additive. But since f is not necessarily nonnegative, μ will no longer be nonnegative or, indeed, monotone. On the other hand, μ is quite well behaved. First, since f is integrable, $\mu(E) \in \mathbb{R}$ for any $E \in \mathcal{M}$; in fact,

$$|\mu(E)| = \left| \int_E f \right| \le \int_E |f| \le \int |f| < \infty.$$

Second, if (E_n) is a sequence of pairwise disjoint, measurable sets, and if $E = \bigcup_{n=1}^{\infty} E_n$, then $\sum_{n=1}^{\infty} \mu(E_n) = \sum_{n=1}^{\infty} \int_{E_n} f$ is *absolutely convergent* with sum $\mu(E) = \int_E f$. In this case, we say that μ is a **signed measure**.

Corollary 18.26 *If $f \in L_1$, then the map $E \mapsto \int_E f$ is a signed measure on $\mathcal{M}$. In particular, if (E_n) is any sequence of pairwise disjoint, measurable sets, and if $E = \bigcup_{n=1}^{\infty} E_n$, then*

$$\sum_{n=1}^{\infty} \left| \int_{E_n} f \right| \le \int_E |f| < \infty \quad \text{and} \quad \sum_{n=1}^{\infty} \int_{E_n} f = \int_E f.$$

EXERCISES

38. If $f \in L_1[\,0,1\,]$, show that $x^n f(x) \in L_1[\,0,1\,]$ for $n = 1, 2, \ldots$ and compute $\lim_{n\to\infty} \int_0^1 x^n f(x)\,dx$.

39. Compute $\sum_{n=0}^{\infty} \int_0^{\pi/2} \left(1 - \sqrt{\sin x}\right)^n \cos x\,dx$. Justify your calculations.

▷ **40.** Let (f_n), (g_n), and g be integrable, and suppose that $f_n \to f$ a.e., $g_n \to g$ a.e., $|f_n| \le g_n$ a.e., for all n, and that $\int g_n \to \int g$. Prove that $f \in L_1$ and that $\int f_n \to \int f$. [Hint: Revise the proof of the Dominated Convergence Theorem.]

41. Let (f_n), f be integrable, and suppose that $f_n \to f$ a.e. Prove that $\int |f_n - f| \to 0$ if and only if $\int |f_n| \to \int |f|$.

42. Let (f_n) be a sequence of integrable functions and suppose that $|f_n| \le g$ a.e., for all n, for some integrable function g. Prove that

$$\int \left(\liminf_{n\to\infty} f_n\right) \le \liminf_{n\to\infty} \int f_n \le \limsup_{n\to\infty} \int f_n \le \int \left(\limsup_{n\to\infty} f_n\right).$$

43. Let f be measurable and finite a.e. on $[\,0,1\,]$.

(a) If $\int_E f = 0$ for all measurable $E \subset [\,0,1\,]$ with $m(E) = 1/2$, prove that $f = 0$ a.e. on $[\,0,1\,]$.

(b) If $f > 0$ a.e., show that $\inf\left\{\int_E f : m(E) \ge 1/2\right\} > 0$.

44. Show that $\lim_{n\to\infty} \int_0^1 f_n = 0$ where $f_n(x)$ is:

(a) $\dfrac{nx}{1+n^2x^2}$

(b) $\dfrac{n\sqrt{x}}{1+n^2x^2}$

(c) $\dfrac{nx\log x}{1+n^2x^2}$

(d) $\dfrac{n^{3/2}x}{1+n^2x^2}$.

[Hint: $1 + n^2x^2 \ge 2nx$.]

45. Find:

(a) $\lim_{n\to\infty} \int_0^\infty \frac{\sin(e^x)}{1+nx^2}\,dx$ (b) $\lim_{n\to\infty} \int_0^1 \frac{n\cos x}{1+n^2x^{3/2}}\,dx.$

46. Fix $0 < a < b$, and define $f_n(x) = ae^{-nax} - be^{-nbx}$. Show that $\sum_{n=1}^\infty \int_0^\infty |f_n| = \infty$ and $\int_0^\infty \left(\sum_{n=1}^\infty f_n\right) \neq \sum_{n=1}^\infty \int_0^\infty f_n$.

▷ **47.** Compute the following limits, justifying your calculations:

(a) $\lim_{n\to\infty} \int_0^\infty \frac{n\sin(x/n)}{x(1+x^2)}\,dx$

(b) $\lim_{n\to\infty} \int_0^1 \frac{1+nx^2}{(1+x^2)^n}\,dx$

(c) $\lim_{n\to\infty} \int_0^\infty \frac{\sin(x/n)}{(1+x/n)^n}\,dx$

(d) $\lim_{n\to\infty} \int_a^\infty \frac{n}{1+n^2x^2}\,dx.$

[The answer in (d) depends on whether $a > 0$, $a = 0$, or $a < 0$. How is this reconciled by the various convergence theorems?]

48. Let $\alpha, \beta \in \mathbb{R}$, and define $f(x) = x^\alpha \sin(x^\beta)$, $0 < x \le 1$. For what values of α and β is f: (i) Lebesgue integrable? (ii) Riemann integrable (in the sense that $\lim_{\varepsilon\to 0^+} \int_\varepsilon^1 f(x)\,dx$ exists)?

49. For which $\alpha \in \mathbb{R}$ is $\sum_{n=1}^\infty xn^{-\alpha}e^{-nx}$ continuous on $[0,\infty)$? in $L_1[0,\infty)$?

50. Let $f(x) = \sum_{n=1}^\infty (1/n)e^{-n(x-n)^2}$ for $x \in \mathbb{R}$. Is f in $L_1(\mathbb{R})$? continuous on $\mathbb{R}$? differentiable on $\mathbb{R}$?

51. Let (f_n) be a sequence of measurable functions with $|f_n| \le g$ for all n, where $g \in L_1$. If $f_n \to f$ a.e., prove that $f_n \to f$ almost uniformly. In other words, show that the conclusion of Egorov's theorem remains valid under the hypotheses of the Dominated Convergence Theorem. [Hint: In the notation of the proof of Theorem 17.13, it is enough to show that, for fixed k, some $E(n,k)$ has finite measure. Show that $E(n,k) \subset \{2g \ge 1/k\}$ in this case and argue that $m\{2g \ge 1/k\} < \infty$.]

Approximation of Integrable Functions

As a final installment in our discussion of the structure of Lebesgue integrable functions, we return to our Basic Construction and uncover a long list of dense subsets of L_1.

Theorem 18.27 *Let f be Lebesgue integrable on $\mathbb{R}$, and let $\varepsilon > 0$. Then:*

(i) *There is an integrable simple function φ with $\int |f - \varphi| < \varepsilon$.*

(ii) *There is a continuous function $g : \mathbb{R} \to \mathbb{R}$ such that $g = 0$ outside some bounded interval and such that $\int |f - g| < \varepsilon$.*

(iii) *There is an (integrable) step function h with $\int |f - h| < \varepsilon$.*

PROOF. From the Monotone Convergence Theorem we can find a compact interval $[a, b]$ such that $\int_{\mathbb{R}\setminus[a,b]} |f| < \varepsilon/4$. We will build all of our approximating functions with support in $[a, b]$; that is, each will be chosen to vanish outside of $[a, b]$.

(i) There is a sequence of (integrable) simple functions (φ_k) with $\varphi_k = 0$ off $[a, b]$, $\varphi_k \to f$ on $[a, b]$, and $|\varphi_k| \le |f|$. It follows from the Dominated Convergence Theorem that $\int_a^b |f - \varphi_k| \to 0$. Now choose k and $\varphi = \varphi_k$ with $\int_a^b |f - \varphi| < \varepsilon/4$ and, hence, $\int_{\mathbb{R}} |f - \varphi| < \varepsilon/2$.

(ii) The function φ is bounded; choose K such that $|\varphi| \le K$. Now, from Theorem 17.20 (and Exercise 17.45), we can find a continuous function g on $\mathbb{R}$, vanishing outside of $[a, b]$, that satisfies $|g| \le K$ and $m\{g \ne \varphi\} < \varepsilon/(8K)$. Thus,

$$\int |\varphi - g| = \int_a^b |\varphi - g| < 2K \cdot \frac{\varepsilon}{8K} = \frac{\varepsilon}{4},$$

and hence $\int |f - g| \le \int |f - \varphi| + \int |\varphi - g| < 3\varepsilon/4$.

(iii) From Lemma 12.2 we know that every continuous function on $[a, b]$ can be uniformly approximated by a step function. In particular, we can find a step function h on $[a, b]$ such that $\|g - h\|_\infty < \varepsilon/[4(b - a)]$. Thus,

$$\int |g - h| = \int_a^b |g - h| < \frac{\varepsilon}{4(b-a)} \cdot (b - a) = \frac{\varepsilon}{4},$$

and hence $\int |f - h| \le \int |f - g| + \int |g - h| < \varepsilon$. □

Corollary 18.28 *$C[a, b]$ is dense in $L_1[a, b]$. In fact, given $f \in L_1[a, b]$ and $\varepsilon > 0$, there exists a polynomial p with rational coefficients satisfying $\int_a^b |f - p| < \varepsilon$. Consequently, $L_1[a, b]$ is separable.*

EXERCISES

▷ **52.** Prove Corollary 18.28.

▷ **53.** Prove that $L_1(\mathbb{R})$ is separable.

54. Given $f \in L_1(\mathbb{R})$ and $\varepsilon > 0$, show that there is an infinitely differentiable function $\psi \in C^\infty(\mathbb{R})$ such that $\psi = 0$ outside some bounded interval, and such that $\int |f - \psi| < \varepsilon$. [Hint: Review the proof of Theorem 11.12.]

▷ **55.** Prove the Riemann–Lebesgue lemma: If f is integrable on $\mathbb{R}$, then $f(x)\cos nx$ is integrable and $\lim_{n\to\infty} \int_{-\infty}^{\infty} f(x) \cos nx\, dx = 0$. The same is true with $\sin nx$ in place of $\cos nx$. [Hint: First try $f = \chi_{[a,b]}$.]

56. Given $f \in L_1(\mathbb{R})$, define $g(x) = \int_{-\infty}^{\infty} f(t) \sin(xt)\, dt$ for $x \in \mathbb{R}$. Show that g is continuous on $\mathbb{R}$ and that $g(x) \to 0$ as $x \to \pm\infty$; hence, g is uniformly continuous on $\mathbb{R}$.

▷ **57.** Prove the following statements, where $f : \mathbb{R} \to \mathbb{R}$.

(a) If f is measurable, then so is $g(x) = f(x + t)$ for any t.

(b) If f is integrable, then so is $g(x) = f(x+t)$ and $\int f = \int g$. [Hint: This is easy if f is a step function.]

(c) If f is integrable, then $\lim_{t\to 0} \int_{-\infty}^{\infty} |f(x) - f(x+t)|\,dx = 0$.

(d) If f is integrable, find $\lim_{t\to\infty} \int_{-\infty}^{\infty} |f(x) - f(x+t)|\,dx$.

58. Prove the following statements, where $f : \mathbb{R} \to \mathbb{R}$.

(a) If f is measurable, then so is $g(x) = f(ax)$ for any a.

(b) If f is integrable, and if $a \neq 0$, then $g(x) = f(ax)$ is integrable and $\int f = |a| \int g$. [Hint: This is easy if f is a step function.]

(c) If f is integrable, then $\lim_{a\to\infty} \int_{-\infty}^{\infty} f(ax)\,dx = 0$.

59. Let $f \in L_1(\mathbb{R})$ and define $\phi(x) = \sum_{n=1}^{\infty} f(2^n x + (1/n))$. Show that φ is integrable and that $\int f = \int \varphi$.

60.

(a) Show that there is a sequence of polynomials (P_n) such that $P_n \to 0$ pointwise on $[0, 1]$, but with $\int_0^1 P_n(x)\,dx \to 3$.

(b) Find $\int_0^1 \sup_n |P_n(x)|\,dx$ for this sequence of polynomials.

▷ **61.** Given $f \in L_1[0, 2\pi]$ and $\varepsilon > 0$, show that there is a trig polynomial T such that $\int_0^{2\pi} |f - T| < \varepsilon$. [Hint: The proof of Theorem 18.27 (ii) shows that there is a continuous function g with $g(0) = 0 = g(2\pi)$ such that $\int_0^{2\pi} |f - g| < \varepsilon/2$. By setting $g(x \pm 2n\pi) = g(x)$ for any $n \in \mathbb{N}$, we may now assume that $g \in C^{2\pi}$.]

Notes and Remarks

While we have chosen an approach to defining the Lebesgue integral that is similar in spirit to Lebesgue's original presentation, there are many other equally viable approaches, including the familiar "area under the graph" approach (see Wheeden and Zygmund [1977]), the "upper and lower integral" approach (see Apostol [1975]), the "limit of step functions" approach (see Chae [1980] or Riesz and Sz.-Nagy [1955]), and at least one approach that avoids measure theory altogether (see Van Daele [1990], for example). But while several authors take the "simple function" approach, not so many bother to check the details. The particulars here are based in part on the painstaking presentations in the books by Folland [1984] and by Royden [1963]. Once the Lebesgue integral has been defined for nonnegative measurable functions, however, the differences between the various approaches tend to fade. Bear this in mind should you consult one of the references given below.

The articles by Bliss [1917], Gillman [1993], Goffman [1953b], Hildebrandt [1917], and Riesz [1920, 1936, 1949], together with Hawkins [1970], and Lebesgue's own *Leçons*, Lebesgue [1928], include discussions of some of the alternative approaches to integration and their history. The books by Folland [1984], Hewitt and Stromberg

[1965], Royden [1963], and Rudin [1966] include various abstractions and generalizations on these themes.

Exercise 44 is taken from De Barra [1974]. Exercises 26 and 47 are taken from Folland [1984]. Exercises 38, 39, and 59 are based on exercises in Torchinsky [1988]. A short proof of Exercise 41 (in a more general setting) is given in Novinger [1992]. Exercises 43 and 60 are taken from notes for a course on real analysis offered by W. B. Johnson at The Ohio State University in 1974–75. The result stated in Exercise 55 is sometimes called *Mercer's theorem*; many authors refer to the result in Exercise 56 as the Riemann–Lebesgue lemma. Lebesgue's version of the lemma appears in Lebesgue [1906].

CHAPTER NINETEEN

Additional Topics

We continue our study of Lebesgue measure and integration by pursuing a few additional topics of interest. Since we have already been afforded some practice with the basic ideas in earlier chapters, the presentation of these extra topics will be streamlined by relegating a larger proportion of the details to the exercises.

Convergence in Measure

We have now seen several modes or types of convergence for sequences of real-valued functions. In this section we will discuss yet another mode of convergence, called convergence in measure. To motivate this new notion, let's begin with a simple observation.

Suppose that (f_n) is a sequence of integrable functions that converges in L_1 to some (integrable) function f. Can we claim that (f_n) converges pointwise a.e. to f? Well, not exactly (see Exercise 1, below), but we can at least make this claim: Given $\varepsilon > 0$, Chebyshev's inequality tells us that

$$m\{|f_n - f| \geq \varepsilon\} \leq \frac{1}{\varepsilon} \int |f_n - f| \to 0$$

as $n \to \infty$. In other words, (f_n) cannot get too far away from f "in measure." Let's give a name to this new phenomenon.

Throughout, we let f and (f_n) be measurable, real-valued functions defined on some common measurable domain $D \subset \mathbb{R}$. We say that (f_n) **converges in measure** to f on the set D if, for each $\varepsilon > 0$, we have

$$m\big\{x \in D : |f_n(x) - f(x)| \geq \varepsilon\big\} \to 0 \qquad \text{as } n \to \infty.$$

Equivalently, (f_n) converges in measure to f if and only if, given $\varepsilon > 0$, there exists an N such that

$$m\{|f_n - f| \geq \varepsilon\} < \varepsilon \qquad \text{for all } n > N$$

(see Exercise 10). We will occasionally abbreviate these statements by using the suggestive shorthand $f_n \xrightarrow{m} f$.

Our goal in this section is to investigate this newest mode of convergence and to answer the question: Does convergence in measure tell us anything about pointwise convergence?

Examples 19.1

(a) Clearly, (f_n) converges in measure to f if and only if $(f_n - f)$ converges in measure to 0. Thus, as with most of the modes of convergence we are familiar with, null sequences are again the general case.

(b) While convergence in measure is implied by convergence in L_1, it is by no means the same thing. Consider, for example, the sequence $f_n = n\chi_{(0,1/n)}$ on $D = [0,1]$. We clearly have $f_n \xrightarrow{m} 0$, as well as $f_n \to 0$ pointwise, but $f_n \not\to 0$ in L_1 since $\int f_n = 1$ for each n. In fact, (f_n) is not even Cauchy in L_1 since $\int |f_{2n} - f_n| = 1$ for every n.

(c) Convergence in measure is not implied by pointwise convergence, in general. The sequence $f_n = \chi_{[n,n+1]}$ converges pointwise to 0 on $[0,\infty)$, for example, while $m\{|f_n| \geq \varepsilon\} = 1$ for any $0 < \varepsilon < 1$. Along similar lines, the sequence $g_n(x) = x/n$ converges pointwise to 0 on $[0,\infty)$, but $m\{g_n \geq \varepsilon\} = \infty$ for any $\varepsilon > 0$.

(d) Nor is pointwise convergence implied by convergence in measure. To see this, we will need to construct a somewhat more elaborate example: For each $n = 0, 1, 2, \ldots$ and each $k = 0, 1, \ldots, 2^n - 1$ we put $E_{k+2^n} = [k2^{-n}, (k+1)2^{-n}]$; in so doing, we enumerate the subintervals of $[0,1]$ with consecutive dyadic rational endpoints as a sequence (E_j). Now the sequence $f_j = \chi_{E_j}$ converges in measure to 0 on $[0,1]$ since $m\{f_{k+2^n} \geq \varepsilon\} = 2^{-n}$ for any $0 < \varepsilon < 1$. But (f_j) does not converge pointwise, or even pointwise a.e., to 0. Indeed, since each $x \in [0,1]$ is the limit of a sequence of dyadic rationals, we have $\limsup_{j\to\infty} f_j(x) = 1$ for every x. (Why?)

The conclusion to be drawn from these few examples is that we have defined a new mode of convergence that is strictly different from any that we have seen thus far. Nevertheless, convergence in measure is more closely related to pointwise convergence than you might imagine. As a first step in this direction, consider the following observation (recall the discussion following Theorem 17.13, Egorov's theorem).

Proposition 19.2. *If (f_n) converges almost uniformly to f on D, then (f_n) converges a.e. and in measure to f on D.*

PROOF. The fact that (f_n) converges a.e. to f follows from Exercise 17.36; we need only show that (f_n) converges in measure to f.

Let $\varepsilon > 0$. Since (f_n) converges almost uniformly to f, there is a measurable subset E of D with $m(E) < \varepsilon$ such that (f_n) converges uniformly to f on $D \setminus E$. Thus, we can find an index N such that $|f_n(x) - f(x)| < \varepsilon$ for all $x \in D \setminus E$ and all $n > N$. In particular, for any $n > N$ we have

$$\begin{aligned} m\{x \in D : |f_n - f| \geq \varepsilon\} &\leq m\{x \in D \setminus E : |f_n - f| \geq \varepsilon\} + m(E) \\ &= m(E) < \varepsilon. \end{aligned}$$

Hence, (f_n) converges in measure to f on D (see Exercise 10). □

By combining this observation with Egorov's theorem, we arrive at a connection between convergence in measure and convergence pointwise a.e. on sets of *finite* measure (Example 19.1 (c) demonstrates the necessity of this extra condition).

Corollary 19.3. *If (f_n) converges pointwise a.e. to f on D, where D has finite measure, then (f_n) also converges in measure to f on D.*

EXERCISES

▷ **1.** Find a sequence of integrable functions (f_n) such that $\int |f_n| \to 0$ but $f_n \not\to 0$ pointwise a.e.

2. Find a sequence of integrable functions (f_n) such that $f_n \to 0$ *uniformly* but $\int |f_n| = 1$ for all n.

3. Prove that $f_n \xrightarrow{m} f$ if and only if $f_n - f \xrightarrow{m} 0$.

▷ **4.** Fill in the missing details in Example 19.1 (d).

▷ **5.** Show that $m\{|f - g| \geq \varepsilon\} \leq m\{|f - h| \geq \varepsilon/2\} + m\{|h - g| \geq \varepsilon/2\}$. Thus, the expression $m\{|f - g| \geq \varepsilon\}$ behaves rather like a metric.

6. Prove that limits in measure are unique up to equality a.e. That is, if (f_n) converges in measure to both f and g, then $f = g$ a.e.

7. If $f_n \xrightarrow{m} f$ and $g_n \xrightarrow{m} g$, prove that $f_n + g_n \xrightarrow{m} f + g$.

8. If $f_n \xrightarrow{m} f$ and $g_n \xrightarrow{m} g$, does it follow that $f_n g_n \xrightarrow{m} fg$? If not, what additional hypotheses are needed?

9. True or false? If $f_n \xrightarrow{m} f$, then $|f_n| \xrightarrow{m} |f|$.

▷ **10.** Prove that $f_n \xrightarrow{m} f$ if and only if, given $\varepsilon > 0$, there exists an N such that $m\{|f_n - f| \geq \varepsilon\} < \varepsilon$ for all $n > N$.

11. If (f_n) converges in measure to f, show that every subsequence of (f_n) converges in measure to f.

▷ **12.** We say that (f_n) is **Cauchy in measure** if, given $\varepsilon > 0$, there exists an N such that $m\{|f_n - f_m| \geq \varepsilon\} < \varepsilon$ whenever $m, n > N$. If (f_n) converges in measure, show that (f_n) is necessarily Cauchy in measure.

▷ **13.** If (f_n) is Cauchy in measure, and if some subsequence (f_{n_k}) converges in measure to f, prove that (f_n) converges in measure to f.

The connection between convergence in measure and pointwise convergence is supplied by the following fundamental result, due to F. Riesz.

Theorem 19.4. *Let (f_n) be a sequence of real-valued measurable functions, all defined on a common measurable domain D. If (f_n) is Cauchy in measure, then there is a measurable function $f : D \to \mathbb{R}$ such that (f_n) converges in measure to f. Moreover, there is a subsequence (f_{n_k}) of (f_n) that converges pointwise a.e. to f.*

PROOF. We first establish the "moreover" claim by showing that (f_n) has a subsequence that is pointwise Cauchy. To accomplish this we appeal to an old trick: Since (f_n) is Cauchy in measure, we may choose a subsequence (f_{n_k}) satisfying

$$m\{x \in D : |f_{n_{k+1}}(x) - f_{n_k}(x)| \ge 2^{-k}\} < 2^{-k}$$

for all k. (How?) In other words, setting $E_k = \{|f_{n_{k+1}} - f_{n_k}| \ge 2^{-k}\}$, we have $m(E_k) < 2^{-k}$ for all k.

Now, since $\sum_k m(E_k) < \infty$, the Borel–Cantelli lemma, Corollary 16.24, tells us that the set

$$E = \limsup_{k\to\infty} E_k = \bigcap_{k=1}^{\infty}\bigcup_{j=k}^{\infty} E_j$$

has measure zero. Notice that for any $x \notin E$ we have $x \notin \bigcup_{j=k}^{\infty} E_j$ for some k sufficiently large, and hence

$$|f_{n_{j+1}}(x) - f_{n_j}(x)| < 2^{-j} \qquad \text{for all } j \ge k. \tag{19.1}$$

In particular, we must have $\sum_j \big(f_{n_{j+1}}(x) - f_{n_j}(x)\big) < \infty$. Thus, for any $x \notin E$, the limit

$$f(x) = \lim_{j\to\infty} f_{n_j}(x) = f_{n_1}(x) + \sum_{j=1}^{\infty}\big(f_{n_{j+1}}(x) - f_{n_j}(x)\big)$$

exists. If we define $f(x) = 0$ for $x \in E$, then we have defined a measurable function f for which $f_{n_k}(x) \to f(x)$ for any $x \notin E$; that is, $f_{n_k} \to f$ a.e.

All that remains is to check that $f_n \to f$ in measure. To this end, first notice that for $x \notin E$ we may write

$$f(x) - f_{n_k}(x) = \sum_{j=k}^{\infty}\big(f_{n_{j+1}}(x) - f_{n_j}(x)\big),$$

and hence, from equation (19.1), for any $x \notin \bigcup_{j=k}^{\infty} E_j$ we have

$$|f(x) - f_{n_k}(x)| \le \sum_{j=k}^{\infty} |f_{n_{j+1}}(x) - f_{n_j}(x)| < \sum_{j=k}^{\infty} 2^{-j} = 2^{-k+1}.$$

(In other words, (f_{n_k}) converges almost uniformly to f.) In particular, we must have

$$m\{|f - f_{n_k}| \ge 2^{-k+1}\} \le m\left(\bigcup_{j=k}^{\infty} E_j\right) \le \sum_{j=k}^{\infty} 2^{-j} = 2^{-k+1}.$$

Thus, (f_{n_k}) converges in measure to f. Since (f_n) is Cauchy in measure, this easily implies that (f_n) itself converges in measure to f. (See Exercise 13.) □

It follows from Riesz's theorem that the collection of measurable, real-valued functions on D is closed under convergence in measure; that is, if (f_n) is a sequence of measurable functions and if, for some function f on D, we have $m^*\big\{x \in D : |f_n(x) - f(x)| \ge \varepsilon\big\} \to 0$ $(n \to \infty)$ for every $\varepsilon > 0$, then f is measurable. (Why?)

Combining Riesz's theorem with our very first observation on convergence in measure yields:

Corollary 19.5. *If (f_n) is a sequence of integrable functions that converges in L_1 to an integrable function f, then some subsequence of (f_n) converges a.e. to f.*

EXERCISES

▷ **14.** Assuming that $m(D) < \infty$, prove that (f_n) converges in measure to f on D if and only if every subsequence of (f_n) has a further subsequence that converges pointwise a.e. to f on D. Is this still true without the requirement that $m(D) < \infty$?

15. If $f_n \to f$ in L_1, prove that there is a subsequence of (f_n) that converges almost uniformly to f. [Hint: By passing to a subsequence we may suppose that $f_n \to f$ a.e., and that $\int |f_n - f| < 2^{-n}$ for all n. Now repeat the proof of Egorov's theorem (Theorem 17.13), arguing that the set $E(1, k)$ has finite measure in this case.]

▷ **16.** Over a set of finite measure we can actually describe convergence in measure in terms of a *metric*. For example, consider

$$d(f, g) = \int_0^1 \min\{|f(x) - g(x)|, 1\}\, dx,$$

where f and g are measurable, real-valued functions on $[0, 1]$.

(a) Check that $d(f, g)$ is a pseudometric, with $d(f, g) = 0$ if and only if $f = g$ a.e. [Hint: $\rho(x, y) = \min\{|x - y|, 1\}$ defines a metric on $\mathbb{R}$; see Exercise 3.5.]

(b) Prove that (f_n) converges in measure to f on $[0, 1]$ if and only if $d(f_n, f) \to 0$ as $n \to \infty$.

(c) Prove that (f_n) is d-Cauchy if and only if (f_n) is Cauchy in measure.

17. We denote the collection of all (equivalence classes of) measurable, finite a.e., extended real-valued functions on $[0, 1]$ by $L_0[0, 1]$, where we identify any two functions that agree a.e. (just as we do for $L_1[0, 1]$). Prove that $(L_0[0, 1], d)$ is a complete metric space, where $d(f, g)$ is the expression defined in Exercise 16.

18. There are a wide variety of (pseudo)metrics that describe convergence in measure. For example, let

$$\tau(f, g) = \int_0^1 \frac{|f - g|}{1 + |f - g|}$$

and verify that (f_n) converges in measure to f on $[0, 1]$ if and only if $\tau(f_n, f) \to 0$ as $n \to \infty$. [Hint: The metric $\sigma(x, y) = |x - y|/(1 + |x - y|)$ is equivalent to the metric ρ of Exercise 16 (a).]

19. In sharp contrast to convergence in measure, the topology of convergence pointwise a.e. cannot, in general, be described by a metric. (And this is precisely why pointwise a.e. convergence is often problematic.) To see this, prove that:

(a) There is a sequence of measurable functions (f_n) on $[0, 1]$ that fails to converge pointwise a.e. to 0, but such that every subsequence of (f_n) has a further subsequence that does converge pointwise a.e. to 0.

(b) There is no metric ρ on $L_0[0, 1]$ satisfying $\rho(f_n, f) \to 0$ if and only if $f_n \to f$ a.e.

20. Note that while convergence in measure can sometimes be described by a metric, and while the collection of measurable functions is clearly a vector space, the topology of convergence in measure is not always "compatible" with the vector space operations. To see this, find a measurable, real-valued function f on $[0, \infty)$, for example, such that $\lambda_n f \not\to 0$ in measure no matter how a sequence of scalars $\lambda_n \to 0$ is chosen. This means that the topology of convergence in measure on $[0, \infty)$ cannot be described by a norm. Why?

▷ **21.** Prove that Fatou's lemma holds for convergence in measure: If (f_n) is a sequence of nonnegative measurable functions and $f_n \xrightarrow{m} f$, show that $f \geq 0$ a.e. and that $\int f \leq \liminf_{n\to\infty} \int f_n$. [Hint: First pass to a subsequence (f_{n_k}) with $\lim_{k\to\infty} \int f_{n_k} = \liminf_{n\to\infty} \int f_n$.]

▷ **22.** Let (f_n) be a sequence of measurable functions with $|f_n| \leq g$, for all n, where $g \in L_1$. If (f_n) converges to f in measure, prove that $|f| \leq g$ a.e. and that (f_n) converges to f in L_1. In other words, prove that the Dominated Convergence Theorem holds for convergence in measure.

The L_p Spaces

In this section we extend our discussion of the space of integrable functions L_1 by introducing an entire scale of spaces L_p, $1 \leq p \leq \infty$. The so-called Lebesgue spaces L_p are the "continuous" analogues of the familiar sequence spaces ℓ_p. Just as with the ℓ_p spaces, we will find that the case $p = \infty$ demands special treatment, and so we begin by focusing on the range $1 \leq p < \infty$.

Given a measurable subset E of $\mathbb{R}$ (with $m(E) > 0$) and a real number $1 \leq p < \infty$, we define the space $L_p(E)$ to be the collection of all equivalence classes, under equality a.e., of measurable functions $f : E \to \bar{\mathbb{R}}$ for which $|f|^p \in L_1(E)$; that is,

$$\int_E |f|^p < \infty.$$

We define a norm on $L_p(E)$ by setting

$$\|f\|_p = \left(\int_E |f|^p \right)^{1/p} \tag{19.2}$$

for $f \in L_p(E)$. This expression is clearly well defined; in other words, if $f = g$ a.e., then $\|f\|_p = \|g\|_p$. Of course, we will want to check that $L_p(E)$ is, indeed, a vector space and that this expression is actually a norm.

Please recall that we have already encountered a relative of the space $L_2(E)$ in Chapter Fifteen. In that chapter we used the symbol L_2 to denote, essentially, the space $L_2[-\pi, \pi]$ (except that we divided the expression in equation (19.2) by $\sqrt{\pi}$ and, of course, we spoke of Riemann integrable functions). For the moment we will ignore this earlier meeting, but we will have more to say about these close cousins later in the chapter.

Just as in the case of L_1, we will turn a blind eye to equivalence classes and simply speak of the elements of $L_p(E)$ as functions, but with the added proviso that statements concerning $L_p(E)$ functions are at best valid almost everywhere. As an example of this, please note that if $f \in L_p(E)$, then f is finite a.e. on E; in other words, f is allowed to take on infinite values at a "few" points.

And, again as in the case of L_1, the underlying set E typically has little bearing on the properties of $L_p(E)$ that are of interest to us. If the discussion at hand does not depend on the set E, we will simply write L_p to denote a typical space $L_p(E)$. For the most part, we will consider only the spaces $L_p[0, 1]$, $L_p[0, \infty)$, and $L_p(\mathbb{R})$. There is no harm here in assuming that the unadorned symbols L_p denote the space $L_p(\mathbb{R})$.

As we have already witnessed with the ℓ_p-norm, the proof that equation (19.2) defines a norm will require a few elementary inequalities. Each of these should look very familiar (if not, you may want to review Lemmas 3.5–3.7 and Theorem 3.8). In what follows, we will concentrate on the range $1 < p < \infty$ (since we already know that L_1 is a normed space). To begin, notice that we certainly have $\|f\|_p = 0$ if and only if $f = 0$ a.e. (Why?) It is also clear that if $f \in L_p$, then $cf \in L_p$ for any scalar $c \in \mathbb{R}$; moreover, $\|cf\|_p = |c|\,\|f\|_p$. As with ℓ_p, the real battle is with the triangle inequality. To strike a first blow in this battle, let's check that L_p is a vector space.

Lemma 19.6. *Let $1 < p < \infty$. If $f, g \in L_p$, then $f + g \in L_p$ and $\|f + g\|_p^p \le 2^p(\|f\|_p^p + \|g\|_p^p)$. Consequently, L_p is a vector space.*

PROOF. The result follows from Lemma 3.5. Given $f, g \in L_p$, we have

$$|f(x) + g(x)|^p \le (|f(x)| + |g(x)|)^p \le 2^p(|f(x)|^p + |g(x)|^p) \quad \text{a.e.} \tag{19.3}$$

and hence

$$\begin{aligned}\int |f(x) + g(x)|^p\,dx &\le \int (|f(x)| + |g(x)|)^p\,dx \\ &\le 2^p \int (|f(x)|^p + |g(x)|^p)\,dx < \infty. \quad \square\end{aligned}$$

Please note the presence of an "a.e." in equation (19.3). Since we only know that f and g are finite a.e. (in fact, f and g may only be defined a.e.), we are only allowed to apply the inequality of Lemma 3.5 for a.e. x.

Next, we have the L_p version of *Hölder's inequality.*

Hölder's Inequality 19.7. *Let $1 < p < \infty$, and let q be defined by*

$$\frac{1}{p} + \frac{1}{q} = 1.$$

If $f \in L_p(E)$ and $g \in L_q(E)$, then $fg \in L_1(E)$ and

$$\left| \int_E fg \right| \le \int_E |fg| \le \|f\|_p \, \|g\|_q.$$

PROOF. We may suppose that $\|f\|_p > 0$ and $\|g\|_q > 0$ (why?); hence, we need to prove that

$$\int_E \frac{|f|}{\|f\|_p} \cdot \frac{|g|}{\|g\|_q} \le 1.$$

We now appeal to Young's inequality (Lemma 3.6): For a.e. x we have

$$\frac{|f(x)|}{\|f\|_p} \cdot \frac{|g(x)|}{\|g\|_q} \le \frac{1}{p} \cdot \frac{|f(x)|^p}{\|f\|_p^p} + \frac{1}{q} \cdot \frac{|g(x)|^q}{\|g\|_q^q},$$

and so integration over E yields

$$\begin{aligned} \int_E \frac{|f|}{\|f\|_p} \cdot \frac{|g|}{\|g\|_q} &\le \frac{1}{p} \cdot \frac{1}{\|f\|_p^p} \int_E |f|^p + \frac{1}{q} \cdot \frac{1}{\|g\|_q^q} \int_E |g|^q \\ &= \frac{1}{p} + \frac{1}{q} = 1. \quad \square \end{aligned}$$

When $p = q = 2$, the conclusion of Hölder's inequality reads

$$\left| \int_E fg \right| \le \left(\int_E |f|^2 \right)^{1/2} \left(\int_E |g|^2 \right)^{1/2},$$

which is an inequality that is usually referred to as the **Cauchy–Schwarz inequality** (and one that we put to good use in Chapter Fifteen).

Finally, we are ready for the L_p version of *Minkowski's inequality* (i.e., the triangle inequality). As an intermediate step, we isolate a key ingredient in the proof that is of independent interest; the proof of the following lemma is left as an exercise (Exercise 23).

Lemma 19.8. *Let $1 < p < \infty$ and let q be defined by $p^{-1} + q^{-1} = 1$. If $f \in L_p$, then $|f|^{p-1} \in L_q$ and*

$$\| \, |f|^{p-1} \, \|_q = \|f\|_p^{p-1}.$$

Minkowski's Inequality 19.9. *Let $1 \le p < \infty$ and let $f, g \in L_p$. Then, $f + g \in L_p$ and $\|f + g\|_p \le \|f\|_p + \|g\|_p$. Consequently, $\|\cdot\|_p$ is a norm.*

PROOF. The theorem is clearly true when $p = 1$, so we will suppose that $p > 1$ here.

The fact that $f + g \in L_p$ follows from Lemma 19.6. To prove the triangle inequality, we next apply Hölder's inequality. If q is defined by $p^{-1} + q^{-1} = 1$, that is, if $q = (p/p - 1)$, then

$$\begin{aligned}
\|f+g\|_p^p = \int |f+g|^p &= \int |f+g|\cdot|f+g|^{p-1} \\
&\le \int |f|\cdot|f+g|^{p-1} + \int |g|\cdot|f+g|^{p-1} \\
&\le \left(\int |f|^p\right)^{1/p}\left(\int |f+g|^{(p-1)q}\right)^{1/q} \\
&\quad + \left(\int |g|^p\right)^{1/p}\left(\int |f+g|^{(p-1)q}\right)^{1/q} \\
&= \|f\|_p\,\|f+g\|_p^{p-1} + \|g\|_p\,\|f+g\|_p^{p-1}.
\end{aligned}$$

Dividing by $\|f+g\|_p^{p-1}$, the result follows. □

EXERCISES

▷ **23.** Prove Lemma 19.8.

24. Show that equality holds in Hölder's inequality if and only if $A|f|^{p-1} = B|g|$ for some nonnegative constants A and B, not both zero, if and only if $C|f|^p = D|g|^q$ for some nonnegative constants C and D, not both zero.

25. When does equality hold in Minkowski's inequality?

26. If $m(E) < \infty$ and $f \in L_q(E)$, show that $\|f\|_p \le (m(E))^{1/p-1/q}\|f\|_q$ for $1 \le p < q < \infty$. Thus, as sets, $L_q(E) \subset L_p(E)$ whenever $m(E) < \infty$. [Hint: Hölder's inequality.] In particular, if $E = [0,1]$, notice that the L_p-norms increase with p; that is, $\|f\|_p \le \|f\|_q$ for $1 \le p < q < \infty$.

27. Given $1 \le p < q < \infty$, show that $L_p(\mathbb{R}) \ne L_q(\mathbb{R})$ by showing that neither containment holds. That is, construct functions $f \in L_p(\mathbb{R}) \setminus L_q(\mathbb{R})$ and $g \in L_q(\mathbb{R}) \setminus L_p(\mathbb{R})$.

28. Given $1 \le p, q, r < \infty$ with $r^{-1} = p^{-1} + q^{-1}$, prove the following generalization of Hölder's inequality: $\|fg\|_r \le \|f\|_p\,\|g\|_q$ whenever $f \in L_p$ and $g \in L_q$.

29. Given $1 \le p, q < \infty$ and $0 \le \alpha \le 1$, let $r = \alpha p + (1-\alpha)q$. Prove *Liapounov's inequality*: $\|f\|_r^r \le \|f\|_p^{\alpha p}\,\|f\|_q^{(1-\alpha)q}$.

▷ **30.** If (f_n) converges to f in L_p, prove that (f_n) converges in measure to f. Thus, some subsequence of (f_n) converges a.e. to f. If (f_n) is Cauchy in L_p, prove that (f_n) is Cauchy in measure.

31. If (f_n) converges to f in L_p, does $(|f_n|^p)$ converge to $|f|^p$ in L_1? in measure?

32. Given $1 \le p < \infty$, construct $f, g \in L_p(\mathbb{R})$ such that $fg \notin L_p(\mathbb{R})$. Thus, although L_p is a vector space and a lattice under the usual pointwise a.e. operations on functions, it is not typically an algebra of functions.

▷ **33.** If f and g are disjointly supported elements of L_p, that is, if $fg = 0$ a.e., show that $\|f+g\|_p^p = \|f\|_p^p + \|g\|_p^p$.

34. Let (A_n) be a sequence of disjoint measurable sets. Show that $\sum_n a_n \chi_{A_n}$ converges in L_p if and only if $\sum_n |a_n|^p m(A_n) < \infty$.

35. Show that the collection of integrable simple functions is dense in L_p, for any $1 \le p < \infty$. [Hint: Repeat the proof of Theorem 18.27 (i).]

36. For any $1 \le p < \infty$, prove that the space $L_p(\mathbb{R})$ is separable. Conclude that $L_p[\,0,1\,]$ is also separable.

37. Given $1 \le p < \infty$, $f \in L_p[\,0,1\,]$, and $\varepsilon > 0$, show that there is a function $g \in C[\,0,1\,]$ such that $\|f-g\|_p < \varepsilon$. Conclude that $C[\,0,1\,]$ is a dense subspace of $L_p[\,0,1\,]$ (where $C[\,0,1\,]$ is embedded into $L_p[\,0,1\,]$ in the obvious way: $f \mapsto [f]$). [Hint: Theorem 18.27 (ii).]

We could now fashion a proof that L_p is, in fact, a complete normed space following Theorem 18.24. Instead, though, we present a proof that uses a little of the machinery that we developed in the previous section.

Theorem 19.10. *L_p is complete for any $1 \le p < \infty$.*

PROOF. Fix $1 \le p < \infty$, and let (f_n) be a Cauchy sequence in L_p. In particular, (f_n) is a bounded sequence in L_p; that is, the sequence $\int |f_n|^p$ is bounded.

Now, (f_n) is also Cauchy in measure (Exercise 30). Thus, by Theorem 19.4, there is a subsequence (f_{n_k}) that converges a.e. to some measurable f. To complete the proof, then, it is enough to show that $f \in L_p$ and that (f_{n_k}) converges to f in L_p-norm. But, since $(|f_{n_k}|^p)$ converges a.e. to $|f|^p$, we may appeal to Fatou's lemma to conclude that

$$\int |f|^p \le \liminf_{k\to\infty} \int |f_{n_k}|^p \le \sup_n \int |f_n|^p < \infty.$$

Hence, $f \in L_p$. The proof that (f_{n_k}) converges to f in L_p-norm follows similar lines: The sequence $(|f_{n_j} - f_{n_k}|^p)_{j=1}^\infty$ converges a.e. to $|f - f_{n_k}|^p$, and so, given $\varepsilon > 0$,

$$\int |f - f_{n_k}|^p \le \liminf_{j\to\infty} \int |f_{n_j} - f_{n_k}|^p < \varepsilon,$$

provided that k is sufficiently large. (Why?) □

EXERCISES

38. Suppose that (f_n) is in L_p, $1 \le p < \infty$, with $\|f_n\|_p \le 1$ and $f_n \to f$ a.e. Prove that $f \in L_p$ and that $\|f\|_p \le 1$.

39. Let $f, f_n \in L_p$, $1 \le p < \infty$, and suppose that $f_n \to f$ a.e. Show that $\|f_n - f\|_p \to 0$ if and only if $\|f_n\|_p \to \|f\|_p$. [Hint: First note that $2^p(|f_n|^p +$

$|f|^p) - |f_n - f|^p \geq 0$ a.e., and then apply Fatou's lemma.] Note that the result also holds if "a.e." is replaced by "in measure."

40. For $1 < p < \infty$ and $a, b \geq 0$, show that $a^p + b^p \leq (a+b)^p \leq 2^{p-1}(a^p + b^p)$, and that the reverse inequalities hold when $0 < p < 1$. [Hint: Consider the function $\varphi(x) = (1+x)^p/(1+x^p)$ for $0 \leq x \leq 1$.]

41. It makes perfect sense to consider the spaces L_p for $0 < p < 1$, too. In this range, expression (19.3) no longer defines a norm; nevertheless, L_p is a complete metric linear space. For $0 < p < 1$, prove that:

(a) L_p is a vector space.

(b) The expression $d(f, g) = \int |f - g|^p$ defines a complete, translation-invariant metric on L_p.

(c) Let $p^{-1} + q^{-1} = 1$ (note that $q < 0$!). If $0 \leq f \in L_p$ and if $g \geq 0$ satisfies $0 < \int g^q < \infty$, then $\int fg \geq \left(\int f^p\right)^{1/p} \left(\int g^q\right)^{1/q}$.

(d) If $f, g \in L_p$ with $f, g \geq 0$, then $\|f + g\|_p \geq \|f\|_p + \|g\|_p$.

(e) If $f, g \in L_p$, then $\|f + g\|_p \leq 2^{1/p}(\|f\|_p + \|g\|_p)$.

At the beginning of this section, the L_p spaces were advertised as analogues of the ℓ_p spaces. As such, the space L_∞, whatever it is, should look like a collection of *bounded* functions. But if L_p functions are allowed to take on infinite values at a "few" points, how are we to make sense of the word "bounded"? The answer is that a "function" in L_∞ is one that is *equivalent* to a bounded measurable function; that is, it is equal a.e. to a bounded function.

We say that a measurable function $f : E \to \bar{\mathbb{R}}$ is **essentially bounded** (on E) if there exists some constant $0 \leq A < \infty$ such that $|f| \leq A$ a.e.; that is, $m\{x \in E : |f(x)| > A\} = 0$. Now there are many choices of the constant A, for if $|f| \leq A$ a.e., then $|f| \leq A + 1$ a.e., too. The smallest constant that works here is called the **essential supremum** of f (over E), which is written

$$\operatorname*{ess.sup}_{x \in E} |f(x)| = \inf\{A \geq 0 : m\{x \in E : |f| > A\} = 0\}. \tag{19.4}$$

Please note that the essential supremum of f would be unchanged even if we were to alter f on a null set. In other words, if f and g are essentially bounded, and if $f = g$ a.e. on E, then we have $\text{ess.sup}_E |f| = \text{ess.sup}_E |g|$.

The essential supremum is not as strange a beast as you might imagine; it is really quite natural to consider almost everywhere boundedness. By way of an example, notice that if $f : [0, 1] \to \bar{\mathbb{R}}$ is measurable and essentially bounded, and if $N \subset [0, 1]$ is a null set, then

$$\int_0^1 |f| = \int_{[0,1]\setminus N} |f| \leq \sup_{x \notin N} |f(x)|.$$

Thus,

$$\int_0^1 |f| \leq \inf\left\{\sup_{x \notin N} |f(x)| : m(N) = 0\right\}.$$

The right-hand side of this last inequality is precisely the essential supremum of f over $[0, 1]$ (see Exercise 45), and it provides a somewhat better upper estimate for $\int_0^1 |f|$ than the uniform norm $\sup_{0 \le x \le 1} |f(x)|$ (see Exercise 44).

Finally, we denote the collection of all equivalence classes, under equality a.e., of essentially bounded measurable functions on E by $L_\infty(E)$, and we define

$$\|f\|_\infty = \operatorname*{ess.sup}_{x \in E} |f(x)| \tag{19.5}$$

for $f \in L_\infty(E)$. By our earlier remarks, this expression is well defined on equivalence classes; in other words, if $f = g$ a.e., then $\|f\|_\infty = \|g\|_\infty$. Just as with L_p, the symbols L_∞ denote a typical space $L_\infty(E)$.

As always, we will want to check that L_∞ is a vector space and that $\|\cdot\|_\infty$ is a legitimate norm. Moreover, since this is nearly the same expression that we have been using for the uniform norm, we will want to check that this new norm coincides with the more familiar sup norm in certain cases. Most of these details will be left as exercises. To avoid potential confusion, though, throughout the remainder of this section the expression $\|\cdot\|_\infty$ will always denote the essential supremum norm (19.4).

EXERCISES

▷ **42.** Let $f : E \to \bar{\mathbb{R}}$ be measurable and essentially bounded, and let $A = \operatorname{ess.sup}_{x \in E} |f(x)|$. Prove that:

(a) $0 \le A < \infty$ and $|f| \le A$ a.e.

(b) $f = 0$ a.e. if and only if $A = 0$.

(c) If $0 \le A' < A$, then $m\{|f| > A'\} \neq 0$.

Thus, $|f| \le \|f\|_\infty$ a.e., where $\|f\|_\infty$ is the L_∞-norm of f and $\|f\|_\infty$ is the smallest constant with this property.

▷ **43.** If $f \in L_\infty$, is $m\{|f| = \|f\|_\infty\} > 0$? Is $\{|f| = \|f\|_\infty\} \neq \emptyset$? Explain.

44. If $f : E \to \mathbb{R}$ is a measurable, (everywhere) bounded function, prove that $\operatorname{ess.sup}_E |f| \le \sup_E |f|$. Give an example showing that strict inequality can occur.

▷ **45.** If $f : E \to \bar{\mathbb{R}}$ is essentially bounded, show that

$$\operatorname*{ess.sup}_{x \in E} |f(x)| = \inf \left\{ \sup_{x \in E \setminus N} |f(x)| : m(N) = 0 \right\}.$$

Moreover, show that this infimum is actually attained; that is, prove that there is a null set N such that $\operatorname{ess.sup}_E |f| = \sup_{E \setminus N} |f|$.

▷ **46.** Let $f \in C[0, 1]$ and $0 \le A < \infty$. If $|f(x)| \le A$ for a.e. $x \in [0, 1]$, prove that, in fact, $|f(x)| \le A$ for all $x \in [0, 1]$. Conclude that

$$\sup_{0 \le x \le 1} |f(x)| = \operatorname*{ess.sup}_{0 \le x \le 1} |f(x)|$$

in this case. In other words, $\|f\|_{C[0,1]} = \|f\|_{L_\infty[0,1]}$.

▷ **47.** If $f, g : E \to \bar{\mathbb{R}}$ are essentially bounded, show that $f + g$ is essentially bounded and that $\|f + g\|_\infty \le \|f\|_\infty + \|g\|_\infty$, where $\|\cdot\|_\infty$ denotes the L_∞-norm. [Hint: It is enough to show that $|f + g| \le \|f\|_\infty + \|g\|_\infty$ a.e.] Conclude that L_∞ is a normed vector space.

48. If $f, g \in L_\infty$, show that $fg \in L_\infty$ and $\|fg\|_\infty \le \|f\|_\infty \|g\|_\infty$. Conclude that L_∞ is a normed algebra. [Compare this with Exercise 32.] Is L_∞ a normed lattice (under the usual pointwise a.e. ordering)?

▷ **49.** Prove that $L_\infty(\mathbb{R})$ is *not* separable. More generally, if $m(E) > 0$, then $L_\infty(E)$ is not separable. [Hint: If A and B are disjoint, notice that $\|\chi_A - \chi_B\|_\infty = 1$.]

50. Show that the collection of all simple functions is dense in L_∞. [Hint: Recall the Basic Construction, Theorem 17.14.] If $m(E) < \infty$, show that the *integrable* simple functions are dense in $L_\infty(E)$. Is this true without the restriction that $m(E) < \infty$? Explain.

▷ **51.** If $m(E) < \infty$, show that, as sets, $L_\infty(E) \subset L_p(\mathbb{R})$, for any $1 \le p < \infty$, and that $\|f\|_p \le (m(E))^{1-1/p}\|f\|_\infty$ for any $f \in L_\infty(E)$. In particular, if $f \in L_\infty[0, 1]$, then $\|f\|_1 \le \|f\|_p \le \|f\|_\infty$ for any $1 < p < \infty$.

52. If $f \in L_\infty[0, 1]$, show that $\lim_{p\to\infty} \|f\|_p = \|f\|_\infty$. [Hint: First note that $\lim_{p\to\infty} \|f\|_p$ exists by Exercise 51. Next, consider the integral of $|f|^p$ over the set $\{|f| > \|f\|_\infty - \varepsilon\}$.]

53. If $m(E) < \infty$, show that $L_\infty(E)$ is a dense subspace of $L_p(E)$, for any $1 \le p < \infty$.

▷ **54.** Given $f \in L_p$, $1 \le p < \infty$, and $g \in L_\infty$, prove that $fg \in L_p$ and that $\|fg\|_p \le \|f\|_p \|g\|_\infty$. [Note that for $p = 1$ this gives Hölder's inequality (for $q = \infty$).]

55. Let $f_n \to f$ in L_p, $1 \le p < \infty$, and let (g_n) be a sequence in L_∞ with $\|g_n\|_\infty \le 1$ and $g_n \to g$ a.e. Show that $f_n g_n \to fg$ in L_p.

Finally, a word or two about convergence in L_∞. We begin with a simple observation: Convergence in L_∞ is the same as uniform a.e. convergence.

Lemma 19.11. *If (f_n) converges to 0 in $L_\infty(E)$, then there is a null set $A \subset E$ such that (f_n) converges uniformly to 0 on $E \setminus A$.*

PROOF. For each n, there is a null set A_n such that $|f_n(x)| \le \|f_n\|_\infty$ for all $x \in E \setminus A_n$. If we set $A = \bigcup_n A_n$, then A is a null set and

$$\sup_{x\in E\setminus A} |f_n(x)| \le \sup_{x\in E\setminus A_n} |f_n(x)| = \|f_n\|_\infty \to 0 \qquad \text{as } n \to \infty. \quad \square$$

Theorem 19.12. *L_∞ is complete.*

PROOF. Let (f_n) be a Cauchy sequence in $L_\infty(E)$. Then, there is a null set A such that (f_n) is uniformly Cauchy on $E \setminus A$. Indeed, for each m and n, we may choose a null set $A_{m,n}$ such that $|f_m(x) - f_n(x)| \le \|f_m - f_n\|_\infty$ for all $x \in E \setminus A_{m,n}$. Putting $A = \bigcup_{m,n} A_{m,n}$ does the trick. Thus, (f_n) converges uniformly on $E \setminus A$. If we define $f(x) = \lim_{n\to\infty} f_n(x)$ for $x \in E \setminus A$ and $f(x) = 0$ for $x \in A$, then f is a bounded measurable function. All that remains is to check that $\|f_n - f\|_\infty \to 0$. But, since A is a null set,

$$\|f_n - f\|_\infty \le \sup_{x \in E\setminus A} |f_n(x) - f(x)|$$

(see Exercise 45), and the right-hand side tends to 0 as $n \to \infty$, since (f_n) converges uniformly to f on $E \setminus A$. □

EXERCISE

56. Under the obvious inclusion (i.e., $f \mapsto [f]$), show that $C[0,1]$ is a closed subspace of $L_\infty[0,1]$.

Approximation of L_p Functions

In analogy with Theorem 18.27, we next discuss the approximation of elements of L_p by simpler functions. As with Theorem 18.27, most of the work here is done by the Basic Construction.

Theorem 19.13. *Let* $1 < p < \infty$, *let* $f \in L_p(\mathbb{R})$, *and let* $\varepsilon > 0$. *Then:*

(i) *There is an integrable simple function* φ *with* $\|f - \varphi\|_p < \varepsilon$.

(ii) *There is a continuous function* $g : \mathbb{R} \to \mathbb{R}$ *such that* $g = 0$ *outside some bounded interval and such that* $\|f - g\|_p < \varepsilon$.

(iii) *There is an (integrable) step function* h *with* $\|f - h\|_p < \varepsilon$.

PROOF. The key observation here is that $|f|^p \in L_1(\mathbb{R})$. Thus, from the Monotone Convergence Theorem, we can find a compact interval $[a,b]$ such that $\int_{\mathbb{R}\setminus[a,b]} |f|^p < (\varepsilon/4)^p$. We will build all of our approximating functions with support in $[a,b]$; that is, each will be chosen to vanish outside of $[a,b]$.

(i) There is a sequence of (integrable) simple functions (φ_k) with $\varphi_k = 0$ off $[a,b]$, $\varphi_k \to f$ on $[a,b]$, and $|\varphi_k| \le |f|$. Using equation (19.3), we have $|f - \varphi_k|^p \le 2^p(|f|^p + |\varphi_k|^p) \le 2^{p+1}|f|^p$, and it now follows from the Dominated Convergence Theorem that $\int_a^b |f - \varphi_k|^p \to 0$. Finally, choose k and $\varphi = \varphi_k$ with $\int_a^b |f - \varphi|^p < (\varepsilon/4)^p$ and, hence, $\int_{\mathbb{R}} |f - \varphi|^p < 2(\varepsilon/4)^p < (\varepsilon/2)^p$.

(ii) φ is bounded; choose K such that $|\varphi| \le K$. Now, from Theorem 17.20 (and Exercise 45), we can find a continuous function g on $\mathbb{R}$, vanishing outside of $[a,b]$, such that $|g| \le K$ and $m\{g \ne \varphi\} < (\varepsilon/8K)^p$. Then,

$$\int |\varphi - g|^p = \int_a^b |\varphi - g|^p < (2K)^p \cdot \left(\frac{\varepsilon}{8K}\right)^p = \left(\frac{\varepsilon}{4}\right)^p,$$

and hence $\|f-g\|_p \le \|f-\varphi\|_p + \|\varphi - g\|_p < 3\varepsilon/4$.

The proof of (iii) is left as an exercise. □

Corollary 19.14. *The integrable simple functions are dense in L_p for $1 \le p < \infty$.*

Corollary 19.15. *$C[a,b]$ is dense in $L_p[a,b]$ for $1 \le p < \infty$. Hence, $L_\infty[a,b]$ is dense in $L_p[a,b]$ for $1 \le p < \infty$.*

Corollary 19.14 and the first statement in Corollary 19.15 do not hold for $p = \infty$. However, as an almost immediate consequence of the Basic Construction, it is true that the simple functions are dense in L_∞ and that the integrable simple functions are dense in $L_\infty(E)$ whenever $m(E) < \infty$ (see Exercise 50).

EXERCISES

57. Prove Theorem 19.13 (iii).

58. Prove Proposition 19.16.

59. Fix $1 \le p \le \infty$. Prove or disprove: When considered as a subset of $L_p[a,b]$, the Riemann integrable functions $\mathcal{R}[a,b]$ are dense in $L_p[a,b]$. Does your answer depend on p? [Hint: Recall that the elements of $\mathcal{R}[a,b]$ are bounded measurable functions.]

▷ **60.** Fix $1 < p < \infty$, $f \in L_p[a,b]$, and $\varepsilon > 0$. Show that there is an algebraic polynomial Q and a trig polynomial T such that $\|f - Q\|_p < \varepsilon$ and $\|f - T\|_p < \varepsilon$.

61. Prove that $L_p(\mathbb{R})$ and $L_p[0,1]$ are separable for any $1 \le p < \infty$. Try to give at least two different proofs.

62. Let $1 \le p < \infty$, and let $f \in L_p(\mathbb{R})$. Given $\varepsilon > 0$, show that there is a $\delta > 0$ such that $\|f\chi_A\|_p < \varepsilon$ whenever $m(A) < \delta$. [Hint: This is easy if f is bounded.] Does this result hold for $p = \infty$? Explain.

63. Fix $1 \le p < \infty$. Given $h \in \mathbb{R}$, define a map T_h on $L_p(\mathbb{R})$ by setting $(T_h(f))(x) = f(x+h)$ for $f \in L_p(\mathbb{R})$ and $x \in \mathbb{R}$.

(a) Show that $T_h(f) \in L_p(\mathbb{R})$ and that $\|T_h(f)\|_p = \|f\|_p$.

(b) Show that the map $f \mapsto T_h(f)$ is linear. Conclude that T_h is an isometry on $L_p(\mathbb{R})$ for any h.

(c) Prove that $\lim_{h\to 0} \|f - T_h f\|_p = 0$. [Hint: This is easy if f is uniformly continuous.]

(d) Does $\lim_{h\to\infty} \|f - T_h f\|_p$ exist? If so, compute it.

64. Let $1 \le p < \infty$, let $f \in L_p$, and let $g \in L_q$, where $p^{-1} + q^{-1} = 1$.

(a) Show that $h(x) = \int_{-\infty}^{\infty} f(t)\, g(x+t)\, dt$ defines a bounded *continuous* function on $\mathbb{R}$ satisfying $\|h\|_\infty \le \|f\|_p \|g\|_q$. [Hint: Exercise 63 (c).]

(b) If one of f or g is differentiable, show that h is also differentiable and find a formula for $h'(x)$ (in terms of either f' or g').

More on Fourier Series

With the Lebesgue integral and the L_p spaces now at our disposal, we take another brief look at Fourier series with an eye toward improving, or at least restating, a few key results from Chapter Fifteen.

Following our earlier notation, we define the Fourier series of a 2π-periodic function $f : \mathbb{R} \to \mathbb{R}$ by

$$\frac{a_0}{2} + \sum_{k=1}^{\infty} \left(a_k \cos kx + b_k \sin kx\right),$$

where the Fourier coefficients a_k and b_k are given by

$$a_k = \frac{1}{\pi} \int_{-\pi}^{\pi} f(t) \cos kt \, dt \qquad \text{and} \qquad b_k = \frac{1}{\pi} \int_{-\pi}^{\pi} f(t) \sin kt \, dt. \tag{19.6}$$

However, we now require that f be Lebesgue integrable on $[-\pi, \pi]$ and we interpret each of the integrals in equation (19.6) as a Lebesgue integral.

Virtually every observation, and every calculation, that we made in Chapter Fifteen will remain valid in this new setting with only a few minor adjustments here and there. A rather obvious modification is that the Riemann integral should everywhere be replaced by the Lebesgue integral, thus providing us with a larger class of functions that admit representation by Fourier series.

On the other hand, there is one major difference here: Because we have assumed that Riemann integrable functions are bounded, we know that f^2 is Riemann integrable whenever f is; in other words, in the context of Chapter Fifteen, the collection $L_2[a, b]$ is simply a new name for the collection $\mathcal{R}[a, b]$. But we make no such boundedness assumption on Lebesgue integrable functions. In particular, the integrability of f now tells us nothing about the integrability of f^2. The Lebesgue spaces $L_2[a, b]$ and $L_1[a, b]$, as presented in this chapter, are quite different from their Chapter Fifteen cousins. Thus, the "L_2-theory" developed in Chapter Fifteen is especially meaningful in the context of the Lebesgue integral: Isolating the collection of Lebesgue square-integrable functions is not only a convenience but a necessity.

As an example of this subtle difference, we first note that Observation 15.1 (b) (as restated in Observation 15.1 (d)) remains true for the Lebesgue integral provided we assume that f^2 is Lebesgue integrable on $[-\pi, \pi]$. In what follows, it will again be helpful to renormalize the L_2-norm by setting

$$\|f\|_2 = \left(\frac{1}{\pi} \int_{-\pi}^{\pi} |f(x)|^2 \, dx\right)^{1/2}. \tag{19.7}$$

We will take this expression to be the norm on $L_2[-\pi, \pi]$ throughout the remainder of the section. It is easy to see that this normalization has no effect on the results for $L_2[-\pi, \pi]$ that were developed earlier in this chapter. In particular, Hölder's inequality, Minkowski's inequality, and Theorem 19.10 all hold in this new setting (see Exercises 65–67).

Proposition 19.16. *If $f \in L_2[-\pi, \pi]$, then $s_n(f)$ is the nearest point to f out of $\mathcal{T}_n$ relative to the L_2-norm. In other words,*

$$\inf_{T \in \mathcal{T}_n} \|f - T\|_2 = \|f - s_n(f)\|_2.$$

Moreover,

$$\begin{aligned} \|f - s_n(f)\|_2^2 &= \frac{1}{\pi} \int_{-\pi}^{\pi} f(x)^2\,dx - \frac{a_0^2}{2} - \sum_{k=1}^{n} \left(a_k^2 + b_k^2\right) \\ &= \|f\|_2^2 - \|s_n(f)\|_2^2. \end{aligned} \tag{19.8}$$

PROOF. The proof of the proposition is identical to that given for Observation 15.1 (b) once we justify the existence of the integrals used in that proof. Of course, if $f \in L_2[-\pi, \pi]$, then f^2 is integrable on $[-\pi, \pi]$. Next, notice that since a trig polynomial $T \in \mathcal{T}_n$ is (continuous and) bounded, it follows that T^2 and, hence, $[f - T]^2$ are Lebesgue integrable on $[-\pi, \pi]$. Finally, if both f and T are in $L_2[-\pi, \pi]$, then Hölder's inequality assures us that the product fT is integrable on $[-\pi, \pi]$. Thus, the various integrals used in the proof of Observation 15.1 (b) exist. □

As a consequence of Proposition 19.16, it is immediate that Bessel's inequality (15.3) holds for $f \in L_2[-\pi, \pi]$. That is, if $f \in L_2[-\pi, \pi]$, then the Fourier coefficients of f are square-summable and satisfy

$$\frac{a_0^2}{2} + \sum_{k=1}^{\infty} \left(a_k^2 + b_k^2\right) \le \frac{1}{\pi} \int_{-\pi}^{\pi} f(x)^2\,dx. \tag{19.9}$$

Hence, Riemann's Lemma 15.4 is also valid in this case. But we can say even more: As evidence that the Lebesgue integral is easy to work with in this regard, we next sketch a direct proof of the *Riemann–Lebesgue Lemma* (Exercise 18.55), Lebesgue's variation on Riemann's Lemma.

Theorem 19.17. *If f is Lebesgue integrable on $[-\pi, \pi]$, then*

$$\lim_{n\to\infty} \int_{-\pi}^{\pi} f(x)\cos nx\,dx = 0 = \lim_{n\to\infty} \int_{-\pi}^{\pi} f(x)\sin nx\,dx.$$

PROOF. First consider the case $f = \chi_{[a,b]}$, where $-\pi \le a < b \le \pi$. Clearly,

$$\int_{-\pi}^{\pi} f(x)\cos nx\,dx = \int_a^b \cos nx\,dx = (\sin b - \sin a)/n \to 0$$

as $n \to \infty$.

Now, given $\varepsilon > 0$, Theorem 18.27 (iii) supplies a step function h, vanishing outside $[-\pi, \pi]$, such that $\int_{-\pi}^{\pi} |f - h| < \varepsilon$. But h is just a linear combination of functions of the form $\chi_{[a,b]}$. Thus, from the first part of the proof (and the linearity of the integral) we may choose n sufficiently large so that $\left|\int_{-\pi}^{\pi} h(x)\cos nx\,dx\right| < \varepsilon$.

The triangle inequality does the rest:

$$\left|\int_{-\pi}^{\pi} f(x)\cos nx\,dx\right| \le \left|\int_{-\pi}^{\pi} h(x)\cos nx\,dx\right| + \left|\int_{-\pi}^{\pi} (f(x)-h(x))\cos nx\,dx\right|$$
$$< \varepsilon + \int_{-\pi}^{\pi} |f-h| < 2\varepsilon.$$

In other words, $\int_{-\pi}^{\pi} f(x)\cos nx\,dx \to 0$ as $n \to \infty$. The proof with $\sin nx$ in place of $\cos nx$ is essentially identical. □

Clearly, Observations 15.1 (e) and 15.1 (f) are unaffected by our choice of integral, so we next revise Observation 15.1 (g). The proof of the following proposition is essentially identical to that given in Observation 15.1 (g) but, since it is an extremely important result, the details bear repeating. It is sometimes referred to as the *Riesz–Fischer theorem.*

Proposition 19.18. *If $f \in L_2[-\pi,\pi]$, then $\|s_n(f)-f\|_2 \to 0$.*

PROOF. Let $\varepsilon > 0$, and choose a function $g \in C^{2\pi}$ satisfying $\|f-g\|_2 < \varepsilon$ (see Exercise 60). Next, since s_n is linear, we have

$$\|f-s_n(f)\|_2 \le \|f-g\|_2 + \|g-s_n(g)\|_2 + \|s_n(f-g)\|_2.$$

From Bessel's inequality, we have $\|s_n(f-g)\|_2 \le \|f-g\|_2 < \varepsilon$ and so, from Observation 15.1 (e), we get

$$\|f-s_n(f)\|_2 \le 2\varepsilon + \|g-s_n(g)\|_2 < 3\varepsilon$$

for all n sufficiently large. □

As an immediate consequence of Proposition 19.18, notice that if $f \in L_2[-\pi,\pi]$, then $s_n(f)$ converges *in measure* to f on $[-\pi,\pi]$ (see Exercise 30). Thus, although the pointwise convergence of Fourier series is a thorny problem in general, every $f \in L_2[-\pi,\pi]$ has a Fourier series that converges in at least this "general" sense. Moreover, Riesz's Theorem 19.4 now supplies a *subsequence* of $(s_n(f))$ that converges pointwise a.e. to f. Better and better! Since $L_2[-\pi,\pi]$ contains the (bounded) Riemann integrable functions on $[-\pi,\pi]$, we have arrived at a simple, general convergence result for the Fourier series of a large class of functions.

Returning to our "surgery" on Observation 15.1, notice that *Parseval's equation* (15.5) follows easily from Proposition 19.18. Specifically, if $f \in L_2[-\pi,\pi]$, then, from equation (19.8) and Proposition 19.18 we have $\|f\|_2^2 = \lim_{n\to\infty}\|s_n(f)\|_2^2$. In other words,

$$\frac{1}{\pi}\int_{-\pi}^{\pi} |f(x)|^2\,dx = \frac{a_0^2}{2} + \sum_{k=1}^{\infty}\left(a_k^2+b_k^2\right) \tag{19.10}$$

for $f \in L_2[-\pi,\pi]$. It now follows, as in Observation 15.1 (i), that distinct elements of $L_2[-\pi,\pi]$ have distinct Fourier series. That is, if $f, g \in L_2[-\pi,\pi]$ satisfy $\int_{-\pi}^{\pi}\big[f(x)-g(x)\big]\cos nx\,dx = 0$ and $\int_{-\pi}^{\pi}\big[f(x)-g(x)\big]\sin nx\,dx = 0$ for all $n = 0, 1, 2, \ldots$, then $\int_{-\pi}^{\pi}\big[f(x)-g(x)\big]^2\,dx = 0$ and, hence, $f-g=0$ a.e.

Our next result is, in a sense, a converse to Bessel's inequality. It is also sometimes referred to as the *Riesz–Fischer theorem.*

Proposition 19.19. *If $(a_n)_{n=0}^{\infty}$ and $(b_n)_{n=1}^{\infty}$ are real sequences with $\sum_{k=1}^{\infty}(a_k^2 + b_k^2) < \infty$, then there is an $f \in L_2[-\pi, \pi]$ satisfying*

$$s_n(f)(x) = \frac{a_0}{2} + \sum_{k=1}^{n}\big(a_k \cos kx + b_k \sin kx\big),$$

for all n, and

$$\frac{1}{\pi}\int_{-\pi}^{\pi}|f(x)|^2\,dx = \frac{a_0^2}{2} + \sum_{k=1}^{\infty}\big(a_k^2 + b_k^2\big).$$

PROOF. Let $T_n(x) = (a_0/2) + \sum_{k=1}^{n}\big(a_k \cos kx + b_k \sin kx\big)$ and notice that for $0 < m < n$ we have

$$\begin{aligned}\|T_n - T_m\|_2^2 &= \left\|\sum_{k=m+1}^{n}\big(a_k \cos kx + b_k \sin kx\big)\right\|_2^2 \\ &= \sum_{k=m+1}^{n}\big(a_k^2 + b_k^2\big) \to 0 \qquad \text{as } m, n \to \infty.\end{aligned}$$

Thus, (T_n) is a Cauchy sequence in $L_2[-\pi, \pi]$ and, as such, converges to some $f \in L_2[-\pi, \pi]$ by Theorem 19.10. Now notice that if $k \le n$, we have

$$\begin{aligned}\left|a_k - \frac{1}{\pi}\int_{-\pi}^{\pi} f(x)\cos kx\,dx\right| &= \left|\frac{1}{\pi}\int_{-\pi}^{\pi}\big(T_n(x) - f(x)\big)\cos kx\,dx\right| \\ &\le \frac{1}{\pi}\int_{-\pi}^{\pi}\big|T_n(x) - f(x)\big|\,dx \\ &\le \sqrt{2}\,\|T_n - f\|_2 \to 0 \qquad \text{as } n \to \infty.\end{aligned}$$

Thus, $a_k = (1/\pi)\int_{-\pi}^{\pi} f(x)\cos kx\,dx$. Similarly, $b_k = \frac{1}{\pi}\int_{-\pi}^{\pi} f(x)\sin kx\,dx$. Since $f \in L_2[-\pi, \pi]$, the rest is easy. □

We can easily collect several of our observations into a single "abstract" formulation: The map that sends an $f \in L_2[-\pi, \pi]$ into its sequence of Fourier coefficients $(a_0, a_1, b_1, a_2, b_2, \ldots)$ is a linear isometry from $L_2[-\pi, \pi]$ onto ℓ_2! Indeed, since the map is clearly linear, Parseval's equation (19.10) tells us that the map is an isometry into ℓ_2, while Proposition 19.19 tells us that the map is, in fact, onto ℓ_2.

This observation, which is itself sometimes called the Riesz–Fischer theorem, is a seminal result in functional analysis. It says that everything we need to know about the "big" space of functions $L_2[-\pi, \pi]$ could be gleaned from the "little" space of sequences ℓ_2. In particular, the proof that $L_2[-\pi, \pi]$ is complete, which would appear to require several measure-theoretic tools, could apparently be deduced from the elementary fact that ℓ_2 is complete. Likewise, Proposition 19.18 (the fact that the trigonometric system is complete in $L_2[-\pi, \pi]$ in the classical sense) should follow from the analogous (and much simpler) result that each element of $x \in \ell_2$ is the norm limit of the sequence of truncated elements $(x_1, \ldots, x_n, 0, 0, \ldots)$. Amazing!

The circle of ideas represented by the various Riesz–Fischer theorems constitutes one of the earliest examples of a functional analytic argument: In this case, a "soft" fact about isometries between abstract spaces yields "hard" information about Fourier series.

EXERCISES

▷ **65.** Suppose that we renormalize $L_p[-\pi, \pi]$ by setting

$$\|f\|_p = \left(\frac{1}{\pi}\int_{-\pi}^{\pi} |f(x)|^p\,dx\right)^{1/p},$$

for $1 \le p < \infty$ (but leave $\|f\|_\infty$ as in equation (19.5)). Check that Hölder's inequality and Minkowski's inequality remain true in this new setting. The renormalized space $L_p[-\pi, \pi]$ is obviously still complete. Why?

▷ **66.** With the L_p-norms defined as in Exercise 65, check that $\|f\|_p \le \|f\|_q$ for any $1 \le p \le q \le \infty$ and any $f \in L_q[-\pi, \pi]$.

67. With the L_p-norms defined as in Exercise 65, prove that we still have $\lim_{p\to\infty} \|f\|_p = \|f\|_\infty$ for $f \in L_\infty[-\pi, \pi]$. (In other words, there is no need to scale the $L_\infty[-\pi, \pi]$-norm.)

◇

Notes and Remarks

Much of the material in this chapter is due to the great Hungarian mathematician Frigyes (Frédéric, Friedrich) Riesz. Riesz introduced convergence in measure in Riesz [1909a], wherein he proved that a sequence converging in measure has a subsequence converging a.e. (Theorem 19.4 and Corollary 19.5). The fact that convergence a.e. over a set of finite measure implied convergence in measure (Theorem 19.3) had already been pointed out by Lebesgue [1906]. As an application, Riesz points out that the Fourier series of a Lebesgue square-integrable function must converge in this "general" sense (combine the result of Exercise 30 with Proposition 19.18). In Riesz [1910a], Riesz points out that Fatou's Lemma and Lebesgue's Dominated Convergence Theorem are valid for convergence in measure (see Exercises 21 and 22).

Fréchet [1921] first proved that convergence in measure could be described by a metric, namely, $d(f, g) = \inf\{\varepsilon + m\{|f - g| > \varepsilon\} : \varepsilon > 0\}$. Another metric (for convergence in probability) is discussed in Dudley [1989; §9.2]. The counterexamples discussed in Exercises 19 and 20 were pointed out to me by D. J. H. Garling and S. J. Dilworth.

Theorem 19.17 is often called *Mercer's theorem* (see also Exercises 18.55 and 18.56, and the notes to Chapter Eighteen). For a discussion of the contributions of Riemann and Lebesgue, see Hawkins [1970].

In 1908, Erhard Schmidt (this is the Schmidt of the "Gram–Schmidt process") introduced what he called "function spaces" (Schmidt [1908]). In modern terminology, Schmidt developed the general theory of the space that we would call ℓ_2, the collection of all sequences (z_j) of *complex* numbers satisfying $\sum_{j=1}^{\infty} |z_j|^2 < \infty$ and endowed with the inner product $(z, w) = \sum_{j=1}^{\infty} z_j \bar{w}_j$. Schmidt further introduced (possibly for the first time) the double bar notation $\|z\|$ to denote the norm of z, defined by $\|z\|^2 = (z, z) = \sum_{j=1}^{\infty} z_j \bar{z}_j = \sum_{j=1}^{\infty} |z_j|^2$. He deduced Bessel's inequality in this generalized setting, went on to consider various types of convergence, and defined the notion of a closed subspace. Schmidt's most important contribution from this work is what we today call *the projection theorem*.

Schmidt [1908] and Fréchet [1907, 1908] remarked that the space $L_2[\,a, b\,]$ supported a geometry that was completely analogous to Schmidt's space of square-summable sequences.

Meanwhile, in a series of papers from 1907, Riesz [1907a, 1907b, 1908, 1910b] investigated the collection of (Lebesgue) square-integrable functions, a space that Riesz would later refer to as L_2 (Riesz [1910b]). Riesz was motivated in this by Hilbert's work on integral equations, and also by the recent introduction of the Lebesgue integral, an important paper of Pierre Fatou that applied the new integral (Fatou [1906]), and Fréchet's work on abstract spaces (Fréchet [1906, 1907]). The main result in Riesz [1907a] states that there is a one-to-one correspondence between Schmidt's space ℓ_2 and the space L_2 (by means of an intermediary orthonormal sequence).

The spaces L_p for $1 < p < \infty$ were introduced in Riesz [1910b]. In fact, the integral versions of Hölder's inequality (Lemma 19.7) and Minkowski's inequality (Theorem 19.9) are due to Riesz. The result in Exercise 39 was first proved by Radon [1913] and, independently, by Riesz [1928a, 1928c] (it is sometimes called the *Radon–Riesz theorem*); see also Novinger [1992]. To better understand the embedding of $C[\,a, b\,]$ into $L_p[\,a, b\,]$, as in Exercises 37, 46, 56 and 60, and Corollary 19.15, see the note by Zaanen [1986].

Independently, and at nearly the same time as Riesz, Ernst Fischer [1907a, 1907b] considered the notion of *convergence in mean* for square-summable functions, that is, convergence in L_2-norm. Fischer's most important result, in modern language, is the fact that L_2 is complete with respect to convergence in mean. From this, Fischer deduced Riesz's result, above, and the combined result is usually referred to as the Riesz–Fischer theorem. Today this result is viewed as a remarkable discovery, but at the time it was considered a mere technical observation in a very specialized area.

The "L_2-theory" was originally introduced using the Lebesgue integral, and was offered as an early application of the power of Lebesgue's new theory. The Riesz–Fischer theorem stands out as an important early contribution to both harmonic and functional analysis. It would ultimately lead to the modern theory of Hilbert spaces, that is, complete normed spaces in which the norm is induced by an inner product, such as ℓ_2 (see Lemma 3.3 and the remarks above) and L_2 (see Observation 15.1 (c)). For a more thorough history of the development of function spaces, the Riesz–Fischer theorem, and the early history of functional analysis, see Bernkopf [1966, 1967], Dieudonné [1981], Dudley [1989], Dunford and Schwartz [1958], Hawkins [1970], Kline [1972], Monna [1973], Nikolśkij [1992], and Taylor [1982].

Although Riesz's observation that a subsequence of $(s_n(f))$ converges pointwise a.e. to $f \in L_2$ is quite general, it would be more satisfying to know that the sequence $(s_n(f))$ itself converged pointwise a.e. to f. Since it is a natural question, Luzin was led to pose this as a problem in 1915. It would go unsolved for over 50 years. That it is, indeed, true that each $f \in L_2[-\pi, \pi]$ is the a.e. limit of its Fourier series is a very deep modern result due to Lennart Carleson [1966]. Carleson's theorem marked the end of a centuries-long search for a general convergence result on Fourier series. Carleson's theorem was later generalized to $L_p[-\pi, \pi]$, $1 < p \leq \infty$, by Hunt [1971]. See Mozzochi [1971], and also Goffman and Waterman [1970] and Halmos [1978].

CHAPTER TWENTY

Differentiation

Lebesgue's Differentiation Theorem

In the last several chapters, we have raised questions about differentiation and about the Fundamental Theorem of Calculus that have yet to be answered. For example:

- For which f does the formula $\int_a^b f' = f(b) - f(a)$ hold? If f' is to be integrable, then at the very least we will need f' to exist almost everywhere in $[a, b]$. But this alone is not enough: Recall that the Cantor function $f : [0, 1] \to [0, 1]$ satisfies $f' = 0$ a.e., but $\int_0^1 f' = 0 \neq 1 = f(1) - f(0)$.
- Stated in slightly different terms: If g is integrable, is the function $f(x) = \int_a^x g$ differentiable? And, if so, is $f' = g$ in this case? For which f is it true that $f(x) = \int_a^x g$ for some integrable g?

In our initial discussion of the Stieltjes integral, we briefly considered the problem of finding the density of a thin metal rod with a known distribution of mass. That is, we were handed an increasing function $F(x)$ that gave the mass of that portion of the rod lying on $[a, x]$, and we asked for its density $f(x) = F'(x)$. We side-stepped this question entirely at the time, defining a new integral in the process, but perhaps it merits posing again.

- Given α increasing, is α differentiable at enough points so as to have $\int_a^b f\,d\alpha = \int_a^b f(x)\,\alpha'(x)\,dx$ hold for, say, all continuous f? That is, is every Riemann–Stieltjes integral a Lebesgue integral? Or even a Riemann integral?
- In particular, if f is of bounded variation, does f' exist? Is f' integrable? If so, is it the case that $V_a^b f = \int_a^b |f'|$? This would give the analogue, in one dimension, of the integral formula for arc length.
- A certain special case is worth considering on its own: Early on in our discussion of Lebesgue measure, we encountered the function $f(x) = m(E \cap (-\infty, x])$, where E is a measurable set of finite measure. We might also write $f(x) = \int_{-\infty}^x \chi_E$, which makes it all the easier to see that f is continuous. The function f represents the distribution of mass of an object whose density is χ_E. The question in this case is whether f is differentiable and, if so, whether $f' = \chi_E$.

In this chapter, thanks to the genius of Lebesgue, we will finally supply answers to several of these questions. Here is the key result:

Lebesgue's Differentiation Theorem 20.1. *If $f : [a, b] \to \mathbb{R}$ is monotone, then f has a finite derivative at almost every point in $[a, b]$.*

That's the good news.... The bad news may come as a surprise to you: Differentiation is hard! It's nothing that we can't handle, mind you, but it is technically more demanding than integration. The reason for this is nothing new; we have already seen that derivatives are harder to come by than integrals. It's easy to see, for example, that every continuous function on $[0, 1]$ is Riemann integrable while, as we now know, the "typical" continuous function fails to have a finite derivative at even a single point. (Recall our discussion at the end of Chapter Eleven.) But, the news isn't all bad: There are only a few hard technical details to sort through. The rest is smooth sailing.

Now, since we want to discuss functions that may not be differentiable in the strict sense, it will help matters if we introduce a "loose" notion of the derivative. An easy choice here is to consider the derived numbers of a function. Given a function $f : \mathbb{R} \to \mathbb{R}$, an extended real number λ is called a **derived number** for f at the point x_0 if there exists a sequence $h_n \to 0$ ($h_n \neq 0$) such that

$$\lim_{n\to\infty} \frac{f(x_0 + h_n) - f(x_0)}{h_n} = \lambda.$$

In other words, λ is a derived number for f at x_0 if some sequence of difference quotients for f at x_0 converges to λ (where we include $\lambda = \pm\infty$ as possibilities). We will abbreviate this lengthy statement using the terse shorthand

$$\lambda = Df(x_0),$$

with the understanding that $Df(x_0)$ denotes just one of possibly many different derived numbers for f at x_0. [In other words, Df is *not* a function.]

Since we permit infinite derived numbers, it is clear that derived numbers exist at every point x_0. (Why?) Of course, if the derivative $f'(x_0)$ exists (whether finite or infinite), then $f'(x_0)$ is a derived number for f at x_0. In fact, in this case, $f'(x_0)$ is the only possible derived number for f at x_0. (Why?)

As an example, consider the function $f(x) = x \sin(1/x)$, $x \neq 0$, $f(0) = 0$, at the point $x_0 = 0$. If we set $h_n^{-1} = (4n - 3)\pi/2$, then

$$\frac{f(x_0 + h_n) - f(x_0)}{h_n} = \frac{h_n \sin(h_n^{-1})}{h_n} = \sin\frac{(4n-3)\pi}{2} = 1$$

for all $n = 1, 2, \ldots$. Thus, $\lambda = 1$ is a derived number for f at 0. It is not hard to see that every number in $[-1, 1]$ is a derived number for f at 0.

EXERCISES

1. Compute the derived numbers for $f = \chi_{\mathbb{Q}}$.

2. Consider $f(x) = x \sin(1/x)$, $x \neq 0$, $f(0) = 0$, at the point $x_0 = 0$. Show that every number in $[-1, 1]$ is a derived number for f at 0.

▷ **3.** Let $f : [a, b] \to \mathbb{R}$. Show that derived numbers for f exist at every point x_0 in $[a, b]$. [Hint: See, for example, Exercise 1.26.]

▷ **4.** If $f : [a, b] \to \mathbb{R}$ is increasing, show that all of the derived numbers for f are nonnegative (i.e., in $[0, \infty]$).

▷ **5.** Let $f : \mathbb{R} \to \mathbb{R}$ and let $x_0 \in \mathbb{R}$. Prove that $f'(x_0)$ exists (as a finite real number) if and only if all of the derived numbers for f at x_0 are equal (and finite). Is this still true when $f'(x_0) = \pm\infty$?

6. Let $f, g : \mathbb{R} \to \mathbb{R}$, let $x_0 \in \mathbb{R}$, and suppose that $g'(x_0)$ exists as a finite real number. Show that λ is a derived number for f at x_0 if and only if $\lambda + g'(x_0)$ is a derived number for $f + g$ at x_0.

7. If $f : (a, b) \to \mathbb{R}$ is differentiable, show that f' is Borel measurable. If f is only differentiable a.e., show that f' is still Lebesgue measurable.

8. If $f'(x)$ exists and satisfies $|f'(x)| \le K$ for all x in $[a, b]$, prove that $m^*\big(f(E)\big) \le K m^*(E)$ for any $E \subset [a, b]$.

With the notion of derived numbers (and Exercise 5) at our disposal, we can now describe our plan of attack on Lebesgue's theorem. To say that a function f has a finite derivative almost everywhere is the same as saying that the set of points x_0 at which f has two different derived numbers, say $D_1 f(x_0) < D_2 f(x_0)$, has measure zero. To address this, we will use a bit of standard trickery and consider instead those derived numbers that satisfy $D_1 f(x_0) \le p < q \le D_2 f(x_0)$, where $p < q$ are real numbers. Thus, we would like to know something about the measure of the set of points at which either $Df(x) \le p$ or $Df(x) \ge q$ occurs.

Now Lebesgue's theorem concerns a monotone function f, but it should be clear that we need only consider the case where f is increasing. In fact, we will first consider the case where f is *strictly* increasing; the general case will follow easily from this. Finally, we can circumvent occasional concerns about the domain of f simply by assuming that every function $f : [a, b] \to \mathbb{R}$ has been extended to all of $\mathbb{R}$ by setting $f(x) = f(a)$ for $x < a$ and $f(x) = f(b)$ for $x > b$.

Lemma 20.2. *Let $f : [a, b] \to \mathbb{R}$ be strictly increasing, let $E \subset [a, b]$, and let $0 \le p < \infty$. If, for every $x \in E$, there exists at least one derived number for f satisfying $Df(x) \le p$, then $m^*\big(f(E)\big) \le p\, m^*(E)$.*

Proof. Let $\varepsilon > 0$, and choose a bounded open set $G \supset E$ such that $m(G) < m^*(E) + \varepsilon$. For each $x_0 \in E$, choose a null sequence (h_n), with $h_n \ne 0$ for all n, such that

$$\lim_{n\to\infty} \frac{f(x_0 + h_n) - f(x_0)}{h_n} = Df(x_0) \le p.$$

Now consider the intervals

$$d_n(x_0) = \begin{cases} [x_0, x_0 + h_n], & \text{if } h_n > 0, \\ [x_0 + h_n, x_0], & \text{if } h_n < 0, \end{cases} \tag{20.1a}$$

and

$$\Delta_n(x_0) = \begin{cases} [\, f(x_0), f(x_0 + h_n)\,], & \text{if } h_n > 0, \\ [\, f(x_0 + h_n), f(x_0)\,], & \text{if } h_n < 0. \end{cases} \tag{20.1b}$$

The intervals $\{d_n(x_0) : x_0 \in E, n \geq 1\}$ cover E while the intervals $\{\Delta_n(x_0) : x_0 \in E, n \geq 1\}$ cover $f(E)$. Notice that since f is strictly increasing, we have $m(\Delta_n(x_0)) > 0$ for any x_0, n.

Since $h_n \to 0$, we may suppose that $d_n(x_0) \subset G$ for all n. We may also suppose that

$$\frac{f(x_0+h_n) - f(x_0)}{h_n} < p + \varepsilon \tag{20.2}$$

for all n. Since

$$m(d_n(x_0)) = |h_n| \qquad \text{and} \qquad m(\Delta_n(x_0)) = |f(x_0+h_n) - f(x_0)|,$$

equation (20.2) can be written as

$$m(\Delta_n(x_0)) < (p+\varepsilon)\, m(d_n(x_0))$$

for all n. In particular, we must have $m(\Delta_n(x_0)) \to 0$ as $h_n \to 0$. Thus, the intervals $\{\Delta_n(x_0) : x_0 \in E, n \geq 1\}$ actually form a Vitali cover for $f(E)$.

By Theorem 16.27, we can find countably many pairwise disjoint intervals $\{\Delta_{n_i}(x_i)\}$ such that

$$m^*\left(f(E) \setminus \bigcup_{i=1}^{\infty} \Delta_{n_i}(x_i)\right) = 0.$$

Thus,

$$m^*(f(E)) \leq \sum_{i=1}^{\infty} m(\Delta_{n_i}(x_i)) < (p+\varepsilon) \sum_{i=1}^{\infty} m(d_{n_i}(x_i)). \tag{20.3}$$

But the intervals $\{d_{n_i}(x_i)\}$ must also be pairwise disjoint. (Why?) Hence,

$$\sum_{i=1}^{\infty} m(d_{n_i}(x_i)) = m\left(\bigcup_{i=1}^{\infty} d_{n_i}(x_i)\right) \leq m(G). \tag{20.4}$$

Combining equations (20.3) and (20.4) yields

$$m^*(f(E)) < (p+\varepsilon)\, m(G) < (p+\varepsilon)(m^*(E) + \varepsilon).$$

Letting $\varepsilon \to 0$, we get $m^*(f(E)) \leq p\, m^*(E)$. □

A similar, but slightly more complicated line of reasoning applies to the set of points where $Df(x) \geq q$.

Lemma 20.3. *Let $f : [a,b] \to \mathbb{R}$ be strictly increasing, let $E \subset [a,b]$, and let $0 < q < \infty$. If, for every $x \in E$, there exists at least one derived number for f satisfying $Df(x) \geq q$, then $m^*(f(E)) \geq q\, m^*(E)$.*

PROOF. Let $\varepsilon > 0$. Since $f(E)$ is bounded, we may choose a bounded open set $G \supset f(E)$ such that $m(G) < m^*(f(E)) + \varepsilon$. For each $x_0 \in E$, choose a null sequence (h_n) such that

$$\lim_{n\to\infty} \frac{f(x_0+h_n) - f(x_0)}{h_n} = Df(x_0) \geq q.$$

As before, we may suppose that

$$\frac{f(x_0 + h_n) - f(x_0)}{h_n} > q - \varepsilon$$

for all n. Thus, if we define the intervals $d_n(x_0)$ and $\Delta_n(x_0)$ exactly as in equation (20.1), then we have

$$m\big(\Delta_n(x_0)\big) > (q - \varepsilon)\, m\big(d_n(x_0)\big)$$

for all n and all $x_0 \in E$.

We would like to argue, as before, that by reducing to countably many intervals we can compare the measures of E and $f(E)$, by way of the open set G. In this case, we want to know when $\Delta_n(x_0)$ is contained in G. But, if $x_0 \in E$ is a point of continuity of f, then $\Delta_n(x_0)$ will be completely contained in G for all n sufficiently large. (Why?) This works at nearly every point $x_0 \in E$: If we let S denote the set of points in E at which f is continuous, then, since f is monotone, the set $E \setminus S$ is at most countable. In summary, we will suppose that $\Delta_n(x_0) \subset G$ actually occurs for all n and all $x_0 \in S$. Now we are ready for Vitali!

The intervals $\{d_n(x_0) : x_0 \in S, n \geq 1\}$ obviously form a Vitali cover for S. Thus, there are countably many pairwise disjoint intervals $\{d_{n_i}(x_i)\}$ such that

$$m^*\left(S \setminus \bigcup_{i=1}^{\infty} d_{n_i}(x_i)\right) = 0.$$

Hence,

$$m^*(S) \leq \sum_{i=1}^{\infty} m\big(d_{n_i}(x_i)\big) < \frac{1}{q - \varepsilon} \sum_{i=1}^{\infty} m\big(\Delta_{n_i}(x_i)\big). \tag{20.5}$$

Now, since f is strictly increasing, the intervals $\{\Delta_{n_i}(x_i)\}$ must also be pairwise disjoint. Consequently,

$$\sum_{i=1}^{\infty} m\big(\Delta_{n_i}(x_i)\big) = m\left(\bigcup_{i=1}^{\infty} \Delta_{n_i}(x_i)\right) \leq m\,(G). \tag{20.6}$$

Combining our observations in light of equations (20.5) and (20.6) yields

$$m^*(E) = m^*(S) < \frac{1}{q - \varepsilon}\big[m^*\big(f(E)\big) + \varepsilon\big].$$

Letting $\varepsilon \to 0$, we get $m^*\big(f(E)\big) \geq q\, m^*(E)$. □

The hard work is (almost) over! Now we sit back and collect the benefits:

Corollary 20.4. *If $f : [\,a, b\,] \to \mathbb{R}$ is increasing, then the set of points at which at least one derived number for f is infinite has measure zero.*

PROOF. This is nearly obvious if f is strictly increasing. In this case, Lemma 20.3 tells us that if the set $E = \{x : Df(x) = +\infty\}$ has nonzero measure, then the set $f(E)$ would have infinite measure. (Why?) This is clearly impossible since $f(E) \subset [\, f(a), f(b)\,]$.

If f is not strictly increasing, we consider instead the function $g(x) = f(x) + x$. Since g is strictly increasing and satisfies

$$\frac{g(x+h) - g(x)}{h} = \frac{f(x+h) - f(x)}{h} + 1,$$

it is clear that $\{x : Df(x) = +\infty\} = \{x : Dg(x) = +\infty\}$. The latter set has measure zero. □

Corollary 20.5. *Let $f : [a, b] \to \mathbb{R}$ be increasing and let $0 \le p < q < \infty$. If at every point x in some set $E_{p,q} \subset [a, b]$ there exist two derived numbers for f satisfying $D_1 f(x) \le p < q \le D_2 f(x)$, then $m(E_{p,q}) = 0$.*

PROOF. If f is strictly increasing, then Lemmas 20.2 and 20.3 imply that

$$q\, m^*(E_{p,q}) \le m^*\big(f(E_{p,q})\big) \le p\, m^*(E_{p,q}),$$

and hence that $m(E_{p,q}) = 0$.

When f is not strictly increasing, we simply apply the first part of the proof to the function $g(x) = f(x) + x$, replacing p by $p + 1$ and q by $q + 1$. □

Finally we are ready for the proof of Lebesgue's theorem.

Theorem 20.6. *If $f : [a, b] \to \mathbb{R}$ is increasing, then f has a finite derivative at almost every point in $[a, b]$.*

PROOF. Let E denote the set of points $x \in [a, b]$ at which $f'(x)$ does not exist. Now a bit of shorthand makes the rest of the proof easy: Let's agree to write $\{x : D_1 f(x) < D_2 f(x)\}$ to denote the set of points x at which f has two different derived numbers $D_1 f(x) < D_2 f(x)$. Then,

$$E = \{x : D_1 f(x) < D_2 f(x)\} = \bigcup_{\substack{p<q \\ p,q\in\mathbb{Q}}} \{x : D_1 f(x) \le p < q \le D_2 f(x)\},$$

where $E_{p,q} \equiv \{x : D_1 f(x) \le p < q \le D_2 f(x)\}$ denotes the set of points x at which f has two different derived numbers satisfying $D_1 f(x) \le p < q \le D_2 f(x)$. From Corollary 20.5, each $E_{p,q}$ has measure zero. There are at most countably many such sets for $p, q \in \mathbb{Q}$ and hence $m(E) = 0$; that is, $f'(x)$ exists at almost every point in $[a, b]$. From Corollary 20.4, we know that the set of points at which $f'(x) = +\infty$ has measure zero; thus, $f'(x)$ exists as a finite real number almost everywhere. □

Corollary 20.7. *If $f \in BV[a, b]$, then f has a finite derivative at almost every point in $[a, b]$.*

EXERCISES

▷ **9.** Consider the Cantor function f on $[0, 1]$. We know that $f'(x) = 0$ when $x \in [0, 1] \setminus \Delta$, the complement of the Cantor set. Compute $f'(x)$, if possible, when

$x \in \Delta$. [Hint: If x is an endpoint, show that $f'(x)$ does not exist; otherwise, show that $f'(x) = +\infty$.]

10. Prove or disprove: If every derived number for f on $[a, b]$ is nonnegative (or $+\infty$), then f is increasing.

11. If $m(E) = 0$, prove that there is a continuous, increasing function $f : \mathbb{R} \to \mathbb{R}$ such that $f'(x) = +\infty$ at each point $x \in E$. [Hint: Let (U_n) be a decreasing sequence of open sets containing E with $m(U_n) < 2^{-n}$. Now let $f_n(x) = m\big((-\infty, x) \cap U_n\big)$ and let $f = \sum_n f_n$.]

Now that we have expanded our collection of differentiable functions, the next item on the agenda is the Fundamental Theorem of Calculus. To address this and other questions raised at the beginning of this chapter, we will first need to discuss the measurability and integrability of derivatives.

Theorem 20.8

(i) *If f is increasing on $[a, b]$, then f' is measurable, $f' \geq 0$ a.e., and $\int_a^b f' \leq f(b) - f(a)$.*

(ii) *If $f \in BV[a, b]$, then $f' \in L_1[a, b]$ and $\int_a^b |f'| \leq V_a^b f$.*

PROOF. Recall our assumption that any function f on $[a, b]$ has been extended to all of $\mathbb{R}$ by setting $f(x) = f(a)$ for $x < a$ and $f(x) = f(b)$ for $x > b$.

The proof of (i) is easier than you might imagine: An increasing function f is measurable, and

$$f'(x) = \lim_{n\to\infty} n\left(f\left(x + \frac{1}{n}\right) - f(x)\right)$$

for almost every x in $[a, b]$. Hence, f' is measurable and $f' \geq 0$ a.e. (Why?) Next we use Fatou's Lemma to estimate $\int_a^b f'$:

$$\begin{aligned}
\int_a^b f' &= \int_a^b \lim_{n\to\infty} n\left(f\left(x + \frac{1}{n}\right) - f(x)\right) dx \\
&\leq \liminf_{n\to\infty} n\left(\int_a^b f\left(x + \frac{1}{n}\right) dx - \int_a^b f(x)\, dx\right) \\
&= \liminf_{n\to\infty} n\left(\int_{a+(1/n)}^{b+(1/n)} f - \int_a^b f\right) \\
&= \liminf_{n\to\infty} n\left(\int_b^{b+(1/n)} f - \int_a^{a+(1/n)} f\right) \\
&\leq f(b) - f(a),
\end{aligned}$$

since f is increasing and since $f(x) = f(b)$ for $x > b$. Please note that the "change of variable" is easily justified here; indeed, since f is monotone, each of the integrals above is actually a Riemann integral.

Now suppose that f is of bounded variation on $[a, b]$, and recall that we may write $f = v - (v - f)$, where $v(x) = V_a^x f$, and where v and $v - f$ are both increasing. Of course, then $f' = v' - (v - f)'$ exists a.e. and is measurable. But, by recalling a basic inequality, we really get something more: For $x < y$ we have

$$|f(y) - f(x)| \le V_x^y f = v(y) - v(x),$$

and it follows that $|f'| \le v'$ a.e. So, from the first part of the proof, f' is integrable and

$$\int_a^b |f'| \le \int_a^b v' \le v(b) - v(a) = V_a^b f. \quad \square$$

We have made some progress on one of our questions: We now know that if f is of bounded variation, then f' exists a.e. and f' is integrable. This still is not enough to make the formula $f(b) - f(a) = \int_a^b f'$ hold (recall the Cantor function). But is it *necessary* to have $f \in BV[a, b]$ in order that the formula hold? The answer is: Yes, and then some. To see this, we will turn the question around: If we set $f(x) = \int_a^x g$, where g is integrable, is f of bounded variation? If so, is $f' = g$ a.e. (in which case, $f(x) = \int_a^x f'$)? The answers are supplied by our next result.

Theorem 20.9. *Let g be integrable on $[a, b]$, and let $f(x) = \int_a^x g$. Then:*
(i) *$f \in C[a, b] \cap BV[a, b]$ and $\int_a^x |f'| \le V_a^x f \le \int_a^x |g|$.*
(ii) *$f \equiv 0$ if and only if $g = 0$ a.e.*
(iii) *$f' = g$ a.e.; hence, $f(x) = \int_a^x f'$ and $V_a^x f = \int_a^x |f'|$.*

PROOF. (i) is very easy. We have already seen that indefinite integrals are continuous (see Corollary 18.21). That f is of bounded variation is surprisingly easy, too. Notice that

$$f(x) = \int_a^x g = \int_a^x g^+ - \int_a^x g^-$$

and both $\int_a^x g^+$ and $\int_a^x g^-$ are increasing. Hence, by the triangle inequality for variations,

$$V_a^x f \le \int_a^x g^+ + \int_a^x g^- = \int_a^x |g|.$$

That $\int_a^x |f'| \le V_a^x f$ is a consequence of Theorem 20.8 (ii).

Next, (ii) follows from considering $\int_a^x g$ as a measure (see Corollary 18.26). If $f \equiv 0$, then

$$\begin{aligned}
\int_a^x g = 0 \text{ for all } x \implies \int_c^d g &= 0 \qquad \text{for all } (c, d) \subset [a, b] \\
\implies \int_U g &= 0 \qquad \text{for all open sets } U \subset [a, b] \\
\implies \int_G g &= 0 \qquad \text{for all } G_\delta\text{-sets } G \subset [a, b] \\
\implies \int_E g &= 0 \qquad \text{for all measurable } E \subset [a, b],
\end{aligned}$$

since every measurable set is, up to a null set, a G_δ-set. Consequently, $g = 0$ a.e. Since $g = 0$ a.e. always forces $f \equiv 0$, this proves (ii).

Finally, we're ready for the proof of (iii). By considering g^+ and g^- separately, we may suppose that $g \geq 0$. Of course, this will make f increasing, and hence $f' \geq 0$ a.e.

Now, let's simplify things further by assuming that g is also bounded, say, $0 \leq g \leq K$. In this case,

$$n\left(f\left(x+\frac{1}{n}\right) - f(x)\right) = n\int_x^{x+(1/n)} g \leq K$$

and $n\,(f\,(x+(1/n)) - f(x)) \to f'(x)$ a.e. So, by the Dominated Convergence Theorem,

$$\begin{aligned}
\int_a^x f' &= \lim_{n\to\infty}\int_a^x n\left(f\left(t+\frac{1}{n}\right) - f(t)\right)dt \\
&= \lim_{n\to\infty}\left[n\int_x^{x+(1/n)} f - n\int_a^{a+(1/n)} f\right] \\
&= f(x) - f(a), \qquad \text{because } f \text{ is continuous,} \\
&= \int_a^x g.
\end{aligned}$$

And now, $\int_a^x f' = \int_a^x g$, for all x, implies that $f' = g$ a.e., from (ii).

In the general case (where g is integrable and nonnegative but not necessarily bounded), we truncate g by defining $g_n(x) = g(x)$ if $g(x) \leq n$ and $g_n(x) = 0$ otherwise; that is, $g_n = g \cdot \chi_{\{g \leq n\}}$. Note that $g_n \to g$ a.e.

Now set $f_n(x) = \int_a^x g_n$. Since $0 \leq g_n \leq g$, we have that $f = (f - f_n) + f_n$, and each of $f - f_n$ and f_n is evidently increasing. But g_n is bounded: $0 \leq g_n \leq n$; thus, by the case just proved, $f_n' = g_n$ a.e. Hence,

$$f' = (f - f_n)' + f_n' \geq f_n' = g_n \to g \text{ a.e.}$$

It follows that $f' \geq g$ a.e., and this turns out to be enough. Since f is increasing, we get

$$f(x) = f(x) - f(a) \geq \int_a^x f' \geq \int_a^x g = f(x).$$

Hence, $f' = g$ a.e. □

Corollary 20.10. *Let E be a measurable subset of $\mathbb{R}$ with finite measure, and consider the "distribution" function $f(x) = m\big(E \cap (-\infty, x\,]\big)$. Then, for almost every x in $\mathbb{R}$, the "density" $f'(x)$ exists and satisfies $f' = \chi_E$ a.e. That is, $f'(x) = 1$ for a.e. $x \in E$ and $f'(x) = 0$ for a.e. $x \in E^c$.*

PROOF. As we have already noted, $f(x) = \int_{-\infty}^x \chi_E$. Thus, since χ_E is integrable, we have

$$f'(x) = \lim_{h\to 0}\frac{1}{h}\int_x^{x+h} \chi_E = \chi_E(x) \qquad \text{for a.e. } x. \quad \square$$

Corollary 20.11. (Lebesgue's Density Theorem) *Let E be a measurable set, and define the* **metric density** *of E at a point $x \in \mathbb{R}$ by*

$$D_E(x) = \lim_{h \to 0} \frac{1}{2h}\, m\big(E \cap [\,x-h, x+h\,]\big),$$

provided that this limit exists. Then, $D_E(x) = 1$ for a.e. $x \in E$ and $D_E(x) = 0$ for a.e. $x \in E^c$. That is, $D_E = \chi_E$ a.e.

PROOF. If $m(E) < \infty$, the conclusion follows immediately from Corollary 20.10. Indeed, in this case, we need only notice that

$$D_E(x) \;=\; \lim_{h \to 0} \frac{1}{2h} \int_{x-h}^{x+h} \chi_E,$$

and that this "two-sided" derivative exists and equals χ_E a.e. (Why?)

Now the limit in question is a *local* property of E: For a given x, the existence of $D_E(x)$ depends only on the set $E \cap [\,x-1, x+1\,]$, for example, which is a set of finite measure. To arrive at a single exceptional set that does not depend on x, where the limit may fail to exist, consider the sets $E_n = E \cap [-n, n\,]$ for $n = 1, 2, \ldots$. We may conclude that the limit exists and equals χ_{E_n} for almost every x in $[-n, n\,]$. By discarding only countably many such exceptional sets, each of measure zero, one for each E_n, we would then have that $D_E(x)$ exists and equals χ_E a.e. □

We extend this result further by considering locally integrable functions, thus taking advantage of the fact that differentiation is a local property. A measurable function $f : \mathbb{R} \to \mathbb{R}$ is said to be **locally integrable** if $\int_a^b |f| < \infty$ for every bounded interval $[\,a, b\,]$.

Corollary 20.12. *Let f be locally integrable. Then, for a.e. $x \in \mathbb{R}$,*

$$\lim_{h \to 0} \frac{1}{h} \int_x^{x+h} f(t)\,dt \;=\; f(x).$$

In fact,

$$\lim_{h \to 0} \frac{1}{h} \int_x^{x+h} |f(t) - f(x)|\,dt = 0$$

for a.e. $x \in \mathbb{R}$.

PROOF. As before, by considering $f\chi_{[-n,n\,]}$ for each n, we might as well suppose that f is integrable and vanishes off some bounded interval. In this case, the first conclusion is an immediate consequence of Theorem 20.9.

The second conclusion takes a bit more work. For each rational r, the first part of the theorem supplies a null set N_r such that

$$\lim_{h \to 0} \frac{1}{h} \int_x^{x+h} |f(t) - r|\,dt \;=\; |f(x) - r| \tag{20.7}$$

for all $x \notin N_r$. Thus, equation (20.7) holds for all r and all $x \notin N$, where $N = \bigcup_{r \in \mathbb{Q}} N_r$ is still a null set.

Now we can make the right-hand side of equation (20.7) arbitrarily small by letting $r \to f(x)$, and so we must have

$$\lim_{h \to 0} \frac{1}{h} \int_x^{x+h} |f(t) - f(x)|\, dt = 0$$

for all $x \notin N$, that is, for a.e. x. □

Let's summarize our progress. Assuming that f' is integrable, then, in order for the formula $f(x) - f(a) = \int_a^x f'$ to hold, it is necessary to have $f \in C[a,b] \cap BV[a,b]$. In fact, if f is the indefinite integral of any $g \in L_1[a,b]$, then we will have to have at least $f \in C[a,b] \cap BV[a,b]$. But, as the Cantor function shows, still more is needed for sufficiency. The missing ingredient is the stronger form of continuity that is typical of the "measure" $\int_a^x g$. Before we formalize this notion, let's take another look at the Cantor function.

Example 20.13

The Cantor function $f : [0,1] \to [0,1]$ cannot be written as the indefinite integral of any $g \in L_1[0,1]$.

PROOF. Recall that $\Delta = \bigcap_{n=1}^{\infty} I_n$, where I_n is the "nth level Cantor set." In particular, the I_n are nested, closed sets satisfying $m(I_n) \to 0$ as $n \to \infty$. More specifically, I_n is the union of 2^n disjoint, closed intervals, each having length 3^{-n}, say, $I_n = \bigcup_{i=1}^{2^n} [x_{n,i}, y_{n,i}]$, where the $x_{n,i}$ and $y_{n,i}$ are "endpoints" of Δ. Since $I_n \supset \Delta$, the Cantor function f maps each I_n onto all of $[0,1]$.

Now suppose that $f(x) = \int_0^x g$ for some $g \in L_1[0,1]$. Then,

$$1 = f(1) - f(0) = \sum_{i=1}^{2^n} [f(y_{n,i}) - f(x_{n,i})] = \int_{I_n} g. \qquad \text{(Why?)}$$

But since $m(I_n) \to 0$, we should also have $\int_{I_n} g \to 0$. Since we are denied this possibility, no such g can exist. □

The problem, in brief, is that $m\big(f(\Delta)\big) = 1$ while $m(\Delta) = 0$, and this (somehow) precludes the possibility of recovering f from f'. We will pursue this idea in detail in the next section.

EXERCISES

12. Find examples of a measurable set E and a point x for which: $D_E(x) = 1/2$; $D_E(x) = 1/3$; $D_E(x)$ does not exist.

13. Fill in the missing details in the proof of Corollary 20.12.

Absolute Continuity

Although we have enumerated various "big questions" several times already, one more incantation couldn't hurt.

Question. Given f, when may we write f as an "indefinite integral"? That is, when does the formula $f(x) = C + \int_a^x g$ hold, for some constant C and some $g \in L_1$?

Question. Given f with $f' \in L_1$, we may consider the function $g(x) = \int_a^x f'$. We know that $g' = f'$ a.e., but does this mean that $f = g + C$ for some constant C? For which f is this true?

The answers to these questions turn out to involve a stronger form of continuity that is satisfied by "indefinite integrals" or "measures" (see Exercise 18.35 or Lemma 20.14, below).

We say that a function $f : [\,a, b\,] \to \mathbb{R}$ is **absolutely continuous** if, for every $\varepsilon > 0$, there exists a $\delta > 0$ such that $\sum_{i\ge1} |f(b_i) - f(a_i)| < \varepsilon$ whenever $\{(a_i, b_i)\}_{i\ge1}$ is any sequence (finite or infinite) of *disjoint* subintervals of $[\,a, b\,]$ satisfying $\sum_{i\ge1}(b_i - a_i) < \delta$.

The requirement that the open intervals $\{(a_i, b_i)\}$ be disjoint is sometimes stated by saying that the corresponding closed intervals $\{[\,a_i, b_i\,]\}$ must be **nonoverlapping**, a self-explanatory nomenclature. However we choose to say it, notice that $m\left(\bigcup_{i\ge1}[\,a_i, b_i\,]\right) < \delta$ is required.

By way of a simple example, notice that every Lipschitz function is absolutely continuous. It is also evident that every absolutely continuous function is (uniformly) continuous. (Why?) If we write $AC[\,a, b\,]$ to denote the collection of all absolutely continuous functions on $[\,a, b\,]$, then, as sets, $\text{Lip}\,1[\,a, b\,] \subset AC[\,a, b\,] \subset C[\,a, b\,]$.

Our goal in this section is to prove that a function f can be written as $f(x) = C + \int_a^x g$, where $g \in L_1$, precisely when f is absolutely continuous. To begin, we prove that an indefinite integral is absolutely continuous.

Lemma 20.14. *If $g \in L_1$, then $f(x) = \int_a^x g$ is absolutely continuous. In fact, given $\varepsilon > 0$, there is a $\delta > 0$ such that $\int_A |g| < \varepsilon$ whenever $m(A) < \delta$.*

Proof. We begin with the proof of the second statement. Given $\varepsilon > 0$, there is a bounded, integrable function h such that $\int |g - h| < \varepsilon/2$. If $0 < K < \infty$ is chosen so that $|h| \le K$, then

$$\int_A |h| \le K m(A) < \frac{\varepsilon}{2} \qquad \text{whenever } m(A) < \frac{\varepsilon}{2K}.$$

Thus,

$$\int_A |g| \le \int_A |g - h| + \int_A |h| < \varepsilon$$

whenever $m(A) < \varepsilon/(2K) \equiv \delta$.

The first conclusion now follows easily from the second: Given nonoverlapping intervals $\{[a_i, b_i]\}$, notice that

$$\sum_{i\geq 1} |f(b_i) - f(a_i)| = \sum_{i\geq 1} \left| \int_{a_i}^{b_i} g \right| \leq \sum_{i\geq 1} \int_{a_i}^{b_i} |g| = \int_A |g|,$$

where $A = \bigcup_{i\geq 1} [a_i, b_i]$, and where the last equation holds by Corollary 18.26. □

The absolute continuity of $f(x) = \int_a^x g$ can be regarded as a condition on the measure $\mu(A) = \int_A |g|$, namely, $\mu(A) < \varepsilon$ whenever $m(A) < \delta$, or $\mu(A) \to 0$ as $m(A) \to 0$. In this sense, absolute continuity is a continuity proper of (certain) measures. (See Exercise 18 for a related condition.)

From Theorem 20.9, indefinite integrals are not only continuous, but also of bounded variation. In fact, the same can be said of any absolutely continuous function.

Proposition 20.15. (i) *If* $f \in AC[a, b]$, *then* $f \in C[a, b] \cap BV[a, b]$. (ii) $f \in AC[a, b]$ *if and only if* $v(x) = V_a^x f \in AC[a, b]$.

PROOF. We have already noted the inclusion $AC[a, b] \subset C[a, b]$. Thus, we first need to show that $AC[a, b] \subset BV[a, b]$. To this end, let $f \in AC[a, b]$, and choose $\delta > 0$ to correspond to the choice $\varepsilon = 1$ in the definition of absolute continuity.

We first note that if $[c, d] \subset [a, b]$ with $d - c < \delta$, then $V_c^d f \leq 1$. Indeed, no matter how we might partition $[c, d] = \bigcup_{i\geq 1} [a_i, b_i]$ into nonoverlapping intervals, we always have $\sum_{i\geq 1} (b_i - a_i) = d - c < \delta$, and hence $\varepsilon = 1$ is always an upper bound for $\sum_{i\geq 1} |f(b_i) - f(a_i)|$. Thus, if we now partition $[a, b]$ into $N = 1 + \left[(b-a)/\delta\right]$ subintervals $\{[c_i, d_i]\}_{i=1}^N$, each of length less than δ, then we would have

$$V_a^b f \leq \sum_{i=1}^N V_{c_i}^{d_i} f \leq N.$$

This proves (i).

But our proof of (i) actually shows much more: If f is absolutely continuous, and if $\{[a_i, b_i]\}_{i\geq 1}$ is any sequence of nonoverlapping intervals with $\sum_{i\geq 1} (b_i - a_i) < \delta$, where $\delta > 0$ corresponds to a given $\varepsilon > 0$ in the definition of absolute continuity for f, then we must have $\sum_{i\geq 1} V_{a_i}^{b_i} f \leq \varepsilon$. Indeed, even if each $[a_i, b_i]$ is further partitioned, the collection of new, smaller subintervals would still have total measure less than δ. Thus, $v(x) = V_a^x f \in AC[a, b]$. That $f \in AC[a, b]$ whenever $v \in AC[a, b]$ is obvious since $|f(b_i) - f(a_i)| \leq V_{a_i}^{b_i} f = |v(b_i) - v(a_i)|$. This proves (ii). □

EXERCISES

▷ **14.** Check that any Lipschitz function on $[a, b]$ is absolutely continuous.

15. Show that the Cantor function is not absolutely continuous on $[0, 1]$. [Hint: Recall Example 20.13.] Conclude that the inclusion $AC[a, b] \subset C[a, b] \cap BV[a, b]$ is proper.

16. Check that $AC[a,b]$ is a subspace and a subalgebra of $C[a,b]$. Is it a sublattice? [Hint: If $f \in AC[a,b]$, is $|f| \in AC[a,b]$?] Is it closed? Explain.

▷ **17.** Prove that $f \in AC[a,b]$ if and only if f can be written as the difference of two increasing, absolutely continuous functions.

18. If $f : [a,b] \to \mathbb{R}$ is increasing and absolutely continuous, prove that $m(f(E)) = 0$ whenever $E \subset [a,b]$ has $m(E) = 0$. [Hint: If $E \subset \bigcup_{i\geq 1}[a_i, b_i]$, then $f(E) \subset \bigcup_{i\geq 1}[f(a_i), f(b_i)]$.]

19. If f is continuous on $[a,b]$, and if $m^*(f(E)) > 0$ for some null set $E \subset [a,b]$, prove that $f(A)$ is nonmeasurable for some (measurable) $A \subset E$.

20. If $f \in C[a,b]$, show that the following are equivalent (for all $E \subset [a,b]$):
(i) $m(E) = 0 \Longrightarrow m(f(E)) = 0$.
(ii) E measurable $\Longrightarrow f(E)$ measurable.
[Hint: For (i) implies (ii), note that f maps F_σ-sets to F_σ-sets. For (ii) implies (i), use Exercise 19.]

21. You will find a variety of seemingly different definitions for absolute continuity in other textbooks. Check that each of the following statements is equivalent to our definition of absolute continuity.
(a) $\forall \varepsilon > 0, \exists \delta > 0$ such that $\sum_{i=1}^n |f(b_i) - f(a_i)| < \varepsilon$ whenever $\{(a_i, b_i)\}_{i=1}^n$ are *finitely many* disjoint subintervals of $[a,b]$ with $\sum_{i=1}^n |b_i - a_i| < \delta$.
(b) $\forall \varepsilon > 0, \exists \delta > 0$ such that $\left|\sum_{i\geq 1}[f(b_i) - f(a_i)]\right| < \varepsilon$ whenever $\{(a_i, b_i)\}$ are disjoint subintervals of $[a,b]$ with $\sum_{i\geq 1} |b_i - a_i| < \delta$.
(c) $\forall \varepsilon > 0, \exists \delta > 0$ such that $\sum_{i\geq 1} V_{a_i}^{b_i} < \varepsilon$ whenever $\{(a_i, b_i)\}$ are disjoint subintervals of $[a,b]$ with $\sum_{i\geq 1} |b_i - a_i| < \delta$.
(d) $\forall \varepsilon > 0, \exists \delta > 0$ such that $\sum_{i\geq 1} \omega(f; [a_i, b_i]) < \varepsilon$ whenever $\{(a_i, b_i)\}$ are disjoint subintervals of $[a,b]$ with $\sum_{i\geq 1} |b_i - a_i| < \delta$. [Recall that $\omega(f; I)$ is the oscillation of f on I.]

It follows from Proposition 20.15 that each absolutely continuous function f is differentiable a.e. and, from Theorem 20.8, that f' is even integrable. Thus it makes sense to ask whether $f(x) = f(a) + \int_a^x f'$ holds. To attack this problem, notice that if we set $g(x) = \int_a^x f'$, then Theorem 20.9 tells us that g is differentiable a.e. and satisfies $g' = f'$ a.e. All that remains is to show that this last condition forces $f = g + C$ for some constant C.

Theorem 20.16. *If $f \in AC[a,b]$ and if $f' = 0$ a.e., then f is constant on $[a,b]$. Thus, if $f, g \in AC[a,b]$ satisfy $f' = g'$ a.e., then $f - g$ is constant on $[a,b]$.*

PROOF. Let $f \in AC[a,b]$ with $f' = 0$ a.e., and fix $a < x < b$. We will prove that $f(x) = f(a)$.

Let $E = \{y \in [a,x] : f'(y) = 0\}$. Please note that E is measurable and that $m([a,x] \setminus E) = 0$. For any point $y \in E$, the fact that $f'(y) = 0$ means that there are arbitrarily small closed intervals $[c, d]$ containing y such that

$$\left|\frac{f(d) - f(c)}{d - c}\right| < \varepsilon.$$

Thus, the collection of closed intervals

$$\mathcal{C} = \left\{[c, d] : [c, d] \subset [a, x] \text{ and } \left|\frac{f(d) - f(c)}{d - c}\right| < \varepsilon\right\}$$

is a Vitali cover for E.

Now, given $\varepsilon > 0$, choose $\delta > 0$ to work in the definition of absolute continuity for f. Then, by Corollary 16.28, there are finitely many disjoint intervals $\{[c_i, d_i]\}_{i=1}^n$ in $\mathcal{C}$ such that

$$m\left(E \setminus \bigcup_{i=1}^n [c_i, d_i]\right) = m\left([a, x] \setminus \bigcup_{i=1}^n [c_i, d_i]\right)$$
$$= (x - a) - \sum_{i=1}^n (d_i - c_i) < \delta.$$

But notice that

$$[a, x] \setminus \bigcup_{i=1}^n [c_i, d_i] = [d_0, c_1) \cup (d_1, c_2) \cup (d_2, c_3) \cup \cdots \cup (d_n, c_{n+1}],$$

where $d_0 = a$ and $c_{n+1} = x$ (if necessary). Hence,

$$\sum_{i=1}^{n+1} (c_i - d_{i-1}) = (x - a) - \sum_{i=1}^n (d_i - c_i) < \delta.$$

That is, we have partitioned $[a, x]$ into two sets of intervals: $\{[c_i, d_i]\}_{i=1}^n$, taken from $\mathcal{C}$, and $\{(d_{i-1}, c_i)\}_{i=1}^{n+1}$, which have small total measure. Now we use the triangle inequality to estimate

$$|f(x) - f(a)| \le \sum_{i=1}^n |f(d_i) - f(c_i)| + \sum_{i=1}^{n+1} |f(c_i) - f(d_{i-1})|$$
$$< \varepsilon \sum_{i=1}^n (d_i - c_i) + \varepsilon < \varepsilon\,((b - a) + 1).$$

Since ε is arbitary, we have $f(x) = f(a)$. □

A function satisfying $f' = 0$ a.e. is called **singular**. Theorem 20.16 says that a function that is simultaneously absolutely continuous and singular must be constant.

Corollary 20.17. *Let* $f : [a, b] \to \mathbb{R}$.

(i) $f \in AC[a, b]$ *if and only if* $f(x) = C + \int_a^x g$ *for some constant* C *and some* $g \in L_1[a, b]$.

(ii) f *is Lipschitz if and only if* $f(x) = C + \int_a^x g$ *for some constant* C *and some* $g \in L_\infty[a, b]$.

(iii) *Each* $f \in BV[a, b]$ *may be written as* $f = g + h$, *where* $g \in AC[a, b]$, $g(a) = 0$, *and where* h *is singular.*

PROOF. We have already talked our way through (i): If $f \in AC[a,b]$, and if we set $h(x) = \int_a^x f'$, then $h' = f'$ a.e. Hence, $f(x) = C + h(x)$. Clearly, $f(x) = f(a) + \int_a^x f'$. The other implication is supplied by Lemma 20.14. The proof of (ii) is left as an exercise (Exercise 22). For (iii), notice that if $f \in BV[a,b]$, then $f' \in L_1[a,b]$. Hence, if we set $g(x) = \int_a^x f'$, then $g \in AC[a,b]$, $g(a) = 0$, $g' = f'$ a.e., and $h = f - g$ satisfies $h' = 0$ a.e. □

When rewritten, Corollary 20.17 (i) will provide a missing detail from Chapter Thirteen along with an alternate version of Proposition 20.15 (ii).

Corollary 20.18. *Let $f : [a,b] \to \mathbb{R}$. Then, the following are equivalent:*

(i) $f \in AC[a,b]$.
(ii) *f' exists a.e., $f' \in L_1[a,b]$, and $f(x) = f(a) + \int_a^x f'$.*
(iii) *f' exists a.e., $f' \in L_1[a,b]$, and $v(x) = V_a^x f = \int_a^x |f'|$.*
(iv) $v \in AC[a,b]$.

PROOF. That (i) implies (ii) is clear. The proof that (ii) implies (iii) follows from Theorem 20.9 (iii) and the fact that $V_a^x f = V_a^x(f - f(a)) = \int_a^x |f'|$. That (iii) implies (iv) is dead easy: If v is an indefinite integral, then $v \in AC[a,b]$. Finally, the fact that (iv) implies (i) is obvious since $|f(x) - f(y)| \le V_y^x f = |v(x) - v(y)|$. □

EXERCISES

▷ **22.** Prove Corollary 20.17 (ii).

23. Prove that the decomposition in Corollary 20.17 (iii) is unique.

24. If $f \in AC[a,b]$ satisfies $f' \ge 0$ a.e., show that f is increasing.

In Chapter Fourteen we raised the question of when a Riemann–Stieltjes integral $\int_a^b f\,dg$ was equal to the Riemann integral $\int_a^b fg'$ (recall Theorem 14.17). We take this one step further and now consider $\int_a^b fg'$ as a Lebesgue integral.

Theorem 20.19. *If $f \in C[a,b]$ and $g \in AC[a,b]$, then*

$$(RS)\int_a^b f\,dg = (L)\int_a^b fg'.$$

PROOF. We want to compare $\int_a^b fg'$ to a typical Riemann–Stieltjes sum for $\int_a^b f\,dg$, say

$$S_g(f,P,T) = \sum_{i=1}^n f(t_i)[g(x_i) - g(x_{i-1})].$$

Since g is absolutely continuous, we may write $g(x_i) - g(x_{i-1}) = \int_{x_{i-1}}^{x_i} g'$, and hence

$$S_g(f, P, T) = \sum_{i=1}^{n} \int_{x_{i-1}}^{x_i} f(t_i)\, g'(x)\, dx.$$

Consequently,

$$\begin{aligned} \left| S_g(f, P, T) - \int_a^b f(x)\, g'(x)\, dx \right| &= \left| \sum_{i=1}^{n} \int_{x_{i-1}}^{x_i} \big[f(t_i) - f(x) \big]\, g'(x)\, dx \right| \\ &\le \sum_{i=1}^{n} \omega(f; [\, x_{i-1}, x_i \,]) \int_{x_{i-1}}^{x_i} |g'(x)|\, dx \\ &\le \alpha \int_a^b |g'(x)|\, dx, \end{aligned}$$

where $\alpha = \max_{1 \le i \le n} \omega(f; [\, x_{i-1}, x_i \,])$. Since f is continuous, $\omega(f; [\, x_{i-1}, x_i \,]) \to 0$ as $x_i - x_{i-1} \to 0$. This proves that the norm integral $(N) \int_a^b f\, dg$ equals $\int_a^b f g'$. Since g is continuous, $(N) \int_a^b f\, dg = (RS) \int_a^b f\, dg$ (see Theorem 14.26). □

As an immediate corollary we get

Corollary 20.20. *If* $f,\, g \in AC[\, a, b \,]$, *then*

$$\int_a^b f g' + \int_a^b g f' = f(b)g(b) - f(a)g(a).$$

Corollary 20.17 and Theorem 20.19 shed new light on the nature of Riemann–Stieltjes integration against integrators of bounded variation. Recall from Chapter Fourteen that each function $g \in BV[\, a, b \,]$ may be written as the sum of a continuous function of bounded variation g_c and a saltus or "pure jump" function g_s. Clearly, any saltus function is also singular. (Why?) Corollary 20.17 tells us that we may further decompose g_c into an absolutely continuous part g_{ac} and a continuous, singular part g_{cs}. That is,

$$g = g_{ac} + g_{cs} + g_s.$$

Theorem 20.19 tells us that integration against g_{ac} reduces to a Lebesgue integral and, as we saw in Chapter Fourteen, integration against g_s reduces to an infinite series. All of the fuss and botheration comes from integration against g_{cs}, the "Cantor function–like" part of g.

Finally, we offer another description of $AC[\, a, b \,]$ that leads to an easy proof that $AC[\, a, b \,]$ is a Banach space under the norm $\| f \|_{BV} = |f(a)| + V_a^b f$. That is, $AC[\, a, b \,]$ is a closed subspace (and even a subalgebra) of $BV[\, a, b \,]$. We will use the characterization given in Corollary 20.17 to write $AC[\, a, b \,] = L_1[\, a, b \,] \oplus \mathbb{R}$.

To simplify the notation, we normalize by considering

$$AC_0[\, a, b \,] = \{ f \in AC[\, a, b \,] : f(a) = 0 \}.$$

Clearly, $AC_0[a,b]$ is a subspace of $AC[a,b]$ and, for $f \in AC_0[a,b]$, the norm simplifies to $\|f\|_{BV} = V_a^b f = \int_a^b |f'|$. If we define a norm on the space $\mathbb{R} \oplus AC_0[a,b]$ by setting $\|(t, f)\| = |t| + V_a^b f$, it then follows that

$$AC[a,b] = \mathbb{R} \oplus AC_0[a,b],$$

isometrically, under the linear map $f \mapsto (f(a), f - f(a))$.

Next we define the map

$$T : L_1[a,b] \to AC_0[a,b] \text{ by } (Tg)(x) = \int_a^x g.$$

That is, $Tg = f$, where $f(x) = \int_a^x g$. Obviously, T is *linear* and *onto* (since $T(f') = f$). Also, T is one-to-one because, in fact, T is an *isometry*:

$$\|Tg\|_{BV} = \|f\|_{BV} = V_a^b f = \int_a^b |f'| = \int_a^b |g| = \|g\|_1.$$

(By the way, what is T^{-1}?) Thus,

$$AC[a,b] = \mathbb{R} \oplus AC_0[a,b] = \mathbb{R} \oplus L_1[a,b],$$

isometrically. Since $L_1[a,b]$ is complete, it follows that $AC_0[a,b]$ must also be complete, and from this it follows easily that $AC[a,b]$ is complete.

Notice, too, that the map T not only preserves the lattice structure of $L_1[a,b]$, but it also carries an extra feature that you might not expect: If $g \geq 0$ a.e., then $Tg \geq 0$, of course, but also

$$g \geq 0 \text{ a.e.} \iff Tg \text{ is } \textit{increasing}.$$

In fact, the lattice decomposition $g = g^+ - g^-$ in L_1 transforms into the Jordan decomposition $f = p - n$, where $f \in AC_0[a,b]$ is written as the difference of its positive and negative variations (see Exercise 26).

Finally, by applying a similar line of reasoning to $BV[a,b]$ we could restate Corollary 20.17 (iii) by writing

$$BV[a,b] = AC_0[a,b] \oplus BV_S[a,b] = L_1[a,b] \oplus BV_S[a,b],$$

where $BV_S[a,b]$ denotes the subspace of singular functions in $BV[a,b]$ (which includes both the constant functions and the saltus functions). That is, each $f \in BV[a,b]$ can be written as $f(x) = \int_a^x f' + h(x)$, where h is singular.

EXERCISES

25. Prove that the Lipschitz functions on $[a,b]$ are dense in $AC[a,b]$ under the variation norm. [Hint: Corollary 20.17 (ii).]

▷ **26.** If $f(x) = \int_a^x g$, where $g \in L_1[a,b]$, prove that the positive and negative variations of f are given by $p(x) = \int_a^x g^+$ and $n(x) = \int_a^x g^-$.

27. Let $PL[a,b]$ denote the subspace of all continuous, piecewise linear functions in $AC[a,b]$ (i.e., the polygonal functions), and let $S[a,b]$ denote the step functions on $[a, b]$. Use the fact that $S[a, b]$ is dense in $L_1[a, b]$ to prove that $PL[a, b]$ is dense in $AC[a, b]$. [Hint: Show that the map $(Tg)(x) = \int_a^x g$ carries $S[a, b]$ onto $PL[a, b]$.]

Notes and Remarks

There is a wealth of literature on differentiation, which is testament to the fact that it is a complex and delicate subject. For an extensive survey of results, see Bruckner [1994]. For more on the history of the results in this chapter, see Hawkins [1970].

The material in this chapter is largely based on the presentation in Natanson [1955]. In particular, we have followed Natanson's lead by opting for the efficacy of derived numbers in our attack on Lebesgue's differentiation theorem rather than the more commonplace *Dini derivatives*. The Dini derivatives of f at x, defined by

$$D_+f(x) = \liminf_{h\to 0^+} \frac{f(x+h)-f(x)}{h}, \qquad D^+f(x) = \limsup_{h\to 0^+} \frac{f(x+h)-f(x)}{h},$$

$$D_-f(x) = \liminf_{h\to 0^-} \frac{f(x+h)-f(x)}{h}, \qquad D^-f(x) = \limsup_{h\to 0^-} \frac{f(x+h)-f(x)}{h},$$

were introduced by (and named after) Ulisse Dini [1878].

Nearly all of the main results in this chapter, including Theorems 20.6, 20.8, 20.9, 20.16 and Corollaries 20.7, 20.11, 20.12, and 20.17, are due to Lebesgue, from roughly 1903 to 1907, and most appeared in the first edition of the *Leçons* in 1904, although not in their current form. Lebesgue's original version of Theorem 20.6, for example, also required that the function f be continuous; this restriction was later shown to be unnecessary (see Lebesgue [1928]). The term "absolutely continuous" was introduced by Vitali [1905b], who published the first proof of Corollary 20.17 (i); Lebesgue [1907] later gave his own proof (essentially the one given here).

The discussion of Corollary 20.17 in terms of Banach space decompositions is based in part on my notes from a course on real analysis offered by W. B. Johnson at The Ohio State University in 1974–1975.

For other presentations of Lebesgue's differentiation theorem see, for example, Riesz and Sz.-Nagy [1955], Taylor [1965], or Chae [1980] (for proofs of Theorem 20.6 not requiring the Vitali Covering Theorem), and Austin [1965] (for a geometric proof of Theorem 20.6). For a proof that $v'(x) = |f'(x)|$ a.e. for $f \in BV[a, b]$, see Wheeden and Zygmund [1977]. For an elementary proof that $m^*(f(E)) \to 0$ as $m^*(E) \to 0$ for $f \in AC[a, b]$, see Łojasiewicz [1988].

For an extensive discussion of absolute continuity and an elementary proof of the *Banach–Zarecki theorem*, which states that a continuous function of bounded variation is absolutely continuous if and only if it maps null sets to null sets, see Varberg [1967] (or Torchinsky [1988]). Varberg's proof is based on the following lemmas, which are of independent interest:

Lemma A. *Let $f : [\,a, b\,] \to \mathbb{R}$. If $f'(x)$ exists and satisfies $|f'(x)| \le K$ for all x in $E \subset [\,a, b\,]$, then $m^*\big(f(E)\big) \le Km^*(E)$.*

Lemma B. *Let $f : [\,a, b\,] \to \mathbb{R}$ be measurable, and let $E \subset [\,a, b\,]$ be measurable. If $f'(x)$ exists (as a finite real number) for all x in E, then $m^*\big(f(E)\big) \le \int_E |f'|$.*

The Banach–Zarecki theorem is immediate from Lemma B (and the ideas found in Exercises 18–20, for example). Lemma B, in turn, is not hard to deduce from Lemma A. The fighting takes place in modifying the proof of Lemma 20.2 to work in the setting of Lemma A. Compare the statement of Lemma A with (the much simpler) Exercise 8.

References

Authors are listed in alphabetical order; each author's works are listed in chronological order by date of first publication (although, in each case, I have actually referred to the most recent edition I could find). Some of the items listed here are not explicitly referred to in the text; nevertheless, all were helpful or influential sources. More than a few of these references are included primarily for their value as historical or expository sources.

Abbott, J. C., Ed. [1978] *The Chauvenet Papers,* 2 vols., The Mathematical Association of America, Washington, DC.

Adams, C. R. [1936] "The space of functions of bounded variation and certain general spaces," *Transactions of the American Mathematical Society,* **40**, 421–438.

Adams, C. R. and Morse, A. P. [1937] "On the space (BV)," *Transactions of the American Mathematical Society,* **42**, 194–205.

Agnew, R. P. [1937] "Convergence in mean and Lebesgue integration," *The American Mathematical Monthly,* **44**, 4–14.

Alas, O. T. [1971] "Metrizable topologies on the real numbers," *The American Mathematical Monthly,* **78**, 773.

Alexandrov, P. and Urysohn, P. [1924] "Theorie der topologischen Räume," *Mathematische Annalen,* **92**, 258–266.

Almkvist, G. and Berndt, B. [1988] "Gauss, Landen, Ramanujan, the arithmetic-geometric mean, ellipses, π, and the *Ladies Diary*," *The American Mathematical Monthly,* **95**, 585–608.

Anderson, R. D. [1962] "Hilbert space is homeomorphic to the countable infinite product of lines," *Bulletin of the American Mathematical Society,* **103**, 249–271.

Apostol, T. [1975] *Mathematical Analysis,* 2nd. ed., Addison-Wesley, Reading, MA.

Aron, R. M. and Fricke, G. H. [1986] "Homomorphisms on $C(\mathbb{R})$," *The American Mathematical Monthly,* **93**, 555.

Arzelà, C. [1889] "Funzioni di linee," *Atti della Reale Accademia dei Lincei Rendiconti Classe di Scienze Fisiche, Matematiche e Naturali,* (5) $\mathbf{5}_{\mathrm{I}}$, 342–348.

——— [1895] "Sulle funzioni di linee," *Memorie della Reale Accademia delle Scienze dell'Istitutio di Bologna Classe di Scienze Fisiche, Matematiche e Naturali,* (5) **5**, 55–74.

Ascoli, G. [1883] "Le curve limiti di una varietà data di curve," *Atti della Reale Academia dei Lincei Memorie della Classe di Scienze Fisiche, Matematiche e Naturali,* (3) **18**, 521–586.

Ash, J. M., Ed. [1976] *Studies in Harmonic Analysis,* Studies in Mathematics v. 13, The Mathematical Association of America, Washington, DC.

Austin, D. G. [1965] "A geometric proof of the Lebesgue differentiation theorem," *Proceedings of the American Mathematical Society,* **16**, 220–221.

Bailey, D. F. [1989] "A historical survey of solution by functional iteration," *Mathematics Magazine,* **62**, 155–166.

Bailey, D. H. [1988] "The computation of π to 29,360,000 decimal digits using Borweins' quartically convergent algorithm," *Mathematics of Computation,* **50**, 283–296.

Baire, R. [1899] "Sur les fonctions de variables réeles," *Annali di Mathematica pura ed applicata,* Series 3, **3**, 1–122.

——— [1905] *Leçons sur les Fonctions Discontinues,* Gauthier-Villars, Paris.

Banach, S. [1922] "Sur les opérations dans les ensembles abstraits et leur application aux équations intégrals," *Fundamenta Mathematicae,* **3**, 133–181.

——— [1923] "Sur le problème de mesure," *Fundamenta Mathematicae,* **4**, 7–33.

——— [1924] "Sur un théorème de M. Vitali," *Fundamenta Mathematicae,* **5**, 130–136.

——— [1930] "Théorème sur les ensembles de première catégorie," *Fundamenta Mathematicae,* **16**, 395–398.

——— [1931] "Über die Baire'sche Kategorie gewisser Funktionmengen," *Studia Mathematica,* **3**, 174–180.

——— [1932] *Théorie des Opérations Linéaires,* 2nd. ed. with corrections, reprinted by Chelsea Publishing Co., New York, 1955.

——— [1987] *Theory of Linear Operations,* North-Holland Mathematical Library v. 38, North-Holland, Amsterdam.

Banach, S. and Kuratowski, K. [1930] "Sur une généralisation du problème de la mesure," *Fundamenta Mathematicae,* **14**, 127–131.

Banach, S. and Tarski, A. [1924] "Sur la décomposition des ensembles des points en parties respectivement congruentes," *Fundamenta Mathematicae,* **6**, 244–277.

Barnsley, M. F. [1988] *Fractals Everywhere,* Academic Press, Boston.

Bartle, R. G. [1964] *The Elements of Real Analysis,* 2nd. ed., Wiley, New York.

——— [1980a] "An extension of Egorov's theorem," *The American Mathematical Monthly,* **87**, 628–633.

——— Ed. [1980b] *Studies in Functional Analysis,* Studies in Mathematics v. 21, The Mathematical Association of America, Washington, DC.

Baten, W. D. [1930] "Simultaneous treatment of discrete and continuous probability by use of Stieltjes integrals," *Annals of Mathematical Statistics,* **1**, 95–100.

Beals, R. [1973] *Advanced Mathematical Analysis,* Springer-Verlag, New York.

Beckenbach, E. F. and Bellman, R. [1961] *An Introduction to Inequalities,* New Mathematical Library v. 3, Random House, New York.

Beer, G. [1988] "UC spaces revisited," *The American Mathematical Monthly,* **95**, 737–739.

Bennett, D. G. and Fisher, B. [1974] "On a fixed point theorem for compact metric spaces," *Mathematics Magazine,* **47**, 40–41; see also "Comment," **48**, 48.

Bernkopf, M. [1966] "The development of function spaces with particular reference to their origins in integral equation theory," *Archive for History of Exact Sciences,* **3**, 1–96.

——— [1967] "A History of Infinite Matrices," *Archive for History of Exact Sciences,* **4**, 308–358.

Bernstein, S. N. [1912] "Démonstration du théorème de Weierstrass fondée sur le calcul des probabilités," *Communications of the Kharkov Mathematical Society,* (2) **13**, 1–2.

Beslin, S. J. [1992] "A note on topological continuity," *Mathematics Magazine,* **65**, 257–258.

Bessaga, Cz. and Pełczyński, A. [1987] "Some aspects of the present theory of Banach spaces," in *Theory of Linear Operations,* S. Banach, Ed., North-Holland, Amsterdam.

Billingsley, P. [1982] "Van der Waerden's continuous nowhere differentiable function," *The American Mathematical Monthly,* **89**, 691.

Birkhoff, G. [1940] *Lattice Theory,* rev. ed., Colloquium Publications 25, American Mathematical Society, New York, 1948.

——— [1943] "What is a lattice?," *The American Mathematical Monthly,* **50**, 484–487.

——— [1973] *A Source Book in Classical Analysis,* Harvard University Press, Cambridge, MA.

Birkhoff, G. and MacLane, S. [1965] *A Survey of Modern Algebra,* 3rd. ed., Macmillan, New York.

Bliss, G. A. [1917] "Integrals of Lebesgue," *Bulletin of the American Mathematical Society,* **24**, 1–47.

Bloom, D. M. [1989] "A pictorial proof of uniform continuity," *The American Mathematical Monthly,* **96**, 250–251.

Boas, R. P. [1960] *A Primer of Real Functions,* The Carus Mathematical Monographs no. 13, 4th. ed., The Mathematical Association of America, Washington, DC, 1996.

——— [1980] "Letter to an author," *The American Mathematical Monthly,* **87**, 376.

Bochner, S. [1979] "Fourier series came first," *The American Mathematical Monthly,* **86**, 197–199.

Borel, E. [1928] *Leçons sur les Fonctions de Variables Réeles,* Gauthier-Villars, Paris.

——— [1950] *Leçons sur la Théorie des Fonctions,* Gauthier-Villars, Paris, 1950.

Botsko, M. W. [1987] "Unified treatment of various theorems in elementary analysis," *The American Mathematical Monthly,* **94**, 450–452.

——— [1988] "An elementary proof that a bounded a.e. continuous function is Riemann integrable," *The American Mathematical Monthly,* **95**, 249–252.

Boyer, C. B. [1968] *A History of Mathematics,* 2nd. ed., rev. by U. Merzbach, Wiley, New York, 1991.

Bray, H. E. [1918] "Elementary properties of the Stieltjes integral," *Annals of Mathematics,* **20**, 177–186.

Briggs, J. M. and Schaffter, T. [1979] "Measure and cardinality," *The American Mathematical Monthly,* **86**, 852–855.

Brooks, F. [1971] "Indefinite cut sets for real functions," *The American Mathematical Monthly,* **78**, 1007–1010.

Brown, A. B. [1936] "A proof of the Lebesgue condition for Riemann integrability," *The American Mathematical Monthly,* **43**, 396–398.

Bruckner, A. M. [1973] "The differentiability properties of typical functions in $C[a, b]$," *The American Mathematical Monthly,* **80**, 679–683.

——— [1978] "Creating differentiability and destroying derivatives," *The American Mathematical Monthly,* **85**, 554–562.

——— [1994] *Differentiation of Real Functions,* 2nd. ed., CRM Monograph Series v. 5, American Mathematical Society, Providence, RI.

Bruckner, A. M. and Ceder, J. [1974] "On improving Lebesgue measure," *Nordisk Matematisk Tidskrift,* **23**, 59–68.

Bruckner, A. M. and Leonard, J. L. [1966] "Derivatives," *The American Mathematical Monthly,* **73**, Part II, 24–56.

Buck, R. C., Ed. [1962] *Studies in Modern Analysis,* Studies in Mathematics v. 1, The Mathematical Association of America; distributed by Prentice-Hall, Englewood Cliffs, NJ.

——— [1967] "Topology and analysis," *Mathematics Magazine,* **40**, 71–74.

Bullen, P. S. [1983] "An inequality for variations," *The American Mathematical Monthly,* **90**, 560.

Bullock, G. L. [1988] "A geometric interpretation of the Riemann-Stieltjes integral," *The American Mathematical Monthly,* **95**, 448–455.

Burgess, C. E. [1990] "Continuous functions and connected graphs," *The American Mathematical Monthly,* **97**, 337–339.

Burkill, J. C. [1964] "Necrology, Charles-Joseph de La Vallée Poussin," *Journal of the London Mathematical Society,* **39**, 165–175.

Burkill, J. C. and Burkill, H. [1970] *A Second Course in Mathematical Analysis,* Cambridge University Press, Cambridge.

Bush, K. A. [1951] "Continuous functions without derivatives," *The American Mathematical Monthly,* **59**, 222–225.

Cajori, F. [1893] *A History of Mathematics,* 3rd. ed., Chelsea Publishing, New York, 1980.

Campbell, S. L. [1986] "Countability of sets," *The American Mathematical Monthly,* **93**, 1986, 480–481.

Cannon, L. and Elich, J. [1993] "Some pleasures and perils of iteration," *The Mathematics Teacher,* **86**, 233–239.

Cantor, G. [1883] "Über unendliche, lineare Punktmannichfaltigkeiten," *Mathematische Annalen,* **20**, 51–58, 545–591.

——— [1915] *Contributions to the Founding of the Theory of Transfinite Numbers,* translated by P. E. B. Jourdain; originally published by Open Court Publishing, 1915; reprinted by Dover Publications, New York, 1955.

Carathéodory, C. [1918] *Vorlesungen über Reelle Funktionen,* Teubner, Leipzig, 1918; 2nd. ed., 1927; reprinted by Chelsea Publishing, New York, 1948.

Carlesson, L. [1966] "On convergence and growth of partial sums of Fourier series," *Acta Mathematica*, **116**, 135–157.

Carlson, B. C. [1971] "Algorithms involving arithmetic and geometric means," *The American Mathematical Monthly*, **78**, 496–505.

Carslaw, H. S. [1930] *Introduction to the Theory of Fourier's Series and Integrals*, 3rd. ed., Macmillan, London.

Cater, F. S. [1982] "Most monotone functions are not singular," *The American Mathematical Monthly*, **89**, 466–469.

——— [1984a] "On van der Waerden's nowhere differentiable function," *The American Mathematical Monthly*, **91**, 307–308.

——— [1984b] "A partition of the unit interval," *The American Mathematical Monthly*, **91**, 564–566.

Chae, S. B. [1980] *Lebesgue Integration*, Marcel Dekker, New York.

Chalice, D. R. [1991] "A characterization of the Cantor function," *The American Mathematical Monthly*, **98**, 255–258.

Chang, G. [1984] "Bernstein polynomials via the shifting operator," *The American Mathematical Monthly*, **91**, 634–638.

Chatterji, S. D. [1988] "A frequent oversight concerning the integrability of derivatives," *The American Mathematical Monthly*, **95**, 758–761.

Chaves, M. A. [1985] "Spaces where all continuity is uniform," *The American Mathematical Monthly*, **92**, 487–489.

Cheney, E. W. [1966] *Introduction to Approximation Theory*, 2nd. ed., Chelsea Publishing, New York, 1982.

Chernoff, P. R. [1980] "Pointwise convergence of Fourier series," *The American Mathematical Monthly*, **87**, 399–400.

Cooke, R. [1979] "The Cantor-Lebesgue theorem," *The American Mathematical Monthly*, **86**, 558–565.

Coppel, W. A. [1969] "J. B. Fourier – On the occasion of his two hundredth birthday," *The American Mathematical Monthly*, **76**, 468–483.

——— [1983] "An interesting Cantor set," *The American Mathematical Monthly*, **90**, 456–460.

Copson, E. T. [1968] *Metric Spaces*, Cambridge University Press, London, 1968; reprinted 1988.

Dauben, J. W. [1971] "The trigonometric background to Georg Cantor's theory of sets," *Archive for History of Exact Science*, **7**, 181–216.

——— [1974] "Denumerability and dimension: The origins of Georg Cantor's theory of sets," *Rete*, **2**, 103–133.

——— [1983] "Georg Cantor and the origins of transfinite set theory," *Scientific American*, June, 122–131.

Davis, P. J. [1963] *Interpolation and Approximation*, Dover Publications, New York, 1975.

——— [1986] "When mathematics says no," *Mathematics Magazine*, **59**, 67–76.

De Barra, G. [1974] *Introduction to Measure Theory*, Van Nostrand Reinhold, New York.

Devaney, R. L. [1992] *A First Course in Chaotic Dynamical Systems*, Addison-Wesley, Reading, MA.

Diamond, H. and Gellès, G. [1984] "Relations among some classes of subsets of $\mathbb{R}$," *The American Mathematical Monthly*, **91**, 19–22.

——— [1985] "Interlaced second category sets," *The American Mathematical Monthly*, **92**, 138–140.

Dieudonné, J. [1981] *History of Functional Analysis*, North-Holland, Amsterdam.

Dini, U. [1878] *Fondamenti per la teorica della funzioni di variabili reali*, Pisa, 1878. Translated and supplemented by J. Lurōth and A. Schepp as *Grundlagen für eine Theorie der Functionen einer veränderlichen rellen Grössen*, Teubner, Liepzig.

Dirichlet, G. L. [1829] "Sur la convergence des séries trigonométriques qui servent à représenter une fonction arbitraire entre des limites données," *Journal für die Reine und Angewandte Mathematik*, **4**, 157–169.

Donner, K. [1982] *Extension of Positive Linear Operators and Korovkin Theorems*, Lecture Notes in Mathematics 904, Springer-Verlag, New York.

Douglas, R. G. [1965] "On lattices and algebras of real valued functions," *The American Mathematical Monthly,* **72**, 642–643.

Du Bois-Reymond, P. [1875] "Versuch einer Classification der willkürlichen Funktionen reeler Argumente nach ihren Änderungen in den kleinsten Intervallen," *Journal für die Reine und Angewandte Mathematik,* **79**, 21–37.

——— [1880] *Zur Geschichte der trigonometrischen Reihen: Eine Entgegnung,* H. Laupp, Tübingen.

Dudley, R. M. [1989] *Real Analysis and Probability,* Wadsworth & Brooks/Cole, Pacific Grove, CA.

Dugac, P. [1981] "Des fonctions comme expressions analytiques aux fonctions représentables analytiquement," in *Mathematical Perspectives: essays on mathematics and its historical development,* J. W. Dauben, Ed., Academic Press, New York.

Dunford, N. and Schwartz, J. T. [1958] *Linear Operators,* Part I, Wiley-Interscience, New York; reprinted, 1988.

Dunham, W. [1990] *Journey Through Genius,* Wiley, New York.

Eberlein, W. F. [1957] "Notes on integration I: The underlying convergence theorem," *Communications on Pure and Applied Mathematics,* **10**, 357–360.

Edgar, G. A. [1990] *Measure, Topology, and Fractal Geometry,* Springer-Verlag, New York.

Edwards, Jr., C. H. [1979] *The Historical Development of the Calculus,* Springer-Verlag, New York.

Egorov, D. F. [1911] "Sur les suites des fonctions mesurables," *Comptes Rendus hebdomadaires des Séances de l'Académie des Sciences, Paris,* **152**, 244–246.

Fatou, P. [1906] "Sèries trigonométrique et sèries de Taylor," *Acta Mathematica,* **30**, 335–400.

Fejér, L. [1900] "Sur les fonctions bornées et intégrables," *Comptes Rendus Hebdomadaries, Séances de l'Académie des Sciences, Paris,* **131**, 984–987.

——— [1904] "Untersuchungen über Fouriersche Reihen," *Mathematische Annalen,* **58**, 51–69.

Ficken, F. A. [1959] "Some uses of linear spaces in analysis," *The American Mathematical Monthly,* **66**, 259–275.

Fischer, E. [1907a] "Sur la convergence en moyenne," *Comptes Rendus hebdomadaires des Séances de l'Académie des Sciences,* **1907**, 1022–1024.

——— [1907b] "Application d'un théorème sur la convergence en moyenne," *Comptes Rendus hebdomadaires des Séances de l'Académie des Sciences,* **144**, 1148–1151.

Fisher, S. D. [1978] "Quantitative approximation theory," *The American Mathematical Monthly,* **85**, 318–332.

Folland, G. B. [1984] *Real Analysis: Modern Techniques and Their Applications,* Wiley, New York.

——— [1992] *Fourier Analysis and Its Applications,* Wadsworth, Belmont, CA.

Fourier, J. B. J. [1878] *The Analytic Theory of Heat,* translated, with notes, by A. Freeman, Dover, New York, 1955.

——— [1888] *Oeuvres de Fourier,* Gauthier-Villars, Paris.

Frank, A. [1980] "The countability of the rational polynomials: A direct method," *The American Mathematical Monthly,* **87**, 810–811.

Franklin, P. [1924] "A simple discussion of the representation of functions by Fourier series," *The American Mathematical Monthly,* **31**, 475–478.

Fréchet, M. [1906] "Sur quelques points du calcul fonctionnel," *Rendiconti del Circolo Matematico di Palermo,* **22**, 1–74.

——— [1907] "Sur les ensembles des fonctions et les opérations linéaires," *Comptes Rendus hebdomadaires des Séances de l'Académie des Sciences,* **144**, 1414–1416.

——— [1908] "Essai de géométrie analytique à une infinité de coordonnées," *Nouvelles Annales de Mathematiques,* (4) **8**, 97–116, 289–317.

——— [1921] "Sur les divers modes de convergence d'une suite de fonctions d'une variable," *Bulletin of the Calcutta Mathematical Society,* **11**, 187–206.

——— [1928] *Les Espaces Abstraits,* Gauthier-Villars, Paris.

——— [1950] "Abstract sets, abstract spaces, and general analysis," *Mathematics Magazine,* **24**, 147–155.

Fredholm, I. [1903] "Sur une classe d'équations fonctionelles," *Acta Mathematica,* **27**, 365–390.

Freilich, G. [1973] "Increasing continuous singular functions," *The American Mathematical Monthly,* **80**, 918–919.

French, R. M. [1988] "The Banach-Tarski theorem," *The Mathematical Intelligencer,* **10** (4), 21–28.

Fulks, W. [1969] *Advanced Calculus,* 2nd. ed., Wiley, New York.

Gaffney, M. P. and Steen, L. A. [1976] *Annotated Bibliography of Expository Writing in the Mathematical Sciences,* The Mathematical Association of America, Washington, DC.

Gelbaum, B. R. [1964] "Banach algebras and their applications," *The American Mathematical Monthly,* **71**, 248–256.

Gelbaum, B. R. and Olmsted, J. M. H. [1964] *Counterexamples in Analysis,* Holden-Day, San Francisco.

Getchell, B. C. [1935] "On the equivalence of two methods of defining Stieltjes integrals," *Bulletin of the American Mathematical Society,* **41**, 413–418.

Gibson, G. A. [1893] "On the history of the Fourier series," *Proceedings of the Edinburgh Mathematical Society,* **11**, 137–170.

Gillman, L. [1993] "An axiomatic approach to the integral," *The American Mathematical Monthly,* **100**, 16–25.

Gillman, L. and Jerison, M. [1960] *Rings of Continuous Functions,* Springer-Verlag, New York.

Goebel, K. and Kirk, W. A. [1990] *Topics in Metric Fixed Point Theory,* Cambridge studies in Advanced Mathematics 28, Cambridge University Press, New York.

Goffman, C. [1953a] *Real Functions,* Rinehart, New York.

——— [1953b] "Definition of the Lebesgue integral," *The American Mathematical Monthly,* **60**, 251–252.

——— [1960] "Functions of finite Baire type," *The American Mathematical Monthly,* **67**, 164–165.

——— [1962] "Preliminaries to functional analysis," in *Studies in Modern Analysis,* R. C. Buck, Ed., The Mathematical Association of America; distributed by Prentice-Hall, Englewood Cliffs, NJ.

——— [1971] "On functions with summable derivatives," *The American Mathematical Monthly,* **78**, 874–875.

——— [1974] "Completeness of the real numbers," *Mathematics Magazine,* **47**, 1–8.

——— [1977] "A bounded derivative which is not Riemann integrable," *The American Mathematical Monthly,* **84**, 205–206.

Goffman, C. and Pedrick, G. [1965] *First Course in Functional Analysis,* 2nd. ed., Chelsea Publishing, New York, 1983.

Goffman, C. and Waterman, D. [1970] "Some aspects of Fourier series," *The American Mathematical Monthly,* **77**, 119–133.

Goldberg, R. R. [1964] *Methods of Real Analysis,* 2nd. ed., Wiley, New York, 1976.

González-Velasco, E. A. [1992] "Connections in mathematical analysis: the case of Fourier series," *The American Mathematical Monthly,* **99**, 427–441.

Grabiner, J. V. [1983] "Who gave you the epsilon? Cauchy and the origins of rigorous calculus," *The American Mathematical Monthly,* **90**, 185–194.

Grattan-Guinness, I. [1970] *The Development of the Foundations of Mathematical Analysis from Euler to Riemann,* MIT Press, Cambridge, MA.

——— Ed. [1972] *Joseph Fourier, 1768–1830,* in collaboration with J. R. Ravetz, MIT Press, Cambridge, MA.

——— Ed. [1980] *From the Calculus to Set Theory, 1630–1910,* Gerald Duckworth & Co., London.

Green, L. C. [1947] "Uniform convergence and continuity," *The American Mathematical Monthly,* **54**, 541.

Gulick, D. [1992] *Encounters With Chaos,* McGraw-Hill, New York.

Hahn, H. [1956a] "Infinity," in *The World of Mathematics,* Vol. 3, J. R. Newman, Ed., Simon and Schuster, New York.

——— [1956b] "The Crisis in Intuition," in *The World of Mathematics,* Vol. 3, J. R. Newman, Ed., Simon and Schuster, New York.

Halmos, P. R. [1960] *Naive Set Theory,* Van Nostrand, Princeton, NJ, 1960; reprinted by Springer-Verlag, New York, 1974.

——— [1978] "Fourier series," *The American Mathematical Monthly,* **85**, 33–34.

——— [1990] "Has progress in mathematics slowed down?," *The American Mathematical Monthly,* **97**, 561–588.

Hardy, G. H. [1916] "Weierstrass' non-differentiable function," *Transactions of the American Mathematical Society,* **17**, 301–325.

——— [1918] "Sir George Stokes and the concept of uniform convergence," *Proceedings of the Cambridge Philosophical Society,* **19**, 148–156.

Hardy, G. H., Littlewood, J. E., and Pólya, G. [1952] *Inequalities,* 2nd. ed., Cambridge University Press, London, reprinted, 1983.

Hartman, S. and Mikusiński, J. [1961] *The Theory of Lebesgue Measure and Integration,* Pergamon Press, Oxford.

Hausdorff, F. [1937] *Grundzüge der Mengenlehre,* 3rd. ed., Von Veit, Leipzig, 1937; published in English as *Set Theory*, 3rd. English ed., Chelsea Publishing, New York, 1978.

Hawkins, T. [1970] *Lebesgue's Theory of Integration: Its Origins and Developments,* 2nd. ed., Chelsea Publishing, New York, 1975.

Hedrick, E. R. [1927] "The significance of Weierstrass's theorem," *The American Mathematical Monthly,* **20**, 211–213.

Heider, L. J. and Simpson, J. E. [1967] *Theoretical Analysis,* Saunders, Philadelphia.

Helly, E. [1912] "Über lineare Funktionaloperationen," *Sitzungsberichte der Kaiserlichen Akademie der Wissenschaften zu Wien. Mathematisch-Naturwissenschaftlichen Klasse,* **121 II A1**, 265–297.

Herivel, J. [1975] *Joseph Fourier: The Man and the Physicist,* Clarendon Press, Oxford.

Hersh, R. and John-Steiner, V. [1993] "A visit to Hungarian mathematics," *The Mathematical Intelligencer,* **15** (2), 13–26.

Hewitt, E. [1960] "The role of compactness in analysis," *The American Mathematical Monthly,* **67**, 499–516.

Hewitt, E. and Stromberg, K. [1965] *Real and Abstract Analysis,* Springer-Verlag, New York.

Hilbert, D. [1952] *Grundzüge einer allgemeinen Theorie der linearen Integralgleichungen,* Chelsea Publishing, New York.

Hildebrandt, T. H. [1917] "On integrals related to and extensions of the Lebesgue integrals," *Bulletin of the American Mathematical Society,* **24**, 113–202.

——— [1926] "The Borel theorem and its generalizations," *Bulletin of the American Mathematical Society,* **32**, 423–474.

——— [1933] "A simple continuous function with a finite derivative at no point," *The American Mathematical Monthly,* **40**, 547–548.

——— [1938] "Definitions of Stieltjes integrals of the Riemann type," *The American Mathematical Monthly,* **45**, 265–278.

——— [1966] "Compactness in the space of quasi-continuous functions," *The American Mathematical Monthly,* **73**, Part II, 144–145.

Hille, E. and Tamarkin, J. D. [1929] "Remarks on a known example of a monotone continuous function," *The American Mathematical Monthly,* **36**, 255–264.

Hirschman, I. I., Ed. [1965] *Studies in Real and Complex Analysis,* Studies in Mathematics v. 3, The Mathematical Association of America; distributed by Prentice-Hall, Englewood Cliffs, NJ.

Hobson, E. W. [1927] *The Theory of Functions of a Real Variable,* 2 vols., Cambridge University Press, London; first American edition, 1950.

Hochstadt, H. [1980] "Eduard Helly, father of the Hahn-Banach theorem," *The Mathematical Intelligencer,* **2** (3), 123–125.

Hoffman, K. [1975] *Analysis in Euclidean Space,* Prentice-Hall, Englewood Cliffs, NJ.

Holbrook, J. [1991] "Stochastic independence and space-filling curves," *The American Mathematical Monthly,* **88**, 426–432.

Hölder, O. [1882] *Beiträge zur Potentialtheorie,* Dissertation, Tübingen.

Horton, G. P. [1918] "Functions of limited variation and Lebesgue integrals," *Annals of Mathematics,* **20**, 1–8.

Hueber, H. [1981] "On uniform continuity and compactness in metric spaces," *The American Mathematical Monthly,* **88**, 204–205.

Hunt, R. A. [1971] "Almost everywhere convergence of Walsh-Fourier series of L^2 functions," in *Actes du Congrès International des Mathématiciens (Nice, 1970),* Tome 2, pp. 665–661, Gauthier Villars, Paris.

Hurwitz, A. [1903] "Über die Fourierschen Konstanten integrirbarer Funktionen," *Mathematische Annalen,* **57**, 425–446.

Jackson, D. [1920] "The general theory of approximation by polynomials and trigonometric sums," *Bulletin of the American Mathematical Society,* **27**, 415–431.

——— [1926] "Note on the convergence of Fourier series," *The American Mathematical Monthly,* **33**, 39–40.

——— [1930] *Theory of Approximation,* Colloquium Publications v. 11, American Mathematical Society, New York.

——— [1934a] "The convergence of Fourier series," *The American Mathematical Monthly,* **41**, 67–84.

——— [1934b] "A proof of Weierstrass's theorem," *The American Mathematical Monthly,* **41**, 309–312.

——— [1941] *Fourier Series and Orthogonal Polynomials,* The Carus Mathematical Monographs no. 6, The Mathematical Association of America, Oberlin, OH.

Jameson, G. J. O. [1974] *Topology and Normed Spaces,* Wiley, New York.

Jeffery, R. L. [1953] *The Theory of Functions of a Real Variable,* 2nd. ed., University of Toronto Press; reprinted by Dover Publications, New York, 1985.

——— [1956] *Trigonometric Series; A Survey,* Canadian Mathematical Congress Lecture Notes Series, No. 2, University of Toronto Press.

Johnsonbaugh, R. and Pfaffenberger, W. E. [1981] *Foundations of Mathematical Analysis,* Marcel Dekker, New York.

Jordan, C. [1881] "Sur la série Fourier," *Comptes Rendus hebdomadaires des Séances de l'Academie des Sciences,* **92**, 228–230.

——— [1892] "Remarques sur les intégrales défines," *Journal de Mathématiques Pures et Appliquées,* (4) **8**, 69–99.

——— [1893] *Cours d'Analyse,* 3 vols., 2nd. ed., Gauthier-Villars, Paris.

Kaplansky, I. [1977] *Set Theory and Metric Spaces,* Chelsea, New York.

Kasriel, R. H. [1971] *Undergraduate Topology,* Saunders, Philadelphia.

Katsuura, H. [1991] "Continuous nowhere differentiable functions – An application of contraction mappings," *The American Mathematical Monthly,* **98**, 411–416.

Kelley, J. L. [1955] *General Topology,* Graduate Texts in Mathematics 27, Springer-Verlag, New York, 1975.

Kestelman, H. [1970] "Riemann integration of limit functions," *The American Mathematical Monthly,* **77**, 182–187.

Kitcher, P. [1983] *The Nature of Mathematical Knowledge,* Oxford, New York.

Klambauer, G. [1973] *Real Analysis,* American Elsevier, New York.

——— [1975] *Mathematical Analysis,* Marcel Dekker, New York.

——— [1979] *Problems and Propositions in Analysis,* Marcel Dekker, New York.

Kleiner, I. [1989] "Evolution of the function concept: A brief survey," *College Mathematics Journal,* **20**, 282–300.

Kline, M. [1972] *Mathematical Thought from Ancient to Modern Times,* Oxford, New York.

Knight, W. J. [1980] "Functions with zero right derivatives are constant," *The American Mathematical Monthly,* **87**, 657–658.

Kolmogorov, A. N. [1930] "Untersuchungen über den Integralbegriff," *Mathematische Annalen,* **103**, 654–696.

Kolmogorov, A. N. and Fomin, S. V. [1970] *Introductory Real Analysis,* rev. English ed., reprinted with corrections by Dover, New York, 1975.

Kominek, Z. [1983] "Measure, category, and the sums of sets," *The American Mathematical Monthly,* **90**, 561–562.

Körner, T. W. [1988] *Fourier Analysis,* Cambridge University Press, New York.

Korovkin, P. P. [1960] *Linear Operators and Approximation Theory,* Hindustan Publishing, Delhi.

Kuller, R. G. [1969] *Topics in Modern Analysis,* Prentice-Hall, Englewood Cliffs, NJ.

Kuratowski, K. [1935] "Quelques problèmes concernant les espaces métriques non-séparable," *Fundamenta Mathematicae,* **25**, 534–545.

——— [1966] *Topology,* Vol. I, Academic Press, New York.

Kuzawa, Sr. M. G. [1970] "Fundamenta Mathematicae: An examination of its founding and significance," *The American Mathematical Monthly,* **77**, 485–492.

La Vallée Poussin, Ch.-J. de [1893] "Sur quelques applications de l'intégrale de Poisson," *Annales de la Société Scientifique de Bruxelles,* **17**, 18–34.

——— [1919] *Leçons sur l'Approximation des Fonctions d'une Variable Réele,* Gauthier-Villars, Paris, 1919; also available in the volume *L'Approximation*, by S. Bernstein and Ch.-J. de La Vallée Poussin, Chelsea Publishing, New York, 1970.

——— [1934] *Intégrales de Lebesgue, Fonctions d'Ensemble, Classes de Baire,* Gauthier-Villars, Paris.

Labarre, A. E. [1965] "Structure theorem for open sets of real numbers," *The American Mathematical Monthly,* **72**, 1114.

Lakatos, I. [1976] *Proofs and Refutations,* Cambridge University Press, New York.

Lance, T. and Thomas, E. [1991] "Arcs with positive measure and a space-filling curve," *The American Mathematical Monthly,* **98**, 124–127.

Langer, R. E. [1947] "Fourier's series, the genesis and evolution of a theory," *The American Mathematical Monthly,* **54**, Suppl., 1–86.

Lebesgue, H. [1898] "Sur l'approximation des fonctions," *Bulletin des Sciences Mathématique,* **22**, 278–287.

——— [1902] "Intégrale, longeur, aire," *Annali di Matematica,* (3) **7**, 231–359.

——— [1906] *Leçons sur les séries trigonométiques,* Gauthier-Villars, Paris.

——— [1907] "Sur la reserche des fonctions primitives par l'intégration," *Atti della Reale Academia dei Lincei. Rendiconti,* (5) $\mathbf{16_1}$, 283–290.

——— [1909] "Sur les intégrales singulières," *Annales de Toulouse,* (3) **1**, 25–117.

——— [1917] "Sur certaines démonstrations d'existence," *Bulletin de la Société Mathématique de France,* **45**, 132–144.

——— [1928] *Leçons sur l'Intégration et la Recherche des Fonctions Primitives,* 2nd. ed., Gauthier-Villars, Paris, 1928; reprinted by Chelsea Publishing, New York, 1973.

——— [1966] *Measure and the Integral,* K. O. May, Ed., Holden-Day, San Francisco.

Levine, L. M. [1977] "On a necessary and sufficient condition for Riemann integrability," *The American Mathematical Monthly,* **84**, 205.

Levine, N. [1960] "Remarks on uniform continuity in metric spaces," *The American Mathematical Monthly,* **67**, 562–563.

Lewin, J. W. [1986] "A truly elementary approach to the bounded convergence theorem," *The American Mathematical Monthly,* **93**, 395–397.

Lipschitz, R. [1876] "Sur la possibilité d'intégrer complètement un système donné d'équations différentielles," *Bulletin de Sciences Mathématiques,* (Ser. 1) **10**, 149–159.

Liusternik, L. A. and Sobolev, V. J. [1961] *Elements of Functional Analysis,* Frederick Ungar Publishing Co., New York.

Łojasiewicz, S. [1988] *An Introduction to the Theory of Real Functions,* Wiley-Interscience, New York.

Long, R. L. [1986] "Remarks on the history and philosophy of mathematics," *The American Mathematical Monthly,* **93**, 609–619.

Lorch, E. R. [1993] "Szeged in 1934," *The American Mathematical Monthly,* **100**, 219–230.

Lorentz, G. G. [1986] *Bernstein Polynomials,* 2nd. ed., Chelsea Publishing, New York.

Luther, N. Y. [1967] "A characterization of almost uniform convergence," *The American Mathematical Monthly,* **74**, 1230–1231.

Luxemburg, W. A. J. [1971] "Arzelà's dominated convergence theorem for the Riemann integral," *The American Mathematical Monthly,* **78**, 970–979.

Lynch, M. [1992] "A continuous, nowhere differentiable function," *The American Mathematical Monthly,* **99**, 8–9.

Maddox, I. J. [1977] "The space of derivatives," *The American Mathematical Monthly,* **84**, 288–289.

——— [1988] *Elements of Functional Analysis,* 2nd. ed., Cambridge University Press, New York.

——— [1989] "The norm of a linear functional," *The American Mathematical Monthly,* **96**, 434–436.

Majumder, N. C. B. [1965] "On the distance set of the Cantor middle third set, III," *The American Mathematical Monthly,* **72**, 725–729.

Maligranda, L. [1995] "A simple proof of the Hölder and the Minkowski inequality," *The American Mathematical Monthly,* **102**, 256–259.

Malik, S. C. [1984] *Mathematical Analysis,* Wiley, New York.

Manheim, J. H. [1964] *The Genesis of Point Set Topology,* Pergamon Press, Oxford; Macmillan, New York.

Mauldin, R. D. [1979] "The existence of non-measurable sets," *The American Mathematical Monthly,* **86**, 45–46.

May, K. O. [1977] *Index of the American Mathematical Monthly Volumes 1 through 80 (1894–1973),* The Mathematical Association of America, Washington, DC.

McCarthy, J. [1953] "An everywhere continuous nowhere differentiable function," *The American Mathematical Monthly,* **60**, 709.

Menger, K. [1943] "What is dimension?," *The American Mathematical Monthly,* **50**, 2–7.

Metzler, R. C. [1971] "On Riemann integrability," *The American Mathematical Monthly,* **78**, 1129–1131.

Miel, G. [1983] "Of calculations past and present: The Archimedean algorithm," *The American Mathematical Monthly,* **90**, 17–35.

Miller, A. D. and Výborný, R. [1986] "Some remarks on functions with one-sided derivatives," *The American Mathematical Monthly,* **93**, 471–475.

Monna, A. F. [1972] "The concept of function in the 19th and 20th centuries, in particular with regard to the discussions between Baire, Borel, and Lebesgue," *Archive for History of Exact Sciences,* **9**, 57–84.

——— [1973] *Functional Analysis in Historical Perspective,* Wiley, New York.

——— [1980] "Hahn–Banach–Helly," *The Mathematical Intelligencer,* **2** (4), 158.

Moore, E. H. [1900] "On certain crinkly curves," *Transactions of the American Mathematical Society,* **1**, 72–90; errata, p. 507.

Moore, E. H. and Smith, H. L. [1922] "A general theory of limits," *American Journal of Mathematics,* **44**, 102–121.

Mozzochi, C. J. [1971] *On the Pointwise Convergence of Fourier Series,* Lecture Notes in Mathematics 199, Springer-Verlag, New York.

Munroe, M. E. [1965] *Introductory Real Analysis,* Addison-Wesley, Reading, MA.

——— [1971] *Measure and Integration,* 2nd. ed., Addison-Wesley, Reading, MA.

Myerson, G. [1991] "First-class functions," *The American Mathematical Monthly,* **98**, 237–240.

Natanson, I. P. [1955] *Theory of Functions of a Real Variable,* 2 vols., Frederick Ungar Publishing Co., New York.

——— [1964] *Constructive Function Theory,* 3 vols., Frederick Ungar Publishing Co., New York.

Neuenschwander, E. [1978] "Riemann's example of a continuous, 'nondifferentiable' function," *The Mathematical Intelligencer,* **1** (1), 40–44.

Newman, D. J. and Parsons, T. D. [1988] "On monotone subsequences," *The American Mathematical Monthly,* **95**, 44–45.

Newman, J. R., Ed. [1956] *The World of Mathematics,* 4 vols., Simon and Schuster, New York.

Newman, M. H. A. [1951] *Elements of the Topology of Plane Sets,* 2nd. ed., Cambridge University Press, London.

Nikolśkij, N. K., Ed. [1992] *Functional Analysis I,* Encyclopedia of Mathematical Sciences, Vol. 19, Springer-Verlag, New York.

Novinger, W. P. [1992] "Mean convergence in L_p spaces," *Proceedings of the American Mathematical Society,* **34**, 627–628.

Osgood, W. F. [1897] "Non-uniform convergence and the integration of series term by term," *American Journal of Mathematics,* **19**, 155–190.

Oxtoby, J. C. [1971] *Measure and Category: A Survey of the Analogies Between Topological and Measure Spaces,* 2nd. ed., Springer-Verlag, New York, 1980.

Peano, G. [1887] *Applicazione Geometriche del Calcolo Infinitesimale,* Torino, Bocca.

Petkovšek, M. [1990] "Ambiguous numbers are dense," *The American Mathematical Monthly,* **97**, 408–411.

Phillips, E. R. [1984] *An Introduction to Analysis and Integration Theory,* Dover, New York.

Picard, E. [1890] "Mémoire sur la théorie des équations aux dérivées partielles et la méthode des approximations successives," *Journal de Mathématique Pures et Appliquées,* **6**, 423–441.

Piranian, G. [1966] "The set of nondifferentiability of a continuous function," *The American Mathematical Monthly,* **73**, Part II, 57–61.

Polubarinova-Kochina, P. Ya. [1966] "Karl Theodor Wilhelm Weierstrass," *Russian Mathematical Surveys,* **21** (3), 195–206.

Pursell, L. E. [1967] "Uniform approximation of real continuous functions on the real line by infinitely differentiable functions," *Mathematics Magazine,* **40**, 263–265.

Radó, T. [1932] "On mathematical life in Hungary," *The American Mathematical Monthly,* **37**, 85–90.

——— [1942] "On semi-continuity," *The American Mathematical Monthly,* **49**, 446–450.

Radon, J. [1913] "Theorie und Anwendungen der absolut additiven Mengenfunktionen," *Sitzungsberichte der Kaiserlichen Akademie der Wissenschaften zu Wien. Mathematisch-Naturwissenschaftlichen Klasse,* **122 IIa**, 1295–1438.

Randolph, J. F. [1940] "Distances between points of the Cantor set," *The American Mathematical Monthly,* **47**, 549–551.

——— [1968] *Basic Real and Abstract Analysis,* Academic Press, New York.

Rao, K. V. R. [1965] "On L- and R-integrals," *The American Mathematical Monthly,* **72**, 1112–1113.

Read, C. B. and Bidwell, J. K. [1976a] "Selected articles dealing with the history of elementary mathematics," *School Science and Mathematics,* **76**, 477–483.

——— [1976b] "Periodical articles dealing with the history of advanced mathematics," *School Science and Mathematics,* **76**, 581–598, 687–703.

Reed, M. and Simon, B. [1972] *Methods of Modern Mathematical Physics, Vol. 1: Functional Analysis,* Academic Press, New York, 1980.

Riemann, B. [1902] *Gesammelte Mathematische Werke und Wissenschaftlicher Nachlass,* 2nd. ed., and *Nachträge,* Teubner, Leipzig, 1902; reprinted by Dover, 1953.

Riesz, F. [1906] "Sur les ensembles des fonctions," *Comptes Rendus hebdomadaires des Séances de l'Académie des Sciences, Paris,* **143**, 738–741.

——— [1907a] "Sur les systèmes orthogonaux de fonctions," *Comptes Rendus hebdomadaires des Séances de l'Académie des Sciences, Paris,* **144**, 615–619.

——— [1907b] "Sur une espèce de géométrique analytique des fonctions sommables," *Comptes Rendus hebdomadaires des Séances de l'Académie des Sciences, Paris,* **144**, 1409–1411.

——— [1908] "Stetigkeitsbegriff und abstrakte mengenlehre," *Atti del IV Congresso Internazionale dei Matematici, Roma,* **2**, 18–24.

——— [1909a] "Sur les suites de fonctions mesurables," *Comptes Rendus hebdomadaires des Séances de l'Académie des Sciences,* **148**, 1303–1305.

——— [1909b] "Les opérations fonctionelles linéares," *Comptes Rendus hebdomadaires des Séances de l'Académie des Sciences,* **149**, 974–977.

——— [1910a] "Sur certains systèmes d'équations fonctionelles et l'approximation des fonctions continues," *Comptes Rendus hebdomadaires des Séances de l'Académie des Sciences,* **150**, 403–406.

——— [1910b] "Untersuchungen über Systeme integrierbarer Funktionen," *Matematische Annalen,* **69**, 449–497.

——— [1911] "Sur certains systèmes singuliers d'équations intégrales," *Annales Scientifiques de l'École Normale Supérieure,* (3) **128**, 33–62.

——— [1913] *Les systèmes d'équations linéaires à une infinité d'inconnues,* Gauthier-Villars, Paris.

——— [1917] "Über Integration unendlicher Folgen," *Jahresbericht der Deutschen Mathematiker-Vereinigung,* **26**, 274–278.

——— [1920] "Sur l'intégrale de Lebesgue," *Acta Mathematica,* **42**, 191–205.

——— [1928a] "Sur la convergence en moyenne," *Acta Scientiarum Mathematicarum,* **4** (1928–1929), 58–64.

——— [1928b] "Elementarer Beweis des Egoroffschen Satzes," *Monatshefte für Mathematik und Physik,* **35**, 243–248.

——— [1928c] "Sur la convergence en moyenne (seconde communication)," *Acta Scientiarum Mathematicarum,* **4** (1928–1929), 182–185.

——— [1930] "Sur la décomposition des opérations fonctionelles linéaires," *Atti del Congresso Internazionale dei Matematici,* Bologna, 1928, **3**, 143–148.

——— [1936] "Sur l'intégrale de Lebesgue comme l'opération inverse de la dérivation," *Annali di Pisa,* (2) **5**, 191–212.

——— [1940] "Sur quelques notions fondamentales dans la théorie générale des opérations linéaires," *Annals of Mathematics,* **41**, 174–206.

——— [1944] "Sur la théorie ergodique," *Commentarii Mathematici Helvetici,* **17**, 221–239.

——— [1949] "L'evolution de la notion d'intégrale depuis Lebesgue," *Annales de l'Institut Fourier,* **1**, 29–42.

——— [1960] *Riesz Frigyes oszegyujtott munkai (Collected Works),* 2 vols., Akademiai Kiado, Budapest.

Riesz, F. and Sz.-Nagy, B. [1955] *Functional Analysis,* Frederick Ungar, 1955; reprinted by Dover, New York, 1992.

Rivlin, T. J. [1981] *An Introduction to the Approximation of Functions,* Dover, New York.

Robbins, H. E. [1943] "A note on the Riemann integral," *The American Mathematical Monthly,* **50**, 617–618.

Rogosinski, W. [1950] *Fourier Series,* Chelsea Publishing, New York.

Ross, K. A. [1980a] "Another approach to Riemann-Stieltjes integrals," *The American Mathematical Monthly,* **87**, 660–662.

——— [1980b] *Elementary Analysis,* Springer-Verlag, New York.

Royden, H. L. [1963] *Real Analysis,* 3rd. ed., Macmillan, New York, 1988.

Rozycki, E. P. [1965] "On Egoroff's theorem," *Fundamenta Mathematicae,* **55**, 289–293.

Rudin, W. [1953] *Principles of Mathematical Analysis,* 3rd. ed., McGraw-Hill, New York, 1976.

——— [1966] *Real and Complex Analysis,* 3rd. ed., McGraw-Hill, New York, 1987.

——— [1983] "Well-distributed measurable sets," *The American Mathematical Monthly,* **90**, 41–42.

Russell, A. M. [1979] "Further comments on the variation function," *The American Mathematical Monthly,* **86**, 480–482.

——— [1980] "A commutative Banach algebra of functions of bounded variation," *The American Mathematical Monthly,* **87**, 39–40.

Sagan, H. [1986] "Approximating polygons for Lebesgue's and Schoenberg's space filling curves," *The American Mathematical Monthly,* **93**, 361–368.

——— [1992] "An elementary proof that Schoenberg's space-filling curve is nowhere differentiable," *Mathematics Magazine,* **65**, 125–128.

Saks, S. [1924] "Sur les nombres dérivés des fonctions," *Fundamenta Mathematicae,* **5**, 98–104.

——— [1937] *Theorie de l'Intégrale,* 2nd rev. ed., G. E. Stechert & Co., Warsaw, 1937; an English translation by L. C. Young appears as *Theory of the Integral,* Dover, New York, 1964.

Satyanarayana, U. V. [1980] "A note on Riemann-Stieltjes integrals," *The American Mathematical Monthly,* **87**, 477–478.

Schaefer, H. H. [1980] "Aspects of Banach lattices," in *Studies in Functional Analysis,* R. G. Bartle, Ed., The Mathematical Association of America, Washington, DC.

Schmidt, E. [1908] "Über die Auflösung linearer Gleichungen mit unendlich vielen Unbekannten," *Rendiconti del Circolo Matematico di Palermo,* **25**, 53–77.

Schoenberg, I. J. [1938] "The Peano curve of Lebesgue," *Bulletin of the American Mathematical Society,* **44**, 519.

——— [1982] *Mathematical Time Exposures,* The Mathematical Association of America, Washington, DC.

——— [1988] *I. J. Schoenberg: Selected Papers,* Birkhauser, Boston.

Seebach, J. A. and Steen, L. A., Eds. [1978] *Mathematics Magazine: 50 Year Index,* The Mathematical Association of America, Washington, DC.

Segal, S. L. [1978] "Riemann's example of a continuous, 'nondifferentiable' function continued," *The Mathematical Intelligencer,* **1** (2), 81–82.

Seidel, P. L. [1847] "Note über eine Eigenschaft der Reihen, welche Discontinuirliche Functionen Darstellen," *Abhandlungen der Mathematisch-Physikalischen Klasse der Könglich Bayerischen Akademie der Wissenschaften,* **5**, 381–393.

Shaskin, Yu. A. [1991] *Fixed Points,* American Mathematical Society, Providence, RI.

Shields, A. [1987a] "Polynomial approximation," *The Mathematical Intelligencer,* **9** (3), 5–7.

——— [1987b] "Luzin and Egorov," *The Mathematical Intelligencer,* **9** (4), 24–27.

——— [1989] "Banach algebras, 1939–1989," *The Mathematical Intelligencer,* **11** (3), 15–17.

Shohat, J. A. [1930] "Stieltjes integrals in mathematical statistics," *Annals of Mathematical Statistics,* **1**, 73–94.

Shohat, J. A. and Tamarkin, J. D. [1943] *The Problem of Moments,* Mathematical Surveys, no. 1, American Mathematical Society, New York.

Sierpiński, W. [1918] "Un théorème sur les ensembles fermés," *Bulletin de l'Académie des Sciences, Cracovie,* Série A, 49–51.

——— [1922] "Démonstration de quelques théorèmes fondamentaux sur les fonctions mesurables," *Fundamenta Mathematicae,* **3**, 314–321.

——— [1974] *Oeuvres Choisies,* 3 vols., PWN, Warsaw.

Simmons, G. F. [1963] *Introduction to Topology and Modern Analysis,* McGraw-Hill, New York, 1963; reprinted by Robert E. Krieger Pub., 1986.

Simon, B. [1969] "Uniform convergence of Fourier series," *The American Mathematical Monthly,* **76**, 55–56.

Singh, A. N. [1969] "The theory and construction of non-differentiable functions," in *Squaring the Circle, and Other Monographs,* Chelsea Publishing, New York.

Sklar, A. [1960] "On the definition of the Riemann integral," *The American Mathematical Monthly,* **67**, 897–900.

Smith, A. [1972] "The differentiability of Riemann's function," *Proceedings of the American Mathematical Society,* **34**, 463–468.

Smith, H. J. S. [1875] "On the integration of discontinuous functions," *Proceedings of the London Mathematical Society,* **6**, 140–153.

Snipes, R. F. [1984] "Is every continuous function uniformly continuous?," *Mathematics Magazine,* **57**, 169–173.

Sprecher, D. A. [1970] *Elements of Real Analysis,* Academic Press, New York, 1970; republished with corrections by Dover Publications, New York, 1987.

Steinhaus, H. [1917] *Nowa vlasnośe mnogości G. Cantor,* Wektor, Poland.

——— [1920] "Sur les distances des points des ensembles de mesure positive," *Fundamenta Mathematicae,* **1**, 93–104 and *Annexe,* 232–233.

——— [1963] "Stefan Banach, 1892–1945," *Scripta Mathematica,* **26**, 93–100.

Stieltjes, T. J. [1894] "Recherches sur les fractions continues," *Annales de la Faculté des Sciences de Toulouse,* **8**, J.1–J.122.

Stokes, G. G. [1848] "On the critical values of the sums of periodic series," *Transactions of the Cambridge Philosophical Society,* **8**, 533–583.

Stone, M. H. [1937] "Applications of the theory of Boolean rings to general topology," *Transactions of the American Mathematical Society,* **41**, 375–481.

——— [1962] "A generalized Weierstrass theorem," in *Studies in Modern Analysis,* R. C. Buck, Ed., The Mathematical Association of America; distributed by Prentice-Hall, Englewood Cliffs, NJ.

Stromberg, K. [1977] "The Banach-Tarski paradox," *The American Mathematical Monthly,* **86**, 151–161.

——— [1981] *An Introduction to Classical Real Analysis,* Wadsworth, Belmont, CA.

Struik, D. J. [1948] *A Concise History of Mathematics,* 4th. rev. ed., Dover, New York.

Suckau, J. W. T. [1935] "On uniform convergence," *American Journal of Mathematics,* **57**, 549–561.

Swift, W. C. [1961] "Simple constructions of non-differentiable functions and space-filling curves," *The American Mathematical Monthly,* **68**, 653–655; see also **69**, 52.

Takács, L. [1978] "An increasing continuous singular function," *The American Mathematical Monthly,* **85**, 35–37.

Taylor, A. E. [1942] "Derivatives in the calculus," *The American Mathematical Monthly,* **49**, 631–642.

——— [1965] *General Theory of Functions and Integration,* 2nd. corrected printing (1966), Blaisdell, Waltham, MA, 1965; reprinted by Dover, New York, 1985.

——— [1982] "A study of Maurice Fréchet: I. His early work on point set theory and the theory of functionals," *Archive for History of Exact Sciences,* **27**, 233–295.

——— [1984] "A life in mathematics remembered," *The American Mathematical Monthly,* **91**, 605–618.

Temple, G. [1981] *100 Years of Mathematics: A Personal Viewpoint,* Springer-Verlag, New York.

Ter Horst, H. J. [1984] "Riemann-Stieltjes and Lebesgue-Stieltjes integrability," *The American Mathematical Monthly,* **91**, 551–559.

Thomas, R. [1985] "A combinatorial construction of a nonmeasurable set," *The American Mathematical Monthly,* **92**, 421–422.

Thurston, H. [1989] "Can a graph be both continuous and discontinuous?," *The American Mathematical Monthly,* **96**, 814–815.

Tolstov, G. P. [1962] *Fourier Series,* Prentice-Hall, Englewood Cliffs, NJ; rev. ed., Dover, New York, 1976.

Tong, J. [1992] "A characterization of continuity," *Mathematics Magazine,* **65**, 255–256.

Torchinsky, A. [1988] *Real Variables,* Addison-Wesley, Redwood City, CA.

Ulam, S. M. [1943] "What is measure?," *The American Mathematical Monthly,* **50**, 597–602.

Van Daele, A. [1990] "The Lebesgue integral without measure theory," *The American Mathematical Monthly,* **97**, 912–915.

Van Dalen, D. and Monna, A. F. [1972] *Sets and Integration. An Outline of the Development,* Wolters-Noordhoff, Groningen.

Van der Waerden, B. L. [1930] "Eine einfaches Beispiel einer nichtdifferenzierbaren stetigen Funktion," *Mathematische Zeitschrift,* **32**, 474–475.

Van Vleck, E. B. [1908] "On non-measurable sets of points with an example," *Transactions of the American Mathematical Society,* **9**, 237–244.

——— [1914] "The influence of Fourier's series upon the development of mathematics," *Science,* **39**, 113–124.

Van Vleck, F. S. [1973] "A remark concerning absolutely continuous functions," *The American Mathematical Monthly,* **80**, 286–287.

Varberg, D. E. [1967] "On absolutely continuous functions," *The American Mathematical Monthly,* **72**, 831–841.

Vitali, G. [1904] "Sulla integrabilità delle funzioni," *Reale Istitutio Lombardo di Scienze e Lettere. Rendiconti,* (2) **37**, 69–73.

——— [1905a] *Sul Problema della Misura dei Gruppi di Punti di una Retta,* Gamberinni e Parmeggiani, Bologna.

——— [1905b] "Sulle funzioni integrali," *Atti della Accademia delle Scienze di Torino. Classe di Scienze Fisiche, Matematiche e Naturali,* **41**, 1021–1034.

Walker, P. L. [1977] "On Lebesgue integrable derivatives," *The American Mathematical Monthly,* **84**, 287–288.

Weierstrass, K. [1885] "Über die analytische Darstellbarkeit sogenannter willkürlicher Functionen einer reellen Veränderlichen," *Sitzungsberichte der Königlich Preussischen Akademie der Wissenshcaften zu Berlin,* 633–639, 789–805.

——— [1886] "Sur la possibilité d'une représentation analytique des fonctions dites arbitraires d'une variable réele," *Journal de Mathématiques Pures et Appliquées,* **2**, 105–138.

——— [1894] *Mathematische Werke von Karl Weierstrass,* 7 vols., Mayer and Müller, Berlin.

Weil, C. E. [1976] "On nowhere monotone functions," *Proceedings of the American Mathematical Society,* **56**, 388–389.

Weir, A. J. [1973] *Lebesgue Integration and Measure,* Cambridge University Press, New York.

——— [1974] *General Integration and Measure,* Cambridge University Press, New York.

Wen, L. [1983] "A space-filling curve," *The American Mathematical Monthly,* **90**, 283.

Weston, D. [1959] "A counter-example concerning Egoroff's theorem," *Journal of the London Mathematical Society,* **34**, 139–140.

——— [1960] "Addendum to a note on Egoroff's theorem," *Journal of the London Mathematical Society,* **35**, 366.

Weyl, H. [1968] *Gesammelte Abhandlungen,* 4 vols., Springer-Verlag, Berlin.

Wheeden, R. L. and Zygmund, A. [1977] *Measure and Integral,* Marcel Dekker, New York.

Whyburn, G. T. [1942] "What is a curve?," *The American Mathematical Monthly,* **49**, 493–497.

Wilansky, A. [1953a] "Two examples in real variables," *The American Mathematical Monthly,* **60**, 317; correction, p. 546.

——— [1953b] "On a nowhere dense set," *The American Mathematical Monthly,* **60**, 411.

Wilder, R. L. [1978] "Evolution of the topological concept of 'connected'," *The American Mathematical Monthly,* **85**, 720–726.

——— [1980] "Correction and addendum to 'Evolution of the topological concept of "connected" '," *The American Mathematical Monthly,* **87**, 31–32.

Wilker, J. B. [1982] "Rings of sets are really rings," *The American Mathematical Monthly,* **89**, 211.

Willard, S. [1970] *General Topology,* Addison-Wesley, Reading, MA.

Williamson, J. H. [1962] *Lebesgue Integration,* Holt, Rinehart and Winston, New York.

Young, W. H. [1904] "A note on the condition of integrability of a function of one variable," *The Quarterly Journal of Pure and Applied Mathematics,* **35**, 189–192.

Zaanen, A. C. [1986] "Continuity of measurable functions," *The American Mathematical Monthly,* **93**, 128–130.

——— [1989] *Continuity, Integration and Fourier Theory,* Springer-Verlag, New York.

Zamfirescu, T. [1981] "Most monotone functions are singular," *The American Mathematical Monthly,* **88**, 47–49.

Zygmund, A. [1935] *Trigonometric Series,* 2nd. ed., Cambridge University Press, London, 1977.

——— [1976] "Notes on the history of Fourier series," in *Studies in Harmonic Analysis,* J. M. Ash, Ed., The Mathematical Association of America, Washington, DC.

——— [1987] "Stanislaw Saks, 1897–1942," *The Mathematical Intelligencer,* **9** (1), 36–43.

Symbol Index

Spaces

Norms

Topic Index